Photorefractive Optics

Materials, Properties, and Applications

Photorefractive Optics

Materials, Properties, and Applications

Editors

Francis Yu and Shizhuo Yin

Department of Electrical Engineering
Pennsylvania State University
University Park, Pennsylvania

ACADEMIC PRESS

A Harcourt Science and Technology Company

San Diego San Francisco New York Boston
London Sydney Tokyo

This book is printed on acid-free paper. ♾

ACADEMIC PRESS
A Harcourt Science and Technology Company
525 B Street, Suite 1900, San Diego, California 92101-4495, USA
http://www.apnet.com

Academic Press
24-28 Oval Road, London NW1 7DX, UK
http://www.hbuk.co.uk/ap/

Library of Congress Catalog Card Number: 99-65284

International Standard Book Number: 0-12-774810-5

Printed in the United States of America
99 00 01 02 03 EB 9 8 7 6 5 4 3 2 1

Contents

Contributing Authors

Numbers indicate which chapter the author worked on.

Partha P. Banerjee (3)
Department of Electrical and Computer Engineering and Center for Applied Optics, University of Alabama in Huntsville, Huntsville Alabama

George Barbastathis (8)
Department of Mechanical Engineering, Massachusetts Institute of Technology, Cambridge, MA

David J. Brady (8)
Department of Electrical and Computer Engineering ,Beckman Institute for Advanced Science and Technology, University of Illinois at Urbana-Champaign, Urbana, IL

Preben Buchhave (1)
Physics Department, Denmark Technical University, Lyngby, Denmark

Karsten Buse (2)
Universität Osnabrück, Fachbereich Physik, Osnabrück, Federal Republic of Germany

Ernest Chuang (12)
Sony Corporation, Tokyo, Japan

John O. Dimmock (3)
Center for Applied Optics and Department of Physics, University of Alabama in Huntsville, Huntsville Alabama

Jean-Jacques P. Drolet (12)
MicroDisplay Corporation, San Pablo, California

Yeshaiahu Fainman (15)
Department of Electrical and Computer Engineering, University of California, San Diego, La Jolla, California

Claire Gu (7)
Department of Electrical Engineering, University of California, Santa Cruz, California

Min Gu (10)
Optoelectronic Imaging Group, School of Communications and Informatics, Victoria University of Technology, Australia

Nicolas Huot (14)
Laboratoire Charles Fabry de l'Institut d'Optique, Unité Mixte du Centre National de la Recherche Scientifique, Orsay Cedex, France

Kazuyoshi Itoh (13)
Department of Applied Physics, Osaka University, Suita, Osaka, Japan

Jean-Michel Jonathan (14)
Laboratoire Charles Fabry de l'Institut d'Optique, Unité Mixte du Centre National de la Recherche Scientifique, Orsay Cedex, France

Satoshi Kawata (9)
Department of Applied Physics, Osaka University, Osaka, Japan

Iam Choon Khoo (4)
Electrical Engineering Department, Pennsylvania State University, University Park, Pennsylvania

Eckhard Krätzig (2)
Universität Osnabrück, Fachbereich Physik, Osnabrück, Federal Republic of Germany

Nickolai V. Kukhtarev (1, 3)
Physics Department, Alabama A & M University, Normal, Alabama

Tatiana Kukhtarev (1)
Physics Department, Alabama A & M University, Normal, Alabama

Kazuo Kuroda (13)
Institute of Industrial Science, University of Tokyo, Komaba, Meguro-ku, Tokyo, Japan

Wenhai Liu (12)
Electrical Engineering Department, California Institute of Technology, Pasadena, California

Osamu Matoba (13)
Institute of Industrial Science, University of Tokyo, Tokyo, Japan

Yuri T. Mazurenko (15)
Department of Electrical and Computer Engineering, University of California, San Diego, La Jolla, California

Andrei L. Mikaelian (6)
Institute of Optical Neural Technologies, Russian Academy of Sciences, Moscow, Russia

Demetri Psaltis (12)
Electrical Engineering Department, California Institute of Technology, Pasadena, California

Gérald Roosen (14)
Laboratoire Charles Fabry de l'Institut d'Optique, Unité Mixte du Centre National de la Recherche Scientifique, Orsay Cedex, France

Pang-chen Sun (15)
Department of Electrical and Computer Engineering, University of California, San Diego, La Jolla, California

Xiangyang Yang (11)
CompuSensor Technology Corporation, Gaithersburg, MD

Puchi Yeh, (7)
Department of Electrical Engineering, University of California, Santa Barbara, California

Xianmin Yi (7)
Department of Electrical Engineering, University of California, Santa Barbara, California

Shizhuo Yin (16)
Department of Electrical Engineering, Pennsylvania State University, University Park, Pennsylvania

Francis T. S. Yu (16)
Department of Electrical Engineering, Pennsylvania State University, University Park, Pennsylvania

Feng Zhao (5)
Jet Propulsion Laboratory, California Institute of Technology, Pasadena, California

Hanying Zhou (5)
Jet Propulsion Laboratory, California Institute of Technology, Pasadena, California

Preface

Photrefrative optics is a rapidly growing, fascinating technology in modern optics which offers a myriad of potential applications. These applications can be envisioned as applied to high-capacity optical memories, dynamic hologram formations, massive interconnections, and high-speed tunable filters, as well as true-time relay lines for phase array antenna processing. This book provides comprehensive coverage of the basic material aspects, optical properties, devices, and numerous ingenious potential applications. This text provides state-of-the-art information on photorefractive optics. The book is written by a collection of world experts in this field, which are well represented by countries such as United States, Japan, Russia, Ukraine, China, France, Germany, Australia, and Denmark.

The book contains 16 selected chapters that begin with the standard photorefractive models, optical properties, wave mixing, hologram formation memories, three-dimensional data storage dynamic, interconnections, space-time processing, application of photoreflective material to wavefront connection and to femtosecond lasers. Finally, this book concludes with a chapter discussing the dynamic process of photoreflective fibers.

Francis T. S. Yu is an Evan Pugh Professor of Electrical Engineering at the Pennsylvania State University, University Park. He has authored/coauthored eight books. He is a Fellow of the Institute of Electrical and Electronics Engineers, the Optical Society of America, and SPIE, the International Society for Optical Engineering. Dr. Yu received his Ph.D. (1964) in Electrical Engineering from the University of Michigan, Ann Arbor.

Shizhuo Yin is an assistant professor of Electrical Engineering at the Pennsylvania State University, University Park. He is a senior member of the Institute of Electrical and Electronics Engineers, and a member of the Optical Society of America, and SPIE. He is a recipient of the 1996 Young Investigator Award for the U.S. Department of the Army. Dr. Yin received his Ph.D. (1993) in Electrical Engineering from the Pennsylvania State University, University Park.

Photorefractive Optics

Materials, Properties, and Applications

Chapter 1

Standard Photorefractive Model as a Foundation of Real-Time Holography

Nickolai V. Kukhtarev

Physics Department, Alabama A & M University, Normal, Alabama

Tatiana Kukhtareva

Physics Department, Alabama A & M University, Normal, Alabama

Preben Buchhave

Physics Department, Denmark Technical University, Lyngby, Denmark

1.1 Introduction (photorefractive "Old Testament")

We were invited by Professor Francis Yu to write an introductory chapter about photorefractive models. It was natural for us to start from the first

PHOTOREFRACTIVE OPTICS
Materials, Properties and Applications

ISBN 0-12-774810-5/$30.00

model, formulated in 1971–1976 by extension of the semiconductor band-transport model. Since its first publication (the best-known theoretical version may be the one published in 1976 [1, 2]) this standard photorefractive model (SPM) has had its own life. Similar to many widely accepted "standard" models SPM was not originally welcomed to the international journals. A detailed paper on SPM with experimental results and analysis of high-contrast gratings was published in *Ferroelectrics* [3] in 1979 after almost 3 years of extensive discussions with negatively minded reviewers. Later this paper [3] was one of the most cited papers as may be easily seen by browsing the *Science Citation Index*. In spite of brief discussion in [3], the case of high contrast, SPM was widely cited as a model of low contrast approximation. This model stems from previous research efforts of one of the authors (N. K.) in semiconductors with periodic distribution of ionized impurity. Motivation for development of this model was inspired by experimental results of A. Semenyuk on galvanic-magnetic properties of the Ge-crystals with stratified distribution of the donor-type impurities [1, 2]. Different types of radiation (neutrons, or γ-rays) introduce compensating (acceptor-type) defects. These radiation-induced defects drastically change electrical macroscopic conductivity—e.g., the Hall effect and magnetoresistance measurements in the inhomogeneous Ge-crystals.

Theoretical description of galvanic-magnetic properties of the inhomogeneous semiconductors resulted in a Ph.D. dissertation for N. K. in 1973. The papers of the investigators from RCA [4] (Amodei and Staebler), on recording of photoinduced gratings in $LiNbO_3$, inspired a lot of interest in holographic-grating technique. It was natural to modify previous models of inhomogeneous photogeneration in semiconductors to holographic grating recording.

Compensating centers are important for the explanation of holographic memory by allowing long-living redistribution of the space-charge field. The major faults of the early models of Deygen *et al.* [5] and Chen [6] were just lack of compensation centers in the description of "optical damage" (early version of the term for the photorefractive effect). In our papers [1, 2] and *Ferroelectrics '79* and papers of Krumins *et al.* [7], SPM with the compensating centers was successfully tested for the description of grating recording in $LiNbO_3$ and $KNbO_3$. Besides optical properties manifested in self-diffraction and energy exchange, it was also predicted that electric current will be modified by the holographic grating in photoconductors [1, 3, 7]. Another prediction was about the generation of photovoltage (photoelectromotive force or photo-EMF) during recording of holographic grating (HG) [8]. Later this idea of photo-EMF was intensively exploited in the series of papers of S.Stepanov *et al.* [9] for the practically important case of transient EMF. It looks like transient EMF effect will be soon implemented in practical devices for sensitive vibrometry [10]. Manifestation of natural symbioses of the electrical and optical features of the dynamic HG were oscillations of the diffraction efficiency and energy exchange [11] during recording in the external electric field. These oscillations were interpreted later as the damping space-charge waves or eigenmodes widely known in semiconductors [12]. For high enough electric fields these damping space-charge waves may become amplifying and may result in the electrical instabilities [13]. An important precursor of instabilities in semiconductors is negative differential conductivity (NDC) [13]. It was predicted in [2] in 1975 that holographic gratings in photoconductive material may lead to inhomogeneous heating of photocarriers, and to stimulation of NDC and electrical instabilities.

Another theoretical paper [14] described the possible use of self-diffraction for monitoring and characterization of radiation defects in the photosensitive materials.

In concluding this historical introduction, let us summarize theoretical results derived in 1971–1980 using SPM:

- Formulation of adequate photorefractive (PR) models with compensating centers
- Formulation of self-consistent models of grating recording and self-diffraction (nonlinear diffusion-drift-recombination-material equation and wave equations)
- Prediction of photo-EMF in HG
- Prediction of photoinduced anisotropic conductivity by HG
- Prediction of instabilities in HG in the external ("heating") E-field

Some of these theoretical results experienced further development with experimental verification (as peculiarities of self-diffraction and phase conjugation, holographic photo-EMF, electrical instabilities in HG, and optical storage based on movement of compensating impurities).

1.2 Basic equations

We will formulate the general form of starting equations describing known and expected results stemming from SPM. Taking into account the growing interest in the spatiotemporal patterns in different fields of modern science, we will write equations in the appropriate three-dimensional (3-D) form.

Originally formulated for semiconductors, SPM has more general fields of application, where the concepts of photogeneration-recombination and drift may be justified. As examples, we can mention the application of SPM to description of grating recording in the PR polymers (and PR liquid crystals [15]), ferroelectric crystals [3], semiconductor materials [2–5], and paraelectric electrooptical crystals [8, 9].

For this more generalized approach SPM may be termed a drift-diffusion-recombination (DDR) model. A system of equations for photogenerated mobile charged carriers with concentration n, photosensitive ionized centers N (which are fixed in space), and electric field E may be written as

$$
\begin{aligned}
&\varepsilon\varepsilon_0 \frac{\partial \overline{E}}{\partial t} + e\hat{\mu} n\overline{E} + e\hat{D}\overline{\nabla} n = \overline{J} \\
&\overline{\nabla}(\hat{\varepsilon}\varepsilon_0 E) = e(N - N_A - n) \\
&\frac{\partial N}{\partial t} = (g + \beta)(N_0 - N) - rnN
\end{aligned}
\tag{1.1}
$$

Here e is the effective charge of the carrier, $\hat{\mu}$ and $\hat{D}$ are mobility and diffusion coefficients of the mobile carriers, N_0 is total concentration of photosensitive centers, N_A is the concentration of compensating centers, g is the optical generation rate, β is the thermal generation rate and r is the recombination coefficient, ε_0 is the dielectric constant of vacuum, and ε stands for the relative dielectric constants. Total current $\overline{J}$ and electric field $\overline{E}$ obey additional equations:

$$\overline{\nabla}\,\overline{J} = 0 \quad \text{and} \quad \overline{\nabla} \times \overline{E} = 0 \tag{1.2}$$

which are important for considering the (3-D) case.

For noncentrosymmetric materials in the left-hand side (LHS) of the first equation in Eq. (1) we need to add photogalvanic current

$$j^{p}i = \gamma_{ijk} F_j F_k{}^{*} \tag{1.3}$$

where γ is the photogalvanic tensor, and $F_{j,k}$ are the electric-field components of the laser irradiation.

Equations (1–3) are the basic material equations for description of the formation of photoinduced gratings in photosensitive materials. Relations to the optical and electrical manifestation of holographic grating, predicted by SPM, may be visualized by the diagram in (Fig. 1.1). The formal relation of SPM to the optical self-diffraction is due to modulation of the refractive index by the photoinduced gratings of the electric field and/or grating of photoconductivity (as shown in Fig. 1.1).

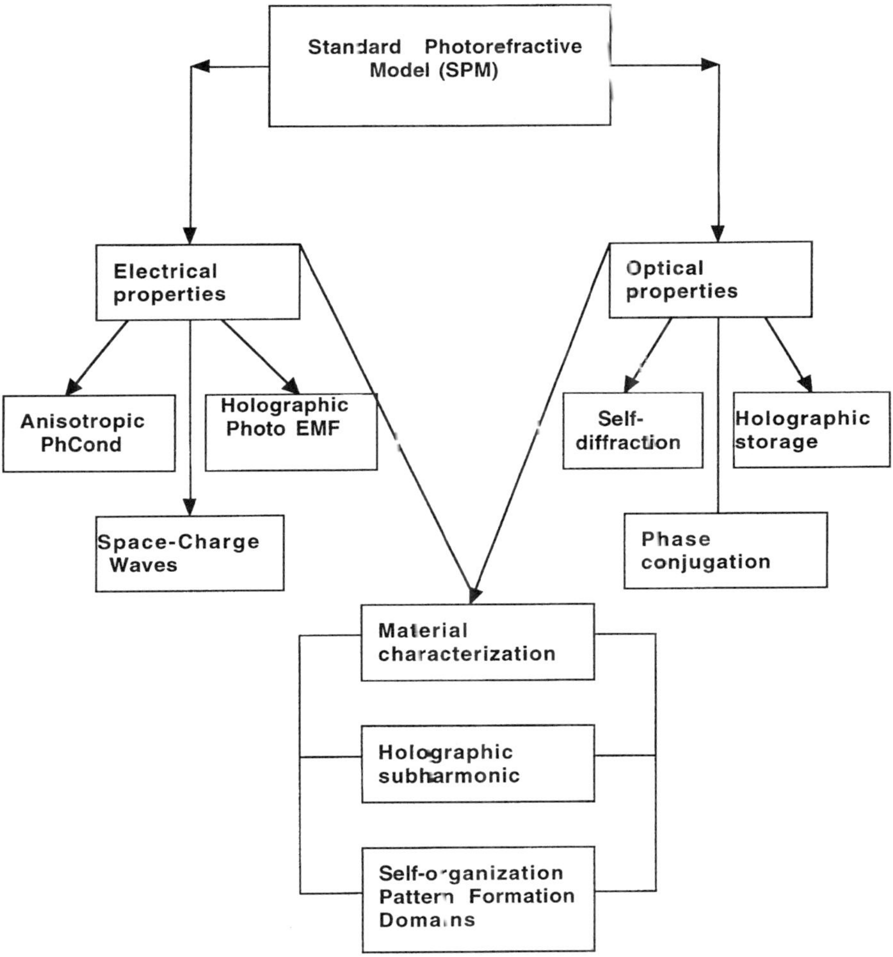

Figure 1.1: Diagram showing relations of SPM to different applications.

1.3 Small-contrast approximation

The equations described in the preceding can be simplified for the case of small contrast of the exciting light interference intensity pattern

$$I(x,t) = I_0\left\{1 + \left[\frac{m}{2}\exp(ikx - i\Omega t) + \text{c.c.}\right]\right\}^{\dagger} \tag{1.4}$$

where m is intensity contrast, k is the grating vector, I_0 is the average intensity, and Ω is the frequency difference between two laser beams. For the one-dimensional case the carrier concentration n and the electric field E in the crystal can be introduced as

$$\begin{aligned} n &= n_0 + (n_1 \exp(iu) + \text{c.c.}) \\ E &= E_0 + (E_1 \exp(iu + \text{c.c.}) \\ u &= kx - \Omega t \end{aligned} \tag{1.5}$$

Here the average electric field E_0 (for short circuit) equals applied voltage, divided by sample thickness. For n_0, n, E we can get from Eqs. (1–5) (assuming linear recombination):

$$\begin{aligned} &\frac{\partial n_0}{\partial t} = g_0(1+\gamma)(N - N_A) - \frac{n}{\tau} \\ &\tau\left(\frac{\partial}{\partial t} - i\Omega\right)n_1 = n_0\left[\frac{m}{2(1+\gamma)} + i\frac{E_1}{E_k} - \frac{E_1}{E_M}\right)\Bigg] - n_1\left(1 + \frac{E_D}{E_M} + i\frac{E_0}{E_M}\right) \\ &\tau_M\left(\frac{\partial}{\partial t} - i\Omega\right)E_1 = (iE_D - E_0)\frac{n_1}{n_0} - E_1 \end{aligned} \tag{1.6}$$

where $\gamma = \sigma_d/\sigma_p = \beta/SI_0$ is the ratio of dark conductivity (σ_d) and photoconductivity (σ_p), $\tau_M = \varepsilon\varepsilon_0/(\sigma_d + \sigma_p)$, $\tau = (rN_A)^{-1}$ is the recombination time, $E_k = eN_A\,(N - \mathrm{N}_A)(\varepsilon\varepsilon_0 kN)^{-1}$ is the so-called limiting space-charge field, $E_D = (D/\mu)k$ is the diffusion field, and $E_M = (k\mu\tau)^{-1}$, $g_0 = SI_0$. In the one-dimensional case, total current does not depend on the coordinate and can be expressed as [16, 19].

$$J = \left\langle n^{-1}\right\rangle^{-1}\left\{e\mu E_0 + \varepsilon\varepsilon_0\left\langle\frac{1}{n}\frac{\partial E}{\partial t}\right\rangle\right\} \tag{1.7}$$

†c.c. = complex conjugated

where $\langle \ldots \rangle$ stands for a spatial averaging. Equation (1.7) accounts for the contributions from both the steady-state current (first term) and the transient (EMF) current (second term). Both types of current can be used for a crystal characterization.

In the low-contrast approximation, the expression for the current can be simplified:

$$J = e\mu n_0 (1 - m^2{}_n)^{1/2} - 0.5\varepsilon\varepsilon_0 \left(m_n \frac{\partial E^*1}{\partial t} + m^*{}_n \frac{\partial E_1}{\partial t} \right) \tag{1.8}$$

where $m_n = 2n_1 / n_0$.

For the steady-state solution, Eq. (6) we have

$$\begin{aligned} n_0 &= g_0\tau(N - N_A)(1 + \gamma); n_1 = \frac{m(iE_D - i\Omega\tau_M)}{2(1 + \gamma)(a + ib)}; \\ E_1 &= \frac{m(iE_D - E_0)}{2(1 + \gamma)(a + ib)}. \end{aligned} \tag{1.9}$$

Here

$$a = 1 + \frac{E_D}{E_k} + \Omega\tau_M\left(\frac{E_0}{E_M} - \Omega\tau\right), b = \frac{E_0}{E_k} - \Omega\tau - \Omega\tau_M\left(1 + \frac{E_D}{E_M}\right),$$

The expression for a total current will be:

$$J = e\mu n_0 \left(E_0 - \frac{m^2 (E_0 + \Omega\tau_M E_D)}{2(1 + \gamma)(a^2 + b^2)} \right) \tag{1.10}$$

From Eq. (1.10) we can see, that even without external electric field, running gratings will introduce electric current (holographic current). The sign of this current will depend on the direction of the running grating.

1.4 Space-charge waves and dispersion relations

Equations for the fundamental spatial harmonics of the electric field E_1 and photoexcited carriers n_1 may be interpreted as equations for the space-charge waves (SCW). This concept of SCW is widely used in semiconductor physics, especially in analysis of stability of the electrical circuits [12, 20, 21]. Even for homogeneous illumination, the external electric field initiates internal SCW (eigenmodes) that propagate between electrodes as a damped wave with frequency ω_k and wavevector k. To find dispersion relations between ω_k and k for the eigenmode we can use Eq. (9) for the spatial harmonic E_1. Formally setting $m \to 0$ (this means that we are considering uniform illumination) from the condition $E_1 \neq 0$ we get the following generation condition for the denominator in Eq. (1.9):

$$a + ib = 0. \tag{1.11}$$

From this generation condition we can get the dispersion relation (for the case $1/\tau >> (\beta + SI_0)N/N_A$)

$$\omega_k = \frac{1 + \dfrac{E_D}{E_k} + i\dfrac{E_0}{E_k}}{\tau_M\left[i\left(1 + \dfrac{E_D}{E_m}\right) - \dfrac{E_0}{E_M}\right]}. \tag{1.12}$$

In an equivalent form this dispersion relation may be rewritten with real and imaginary parts $\omega_k = \omega_R - i\Gamma$

$$\omega_R - i\Gamma = \omega_0 \frac{E_k + E_D + iE_0}{i(E_M + E_D) - E_0} \tag{1.13}$$

with $\omega_0 = E_M(\tau_M E_k)^{-1} = (\beta + sI_0)\, NN_A{}^{-1}$ which designates the ionization rate.

Introduction of the eigenmode frequencies allows us to rewrite amplitude E_1 in the form:

$$E_1 = \frac{m\omega_0\,(iE_D - E_0)\,E_k}{2(1 + \gamma)\,[E_0 - i(E_M + E_D)]\,(\Omega - \omega_R + i\Gamma)} \tag{1.14}$$

with traditional "resonant" dependence of the amplitude E_1 on the detuning frequency.

Interesting features of the dispersion relations (Eqs. 12–13) are:

1. Antisymmetric form $\omega_k(-k) = -\omega_k(k)$
2. Inverse dependence of ω_k on k (for $E_k >> E_0 >> E_M, E_D$)

1.5 High-contrast gratings

Special considerations are deserved in the case of the high-contrast gratings, when two recording beams have near-equal intensities. It means that minimum intensity of light is near zero. It results in stratified distribution of photoconductivity with near-dark conductivity minima regions. Resulting macroscopic steady-state current along holographic grating J_{11} will be $\sqrt{1 - m_n^2}$ times less than the photocurrent with homogeneous illumination with the same intensity I_0. For the high-contrast limit $m_n \to 1$ [19] and current $J_n \to 0$. This simplified approach shows the critical dependence of electrical current on the contrast. Because of the interconnection of the optical and electrical properties, high-contrast effects are also pronounced in self-diffraction (with higher harmonics and subharmonics generation [20, 21]). In the following sections we will discuss electrical and optical effects separately in high-contrast gratings, emphasizing anisotropic photoconductivity and subharmonic generation.

1.6 Photoinduced anisotropic photoconductivity for optical interconnection of two electric circuits

The anisotropic photoconductivity method was originally suggested in [1–3] and tested for $KNbO_3$ [7] and for detection of the so-called reduction factor [19, 22]. This method is based on the measurements of the holographic current in the photorefractive crystal under applied voltage with and without a high-contrast interference pattern. By measuring the current change and using the analytical solution of the SPM for arbitrary contrast it is possible to find modulation of photoconductivity and the carrier diffusion length (as was illustrated in [16] for $B_{12}SiO_{20}$ [BSO] crystals). Figure 1.2 depicts the geometry of the incident intensity fringes and the conductivity grating formation. A constant electric field E_x is applied to the crystal along the x-axis, and a probe field E_y is applied along the y-axis. The tilt of the fringes is measured by the angle φ between the k vector of the grating and the x-axis. The x and y components of the photocurrent density can be found as [16]:

$$
\begin{aligned}
J_x &= E_x(\sigma_\perp \cos^2\varphi + \sigma_\parallel \sin^2\varphi) + E_y(\sigma_\perp - \sigma_\parallel)\frac{\sin^2\varphi}{2} \\
J_y &= E_y(\sigma_\parallel \cos^2\varphi + \sigma_\perp \sin^2\varphi) + E_x(\sigma_\perp - \sigma_\parallel)\frac{\sin^2\varphi}{2}
\end{aligned}
\tag{1.15}
$$

with $\sigma_\parallel = \sigma = e\mu n_0$, $\sigma_\perp = (1 - a^2)^{1/2}\delta$.

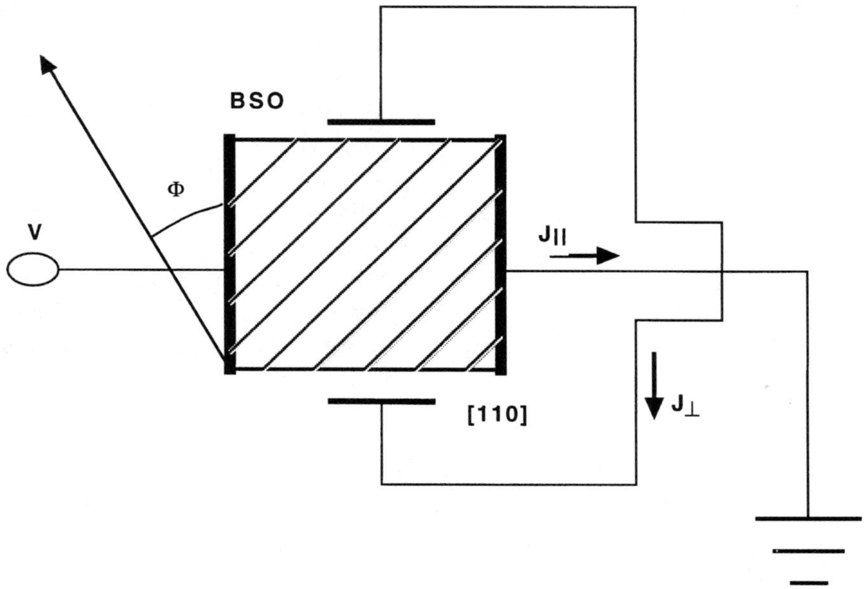

Figure 1.2: Experimental setup for optical interconnection in BSO.

Here modulation of conductivity a is related to the interference fringe contrast m by [19]:

$$a = \frac{M}{1 + (M\delta)^2}\{1 - [1 + M^2(\delta^2 - 1)]^{1/2}\}. \tag{1.16}$$

Here M is the dressed modulation of the free-carrier distribution (or photoconductivity):

$$M = \frac{m(1 + \lambda)}{(1 + \gamma)[(1 + \lambda)^2 + \xi^2]^{1/2}} \tag{1.17}$$

where $\delta = \lambda/(1 + \lambda)$, $\lambda = E_D/E_k$, $\xi = E_0/E_k$.

We can see from Eq. (15), that two electrical circuits (J_x and J_y) may be interconnected by the inteference pattern, using the photoconductor. The strength of interconnection depends on the grating contrast as well as on the angle of the grating rotation.

It is important to notice that for this optical interconnection incoherent illumination can be used. The interference pattern can be imitated with the help of the spatial-light modulator (electrically or optically addressed).

1.7 Photoconductivity grating as an optically scanning antenna

The grating of photocarriers, in addition to effecting photoconductivity modulation, also initiates optical gratings due to modulation of optical parameters. In the noncentrosymmetric materials electric field grating (initiated by photocarrier grating) modulates refractive index via the linear electrooptic effect. This is the standard photorefractive effect, and it is usually a relatively slow process of grating formation and decay (with Maxwell relaxation time).

In the photosensitive semiconductors (as Si) photogenerated carriers can modulate refractive index χ via the Drude effect

$$\chi = \chi_0 \left(1 - \frac{w_p^2}{w^2}\right) \tag{1.18}$$

where $w_p = \frac{e}{2}\sqrt{\frac{n}{m_e \varepsilon \varepsilon_0}}$ is the plasma resonance frequency, m_e is effective mass of the carrier, and w is the diffracting wave frequency. The electromagnetic wave (the light wave or microwave) may be effectively diffracted on this photoinduced plasma grating. This application was suggested and realized in [23] for optically scanning microwaves (millimeter waves) (Fig. 1.3). The angle of diffraction depends on the period controlled by the liquid crystal display. This optically scanned antenna with mirror modification can be also used as an optically controlled optical or microwave interconnection. It is important that the same photoinduced grating of photoconductivity (or plasma grating) can be used simultaneously as coupler between two electrical circuits and as antenna-interconnect. These double-functional features of the photoinduced gratings may be used for the next-generation devices with improved reliability.

1.8 Subharmonic domains of the space-charge waves

The interesting phenomenon of generation of the subharmonic wave (diffracted beam between two recording beams) was observed in $Bi_{12}SiO_{20}$ (BSO) crystals [20, 21]. This effect may be explained as instability of space-charge waves (eigenmodes) naturally generated in the photoconductive crystals by an applied electric field. These eigenmodes can be resonantly

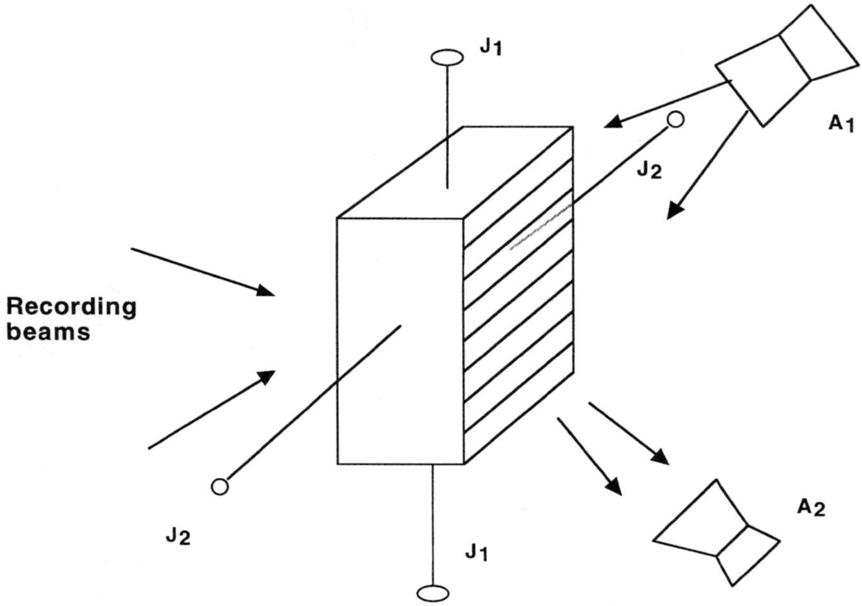

Figure 1.3: Multifunction optically scanning antenna and coupler-interconnect between electrical circuits (A_1 = transmitter; A_2 = receiver; J_1, J_2 = coupled electrical circuits.)

amplified by the moving light-intensity patterns [21]. Another remarkable property of the subharmonic beam is the appearance of the well-defined domain structure [24, 25] in the near-field transversal cross section. It was found that domains move from one electrode to another with the complex dynamic of aggregation and destruction. We need to use a three-dimensional version of SPM to describe these spatiotemporal dynamic patterns. For the quasistationary approximation, valid when recombination of carriers is much faster than Maxwell space-charge relaxation, and for linear recombination we can exclude N from Eq. (1.1). In this case reduced equations for n and E will be

$$nr\left(N_A + \frac{\overline{\nabla}(\varepsilon\varepsilon_0\overline{E})}{e}\right) = G(N_0 - N_A) - \frac{1}{e}\left(G + \frac{\partial}{\partial t}\right)\overline{\nabla}(\varepsilon\varepsilon_0\overline{E})$$

$$\left(\varepsilon\varepsilon_0\frac{\partial}{\partial t} + e\hat{\mu}n\right)\overline{E} + eD\overline{\nabla}n = \overline{J} \tag{1.19}$$

where $G \equiv g + \beta$.

Excluding n, we will get finally only one reduced equation for the electric field E:

$$\varepsilon\varepsilon_0 \frac{\partial \overline{E}}{\partial t} + e(\hat{\mu}\overline{E} + D\overline{\nabla}) \left[\frac{eG(N_C - N_A) - \left(G + \frac{\partial}{\partial t}\right) \overline{\nabla}(\varepsilon\varepsilon_0 \overline{E})}{r(eN_A + \overline{\nabla}(\varepsilon\varepsilon_0 \overline{E}))} \right] = \overline{J}. \qquad (1.20)$$

This equation for total current $\overline{J}$, together with Eqs. (2), may be used as a starting fundamental system for considering the three-dimensional case of subharmonic domain generation. For the important special case of "weak quasineutrality" (when $eN_A > |\overline{\nabla}\, \varepsilon\varepsilon_0 \overline{E}|$), Eq. (1.20) will be simplified [neglecting quadratic terms of (∇E)]:

$$\varepsilon\varepsilon_0 \frac{\partial \overline{E}}{\partial t} + e(\mu\overline{E} + D\overline{\nabla})\tau_r \left[G(N - N_A) - \frac{1}{e}\left(G\frac{N_0}{N_A} + \frac{\partial}{\partial t}\right) \overline{\nabla}(\varepsilon\varepsilon_0 \overline{E}) \right] = \overline{J} \qquad (1.21)$$

Some special cases of the system of Eqs. (2) and (20) were discussed [21, 26], in relation to the subharmonic generation but without addressing domain structure.

The nonlinear system of Eqs. (2) and (20) include parametric generation terms (with G) from the point of view of nonlinear dynamics. This system may parametrically generate subharmonics of the space-charge waves. A more detailed paper on the parametric generation of subharmonics will be published elsewhere.

1.9 Formation of the spatiotemporal patterns and domains, optical channeling

Recent experiments on laser light scattering in photorefractive materials reveal another interesting effect: self-organization of near-field scattering in the regular arrays (hexagon, squares, rolls, etc.) [27, 28]. An example of the hexagonal pattern formation in the photorefractive crystal $KNbO_3$:Fe is shown in Fig. 1.4 [28]. For more details about self-organization in hexagonal arrays see Banergee *et al.* [29]. Here we wish to emphasize one innovative view on near-field pattern formation in transparent materials with phase (not absorption!) gratings. Transversal modulation of the refractive index (phase gratings) of a thick photorefractive crystal may be described also as formation of a bunch of optical

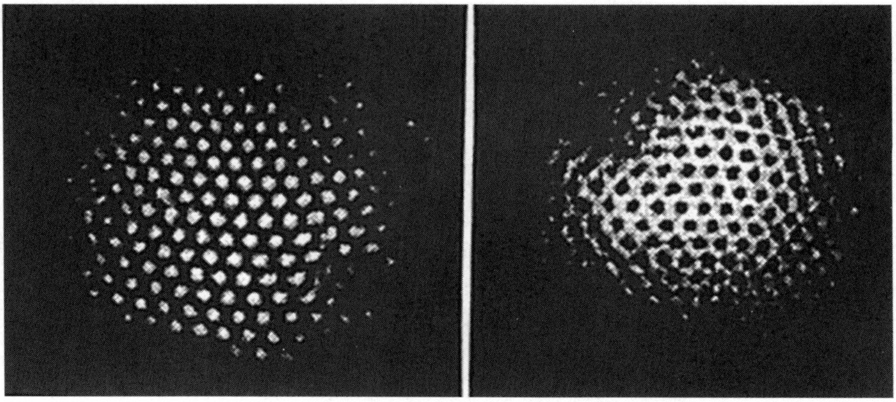

(a) (b)

Figure 1.4: Near-field patterns showing contrast inversion. (a) Positive image. (b) Negative image showing contrast inversion.

waveguides or channels. As it was shown in [30, 31], these transversal modulations of the refractive index may be visualized in the near-field by an optical channeling.

As example, we will consider two-dimensional modulation of the optical dielectric constant

$$\varepsilon(x, y) = \varepsilon_0 + \varepsilon_x \cos k_x x + \varepsilon_y \cos (k_y y + \varphi) \tag{1.22}$$

where ε_0 denotes the average value of the permittivity, and $\varepsilon_{x,y}$ are the amplitudes of the modulation along the transverse x- and y-axis. Introducing the function $\varepsilon(x, y)$ in the Maxwell's equations, we can get the following results for near-field intensity:

$$\begin{aligned} I(x, y) = I_0 \{1 &+ \varepsilon_x (L_x / \lambda)^2 \sin^2 (\pi\lambda^2 / 2L_x^2) \cos k_x \lambda \\ &+ \varepsilon_y (L_y / \lambda)^2 \sin^2 (\pi\lambda^2 / 2L_y^2) \cos (k_y y + \varphi) \end{aligned} \tag{1.23}$$

where $L_{x,y} = 2\pi / k_{x,y}$ and λ is the wavelength. This solution is valid for small modulation, describes longitudinal modulation with the periods

$$z_{x,y} = 2L_{x,y}^2 / L \tag{1.24}$$

and also describes the experimentally observed phenomenon of contrast inversion [28]. For experimental data with $KNbO_3$ ($\lambda = 0.514$ μm, $L_x =$

$L_y = 35$ μm) we get for a longitudinal period $z_x = z_y = 0.49$ cm, that is close to the experimental value of 0.5 cm [28].

In Fig. 1.5 the phenomenon of optical channeling is shown schematically with optical rays bunching together in different transversal planes, forming contrast inversion. We should note that the effect of channeling is important for the thick gratings, and for large enough modulation of refractive index. For optically thin-phase gratings, the description of visualization by Talbot effect is more adequate. In the case of Talbot effect, contrast inversion is due to the propagation in free-space, while channeling is possible only in the thick photorefractive materials.

Another example of visualization by channeling is domain pattern in the photorefractive subharmonics, observed in the near field. In this case the subharmonic beam propagates between recording beams. This geometry is favorable for channeling provided that modulation of refractive index is large enough for wave guiding. In experiments with subharmonic waves, fundamental gratings run with the resonance speed $V_r = \omega_r / k$, while domains move in the opposite direction with the same speed. This interesting result is a manifestation of a specific dispersion relation for the space-charge waves that for $E_k >> E_0 >> E_m$ takes a simple form (see Eq. 1.13):

$$\omega = \tau_M \tau^{-3} (\mu E_0 k)^{-1}$$

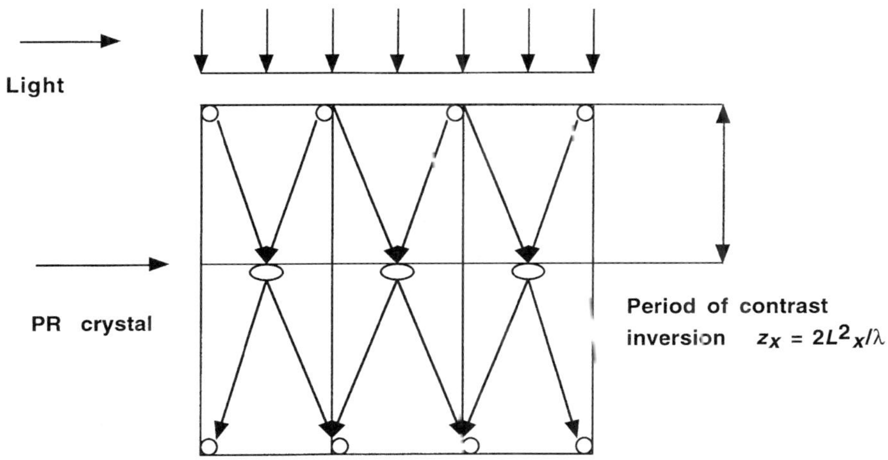

Figure 1.5: Optical channeling by photoinduced gratings with the contrast inversion effect.

where τ_M and τ are Maxwell and recombination times, and μ is the photocarrier mobility. With this "inversion k" dispersion law for the amplitude of the wave package, formed by the two subharmonic waves [21], group velocity for the wave package will be

$$V_g = \frac{\partial \omega}{\partial k} = -\frac{\omega}{k}$$

in accordance with experimental results.

1.10 Conversion of heat into electric current by moving gratings

We will now discuss modification of SPM for the description of thermoelectric conversion of energy of infrared radiation into electric current.

Many technological processes are accompanied by excessive heat. For example, the technology of pyrolitic treatment of different types of materials, including hazardous wastes [32], that use thermochemical reactions, waste heat from vehicle propulsion engines, and nuclear reaction in nuclear power plants. For environmental safety it is important to solve the problem of utilization of the excessive heat. We will address the problem of excessive heat utilization by suggesting a new method of heat-to-electricity conversion based on the moving-grating technique. Recently this method was applied for the creation of photo-EMF in photorefractive crystals, which can be used as sensitive vibration sensors [9, 10]. The same method applied to solid material can generate electrical current by photoelectromotive force (EMF) [9, 10]. It is interesting to note that this method allows the use of the infrared part of solar irradiation spectra, which is a longstanding problem in solar power technology.

The method of photoinduced running grating was suggested for optical decontamination of photosensitive solutions and aerosols [33]. It was supposed that running grating formed by laser irradiation creates photoconductivity and electric field grating through photoionization. The goal of this chapter is to expand this method of running grating for nonionizing radiation such as infrared or heat irradiation. The basic idea of this approach is the use of a spatially modulated heat pattern (temperature grating, or TG) to create in materials periodic distribution of temperature T, that modulates charge distribution through thermodiffusion. These thermoinduced processes lead to the appearance of gratings of electric conductivity σ and temperature T, which may produce electric current with the proper spatial matching. Optimal spatial matching may be achieved by the moving-grating technique described in the basic model.

1.10.1 Basic model of thermoelectric transient current

We will consider, as a specific example, the case when infrared irradiation with spatially modulated intensity $I(x, t)$

$$I(x) = I_0(1 + m\cos(kx - \Omega t)) \tag{1.25}$$

is heating a thin sample layer ((Fig. 1.6), thermoconductive polymer or crystal, etc.), k is the interference pattern grating vector, m is the amplitude of modulation, and Ω is the frequency of running grating. The speed of the running grating V_g is simply related to Ω and k as $V_g = \Omega k^{-1}$. The

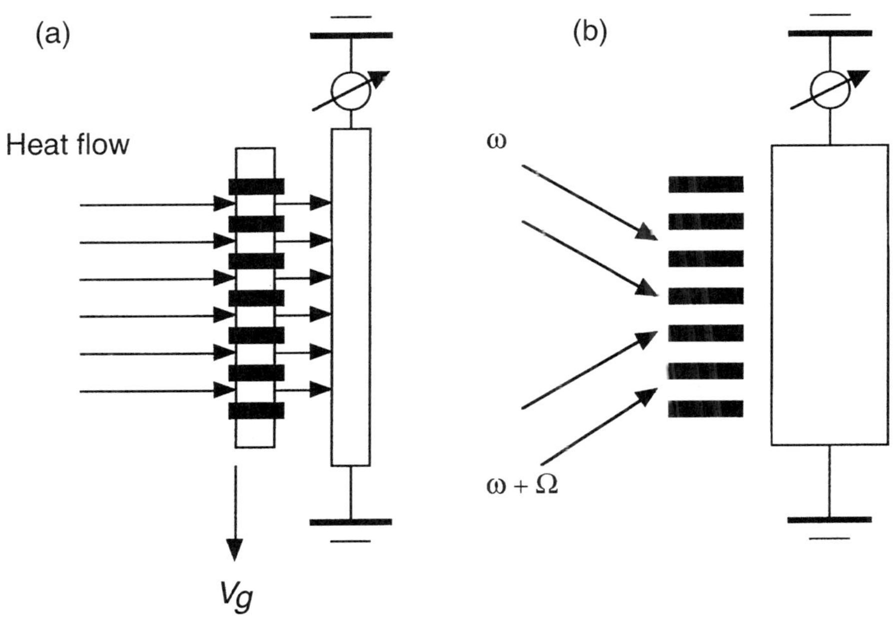

Figure 1.6: Method of creating moving gratings. (a) With incoherent radiation using moving or rotating periodic mask. (b) With coherent frequent-shifted beams ω and $\omega + \Omega$.

kinetic of the temperature T distribution will be described by the heat equation

$$\frac{\partial T}{\partial t} = aI(x) - \frac{T - T^0}{\tau} + D_T \frac{d^2 T}{dx^2} \tag{1.26}$$

where α is the "effective" absorption coefficient, τ is heat relaxation time, T^0 is the initial sample temperature (without radiation), and D_T is the coefficient of thermal diffusion. We consider diffusion only for the x-axis, which is justified for thin film and homogeneous illumination along the y-axis. We will consider two charged components: a negatively charged light component in concentration n and a positively charged heavy component in concentration N. The light component n can be moved by diffusion and thermodiffusion creating an electric current. The resulting redistribution of negative charges will induce diffusion and drift currents and electric field. For determination of n and electric field E we will have the Poisson's equation and conservation law for concentration

$$\nabla(\varepsilon E) = e(N - n) \tag{1.27}$$

$$e\frac{\partial n}{\partial t} = \overline{\nabla} j \tag{1.28}$$

with current

$$j = e\mu n\left\{E + \frac{k_B T}{en}\nabla n - \beta\nabla T\right\}$$

where μ is mobility of the mobile charged particles, ε is the static dielectric constant, β is the thermoelectric (Seebeck's) coefficient, and k_B is the Boltzmann's constant.

Total current J will be the sum of the diffusion-drift current and displacement current

$$\varepsilon\frac{\partial E}{\partial t} \tag{1.29}$$

$$J = j + \varepsilon\frac{\partial E}{\partial t}$$

With Kirchhoff's law for the external electric circuit with resistance R, electrode area S, and interelectrode distance d, and external power supply of voltage V

$$V = E_0 d + JRS \tag{1.30}$$

(here E_0 is an average electric field). We have the basic set of equations needed for the calculation of the thermocurrent.

1.10.2 Solution of the basic equations

We are seeking the steady-state solution of the basic equations for T, n, E in the form of running gratings

$$T = T_0 + (T_1 e^{iu} + \text{c.c.}) \tag{1.31}$$

where $U \equiv \vec{k}\vec{r} - \Omega t$. Similar relations are also valid for n and E. For small contrast approximation ($m < 1$), neglecting dependence of μ and D_T on T we can get

$$\begin{aligned}
T_0 &= T^0 + \Delta T \\
\Delta T &= a\tau I_0; \qquad n_0 = N; \\
T_1 &= \frac{m\Delta T}{2\left(1 + \dfrac{\tau}{\tau_D} - i\tau\Omega\right)}; \qquad n_1 = -\frac{ik\varepsilon}{e}E_1 \\
E_1 &= \frac{i\beta k T_1}{1 + \tau_\mu \Omega_D - i\tau_\mu(\Omega + \Omega_E)}
\end{aligned} \tag{1.32}$$

$$\text{here } \tau_\mu{}^{-1} = \frac{e\mu n_0}{\varepsilon}; \quad \Omega_D \equiv Dk^2; \quad \tau_D{}^{-1} = D_T k^2; \quad \Omega_E = \mu k E_0$$

$$D = \frac{\mu}{e} k_B T$$

From Eq. (32) we can see that for standing gratings ($\Omega = 0$) in the absence of external electric field ($\Omega_E = 0$), temperature grating and grating of n will be in phase (unshifted) in respect to the light intensity pattern, while E-field grating will be $\pi/2$-shifted.

For running light intensity grating ($\Omega \neq 0$) a phase shift depending on Ω will be introduced in gratings of T, n, and E. Macroscopic current in the external circuit can be found by averaging microscopic current (28) in space:

$$J = \sigma\left\{E_0 - \frac{E_T^2(1+f)(\Omega + \Omega_E)m^2}{2E_K[(1 + \tau_\mu\Omega_D)^2 + (\Omega + \Omega_E)^2\tau_\mu^2][(1 + \tau\tau_0{}^{-1})^2 + (\tau\Omega)^2]}\right\} \tag{1.33}$$

where $E_T = \Delta T\, \beta k$ is the thermoelectric field, $\sigma = e\mu n_0$, $E_K = \mu K \tau_\mu$, $f = k_B/\beta e$. With no external voltage $E_0 = -RSJ/d$ and choosing external resistance $R = d/\sigma S$, we get for $\Omega >> \Omega_E$

$$J = -\frac{\sigma}{4}\frac{E_T^2(1+f)\Omega\tau_\mu E_K{}^{-1}m^2}{\left[(1+\Omega_D\tau_\mu)^2 + (\Omega\tau_\mu)^2\right]\left[\left(1+\frac{\tau}{\tau_D}\right)^2 + (\Omega\tau)^2\right]} \tag{1.34}$$

For the optimized value of current, assuming resonance frequency shift $\Omega\tau_\mu = 1$ and $1 >> \Omega_D\tau_\mu\ \tau/\tau_D$, τ/τ_μ we get

$$J = -\frac{\sigma(1+f)}{8E_K}E_T^2 m^2 \tag{1.35}$$

In Eq. (35) current is proportional to the square of radiance intensity $\sim I_0^2$, with a limiting value $E_T \leq E_K$. This limit follows from the criteria of small contrast $\left|\frac{n_1}{n_0}\right| \leq 1$, and not from physical restrictions. For maximal value of current from Eq. (1.35) we get

$$J \leq \frac{\sigma(1+f)E_T^2}{16E_k} = 0.0625\mu k\varepsilon\varepsilon_0(1+f)E_T^2 \tag{1.36}$$

For the typical values for semiconductors $\beta = 1$ mV/^{0}K, $\mu = 10$ cm^2/Vs†, $\sigma = 1.6*10^{-2}$ S/m‡ (for $n_0 = 10^{14}$ cm^{-3}), $\tau_\mu = 10^{-8}$ s, $K = 3*10^3$ cm^{-1}, $E_k = 3.3$ kV/cm, $J_M = 0.03$ A/cm^2, $P_M = J^2/\sigma = 8$ W/cm^3.

To reach these maximal values of J and P, we need a large input intensity of heating radiation I_0. For $\alpha = 10^3$ (cm^{20}K)/Ws and $\tau = 10^{-9}$s, we will get for the critical value of input intensity $I_0 = 3 \cdot 10^7$ W/cm^2.

For the optimized values of parameters with smaller periods and larger intensities: $k{=}10^4$ cm^{-1}, ($L = 6.3$ μm), $\mu = 1$ cm^2/(vs), ($n_0 = 10^{14}$ cm^{-3}), $\sigma = 1.6*10^{-3}$ S/m; $\tau_\mu = \tau = 6*10^{-8}$ s, we get for $I_0 = 3.9*10^8$ W/cm^2, $J = 5*10^3$ A/cm^2, and $P = 10^{11}$ W/cm^3 In this optimal case efficiency of energy conversion

$$\eta = P/aI_0 = 27\%$$

These estimations show only the order of magnitude of the input and

†Vs = volt · second

‡S/m $\frac{\text{siemence}}{\text{meter}}$

output energies. An important result is that for high input intensities, efficiency of energy conversion may be large enough for practical applications.

1.11 Conclusions

We have briefly discussed the evolution of the standard photorefractive model (SPM), starting from the original semiconductor version. Being based on the description of photoionization and drift-diffusion recombination, SPM is valid not only for semiconductors but also for a variety of other photosensitive materials: ferroelectrics, photorefractive polymers, and liquid crystals. In general, SPM is valid for those materials where we can introduce drift-diffusion approximation together with photogeneration-recombination. This model was successful not only to explain the self-diffraction, holographic recording, and storage but also predicted holographic EMF and anisotropic photoconductivity. The connection of SPM to the description of the optical and electrical properties of the photosensitive materials is shown in Fig. 1.1. It is interesting to note that promising features of the conductivity modulation were stated in the early papers [2, 7] and later were "rediscovered" in semiconductor physics [34]. New pragmatic interest in the electrical effects in dynamic gratings was inspired by the concentrated efforts of S. Stepanov's group [9] on application of the transient EMF for ultrasensitive vibrometry. The applicability of the transient EMF method for characterization of the laser crystals [35] was a remarkable surprise for us.

Interest in the optical properties of dynamic gratings was permanently high during more than two decades, due to attractive manifestation of self-diffraction, phase conjugation, and holographic storage. Experimental discovery of the holographic subharmonics in BSO crystals [24, 25] and domain pattern in near field of subharmonics [24, 25] renewed interest in the electrical instabilities and the space-charge waves. At the same time in holographic subharmonic domains both optical and electrical properties of the dynamic gratings were combined together. Another challenge for SPM was explanation of self-organization of the light scattering in the spatiotemporal patterns [27, 28] including photorefractive solitons [36]. It is interesting to note that ideas of real-time holography tested in physically unanimated objects now spread to the biological sciences. The phenomenon of self-diffraction was observed in suspensions of living microorganisms [37].

A promising new application of real-time gratings was suggested recently for the problem of high-density magnetic data storage [38]. From the proposed model of temperature-induced current using moving illumination grating we can see that current can be generated without any external electric field. The direction of this current depends on the direction of the running grating and may be controlled from a distance. Running temperature grating may be generated by the intersection of two coherent infrared (IR) or visible laser beams with frequency shift Ω. Another possibility is using a moving periodic binary mask (such as a radial grating; see Fig. 1.6) before the sample is investigated. In this case, the radiation source may be incoherent, which makes it more attractive for practical applications.

We can conclude that SPM (including material equations and Maxwell's wave equations) may be regarded as a foundation for real-time holography in a variety of different materials including animated biological objects.

A review of applications of the photorefractive gratings for optical storage, interconnection, and processing can be found in [39].

Acknowledgments

This work was done with support from AAMU/SHBCO/MI Environmental Consortium and National Science Foundation (NSF) CREST grant No. HRD-9353548 (for N. K. and T. K.).

We would like to thank Professor F. Yu for his encouraging suggestions during preparation of this paper.

References

1. N. Kukhtarev, *Sov. Tech. Phys. Lett.*, **2**, 438, 1976; N. Kukhtarev and A. Semenyuk, *Radiation Physics of Nonmetallic Crystals* [in Russian], Vol.3, Chap.1, Naukova Dumka, Kiev (1971).
2. N. Kukhtarev, *Sov. Tech. Phys. Lett.*, **1**, 155, 1975; V. Vinetskii, N. Kukhtarev, and A. Semenyuk, *Sov. Phys.-Semicond.*, **6**(18), 879, 1972; V. Vinetskii and N. Kukhtarev, *Sov. Phys. Solid State Phys.*, **16**, 3714, 1974.

3. N. Kukhtarev, V. Markov, S. Odulov, M. Soskin, V. Vinetskii, *Ferroelectrics,* **23**, 949, 1979.
4. J. Amodei and D. Staebler, *J. Appl. Phys.,* **43**, 1042, 1972.
5. M. Deigen, S. Odulov, M. Soskin, and B. Shanina, *Sov. Phys. Solid State,* **16**, 1237, 1975.
6. F. Chen, J. LaMacchia, and D. Fraser, *Appl. Phys. Lett,* **13**, 223, 1968.
7. A. Krumins and P. Gunter, *Phys. Status Solidi,* **A63**, K111, 1981.
8. V. Vinetskii and N. Kukhtarev, *Sov. Tech. Phys. Lett.,* **1**, 84, 1975.
9. S. Trofimov, S. Stepanov, M. Petrov, and M. Krasinkova, *Sov. Tech. Phys. Lett.,* **13**, 108, 1987; I. Sokolov and S. Stepanov, *Optik 93,* 175, 1993.
10. D. Pepper, *Optics & Photonics News,* May 1997, p. 33.
11. N. Kukhtarev, V. Markov, and S. Odulov, *Opt. Comm.,* **23**, 338, 1977.
12. V. Alimpiev and I. Gural'nik, *Sov. Phys., Semicond.* **18**(9), 978, 1984.
13. E. Convell, *Transport Properties,* Academic Press, 1967.
14. N. Kukhtarev and A. Kondrachuk, *Sov. Tech. Phys. Lett.,* **6**(11), 1980.
15. B. Volodin, B. Kippelen, K. Meerholz, N. Kukhtarev, H. J. Caulfield, and N. Peghambarian, *Opt. Lett.,* **21**, 519, 1996.
16. N. Kukhtarev, S. Lyuksyutov, P. Buchhave, H. J. Caulfield , and M. Vasnetsov, *Opt.Lett.,* **21**, 1891, 1996.
17. B. Belinicher and B. Sturman, *Sov. Phys. Usp.,* **23**, 109, 1980.
18. A. Glass, D. Von der Linde, and T. Negram, *Appl. Phys. Lett.,* **25**, 233, 1974.
19. N. Kukhtarev, P. Buchhave, and S. Lyuksyutov, *Phys. Rev A,* **33**, 3133, 1997.
20. C. Kwak, J. Takacs, and L. Solymar, *Opt. Comm.,* **96**, 278, 1993.
21. B. Sturman, M. Mann, J. Otten, and K. Ringhofer, *J. Opt. Soc. Am.* **B12**, 1642, 1995.
22. P. Buchhave, S. Lyuksyutov, M. Vasnetsov, *Opt. Lett.,* **20**, 2363, 1995.
23. V. Manasson, L. Sadovnik, P. Shnitser, R. Mino, and J. Kniger, *SPIE,* **2532**, 290, 1995.
24. A. Grunnet-Jepsen, S. Elston, I. Richter, J. Takacs, and L. Solymar, *Opt. Lett.,* **18**, 1993.
25. P. Buchhave, S. Lyuksyutov, M. Vasnetsov, and C. Heyde, *J. Opt. Soc. Am.,* **B13**, 2595, 1996.
26. S. Shim, C. Kwak, and E. Lee, *Appl. Phys. Lett.,* **66**, 405, 1995.
27. T. Honda, *Opt. Lett.,* **18**, 598, 1993.
28. N. Kukhtarev, T. Kukhtareva, H. J. Caulfield, P. Banerjee, H. Yu, and L. Hesselink, *Opt. Eng.,* **34**, 2261, 1995.

29. P. Banerjee, Chapter 3 in this book, *Photorefractive Optics: Materials, Properties, and Applications,* Academic Press, Boston, 2000.
30. N. Kukhtarev, T. Kukhtareva, H. J. Caulfield, and A. Knyaz'kov, *Optik,* **8**, 1154, 1994.
31. R. Rupp and E. Kratzig, *Phys. Stat. Sol.,* **A72**, K5, 1982.
32. K. Tyson, M. Rymos, E. Hamond, *Biomass and Biochemistry,* Vol.10. 1996, pp. 2–10.
33. N. Kukhtarev, T. Kukhtareva, E. Ward. J. Jones, H. J. Caulfield, *Opt. Eng.,* **37**(9), 2597–2600, 1998.
34. V. Haken, M. Hunhaussen, and Ley, *Phys. Rev,* **B51**, 10579, 1995; D. Ritter, E. Zeldov, and K. Weser, *J. Appl. Phys.,* **62**, 4563, 1987.
35. N. Noginova, N. Kukhtarev, T. Kukhtareva, M. Noginov, H. J. Caulfield, P. Venkatesvarlu, D. Parker, and P. Banerjee, *J. Opt. Soc. Am.,* **B14**, 1390, 1997.
36. M. Saffman, A. Zozulya, and D. Anderson, *J. Opt. Soc. Am,* **B11**, 1409, 1994.
37. K. M. Brown, K. Lamb, D. L. Russel, I. S. Ruddock, A. Cunningham, R. Illingworth, *Opt. Comm.,* **126**, 202, 1996.
38. Yu. Verevkin, V. Petryakov and N. Polushkin, *Techn. Phys. Lett.,* **24**(6), 460, 1998.
39. S. Stepanov, *Appl. Opt.* **33**, 915, 1994.

Chapter 2

Light-Induced Charge Transport in Photorefractive Crystals

Karsten Buse and Eckhard Krätzig

Universität Osnabrück, Fachbereich Physik, Osnabrück, Federal Republic of Germany

Summary

Tailoring of photorefractive crystals for various applications requires detailed knowledge of the light-induced charge transport. In this chapter, in addition to the well-known one-center model, two-center, and three-valence models are discussed which have been developed to describe the transport properties of lithium niobate and lithium tantalate at high light intensities, as well as perovskite, sillenite, and tungsten-bronze-type crystals.

2.1 Introduction

Photorefractive crystals like $LiNbO_3$, $KNbO_3$, or $BaTiO_3$ offer fascinating possibilities in the fields of optical information processing, storage of holograms, laser beam manipulation, or holographic interferometry [1].

PHOTOREFRACTIVE OPTICS
Materials, Properties and Applications

ISBN 0-12-774810-5/$30.00

Large material nonlinearities are the basis of many unique devices, among them self-pumped phase-conjugating mirrors and parametric amplifiers and oscillators. But in most cases further improvements of photorefractive properties — faster response or higher efficiency — are absolutely necessary and require detailed knowledge of the processes involved [2,3].

Photorefractive effects in these crystals are based on the transposition of a light pattern into a refractive index pattern. Under nonuniform illumination charge carriers — electrons or holes — are excited and trapped at new sites. By these means electrical space-charge fields are set up which give rise to a modulation of refractive index. The trapped charge can be released and the field pattern erased by uniform illumination or by heating.

In this chapter, we review recent advances concerning the light-induced charge transport. We start with the so-called one-center model which describes the light-induced charge transport properties of transition metal-doped $LiNbO_3$ and $LiTaO_3$ at usual continuous-wave (cw) laser intensities ($\leq 10^5$ Wm^{-2}). Then we explain two-level models, the two-center and the three-valence model, which have to be applied to other photorefractive crystals. The validity of these models for $LiNbO_3$ and $LiTaO_3$ at high light intensities ($> 10^5$ Wm^{-2}), for ferroelectric perovskites, for sillenite-type, and for tungsten-bronze-type crystals is discussed.

2.2 One-center model

This model allows the description of photorefractive properties of doped $LiNbO_3$ and $LiTaO_3$ at usual cw laser intensities ($\leq 10^5$ Wm^{-2}). The pure crystals are transparent in the visible and near-infrared range. For this reason the crucial influence of transition metal dopants in the crystals was discovered already very early [4,5]. Transition metal ions may occur in two different valence states, making possible the generation of space-charge fields. A particular role is played by Fe impurities which are often present in the ppm range even in undoped (nominally pure) samples. But other transition metals, e.g., Cu, Mn, or Cr, cause similar effects.

Two valence states of transition metal ions in $LiNbO_3$ have been identified by various methods, among them electron-spin resonance (ESR), Mössbauer spectroscopy and optical absorption. In the case of Fe impurities, Mössbauer spectroscopy is of special importance. Both, Fe^{2+}

and Fe^{3+} spectra are observed [6], thus yielding the concentration ratio $c_{Fe^{2+}}/c_{Fe^{3+}}$. These investigations also demonstrate that only Fe^{2+} and Fe^{3+} centers are present. The valence states of impurities in $LiNbO_3$ can be greatly influenced by suitable thermal annealing treatments [5,6].

The light-induced charge transport in doped $LiNbO_3$ is mainly determined by a bulk photovoltaic effect [7], characteristic for pyroelectric crystals, and by photoconductivity. The photovoltaic current density $\vec{j}^{pv}$ was found to depend linearly on light intensity I and hence sesquilinearly [8] on the polarization (unit) vector $\vec{\mathbf{e}}$:

$$j_i^{pv} = \beta_{imn} I e_m e_n^*.$$

For $LiNbO_3$ (point group 3m) there exist four independent nonvanishing tensor elements β_{33}, β_{31}, β_{22}, and β_{15} (contracted indices). These elements are proportional to the concentration of filled traps [9], Fe^{2+} ions in the case of $LiNbO_3$:Fe.

The photoconductivity σ_{ph} is given by $\sigma_{ph} = N_e e\mu$, where N_e is the density, e the charge, and μ the mobility of excited electrons in the conduction band. The density N_e can be derived from the rate equation and the equations for constant trap density and for charge conservation:

$$dN^+/dt = -rN^+N_e + (qSI + \beta)N^0, \qquad N^0 + N^+ = N, \qquad N^+ + N_e = N_c.$$

Here N^+ and N^0 denote the concentrations of empty and filled traps, r is the recombination coefficient, q the quantum efficiency for generating a mobile electron upon absorption of a photon, S the absorption cross section, β the thermal generation rate, N the concentration of filled and empty traps, and N_c a constant concentration of compensation charge to maintain overall charge neutrality. If space-charge limiting effects and thermal excitations are neglected, we obtain in equilibrium: $N_e = q\,SIN^0/(rN^+)$ and $\sigma_{ph} \sim IN^0/N^+$, i.e., for Fe impurity centers $\sigma_{ph} \sim c_{Fe^{2+}}/c_{Fe^{3+}}$. This relation is confirmed experimentally for light of the visible spectral region [9, 10]. But it should be kept in mind that these experiments were carried out at usual cw laser intensities.

Photorefractive effects may be characterized by the saturation value Δn_s of refractive index change and the sensitivity $S = d(\Delta n)/d(It)|_{t=0}$. These quantities may be approximated by the expressions [6] $\Delta n_{so} = \frac{1}{2} n_0^3 r_{13} m \beta_{31} I / \sigma_{ph}$ and $S = \frac{1}{2} n_o^3 r_{13}\, m\, \beta_{31}/(\varepsilon\varepsilon_0)$. Here o refers to ordinary polarization, r_{13} denotes the effective element of the electrooptic tensor, m the modulation index, and ε the static dielectric constant.

From the above results $\beta_{31} \sim c_{Fe^{2+}}$ and $\sigma_{ph} \sim c_{Fe^{2+}} / c_{Fe^{3+}}$ we expect $S \sim c_{Fe^{2+}}$ and $\Delta n_{so} \sim c_{Fe^{3+}}$. This is again confirmed experimentally [10]. Only very strongly oxidized samples with a very small Fe^{2+} concentration exhibit smaller Δn_s values than expected [11] because of space-charge field limitations by depletion of filled traps (Fe^{2+} ions).

2.3 Two-center model

An important step for the understanding of photorefractive properties of ferroelectric perovskites was the discovery of light-induced absorption in $BaTiO_3$ by Motes and Kim [12]. This increase of absorption under illumination was interpreted in terms of two kinds of centers involved [13], each of them occurring in two different states. Holtmann [14] and later Mahgerefteh and Feinberg [15] successfully applied this two-center model to describe the transport properties of $BaTiO_3$. Light-induced absorption was also observed for other perovskites, e.g., $KNbO_3$ [16] and $KTa_{1-x}Nb_xO_3$ ($0 \le x \le 1$, KTN) [17], for sillenites, e.g., $Bi_{12}TiO_{20}$ (BTO) [18], or for $Sr_xBa_{1-x}Nb_2O_6$ ($0 \le x \le 1$, SBN) [19].

Because the photoconductivity of ferroelectric perovskites is mostly dominated by holes in the valence band [14,17], we assume in the following discussion of the two-center model only hole transport. (For electron transport an analogous argumentation holds.) With the help of Fig. 2.1 the transport of charge may be described as follows: We consider two different photorefractive centers C_1 and C_2. For each species, $i = 1, 2$, there are hole sources and traps. We denote the concentration of sources C_i^+ by N_i^+ and the concentration of traps C_i^0 by N_i^0. The total concentration of centers of type i is $N_i = N_i^+ + N_i^0$. Charge conservation requires $N_1^+ + N_2^+ + N_h = N_c$, where N_h is the concentration of holes in the valence band and N_c a constant concentration. The first center has to be a deep-level impurity, e.g., iron, and the second one should be a more shallow trap with respect to the valence band edge. The center C_2 has a relatively low thermal activation energy, such that $N_2^0 >> N_2^+$ holds in the dark. Charge is transported as follows: Upon illumination holes are generated by excitation of electrons from the valence band into C_1^+ centers. The holes migrate in the valence band and are trapped either by C_1^0 or by C_2^0 centers. Trapping at the latter creates C_2^+ centers. With increasing light intensity, more and more holes are generated and N_2^+ grows, too. By these means absorption

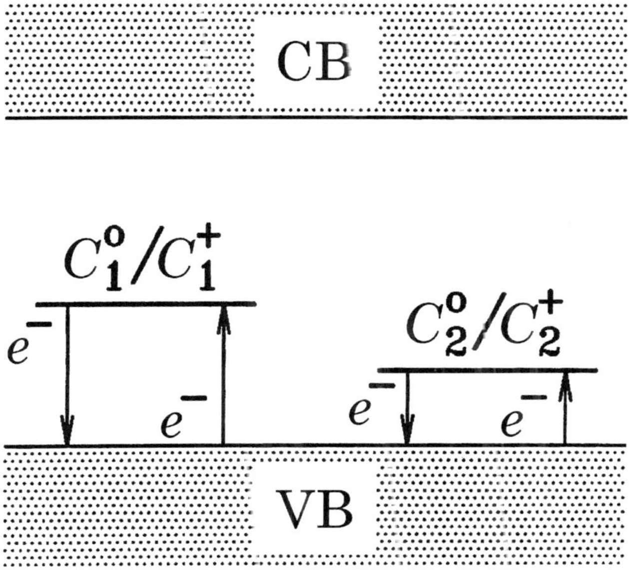

Figure 2.1: Band diagram of the two-center charge-transport model. (CB = conduction band, VB = valence band, C_1 = center 1, C_2 = center 2.)

processes become possible, which result from optical excitations of valence band electrons to C_2^+ centers. This leads to light-induced absorption changes, if C_1^+ and C_2^+ have different photon absorption cross sections. The rate equations read:

$$dN_i^+/dt = -(q_i S_i I + \beta_i)N_i^+ + r_i(N_i - N_i^+)N_h,\ i = 1, 2.$$

Here q_i again denotes the quantum efficiencies for generating a hole upon absorption of a photon, S_i the absorption cross sections, β_i the thermal generation rates, and r_i the recombination coefficients.

Besides the appearance of light-induced absorption the two-center model also explains the nonlinear dependence of photoconductivity σ_{ph} on light intensity [14]. The experimentally often observed relationship for perovskites, $\sigma_{\mathrm{ph}} \sim I^x$ with $0.5 \le x \le 1$ [16, 20], is confirmed by the model. Furthermore, the light-induced absorption coefficient α_{li} is predicted to decrease and the coefficient x to approach one with increasing temperature. This is due to the fact that thermal excitations become more important at higher temperatures; the second center occurs essentially as C_2^0 and the transport properties are mainly determined by center C_1 (one-center situation). These predictions are also verified experimentally as illustrated in Figs. 2.2 and 2.3 for $BaTiO_3$.

When a suitable set of parameters is chosen, the two-center model describes the intensity dependences of σ_{ph} and α_{li} well. Furthermore, a relation $1/\alpha_{li} = 1/[(S_2 - S_1)N_2] + \beta_2 e\mu/[(S_2 - S_1)N_2 r_2 \sigma_{ph}]$ was derived [20] (μ is the mobility of the holes in the valence band). This relation is in perfect agreement with the experimental results as shown in Fig. 2.4. For the thermal activation energies E_1 and E_2 of the centers C_1 and C_2 the values $E_1 = 1.04$ eV and $E_2 = 0.74$ eV were obtained for a $BaTiO_3$ sample [20] and $E_2 = 0.8$ eV for a $KNbO_3$ sample [6]. At very high intensities, experimentally realized by frequency-doubled pulses of a Q-switched neodymium-yttrium-aluminum-garnet (Nd:YAG) laser, additional states

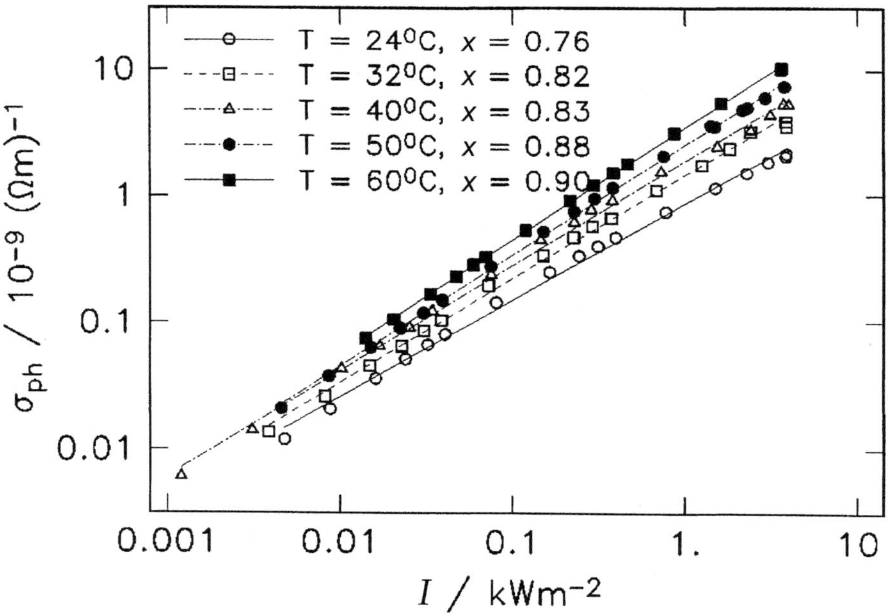

Figure 2.2: Photoconductivity σ_{ph} of a $BaTiO_3$ sample versus light intensity I for different temperatures T. Exponents x are calculated according to $\sigma_{ph} \sim I^x$.

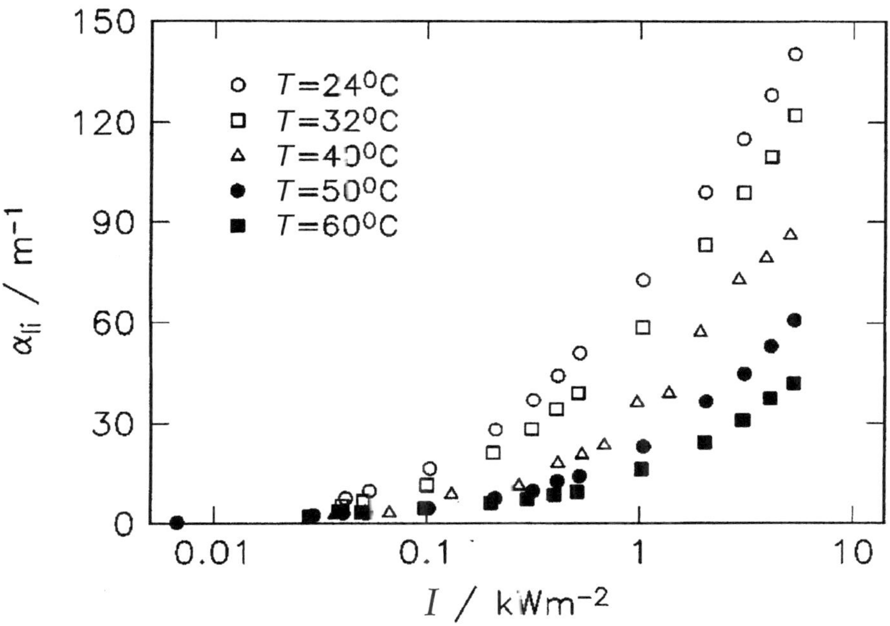

Figure 2.3: Light-induced absorption coefficient α_{li} of a $BaTiO_3$ sample versus light intensity I for different temperatures T.

with activation energies of 0.5 eV ($BaTiO_3$) and 0.35 eV ($KNbO_3$) become involved [21].

Further experimental support for the two-center model was discovered: Activation for infrared (IR) recording [22] and dark buildup of holograms after recording [23]. When a $BaTiO_3$ crystal is illuminated homogeneously with a green light pulse, subsequent recording of a holographic grating with infrared (IR) pulses is possible. After illumination with a green pulse, empty shallow traps C_2^+ exist. Then IR recording is based on excitations of electrons from the valence band to empty shallow traps C_2^+. The generated holes migrate and capture electrons of deep-level impurities C_1^0. Thus the light pattern is transposed to a space-charge pattern according to the population of deep-level impurities C_1. Still-existing empty shallow traps C_2^+ are filled by electrons due to thermal excitations. The remaining grating can be read nondestructively with IR light and erased by illumination with green light.

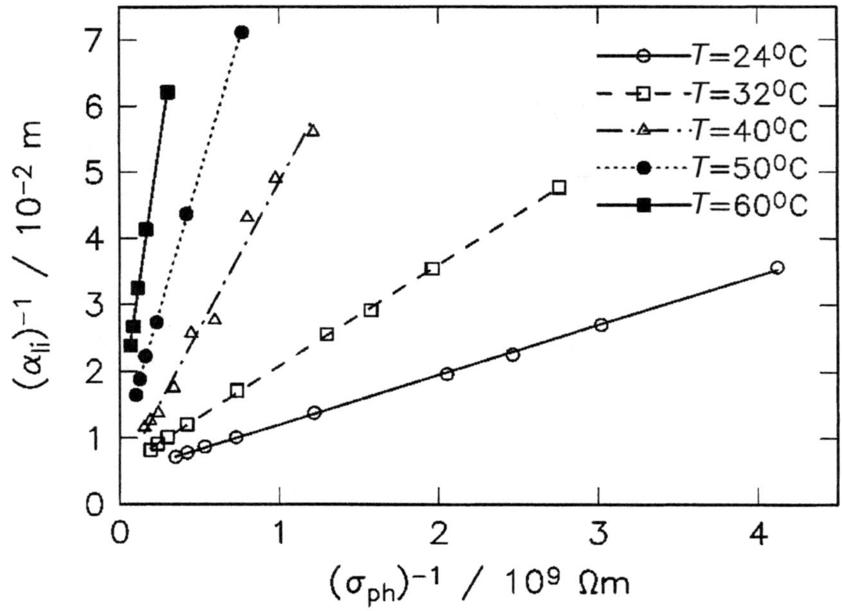

Figure 2.4: Inverse light-induced absorption coefficient α_{li}^{-1} versus inverse photoconductivity σ_{ph}^{-1} at different temperatures T. The light intensity I was varied from 300 Wm^{-2} to 4 kWm^{-2}.

Holograms recorded under suitable conditions in $BaTiO_3$ exhibit an unusual dark buildup [23]. The diffraction efficiency may increase by orders of magnitude after the recording beams are switched off, and then steadily decreases afterwards. The two-center model enables an explanation: Two mutually π-shifted gratings are recorded. They result from modulation of deep traps C_1 and shallow traps C_2, respectively. Both gratings largely compensate each other. In the dark, however, after switching-off illumination, the grating in the shallow traps C_2 relaxes faster than the grating in the deep traps C_1 and therefore holographic diffraction increases. Model calculations yield perfect agreement with the experimental results.

2.4 Three-valence model

But there exists a further possibility to explain the charge transport properties of perovskites. As we pointed out [24], the assumption of one impurity center occurring in three different valence states — the so-called three-valence model — leads to similar conclusions as the two-center model.

The situation is illustrated in Fig. 2.5. The three valence states of the center X are denoted by 0, +, and 2+. The arrows indicate the considered excitation and recombination processes of electrons. At low intensities only X^0 and X^+ states are present, because thermally excited valence band electrons fill X^{2+}. Illumination excites electrons from the valence band into X^+ and generates holes which are annihilated by electrons from X^0. For sufficiently high light intensities the hole concentration becomes large enough that an appreciable number of electrons from X^+ can recombine with holes and generate X^{2+} contributing to absorption. Thus light-induced absorption changes appear. Furthermore, participation of X^{2+} in

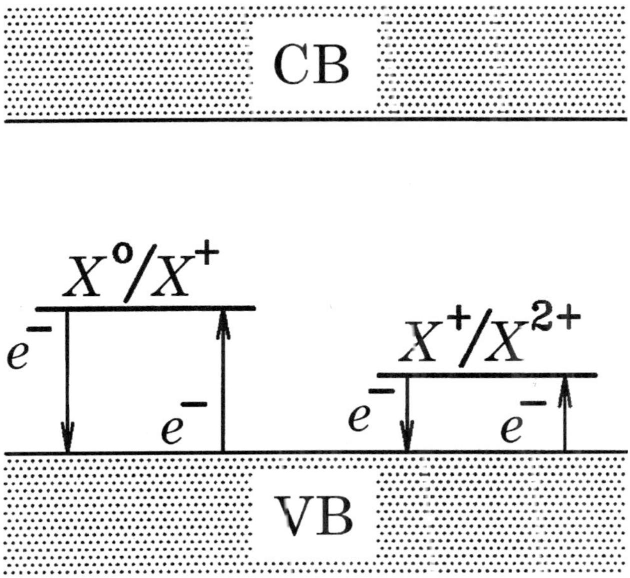

Figure 2.5: Band diagram of the three-valence model (CB = conduction band, VB = valence band; the valence states of the center X are indicated by 0,+, and 2+.)

the charge transport may provide a photoconductivity increasing nonlinearly with light intensity.

Rate equations, charge conservation, and constant trap density may be written as:

$$dN^+/dt = r^0N^0N_h - (\beta^+ + q^+S^+I)N^+ - r^+N^+N_h + (\beta^{2+} + q^{2+}S^{2+}I)N^{2+},$$
$$dN^{2+}/dt = r^+N^+N_h - (\beta^{2+} + q^{2+}S^{2+}I)N^{2+},$$
$$2N^{2+} + N^+ + N_h = N_c, \quad N^0 + N^+ + N^{2+} = N.$$

Here again N^0, N^+, and N^{2+} are the concentrations of X^0, X^+, and X^{2+}, N is the whole impurity concentration, N_h is the concentration of holes in the valence band, β^+ and β^{2+} are the thermal generation rates, q^+ and q^{2+} are the quantum efficiencies for hole generation upon absorption of a photon, S^+ and S^{2+} are the photon absorption cross sections, r^0 and r^+ are the recombination coefficients, and N_c is a constant concentration. We would like to emphasize that the above equations of the three-valence model cannot be derived from those of the two-center model introducing a special relation between the concentrations of deep and shallow traps. But both models lead to similar conclusions:

The three-valence model yields a relation between photoconductivity σ_{ph} and light intensity I which may again be approximated by $\sigma_{ph} \sim I^x$ with $0.5 \le x \le 1$. Light-induced absorption is explained and a linear relationship between α_{li}^{-1} and σ_{ph}^{-1} is deduced, too [24]. Furthermore, activation for IR recording and dark buildup of holograms after recording are interpreted in a similar way as in the case of the two-center model. A quantitative description of the transport properties is also possible [24].

2.5 Charge transport in different crystals

For the two-center and three-valence model the common notation two-level model may be used, but for a special crystal it is of course important to know whether two interacting centers or three valence states of one center are involved. For example, in the case of two centers the charge transport properties may be greatly influenced by elimination of the shallow traps. Such a possibility is lacking for three valence states.

Relatively well understood is the situation in the case of transition metal-doped $LiNbO_3$ and $LiTaO_3$ at high light intensities. Several authors observed deviations [25, 26] from the one-center model which is valid at low light intensities ($\leq 10^5\,\mathrm{Wm^{-2}}$). A detailed investigation of $LiNbO_3$:Fe with frequency-doubled pulses of a Q-switched Nd:YAG laser demonstrates additional contributions to the charge transport properties pointing to a two-level model [27, 28].

The congruently melting composition of $LiNbO_3$ occurs at a relatively large niobium excess (c_{Nb} = [Nb mol]/([Li mol] + [Nb mol]) = 51.6%) and in the crystals Nb^{5+} ions on Li^+ sites (${Nb_{Li}}^{5+}$) are present. These Nb^{5+} centers may trap electrons, form ${Nb_{Li}}^{4+}$ polarons, and act as shallow centers. This interpretation in terms of a two-center model is supported by further arguments: Three valence states of impurities have never been detected in $LiNbO_3$, even after strong thermal annealing treatments. In addition, model calculations [28] indicate that the number of shallow levels has to be much larger than that of deep levels. Finally, if $LiNbO_3$ crystals are doped with Mg or Zn, the number of Nb^{5+} ions on Li sites is decreased and all effects pointing to a two-level model are diminished [29, 30]. As an example, Fig. 2.6 shows the reduction of light-induced absorption with increasing Mg or Zn content.

In strongly reduced nominally pure $LiNbO_3$ crystals bipolarons were observed which consist of a Nb^{4+} ion on a Li^+ site and a Nb^{4+} ion on a Nb^{5+} site [31, 32]. By laser pulses these bipolarons can be dissociated at room temperature and light-induced absorption changes resulting from the polarons appear [33]. A square root intensity dependence is characteristic for this process.

In the case of the ferroelectric perovskites $BaTiO_3$ and $KNbO_3$ (nominally pure) we see some arguments in favor of the three-valence model. Numerous investigations were carried out with different samples, but always light-induced absorption and a nonlinear photoconductivity were observed. By doping, e.g., with Fe, these effects are considerably increased [34]. Thus the interpretation in terms of a two-center model yields a correlation between shallow and deep centers which is difficult to understand. The three-valence model, however, leads to an obvious correlation between different valence states.

Three valence states of impurity centers in perovskites have indeed been found experimentally. Possenriede et al. [35] investigated different $BaTiO_3$ samples by a combination of ESR and optical absorption measurements. They showed that many different centers exist which may contribute to the photorefractive process. In all crystals they observed that Fe^{3+} and Fe^{4+} play a dominant role. Furthermore, they discovered in all samples that illumination generates Fe^{5+}. Perhaps $X^{0,+,2+}$ may be identified

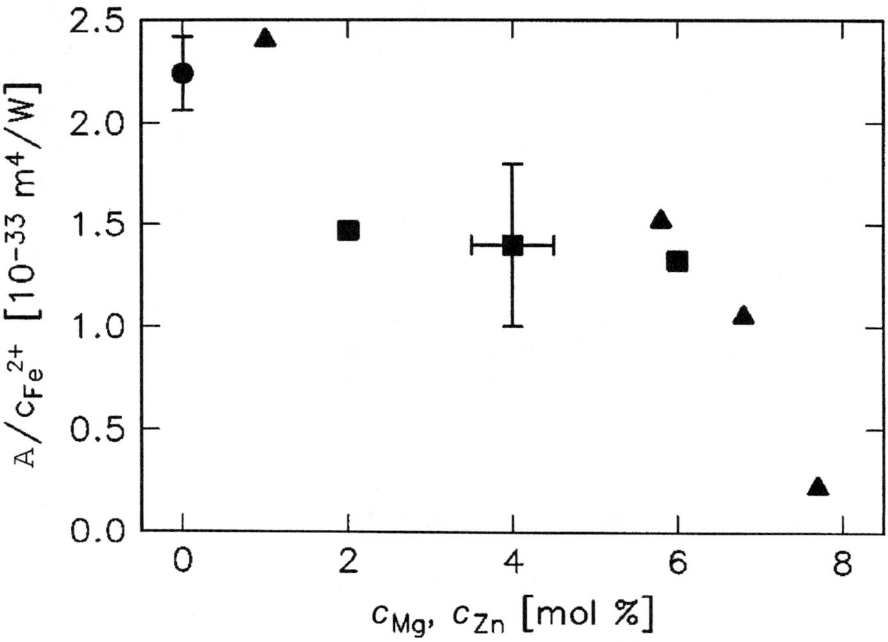

Figure 2.6: Slope $A = d\alpha_{li}/dl|_{l=0}$ of light-induced absorption coefficient α_{li} normalized to $c_{Fe^{2+}}$ versus concentration of Mg (squares) and Zn (triangles) co-dopants (in the melt) at the probe laser wavelength $\lambda = 785$ nm.

by $Fe^{3+,4+,5+}$. An additional center which provides a three-valence system is Cr, occurring as $Cr^{3+,4+,5+}$ in $BaTiO_3$ [35].

Similar arguments pointing to a three-valence model hold for $KNbO_3$ and KTN crystals. With increasing Fe content the photoconductivity of $KNbO_3$:Fe becomes more and more sublinear in intensity and the diffraction efficiency becomes more strongly intensity dependent [36]. For KTN an interpretation in terms of a two-center model leads to the conclusion that the total concentration of shallow traps (filled and empty) is influenced by thermal annealing. This is hard to imagine. Furthermore, the density of shallow centers is correlated with the concentration of deep traps filled with a hole which again seems to be unlikely [37].

The situation becomes still more complicated if perovskites are doped with more than one impurity. An investigation of the light-induced charge transport of $BaTiO_3$:Rh,Fe indicates that several interacting centers in several valence states contribute [38].

There are experimental observations that the charge transport in sillenite-type crystals [$Bi_{12}SiO_{20}$, $Bi_{12}TiO_{20}$, $Bi_{12}GeO_{20}$] has also to be explained by a two-level model [39, 40]. Most probably the intrinsic defect Bi^{3+} on a metal site (Si, Ti, Ge) is involved [41]. Up to now a decision between the two-center and three-valence model seems to be not possible. After excitation with high-intensity laser pulses the simultaneous influence of more than two levels has to be considered [42].

For SBN:Ce crystals with tungsten-bronze structure a description by the two-center model seems to be appropriate [43]. The experiments yield that the concentrations of deep and shallow levels increase linearly with Ce concentration. Below a Ce concentration of 0.1 wt% practically no shallow levels are observed. The two-center model can explain this easily by a small concentration of shallow centers. Explanation with the three-valence model requires a small concentration ratio between donors N^- and traps N°. In this case the photoconductivity should be small; measurements of crystals with small Ce concentration, however, yield just the opposite.

A complete set of charge-transport parameters of photorefractive SBN:Ce has been determined recently [44]: The concentrations of filled and empty traps, the photon-absorption cross section, the quantum efficiency for excitation of an electron upon absorption of a photon, the recombination coefficient and the charge-carrier mobility. By these means a fascinating insight into the microscopic processes is obtained: Of 2000 incident photons which pass through a Ce^{3+} ion only 1 is absorbed, and 100 absorbed photons yield only six mobile electrons. These electrons recombine with shallow levels, and they are 33 times thermally reexcited before they finally recombine with the trap Ce^{4+}. An excited electron is trapped if the distance to a Ce^{4+} ion is smaller than 100 times the ionic radius.

2.6 Conclusions

Tailoring of photorefractive crystals for various applications requires detailed knowledge of the light-induced charge transport. Already rather early the so-called one-center model was developed, which well describes the transport properties of transition metal-doped $LiNbO_3$ and $LiTaO_3$ crystals at usual cw laser intensities; e.g., in $LiNbO_3$:Fe electrons are transferred from Fe^{2+} to Fe^{3+} ions.

But in other cases the situation is much more complicated, e.g., for $LiNbO_3$ and $LiTaO_3$ at high light intensities ($>10^5\,Wm^{-2}$), for ferroelectric perovskites and for tungsten-bronze type crystals. The interpretation of the experimental results for the charge transport requires the assumption of two or more interacting centers (two-center model, many-center model) and / or of more than two valence states of the centers involved (three-valence model, many-valence model). In general we speak of many-level models. Light-induced absorption effects and a nonlinear dependence of photoconductivity on light intensity are typical consequences of these models.

For $LiNbO_3$ and $LiTaO_3$ at high intensities the two-center model must be applied. The second center has been identified as Nb_{Li} polaron. Two centers seem to be involved in the case of SBN:Ce, too. For nominally pure $BaTiO_3$, $KNbO_3$, and KTN crystals the three-valence model seems to be more appropriate. For perovskites doped with several impurities, e.g., for $BaTiO_3$:Rh,Fe, interactions of several centers in several valence states have be taken into account.

Acknowledgment

The authors thank H. Hesse, L. Holtmann, F. Jermann, O. F. Schirmer, M. Simon and U. van Stevendaal for helpful discussions. Financial support of the Deutsche Forschungsgemeinschaft (SFB 225, C5) is gratefully acknowledged.

References

1. P. Günter and J. P. Huignard (eds.), *Photorefractive Materials and Their Applications I and II, Topics in Applied Physics,* Vols. 61 and 62, Springer Verlag, 1988, 1989.
2. K. Buse, "Light-induced charge transport processes in photorefractive crystals I: Models and experimental methods," *Appl. Phys.,* **B64**, 273–291, 1997.
3. K. Buse, "Light-induced charge transport processes in photorefractive crystals II: Materials," *Appl. Phys.,* **B64**, 391–407, 1997.
4. G. E. Peterson, A. M. Glass, T. J. Negran, "Control of the susceptibility of lithium niobate to laser-induced refractive index change," *Appl. Phys. Lett.,* **19**, 130–132, 1971.

5. J. J. Amodei, W. Phillips, D. L. Staebler, "Improved electrooptic materials and fixing techniques for holographic recording," *Appl. Opt.*, **11**, 390–396, 1972.

6. H. Kurz, E. Krätzig, W. Keune, H. Engelmann, U. Gonser, B. Dischler, A. Räuber, "Photorefractive centers in $LiNbO_3$, studied by optical, Mössbauer- and EPR-methods," *Appl. Phys.*, **12**, 355–368, 1977.

7. M. Glass, D. von der Linde, T. J. Negran, "High-voltage bulk photovoltaic effect and the photorefractive process in $LiNbO_3$," *Appl. Phys. Lett.*, **25**, 233–235, 1974.

8. V. I. Belinicher, V. K. Malinowski, B. I. Sturman, "Photogalvanic effect in a crystal with polar axis," *Sov. Phys. JETP,* **46**, 362–366, 1977.

9. E. Krätzig, "Photorefractive effects and photoconductivity in $LiNbO_3$:Fe," *Ferroelectrics,* **21**, 635–636, 1978.

10. R. Sommerfeldt, L. Holtmann, E. Krätzig, B. Grabmaier, "The light-induced charge transport in $LiNbO_3$:Mg,Fe crystals," *Ferroelectrics,* **92**, 219–225, 1989.

11. R. A. Rupp, R. Sommerfeldt, K. H. Ringhofer, E. Krätzig, "Space charge field limitations in photorefractive $LiNbO_3$:Fe crystals," *Appl. Phys.*, **B51**, 364–370, 1990.

12. A. Motes and J. J. Kim, "Intensity-dependent absorption coefficient in photorefractive $BaTiO_3$ crystals," *J. Opt. Soc. Am.* **B4**, 1379–1381, 1987.

13. G. A. Brost, R. A. Motes, J. R. Rotgé, "Intensity-dependent absorption and photorefractive effects in barium titanate," *J. Opt. Soc. Am.*, **B5**, 1879–1885, 1988.

14. L. Holtmann, "A model for the nonlinear photoconductivity of $BaTiO_3$," *Phys. Stat. Sol.*, **A113**, K89–K93,1989.

15. D. Mahgerefteh and J. Feinberg, "Explanation of the apparent sublinear photoconductivity of photorefractive barium titanate," *Phys. Rev. Lett.*, **64**, 2195–2198, 1990.

16. L. Holtmann, K. Buse, G. Kuper, A. Groll, H. Hesse, E. Krätzig, "Photoconductivity and light-induced absorption in $KNbO_3$:Fe," *Appl. Phys.*, **A53**, 81–86, 1991.

17. S. Loheide, S. Riehemann, E. Krätzig, "Light-induced absorption in Fe-doped potassium tantalate-niobate crystals," *Phys. Stat. Sol.*, **A142**, K99–K101, 1994.

18. F. Mersch, K. Buse, W. Sauf, H. Hesse, E. Krätzig, "Growth and characterization of undoped and doped $Bi_{12}TiO_{20}$ crystals," *Phys. Stat. Sol.*, **A140**, 273–281, 1993.

19. M. Simon, A. Gerwens, E. Krätzig, "Light-induced absorption generated with high intensity laser pulses in strontium-barium niobate," *Phys. Stat. Sol.,* **A143**, K125–K128, 1994.

20. L. Holtmann, M. Unland, E. Krätzig, G. Godefroy, "Conductivity and light-induced absorption in $BaTiO_3$," *Appl. Phys.,* **A51**, 13–17, 1990.

21. K. Buse and E. Krätzig, "Light-induced absorption in $BaTiO_3$ and $KNbO_3$ generated with high intensity laser pulses," *Optical Materials,* **1**, 165–170, 1992.

22. K. Buse, L. Holtmann, E. Krätzig, "Activation of $BaTiO_3$ for infrared holographic recording," *Optics Commun,* **85**, 183–186, 1991.

23. K. Buse, J. Freijlich, G. Kuper, E. Krätzig, "Dark build-up of holograms in $BaTiO_3$ after recording," *Appl. Phys.,* **A57**, 437–440, 1993.

24. K. Buse and E. Krätzig, "Three-valence charge-transport model for explanation of the photorefractive effect," *Appl. Phys.,* **B61**, 27–32, 1995.

25. C.-T. Chen, Dae M. Kim, D. von der Linde, "Efficient hologram recording in $LiNbO_3$:Fe using optical pulses," *Appl. Phys. Lett.,* **34**, 321–324, 1979.

26. P. A. Augustov and K. K. Shvarts, "The temperature and light intensity dependence of photorefraction in $LiNbO_3$," *Appl. Phys.,* **21**, 191–194, 1980.

27. F. Jermann and E. Krätzig, "Charge transport processes in $LiNbO_3$ at high intensity laser pulses," *Appl. Phys.,* **A55**, 114–118, 1992.

28. F. Jermann and J. Otten, "The light-induced charge transport in $LiNbO_3$:Fe at high light intensities," *J. Opt. Soc. Am.,* **B10**, 2085–2092, 1993.

29. M. Simon, F. Jermann, E. Krätzig, "Light-induced absorption changes in iron-doped $LiNbO_3$," *Optical Materials,* **3**, 243–250, 1994.

30. M. Simon, F. Jermann, E. Krätzig, "Intrinsic photorefractive centers in $LiNbO_3$:Fe," *Appl. Phys.,* **B61**, 89–93, 1995.

31. O. F. Schirmer, S. Juppe, J. Koppitz, "ESR-, optical and photovoltaic studies of reduced undoped $LiNbO_3$," *Cryst. Latt. Def. and Amorph. Mat.,* **16**, 353–357, 1987.

32. O. F. Schirmer, O. Thiemann, M. Wöhlecke, "Defects in $LiNbO_3$ — I. experimental aspects," *J. Phys. Chem. Sol.,* **52**, 185–200, 1991.

33. F. Jermann, M. Simon, R. Böwer, E. Krätzig, O. F. Schirmer, "Light-induced absorption changes in reduced lithium niobate," *Ferroelectrics,* **165**, 319–327, 1995.

34. L. Holtmann, A. Groll, M. Unland, E. Krätzig, A. Maillard, G. Godefroy, "Light-induced charge transport in $BaTiO_3$," *Tech. Dig. Topical Meeting on Photorefractive Materials, Effects, and Devices,* **II**, 83–86, 1990.

35. E. Possenriede, P. Jacobs, M. Kröse, O. F. Schirmer, "Paramagnetic defects in $BaTiO_3$ and their role in light-induced charge transport — optical studies," *Appl. Phys.*, **A55**, 73–81, 1992.

36. P. Günter and F. Micheron, "Photorefractive effects and photocurrents in $KNbO_3$:Fe," *Ferroelectrics,* **18**, 27–38, 1977.

37. K. Buse, S. Loheide, D. Sabbert, E. Krätzig, "Photorefractive properties of tetragonal $KTa_{0.52}Nb_{0.48}O_3$:Fe crystals and explanation by the three-valence charge-transport model," *J. Opt. Soc. Am.*, **B13**, 2644–2651, 1996.

38. U. van Stevendaal, K. Buse, S. Kämper, H. Hesse, E. Krätzig, "Light-induced charge transport processes in photorefractive barium titanate doped with rhodium and iron," *Appl. Phys.*, **B63**, 315–321, 1996.

39. G. Lesaux, G. Roosen, A. Brun, "Observation and analysis of the fast photorefractive process in BSO," *Opt. Commun.*, **56**, 374–378, 1986.

40. F. P. Strohkendl, "Light-induced dark decays of photorefractive gratings and their observation in $Bi_{12}SiO_{20}$," *J. Appl. Phys.*, **65**, 3773–3780, 1989.

41. H.-J. Reyher, U. Hellwig, O. Thiemann, "Optically detected magnetic resonance of the bismuth-on-metal-site intrinsic defect in photorefractive sillenite crystals," *Phys. Rev.*, **B47**, 5638–5645, 1993.

42. K. Okamoto, T. Sawada, K. Ujihara, "Transient response of a photorefractive grating produced in a BSO crystal by a short light pulse," *Opt. Commun.*, **99**, 82–88, 1993.

43. K. Buse, U. van Stevendaal, R. Pankrath, E. Krätzig, "Light-induced charge transport properties of $Sr_{0.61}Ba_{0.39}Nb_2O_6$ crystals," *J. Opt. Soc. Am.*, **B13**, 1461–1467, 1996.

44. K. Buse, A. Gerwens, S. Wevering, E. Krätzig, "Charge transport parameters of photorefractive strontium-barium niobate crystals doped with cerium," *J. Opt. Soc. Am.*, **B15**, 1674–1677, 1998.

Chapter 3

Nonlinear Self-Organization in Photorefractive Materials

Partha P. Banerjee

Department of Electrical and Computer Engineering and Center for Applied Optics, University of Alabama in Huntsville, Huntsville Alabama

Nickolai V. Kukhtarev

Physics Department, Alabama A&M University, Normal Alabama

John O. Dimmock

Center for Applied Optics and Department of Physics, University of Alabama in Huntsville, Huntsville Alabama

3.1 Introduction

It is indeed intriguing that many natural phenomena as well as the brain or animal behavioral patterns exhibit self-organization. The convective

PHOTOREFRACTIVE OPTICS
Materials, Properties and Applications

ISBN 0-12-774810-5/$30.00

rolls in a liquid when it is heated beyond the Rayleigh-Bernard instability point is an example of pattern formation in hydrodynamics due to self-organization. Here the temperature difference is the driving force or input parameter. Below the critical or threshold temperature, one can only observe random motion of the liquid particles. Besides, in an open container containing the fluid, surface tension can also affect the flow, causing tesselation of the surface and formation of hexagonal cells. Such spontaneous pattern formation is exactly what is termed self-organization, but there is no agent inside the system that does the organizing. The motion of the whole is no longer the sum of the motion of the parts, due to nonlinear interactions between the parts and the environment. Another example of pattern formation is a "wave" among spectators in a stadium — individual spectators communicate and cluster together in groups to create a nearly synchronized pattern that spreads throughout the stadium. Speaking of which, there is enough evidence that human behavioral patterns are self-organized. The human body, for example, is a complex system comprising about 10^2 joints, 10^3 muscles, 10^3 cell types, and 10^{14} neurons or neuron connections. The actions of communication, body movement etc. are the result of self-organization of this complex system pertaining to a certain control or input parameter (stimulus). In a similar fashion it has been shown that the brain itself is an active, dynamic self-organizing system. For more on the self-organizing aspects of the brain and human behavior, the readers are referred to Kelso [1], Kohonen [2], and Haken [3].

Some of the elementary concepts and conditions for self-organization are as follows [1]:

1. Patterns arise spontaneously as a result of nonlinear coupling between large numbers of interacting components.
2. The system must be far from equilibrium. Due to nonlinear interactions, energy is not distributed evenly but coalesces into patterns or flows.
3. Relevant degrees of freedom, or order parameters, must exist near nonequilibrium phase transitions, where loss of stability gives rise to new patterns and / or switching between patterns.
4. Noise must be present in the system, so that fluctuations can "feel" the system stability and provide for the system to self-organize into different patterns.

In this chapter we will discuss self-organization and its effects in optics. In fact, one of the most exciting and potentially useful areas of current research in optics involves the understanding and exploitation of self-organization in nonlinear optical systems. This self-organization may sometimes lead to the evolution of complex spatial patterns which can be regarded as the nonlinear eigenmodes of the system. Generation of these patterns is characteristically marked by the presence of intensity thresholds. The detailed study of the self-organization process, including the spatiotemporal evolution, is needed in order to harness these effects for potential practical applications.

For a long time in nonlinear optics, only problems of temporal dynamics were investigated. However, spatial distributions were only assumed, without regard to their time evolution and hence relationship with temporal instabilities. However in a nonlinear system with complicated temporal dynamics, it turns out that one cannot retain purity in spatial dimensionality. It is therefore equally important to investigate the dynamics of the transverse spatial variations which in fact give rise to very interesting patterns due to self-organization. A vast wealth of patterns can be achieved by using a nonlinear optical element with feedback that has the capability to provide for field transformation, e.g., by spatial filtering. These types of systems have been called optical kaleidoscopes, simply because of the different self-organized patterns that they can generate. Examples of nonlinear self-organized kaleidoscopic patterns are:

Rolls

Rotatory waves

Optical spirals

Hexagonal patterns

Patterns with more complicated geometry

Pattern hopping

An excellent reference for this as well as self-organization in different nonlinear optical systems is the book by Vorontsov and Miller [4].

Information processing applications of nonlinear optics are closely linked to the ability to control nonlinear optical systems which can self-organize in different ways. For instance, different patterns formed through self-organization can be used for coding and processing of optical information [4]. It has been proposed that the existence of several modes in a laser can be used as a base for synergetic computing [5]. Fourier filtering techniques have been used in conjunction with nonlinear optical systems for information processing. Degtiarev and Vorontsov [6] used Fourier filtering in the path of a liquid crystal light valve (LCLV) system with feedback for phase distortion suppression. Such nonlinear optical systems with feedback have also been used for various kinds of pattern generation such as hexagons, rolls, etc. [7]. The dynamics of pattern formation in a coupled LCLV system with feedback have been studied by Thuring *et al.* [8].

Photoinduced scattering of laser radiation into self-organizing patterns has been observed over the past several years in a substantial number of nonlinear materials including gases and liquids [9–14]. Among solids, photorefractive materials such as $KNbO_3$ have been observed to exhibit a rich variety of such scattering including hexagonal pattern formation and rotation, as well as other patterns depending on the experimental conditions [15, 16]. Furthermore simultaneous pattern generation and self-phase conjugation have been observed due to self-organization in this material under other conditions [17, 21]. Hexagon formation has also been observed in other photorefractive materials as well, such as $BaTiO_3$ [22, 23].

Because of the richness of the scattering phenomena observed in $KNbO_3$, we anticipate that the understanding of the origins of photorefractivity and the nature of the self-organization phenomena will lead to novel and substantially enhanced nonlinear applications of this material such as set forth in the following.

1. We anticipate that the self-organization can readily be used to intelligently manufacture diffractive optical elements, such as hexagonal arrays, gratings etc. In this case, one can use the nonlinear properties of the active material to create diffractive optic elements, rather than rely on complicated geometrical processing. In the long run, these patterns can be generated and stored in thin-film photorefractive polymers. In the shorter term, one can image a plane inside of a thick crystal on a film and thereby make such diffractive elements.

2. The near-field pattern is observed to be composed of hundreds of phase-related spots in a hexagonal array which can be caused to shift or move across the face of the crystal. We believe that this can, in principle, be used for hexagonal sampling of images in digital image processing which offers spatial bandwidth savings [24]. Hexagonal array generation

has been traditionally done by fabricating binary phase gratings [25]. These hexagonal arrays can also be used to effectively couple light into a fiber bundle, which may eventually feed into adaptive antenna array structures.

3. The far-field pattern can be used to broadcast separate images of an input pattern in different directions. Further, because these separate images have specific phase relations, unique image processing can be performed by interfering these separate images with each other or with the original beam. It is also conceivable that the far-field pattern comprising six peripheral spots and the central spot can be used to monitor velocity and acceleration of a moving body.

4. As both near- and far-field pattern rotations are extremely sensitive to small misalignments of the pump beam with respect to the crystal surfaces and axes [16, 18], we anticipate that this material can be employed as an integral element in misalignment detection or rotation sensing devices.

5. The self-phase conjugation can be used to form conjugate images in both forward and backward directions without the need of complex additional optics. Edge enhancement, an important aspect of image processing, has been demonstrated using this material and there is potential to develop real-time optically edge-enhanced correlators using this concept.

6. Finally, since we can observe and measure holographic currents during grating recording in the photorefractive material [26, 27], we anticipate that the self-organization effects and their time dependencies can be modified and indeed controlled by application of external electrical fields to the $KNbO_3$ crystal. In the long term, the possibility of superposing external electrical modulation to change the holographic current and hence the diffraction pattern in the near- and far-fields seems feasible. This will open the door to exciting applications of the crystal in nonlinear information and image processing which may be electronically controllable.

Before closing this section, we would like to point out that photoinduced scattering in the $KNbO_3$ crystal has also been observed over a wide range of the visible spectrum [17]. This makes the crystals particularly attractive in low-cost applications using, for instance, portable He-Ne lasers. Not only is the understanding of self-organization important for possible applications, but also the optical and electrical measurements on the crystal are essential in order to characterize the physical properties of the crystal related to charge transport and the nature of the nonlinearity. This will enhance the knowledge base for the crystal, useful for rigorous analysis of the self-organization phenomenon, as well as for other applications. Finally, the optical and electrical measurements can be readily used to characterize other photorefractive crystals as well.

3.2 Basic experimental observations

Self-organization leading to hexagon formation using photorefractive materials such as $KNbO_3$ can be heuristically explained as due to a photoinduced holographic scattering which develops in two stages [16]. In the first stage, scattered light is rearranged into a cone which corresponds to a Fabry-Perot mode of the nonlinear cavity formed by the refractive index mismatch at the crystal interfaces. Reflection gratings, sometimes aided by transmission gratings, may nonlinearly modify the cavity characteristic and the cone angle. At the second stage, waves scattered in the cone write new holographic gratings (second generation gratings), and those among them that have holographic grating vectors equal to the strongest gratings from the set of first generation gratings are enhanced, following a winner-take-all route. This holographic self-organization model conceptually explains the appearance of a hexagonal spot structure around the transmitted beam. Other heuristic explanations are based on the Talbot effect; this was enunciated by Tamburrini *et al.* [13] for a liquid crystal and extended to the case of $KNbO_3$ by Honda and Matsumoto [22]. Other simplified explanations of hexagon formation also exist in the literature [28]. In this case, the authors use a simplified although, maybe, unrelated model of nonlinear susceptibility in the understanding of hexagonal pattern formation in photorefractives. The detailed physics of hexagonal pattern generation in photorefractives in our opinion is complicated and not yet well understood.

$KNbO_3$ is a biaxial electrooptic material with orthorhombic symmetry and has excellent photorefractive properties marked by large beam coupling gains [29], fast buildup times, and large anisotropies [30]. Furthermore, Fe doping in $KNbO_3$ is known to increase the maximum value of the two-beam coupling gain [31]. $KNbO_3$ based phase conjugators have been implemented in various configurations [31–33] and material properties of the crystal have been extensively studied [34, 35]. The net optical nonlinearity of $KNbO_3$ has been studied using standard z-scan techniques [16, 36]. Electrical measurements have been also performed and give valuable information about the Maxwell relaxation time, screening length, and photogalvanic current [26].

Self-organization of an Ar laser single-beam scattering in a photorefractive $KNbO_3$:Fe crystal, first into a scattering cone, and then into a hexagonal pattern was observed by Honda [15] and Banerjee *et al.* [16]. Furthermore, these spots may be made to rotate about the center, and the rotation speed depends on the misalignment of the incident beam from the *c*-axis and the power of the beam. The hexagonal pattern is also influenced in real time by a low-power He-Ne laser (wavelength 632 nm): the spot pattern erases in about a second after the He-Ne laser is turned on, leaving only the scattering cone, and reappears a second after the He-Ne is turned off.

In the simplest experimental setup, an Ar laser (wavelength 514 nm) with horizontal polarization and with initial beam diameter 1 mm is reduced to a beam diameter of 0.5 mm using a confocal lens combination, and illuminates a $KNbO_3$:Fe crystal of dimensions $6 \times 6 \times 7$ mm^3 (Fig. 3.1) [16]. A slightly converging beam may also be used [15]. When the beam is normal to the incident surface, the far-field pattern is stationary in time and comprises a strong central spot with a peripheral ring which appears instantaneously, and thereafter evolves into six symmetrically spaced spots on the scattering cone (Fig. 3.2a). This far-field pattern is observed simultaneously both in the forward and backward directions; however, the diffraction efficiencies (discussed in more detail below) are not identical.

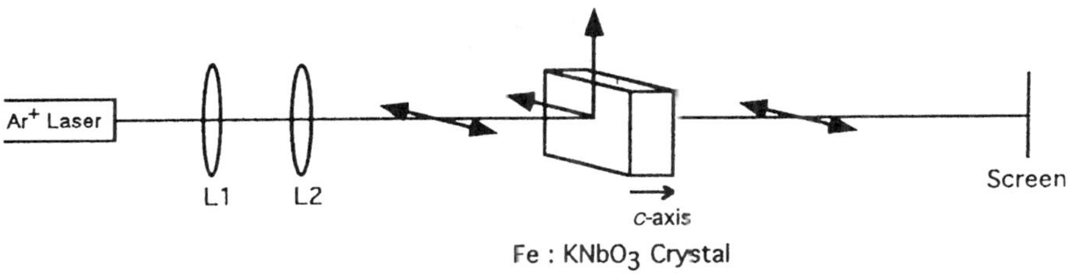

Figure 3.1: Experimental setup to observe hexagon formation in potassium niobate. (L1, L2-lenses.)

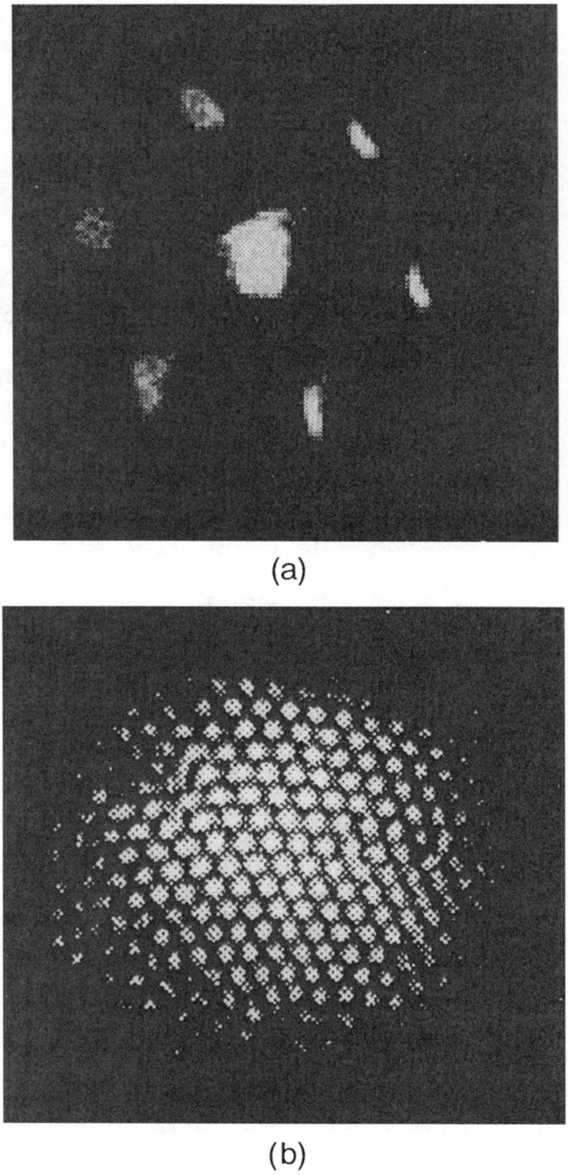

Figure 3.2: (a) Far-field transmission pattern showing central spot and hexagonal pattern. (b) Near-field pattern showing hexagonal spot array. (Source: Ref. [16]. Reprinted with permission.)

The semiangle of divergence θ of the peripheral cone is approximately $0.8°$ in air and is independent of the incident power. The time taken to form the spots is a few seconds for an incident power of 7.5 mW, although the spots may be formed for lower incident powers as well, with a longer formation time. The ring and all spots (central and peripheral) are also predominantly horizontally polarized. The diffraction efficiency for the spots in the forward direction is large: the intensity ratio of each transmitted peripheral spot to the transmitted central spot, which we term the forward diffraction efficiency per spot, is over 7%, for a total forward scattering efficiency into all six spots of 42%. The corresponding diffraction efficiency in the backward direction is about 4% per spot. Finally, the diffraction efficiencies seem to be relatively independent of the incident power over the range of powers investigated (7.5–30 mW).

Upon imaging different planes in the crystal (including the exit face) by a lens for the sake of visualization of the transverse nature of the optical fields, we have found, as shown in Fig. 3.2b, a periodic transverse hexagonal pattern at approximately the exit face of the crystal. Moreover, when the crystal is moved longitudinally by 0.5 cm, the same transverse pattern repeated, indicating a (nonlinearly modified) Talbot-type effect [37] (see Fig. 3.3), with contrast reversal occurring halfway between the Talbot imaging planes. The transverse period, calculated from the longitudinal period, is of the order of 30 μm, indicating a far-field diffraction angle in agreement with our observed value.

If the incident beam is slightly off-normal to the interface (typically by $0.04°$), and the power is increased, the entire hexagonal pattern rotates [16]. The sense of rotation depends on the sense of the angular misalignment; thus, both clockwise and counterclockwise rotations of the pattern are possible through positive and negative angular misalignments. A typical value for the rotation speed in the steady state is 100 degrees per minute for an incident power of 30mW. The rotation speed is smaller for lower powers and smaller misalignments; however, we have observed that a minimum critical power was required to achieve constant-speed rotation in the steady state. Figure 3.4 shows the sense of rotation for various values and signs of the angular misalignment. In fact, this phenomenon of rotation may be effectively used to reconfigure the hexagonal spot array pattern to any desired orientation by increasing the beam power for a finite length of time. The existence of a minimum threshold may suggest a secondary bifurcation [38], rotation is one of the routes by which such patterns may lose stability. Rotation of the far-field could imply that

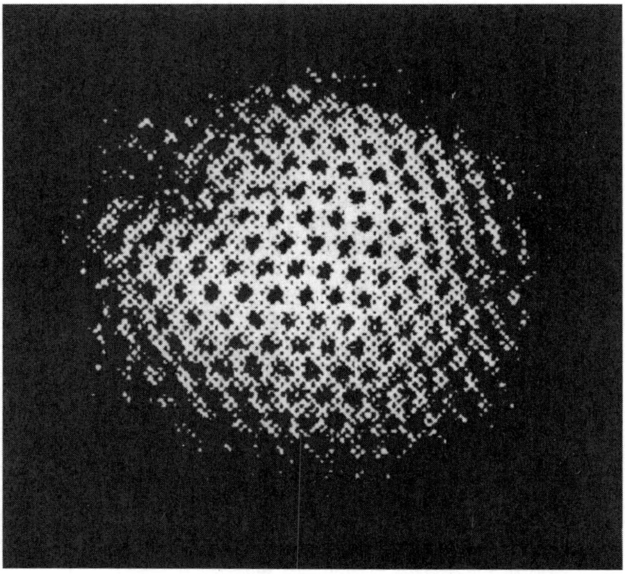

Figure 3.3: Contrast inversion of hexagonal spot array due to Talbot imaging.

individual spots in the near-field (Fig. 3.2b) undergo a change in phase as a function of time; however, this is still under investigation.

For larger angular misalignments (typically 0.3°), the far-field pattern resembles a hexagonal structure where the six spots are located on an ellipse instead of on a circle as in Fig. 3.2a. A larger misalignment (of 0.4°) destroys the hexagonal pattern altogether. What remains are two primary diffracted spots on opposite sides of the main beam orthogonal to the misalignment direction, in agreement with the experimental results of Grynberg *et al.* [9].

In related experiments, Honda obtained similar results by using a $KNbO_3$ crystal in index matching oil along with an external $BaTiO_3$ self-pumped phase conjugate mirror [15]. Also, a slightly converging beam was used, by using a convex lens of focal length 300 mm. The reflectivity of the phase conjugate mirror was about 50%. Pattern formation has also been observed using a $KNbO_3$ crystal along with a plane feedback mirror [39]. As will be discussed in the following, the angle of divergence depends on the length of the feedback path, thus it is possible to change the cone angle by adjusting the position of the external feedback mirror. In this

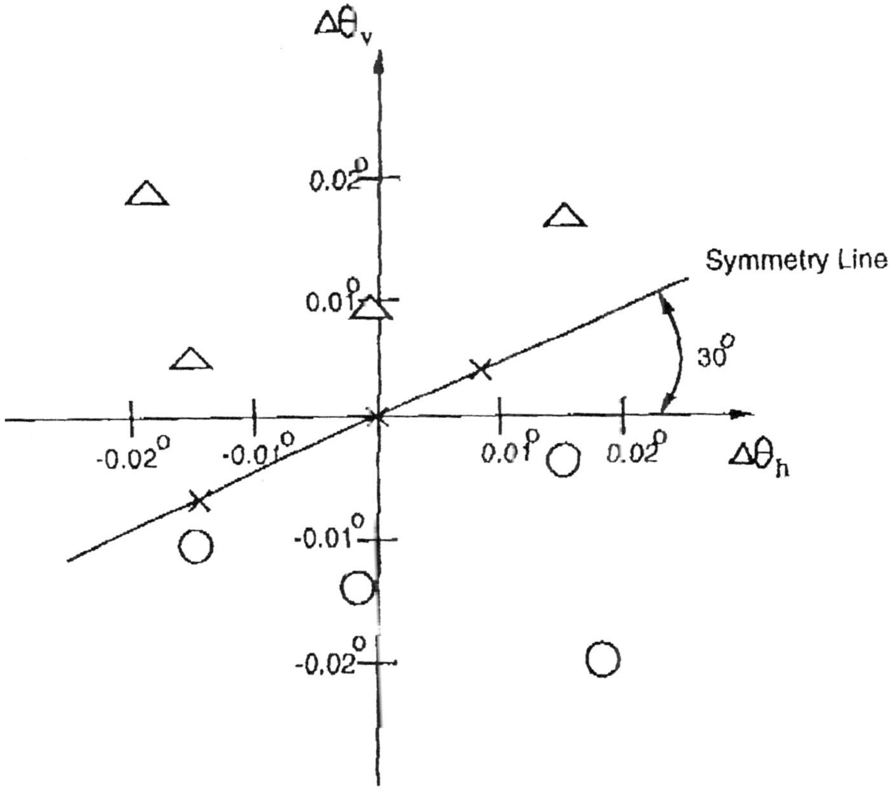

Figure 3.4: Experimental data for direction of rotation and angular misalignment. (Δ = hexagon rotates clockwise; O = hexagon rotates counterclockwise; X = hexagon is stationary.)

case, the crystal *c*-axis should be slightly tilted from the beam axis to reduce the influence of the beam reflected from the back surface of the crystal. It has been also shown that pattern rotation can be achieved by using an additional erase beam, making a small angle with respect to the direction of propagation of the pump beams in the crystal [18, 23]. The speed and sense of rotation of the hexagonal pattern in the far-field may be also controlled with the erase beam. Hexagonal pattern formation has been observed in photorefractive materials other than $KNbO_3$. Hexagonal pattern generation in Co-doped $BaTiO_3$ with an external feedback mirror has been observed by Honda and Matsumoto [22] and by Uesu *et al.* [23]. In all of the preceding experiments, higher-order hexagonal patterns have been observed in the far-field with an increase of incident intensity. Also, interesting patterns have been observed using a single feedback system with a virtual feedback mirror. A virtual feedback mirror is achieved by inserting a lens between the exit plane of the crystal and the external feedback mirror [19, 20]. The lens images the mirror at a certain distance from the exit face of the crystal. Depending on the location of the lens, the image location could be outside or even inside the crystal. Square patterns have been observed using this arrangement. For an appropriate choice of the virtual feedback mirror (typically located inside the crystal), the hexagonal and square patterns have been shown to alternate with time, demonstrating "pattern-hopping," which is testimony to criterion #3 for self-organization in the Introduction.

In a related experiment, self-phase conjugation similar to what was observed in SBN [40] has been observed in $KNbO_3$ [17]. This configuration has applications in image processing as well, as recently shown by Banerjee *et al.* [21]. In a typical experimental setup involving $KNbO_3$: Fe [17], a wave C_1 incident at about 10° to the normal to the crystal surface is reflected from the crystal, producing C'_{-1} (Fig. 3.5). Due to scattering, additional waves C_0 and C'_0 develop, which propagate almost normal to

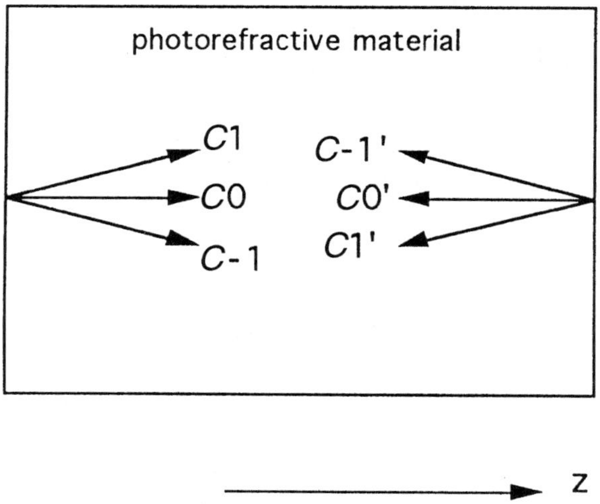

Figure 3.5: Six-wave coupling in potassium niobate.

the crystal surface. These represent concentric Fresnel rings, and are analogous to Fabry-Perot modes in a resonator. With time, the inner ring may decompose into a hexagonal pattern. Furthermore, interaction of the four waves C_0, C'_0, C_1, C'_{-1} gives rise to additional waves C_{-1} (counterpropagating to C'_{-1} and phase conjugate of C_1) and C'_1 (counterpropagating to C_1 and its phase conjugate).

A variation of the above experiment involves interactions initiated by two beams C_1 and C_0, and supported by reflections C'_{-1} and C'_0, to eventually produce C_{-1} and C'_1. In this case we have found that if C_1 represents the field from a point source, the phase conjugate C_{-1} images a certain distance behind the photorefractive crystal. Furthermore, if C_1 is the Fourier transform of an object, the phase conjugate of the object is recovered in the far-field, traveling nominally in the direction of C_{-1}. By changing the position of the object with respect to the front focal plane of the Fourier transform lens, edge enhancement can be achieved [21].

3.3 Theory

3.3.1 Fabry-Perot modes

Assume that a radially symmetric beam $C(r)$, where r represents the radial distance in the transverse plane, is nominally normally incident onto a Fabry-Perot cavity formed by the parallel faces of the photorefractive material. The far-field intensity profile can be shown to be given by $|\hat{C}(\theta)|\, S(\theta)$, where $S(\theta)$ is a shaping function which, to a first approximation, can be shown to be

$$S(\theta) \propto \frac{1}{1 + F(\theta)\sin^2 k_0\theta^2 L/2}. \tag{3.1}$$

Also, $\hat{C}(\theta)$ represents the Fourier transform of $C(r)$, with $\theta = k_r/k_0$, where k_r is the spatial frequency corresponding to r and k_0 is the propagation constant of the light in the medium. In the above relation F is the cavity finesse and L is the thickness of the material. From (3.1), upon setting $k_0\theta^2 L/2 = \pi$, the semiangle of the first ring can be calculated to be approximately 0.4° in the material, which is of the order of our observed value of 0.8° in air. With time, the ring may break up into hexagonal spots, as observed experimentally. Note also that in the experiments, secondary (or higher-order diffraction) rings, and sometimes higher-order hexagonal spots, are also observed. We have observed in our experiments that the radius of the second ring is observed to be $\sqrt{3}$ times that of the first, which can be also derived from (3.1) by setting $k_0\theta^2 L/2 = 3\pi$.

We would like to point out that the existence of Fabry-Perot modes in the crystal cavity supports the concept of periodic imaging during propagation in the crystal. In so-called "open" cavities consisting of a matched or misaligned photorefractive crystal and an external feedback mirror, the concept of Talbot imaging has been used to determine the scattering angle [22, 23]. However we feel that the concept of Talbot imaging as discussed in [22, 23] can only be applied to the case of propagation in the "cavity" between a "thin" slice of the photorefractive material containing the induced reflection grating and the external mirror. More on this is discussed in the following.

In a nonlinear system where the incident beam may originate from light scattering, the coupling between forward and backward traveling waves may be provided by transmission and / or reflection gratings. From experimental results on beam coupling, it has been shown that reflection gratings are dominant. In the remainder of this chapter, we will assume only reflection gratings to be present. It turns out that the scattering angle will be nonlinearly modified depending on the strength of the reflection grating.

3.3.2 Model equations

We represent the forward and backward traveling waves in the nonlinear photorefractive material as

$$E = \mathrm{Re}[(C_e \exp - jk_0 z + C_{e'} \exp jk_0 z) \exp j\omega_0 t] \tag{3.2}$$

where $C_{e,e'}$ denote the forward and backward traveling wave amplitudes, respectively, and ω_0 is the angular frequency of the light in the medium. We also assume that the material has light-induced changes in the refractive index due to reflection gratings formed in the material, with spatial frequency $2k_0$. The spatial evolution of the forward and backward traveling envelopes can be then written as

$$\begin{aligned} L_e C_e &= -jk_0 \delta n C_{e'}, \\ L_{e'} C_{e'} &= -jk_0 \delta n^* C_e, \end{aligned} \tag{3.3}$$

where $L_{e,e'}$ are linear operators given by the relation

$$L_{e,e'} = \partial / \partial z \mp j(1/2k_0) \nabla_\perp^2 . \tag{3.4}$$

The variable δn represents the fractional change of refractive index due to the induced reflection grating and evolves according to

$$\tau \partial \delta n / \partial t + \delta n = \gamma \frac{C_e C_{e'}^*}{|C_e|^2 + |C_{e'}|^2}. \tag{3.5}$$

We consider reflection gratings only for now because they are dominant in photorefractive potassium niobate [15]. Transmission gratings have been assumed in other analyses, such as for the determination of the onset of instability [41]. By using the model of excitation of satellite beams due to propagation of contrapropagating primary or pump beams, and transmission gratings, so-called *spatial dispersion curves* for the onset of instabilities leading to satellite beam formation have been derived. The plots show the dependence of the minimum threshold gain as a function of the angle between the pump and the spatial sidebands [41]. Dispersion curves assuming predominantly transmission gratings and aided by reflection gratings have been also derived by Kukhtarev *et al.* [17]. Later on in the chapter, we will provide the results of such dispersion curves but using reflection gratings in the model, since it pertains more closely to spontaneous pattern generation in photorefractive potassium niobate.

We would like to point out that pattern dynamics have been extensively studied in a bidirectional photorefractive ring resonator assuming transmission grating approximation and four-wave mixing in the active photorefractive medium [42]. Spontaneous symmetry breaking, dynamical oscillations, vortex formation and complex pattern development are predicted for large Fresnel numbers. A photorefractive oscillator with a stable resonator has been used to model a nonlinear dynamical system in which transverse mode patterns have been observed [43].

3.3.3 Instability criterion and the dispersion relation

There is considerable work done on the onset of instabilities in a photorefractive medium with reflection gratings due to counterpropagation of pump beams. The analysis of Sturman and Chernykh [44] assumes a medium in which there is no energy coupling. Saffman *et al.* [45] has performed a more detailed analysis assuming both real and complex coupling coefficients. Honda and Banerjee [39] have improved on their analysis, by showing that pattern generation can occur even for purely energy coupling.

We now present the threshold condition for instabilities derived for the experimental arrangement in [39] with the photorefractive crystal and a feedback mirror. We use the relations in Eqs. (3.3)–(3.5) and substitute

$$C_{e,e'} = C_{0,0'}\,[1 + c_{1,1'} \exp - jK\bullet r + c_{-1,-1'} \exp jK\bullet r],\, c_{i,i'} = C_{i,i'} / C_{0,0'} \tag{3.6}$$

where K is the transverse wavenumber and r denotes the transverse coordinate to get

$$\begin{aligned}
(\partial / \partial z - jk_d)c_1 &= jA\gamma(c_1 + c_{-1}{}^* - c_{1'} - c_{-1'}),\\
(\partial / \partial z + jk_d)c_{-1}{}^* &= -jA\gamma^*(c_1 + c_{-1}{}^* - c_{1'} - c_{-1'}),\\
(\partial / \partial z + jk_d)c_{1'} &= jA\gamma^*(c_1 + c_{-1}{}^* - c_{1'} - c_{-1'}),\\
(\partial / \partial z - jk_d)c_{-1} &= -jA\gamma(c_1 + c_{-1}{}^* - c_{1'} - c_{-1'}),
\end{aligned} \tag{3.7}$$

where $k_d = K^2 / 2k_0$ and $A = A(z) = |C_0|^2\, |C_{0'}|^2 / [|C_0|^2 + |C_{0'}|^2$. Note that since A is a function of z, Eqs. (3.7) cannot be solved analytically. However, when the reflectivity of the feedback mirror is unity or the reflection from the back surface of the crystal is considerable, we can approximate $A \sim 1/4$ [39].

For the case of a feedback mirror placed behind the crystal, the boundary conditions can be written as

$$\begin{aligned}
&c_{1,-1}(0) = 0,\\
&\quad c_{1'}(L) = \exp(-2jk_d l)c_1(L),\, c_{-1'}(L) = \exp(2jk_d l)c_{-1}{}^*(L)
\end{aligned} \tag{3.8}$$

where L is the crystal thickness and l denotes the distance between the photorefractive medium and the feedback mirror. The threshold condition can be found using (3.7) and (3.8) and using the Laplace transformation to solve. Assuming that the mirror is placed against the back surface of the crystal, the "dispersion relation" can be written as

$$\begin{aligned}
\cos wL \cos k_d L &+ (\gamma / 2w) \sin wL \cos k_d L\\
&+ (k_d / w) \sin wL \sin k_d L = 0,
\end{aligned} \tag{3.9}$$

where we have assumed the coupling constant γ to be purely imaginary ($\gamma \to -j\gamma$) and $w^2 = k_d^2 - \gamma^2/4$. Figure 3.6 shows the dispersion curve for this case. When γ is just above the threshold for spatial sideband generation, the direction of the sidebands will correspond to k_d which gives the minimum of the dispersion curve. For other mirror locations the angle between the carrier and the spatial sidebands decreases as shown in [23, 39].

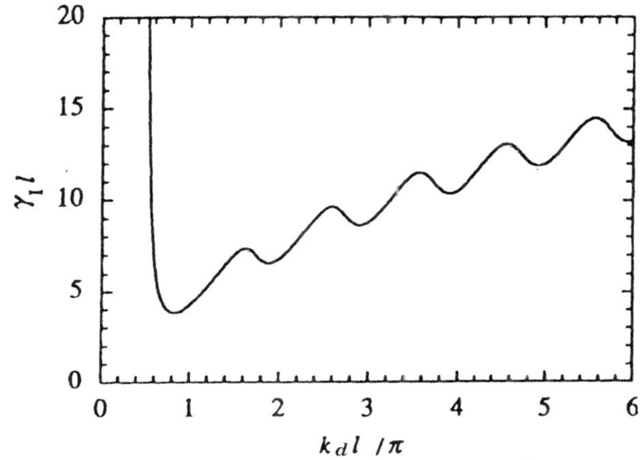

Figure 3.6: Threshold condition assuming purely energy coupling and for mirror feedback. Mirror is located at the back surface of the sample. (Source: Ref. [39]. Reprinted with permission.)

3.3.4 Nonlinear eigenmodes in the steady state

The formulation stated above through Eqs. (3.3–3.5) can also be used to study the exact spatial behavior of the carriers (contrapropagating pumps) and the spatial sidebands. In the steady state, the spatial evolution of the carriers and the spatial sidebands lying on a scattering ring can be studied by solving the system of equations [46–48]

$$\begin{aligned} L_i C_i &= \sum_{j'} \delta n_{ij'} C_{j'} = \gamma \sum_{j'kl'} C_k C_{l'}^* C_{j'} \Big/ \left[\left| \sum_i C_i \right|^2 + \left| \sum_i C_{i'} \right|^2 \right] \\ L_{i'} C_{i'} &= \sum_{j} \delta n_{ji'}^* C_j = \gamma^* \sum_{j'kl'} C_{k'} C_l^* C_j \Big/ \left[\left| \sum_i C_i \right|^2 + \left| \sum_i C_{i'} \right|^2 \right] \end{aligned} \tag{3.10}$$

where we have assumed the optical properties of the photorefractive material to be isotropic. As seen from Eq. (3.10), coupling will occur only between waves whose transverse wavevectors satisfy the general relation $K_i + K_{l'} = K_k + K_{j'}$. An example of a set of contrapropagating pumps and a set of six forward and backward propagating scattered sidebands is shown on the transverse $\boldsymbol{K}$-plane in Fig. 3.7. It can be shown that seven different types of couplings may occur. For example, $K_{l'} = 0$, $K_i = K_k + K_{j'}$ couples the main forward traveling beam with three waves that are

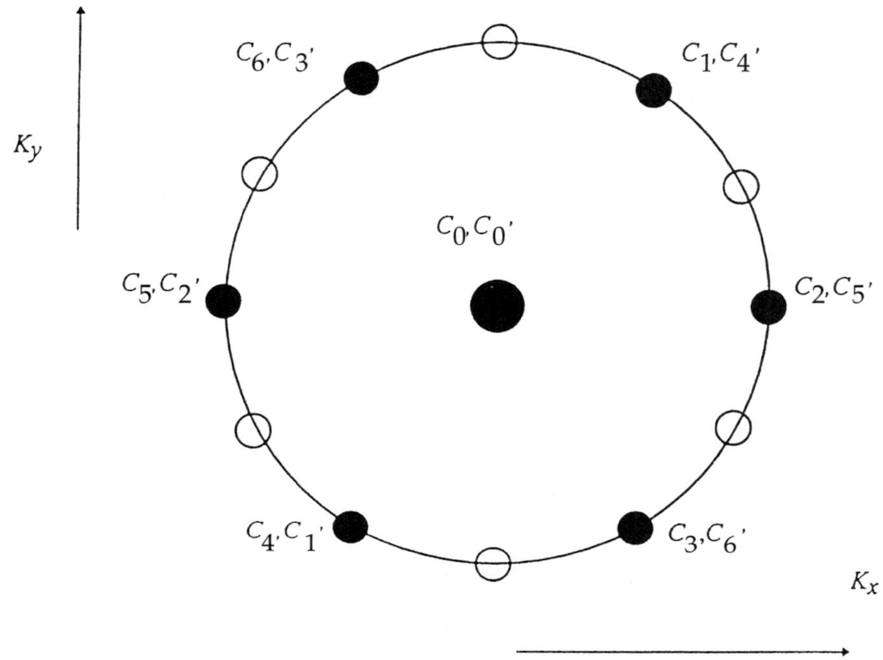

Figure 3.7: Transverse **k**-vectors of hexagonally related scattered waves with regard to that of the forward- and backward-propagating beams. Two sets of hexagonally related scattered beams are shown.

hexagonally related. The interaction $K_i = -K_{l'}$, $K_k = -K_{j'}$ couples sets of hexagonally related waves together.

In what follows, we assume a geometry identical to the experimental arrangement in [16] with only the unmatched crystal and no feedback mirror. Using Eq. (3.10) as a model, the spatial evolution of the carriers and the sidebands have been analyzed for the case when there are 72 sidebands symmetrically distributed on the scattering ring [47, 48]. The preliminary results which were performed using a purely imaginary γ shows the general nature of the "modes" in the steady state that can exist within the interaction region in the photorefractive material. These modes show the permissible values of the phase difference between the pump and the sidebands at the front surface of the material for different values of the gain parameter proportional to γ. Furthermore, one can simultaneously get the spatial variations of the contrapropagating pumps and the spatial sidebands, assumed equal in magnitude for simplicity. The results therefore define the conditions needed for self-organization of the laser beam into a scattering ring, starting from fanning noise in the material. However, it turns out that the ratio of scattered to pump intensities both in the forward and the backward directions are not exactly similar to experimentally observed results [16].

We would like to point out that a simple time evolution simulation to illustrate the basic principle of the formation of the scattering ring and hexagon formation can be performed by starting from the model shown in Eqs. (3.3)–(3.5) and even assuming a constant (intensity independent) imaginary coupling constant, constant amplitudes for the interacting pumps and spatial sidebands in the photorefractive material, and assuming a thin sample [46]. Taking an initial linear scattering from beam fanning, the evolution of the spatial sidebands into a scattering cone and eventually into hexagonal pattern in the far-field is shown in Fig. 3.8.

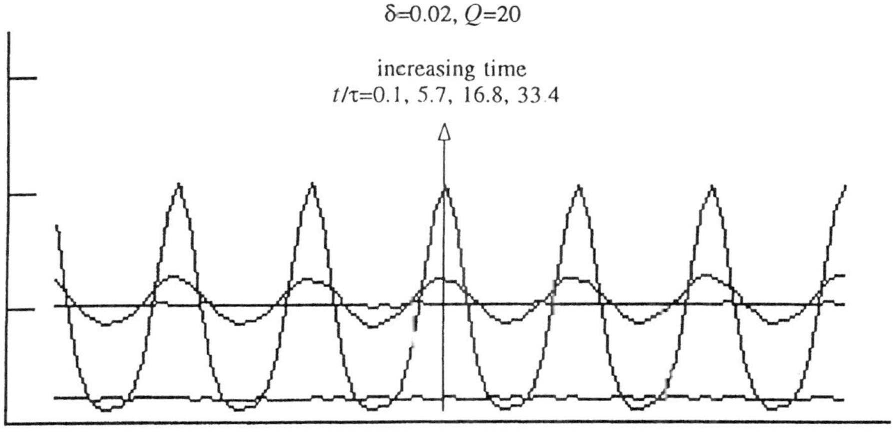

Figure 3.8: Time evolution of scattering around a circle, showing the growth of the scattering ring and eventual formation of the hexagonal spot pattern. The normalized coupling parameter Q showing coupling between hexagonally related points on the scattering ring and the pump is taken to be 20, and the linear scattering coefficient δ from the pump (to initiate the self-organization process) is taken to be 0.02 [46].

The plots show that the energy scattered into the ring as the first stage of the self-organization process essentially later redistributes into the hexagons. The plots are quantitatively modified slightly if transmission gratings are also incorporated into the simulations. All simulation results are in qualitative agreement with experimental observations [16]. If one monitors the minimum value of the gain needed for the onset of instabilities as a function of the linear scattering parameter, it is observed that the threshold gain decreases sharply with increasing initial linear scattering, as expected, and tends to slightly decrease for very high values of the scattering parameter. This decrease can be attributed to the fact that excessive linear scattering tends to deplete the pumps of their initial energy, thus inhibiting the formation of reflection gratings and eventual transfer of pump energies into the spatial sidebands. This simple simulation also demonstrates the justification for looking for exact spatial eigenmodes which depict the spatial variation of the pumps and the scattering ring as the first stage of the self-organization process.

As stated in the preceding discussion, the discrepancy between numerical simulations for the nonlinear eigenmodes and experimental results of the energy scattered into the ring and eventually into the hexagon can be resolved by assuming a complex coupling constant. Possible reasons for the nonideal phase of the coupling coefficient are as follows. As in any photorefractive material, the contribution to photorefractivity can come from both diffusion and photovoltaic contributions. While diffusion creates a space-charge field which is out of phase with the intensity profile, photovoltaic effects give rise to space-charge fields which are in phase with the intensity [49]. In general, therefore, an arbitrary phase difference may exist. Furthermore, even for a purely photovoltaic material, it has been shown that there can exist a phase difference between the intensity grating and the fundamental spatial frequency component of the space-charge field, for large modulation depths. This can also give rise to a complex coupling coefficient [50].

Starting from Eq. (3.10) and setting $\gamma \to \gamma \exp j\phi$ with

$$C_{i,i'} = S_{i,i'}(z)\exp(-jK_{i,i'} \bullet r)\exp(\mp j\pi z \ / \ L)\exp(j\phi_{i,i'}(z)) \qquad (3.11)$$

we can derive the spatial evolution equations for the amplitudes and phsaes of the interacting waves. If we assume that the amplitudes and phases $S_{i,i'}$, $\phi_{i,i'}$, $i,i' \neq 0$ of the interacting waves on the scattering ring are identical for simplicity, we get after extensive algebra, coupled differential equations which have the functional forms [51]

$$\begin{aligned}\partial S_0^2/\partial z &= (\gamma/I)F_0[S_0^2, S_1^2, S_{0'}^2, S_{1'}^2 b, c, \phi, N] - N\delta S_0^2,\\ \partial S_1^2/\partial z &= (\gamma/I)F_1[S_0^2, S_1^2, S_{0'}^2, S_{1'}^2 b, c, \phi, N] + \delta S_0^2,\\ \partial b/\partial z &= (\gamma/I)F_b[S_0^2, S_1^2, S_{0'}^2, S_{1'}^2 b, c, \phi, N] + \pi/L,\end{aligned} \tag{3.12}$$

where

$$b = \phi_{1'} - \phi_{0'} + \pi z/L, c = \phi_1 - \phi_0 - \pi z/L \tag{3.13}$$

The corresponding equations for $S^2_{0',1'}$, c can be found by interchanging the primed and unprimed variables, interchanging b and c in the above equations and replacing L by $-L$. In Eq. (3.12), I is the incoherent intensity and we should point out that Eq. (3.12) is valid assuming up to third-order interactions. The constant δ is a linear scattering parameter. N is the number of interacting waves on the scattering ring, taken here to be equal to 72. The exact expressions for F_i in (3.12) are given in [47, 48, 51]. Conservation rules for waves interacting through the formation of reflection gratings hold, and Eq. (3.12) is solved numerically assuming boundary conditions pertinent to the front and back surfaces of the crystal which generate the counterpropagating waves in a truly mirrorless configuration.

The numerical results (not shown here) show the existence of multiple eigenmodes which are possible in the photorefractive medium. Each eigenmode is characterized by a value of $b(0)$ [assumed equal to $c(0)$] and ϕ, and is the locus of permissible solutions on the $b(0) - \phi$ plane. If the forward and backward scattering ratios, defined as the fraction of the pump energy scattered onto the ring, are monitored, it follows that by relaxing the condition on ϕ, namely, making it arbitrary, it is possible to attain values similar to experimental observations. For instance, for a value of $\phi = 230°$, about 40% of the energy is scattered into the ring in the forward direction, with about 30% in the backward direction, in close agreement with experimental observations [16]. The fact that a complex coupling constant is required to achieve the expected forward and backward scattered energies corroborates the fact that the ideal phase difference (namely 270°) between the intensity grating and the induced refractive index profile is probably changed due to contributions from diffusion from the finite modulation depth of the intensity grating, as explained in the preceding discussion. The phase difference between the pump and the sidebands at the front surface is close to 270° for this case, which implies phase modulation of the profile of the total beam at this plane (and also at the exit plane), with amplitude modulation in the center of the photorefractive material. The analysis also enables us to track the exact spatial evolution of the pumps and the spatial sidebands; this is shown in Fig. 3.9. We would also like to point out that agreement between theory and experiment is only observed for the above value of ϕ, which explains why self-organization is not observed when the experiment is performed with the c-axis of the crystal turned in the reverse direction [16].

Finally we would like to point out other related analytical work in the area of transverse instabilities. The effect of crystal symmetry on the

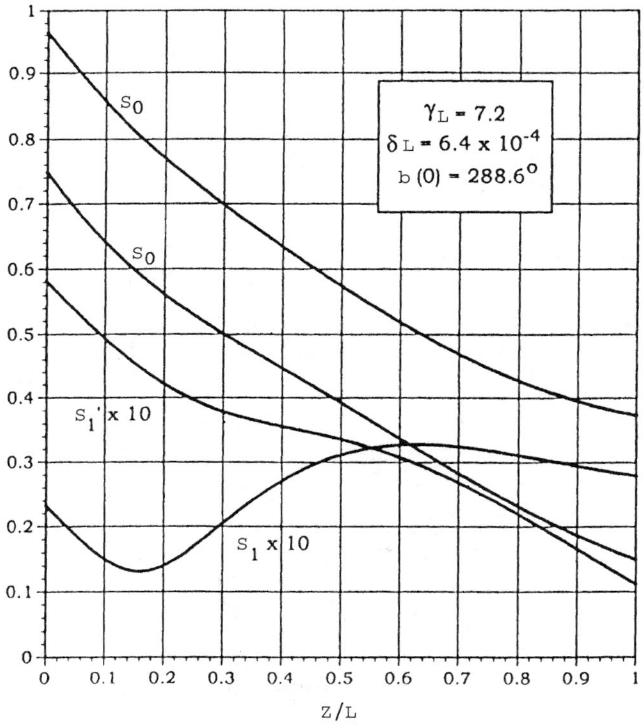

Figure 3.9: Spatial variation of the forward- and backward-scattered and main beam amplitudes during propagation through the crystal.

formation and rotation of hexagonal scattering patterns in photorefractive materials has been discussed by Dimmock [52].

For the orthorhombic, C_{2v}, symmetry of the $KNbO_3$ crystal, the two crystallographic planes perpendicular to the a- and b-axes are planes of reflective symmetry and the c-axis is a two-fold rotation axis. If we consider that the incident laser beam in Fig. 3.1 is polarized along the a-axis, and that the laser beam, and its reflected beams are directed exactly along the c-axis, then the experiment will also possess the C_{2v} symmetry. Namely, it is invariant with respect to twofold rotations about the c-axis, and reflections in the two symmetry planes. Such a configuration can show no right- or left-handedness. The constraints of symmetry on the possible rotations are shown in Fig. 3.10. Let K_x and K_y represent the transverse K vectors of the laser beam with respect to the crystallographic c-axis direction. If the crystal reflection planes are perpendicular to the c-axis then K_x and K_y are proportional to $\Delta\theta_h$ and $\Delta\theta_v$. If the laser beam propagates parallel to either of the two symmetry planes the orthorhombic symmetry is preserved and no rotation can occur. It is only when the laser beam propagates in a direction corresponding to one of the four quadrants that predictable and stable rotation is allowed. Figure 3.4 shows how the direction of rotation experimentally depends on the beam direction. This is not exactly what is predicted in Fig. 3.10. The discrepancy can be explained by postulating additional asymmetries in the material, and

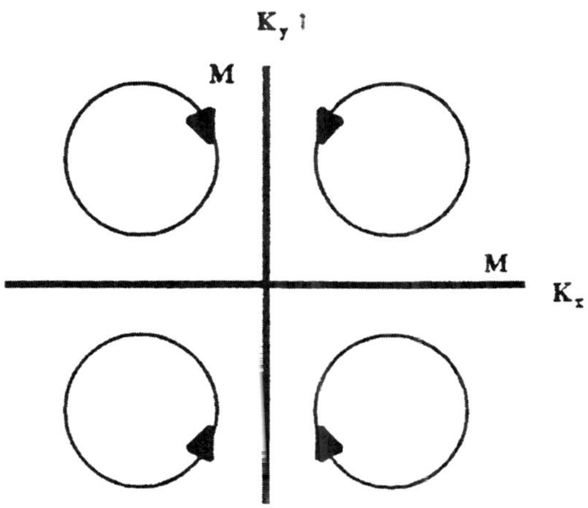

Figure 3.10: Dependence of rotational sense on beam direction.

misorientations of the front and back surfaces with respect to the c-axis of the crystal. The effect of orthorhombic anisotropies of the index of refraction and the electrooptic coefficients has been discussed in detail in [52].

Sandfuchs *et al.* have determined the instability criteria for the case when a voltage is applied across the photorefractive material, and assuming reflection gratings and finite modulation index for the intensities [53]. For related work in Kerr media, the reader is also referred to [54–56].

3.3.5 Self-phase conjugation

As described in Section 3.2, interactions initiated by incident beams $C_{0,1}$ and supported by internal reflections $C_{0',-1'}$ eventually generate $C_{-1,1'}$ [17]. Using equations analogous to Eq. (3.10) for the interfering waves and in the steady state we can write, approximately [46],

$$L_e C_{-1} \approx n_{0,-1'} C_0', \; L_e' C_1' \approx n_{0',1} C_0, \\ n_{0,-1'} \propto \gamma C_0 C_1'^{*}, \; n_{0',1} \propto \gamma C_0' C_1{}^{*}. \tag{3.14}$$

Using the preceding relations, the phase conjugate intensities $I_{-1,1'}$ can be found as

$$I_{1'} \propto TRI_{1inc} I_{0inc}^{2}{}^{2} / (1 + R)^2 (I_{1inc} + I_{0inc})^2 ; I_{-1} \propto RI_{1'} \tag{3.15}$$

where $I_{1inc,0inc}$ are the input intensities of the waves 1 and 0; T and R represent the transmittivity and reflectivity of the interface, respectively. The derivation above can be modified to include the effects of transmission gratings as well. A detailed description of the evolution of the conjugates is important and will be performed in the future, especially in light of recent experimental observations that the time dynamics of the different phase conjugates formed are different and could hence originate from transmission or reflection gratings.

3.3.6 Model of hexagonal formation based on transverse electrical instability

In what follows, we will discuss the contribution of electrical instabilities to the formation of hexagonal structures. An adequate description of self-organized pattern in $KNbO_3$ and other thick photorefractive materials includes material equations (like diffusion-drift model) and Maxwell's equations. Both material and optical equations are nonlinear and are potentially capable of describing formation of spatial-temporal patterns. As an example, we can mention the problem of the holographic subharmonic [57], observed during self-diffraction of two beams with slightly different frequencies, and with an external electric field in $Bi_{12}SiO_{20}$ crystals. The appearance of an additional beam between intersecting "pump beams" was originally explained by optical nonlinearities [58]. Later it was realized that instabilities of material equations, like period doubling, may lead to formation of a subharmonic component in the space-charge field and in the refractive index [59]. Similar trends are visible in the explanation of hexagon patterns in the thick photorefractive materials.

All previous explanations of pattern formation in photorefractive crystals were based on instabilities of optical equations (Maxwell's equations) where material equations play only an insignificant role. Only recently it was realized that photogalvanic currents may be responsible for contrast enhancement, and may result in space-charge instabilities [50]. Quantitatively, formation of the spatial patterns due to photogalvanic current may be explained taking into account the relation $\nabla \cdot J = 0$. This equation implies that the current has a vortex structure and forms closed loops. Detailed calculations of the transversal structure caused by photogalvanic nonlinearity is beyond the scope of this discussion. The ansatz that photogalvanic instabilities lead to transverse patterns of the E-field and refractive index lets us discuss experimental results in the near field.

Transversal modulation of the refractive index of a thick photorefractive crystal may be regarded as recording of a bunch of optical channels or waveguides. As was shown in [60], modulation of the refractive index may be visualized in the near-field as optical channeling. We can thus model the transversal modulation of the dielectric constant by the function

$$\varepsilon(x,y) = \varepsilon_0 + \varepsilon_x \cos K_x x + \varepsilon_y \cos(K_y y + \varphi) \tag{3.16}$$

where ε_0 denotes the average value of the permittivity, and $\varepsilon_{x,y}$ are the amplitudes of the modulation along the transverse x- and y-axis, with wave numbers K_x, K_y and proper phase shift φ. Introducing the function $\varepsilon(x,y)$ in Maxwell's equations, we can get the following result for the near-field intensity:

$$\begin{aligned} I(x,y) = I_0[1 &+ \varepsilon_x(L_x/\lambda)^2 \sin^2(\pi\lambda z/2L_x^2)\cos K_x x \\ &+ \varepsilon_y(L_y/\lambda)^2 \sin^2(\pi\lambda z/2L_y^2)\cos(K_y y + \varphi)] \end{aligned} \tag{3.17}$$

where $L_{x,y} = 2\pi / K_{x,y}$ and λ is the wavelength. The solution [Eq. (3.17)] is valid for small modulation and includes longitudinal modulation with the periods

$$z_{x,y} = 2L_{x,y}^2 / \lambda. \tag{3.18}$$

We can see that Eq. (3.17) also describes contrast inversion. As described before, for experimental values with $KNbO_3$ ($\lambda = 0.514$ μm, $L_x = L_y = 30$ μm), we can get for longitudinal period $z_x = z_y = 0.49$ cm that is close to the experimental value of 0.5 cm [16].

We would like to reiterate that the explanation of hexagonal structures by Talbot effect imaging is valid only for optically thin gratings, where Talbot effect description during propagation in free-space is justified. In our case we have used a thick crystal (1 cm thick) and we should use an adequate model of thick holographic gratings. The channeling effect is pronounced for thick gratings, and naturally describes the effect of contrast inversion. In contrast to Bragg diffraction that normally needs coherent light, channeling may be observed also in incoherent illumination.

3.5 Conclusion

We have summarized, using minimal mathematics, some important aspects of an area which is complicated for two reasons: (1) because of the nonlinear and spatiotemporal nature of the problem, and (2) because the response of a photorefractive material to incident light is a complicated phenomenon, governed by a set of nonlinear coupled differential equations. Wherever possible, experimental results have been quoted or referred to to assure readers that there is some connection to reality behind the complicated mathematics. The list of potential applications given in the Introduction is also meant to excite the reader to future possibilities. We hope the summary of self-organization in photorefractives, as presented, will interest readers to undertake the challenging and unfinished work in the area. Finally, we have tried to compile the important references in the general area, and although undoubtedly some have been left out for brevity and due to oversight, cross-referencing should prove valuable in finding all necessary citations to this rapidly growing field.

Acknowledgment

P. P. B. thanks Prof. K. Matsushita of Osaka City University for his help during preparation of the manuscript. P. B. B. also acknowledges collaboration with Drs. T. Honda (NRLM, Tsukuba, Japan), F. Madarasz (CAO, UAH), and H-L Yu.

References

1. J. A. Scott Kelso, *Dynamic Patterns: The Self Organization of Brain and Behavior*, MIT Press, Cambridge MA, 1995.
2. T. Kohonen, *Self-organization and Associative Memory*, Springer-Verlag, New York, 1984.
3. H. Haken, *Advanced Synergetics: instability hierarchies of self-organizing systems and devices*, Springer-Verlag, New York, 1983.
4. M. A. Vorontsov and W. B. Miller, *Self-Organization in Optical Systems and Applications in Information Technology*, Springer, Berlin, 1995.
5. A. Fuchs and H. Haken, *Neural and Synergistic Computers*, Springer, Berlin, 1988.
6. E. V. Degtiarev and M. A. Voronotsov, "Spatial filtering in nonlinear two-dimensional feedback systems: phase distortion suppression," *J. Opt. Soc. Amer.*, **B12**, 1238, 1995.
7. M. A. Vorontsov and W. J. Firth, "Pattern formation and competition in nonlinear feedback systems with two-dimensional feedback," *Phys. Rev.*, **A49**, 2891, 1994.
8. B. Thuring, A. Schreiber, M. Kreuzer and T. Tcshudi, "Spatio-temporal dynamics due to competing spatial instabilities in a coupled LCLV feedback system," *Physica,* **D96**, 282, 1996.
9. G. Grynberg, E. LeBihan, P. Verkerk, P. Simoneau, J. R. Leite, D. Bloch, S. LeBoiteux and M. Ducloy, "Observation of instabilities due to mirrorless four-wave mixing oscillations in sodium," *Opt. Comm.*, **67**, 363, 1988.
10. J. Pender and L. Hesselink, "Degenerate conical emission in atomic sodium vapor," *J. Opt. Soc. Amer.*, **B7**, 1361, 1990.
11. G. Grynberg, A. Maitre and A. Petrosian, "Flowerlike patterns generated by a laser beam tansmitted through a rubidium cell with a single feedback mirror," *Phys. Rev. Lett.*, **72**, 2379, 1994.

12. B. Thuring, R. Neubecker and T. Tschudi, "Transverse pattern formation in an LCLV feedback system," *Opt. Comm.,* **102**, 111, 1993.

13. M. Tamburrini, M. Bonavita, S. Wabnitz and E. Santamato, "Hexagonally patterened beam filamentation in a thin liquid crystal film with a single feedback mirror," *Opt. Lett.,* **18**, 855, 1993.

14. J. Gluckstad and M. Saffman, "Spontaneous pattern formation in a thin film of bacteriorhodopsin with mixed absorptive nonlinearity," *Opt. Lett.,* **20**, 551, 1995.

15. T. Honda, "Hexagonal pattern formation due to counterpropagation in $KNbO_3$," *Opt. Lett.,* **18**, 598, 1993.

16. P. P. Banerjee, H-L Yu, D. A. Gregory, N. Kukhtarev, and H. J. Caulfield, "Self-organization of scattering in photorefractive $KNbO_3$ into a reconfigurable hexagonal spot array," *Opt. Lett.,* **20**, 10, 1995.

17. N. Kukhtarev, T. Kukhtareva, P. P. Banerjee, H-L Yu and L. Hesselink, "Broad-band, dynamic, holographically self-recorded, and static hexagonal scattering patterens in photorefarctive $KNbO_3$," *Opt. Engr.,* **34**, 2261, 1995.

18. T. Honda, "Flow and controlled rotation of the spontaneous optical hexagon in $KNbO_3$," *Opt. Lett.,* **20**, 851, 1995.

19. T. Honda, H. Matsumoto, M. Sedlatschek, C. Denz and T. Tschudi, "Spontaneous formation of hexagons, squares and sqeezed hexagons in a photorefractive phase conjugator with virtually internal feedback mirror," *Opt. Comm,* **133**, 293, 1997.

20. C. Denz, M. Schwab, M. Sedlatschek, T. Tschudi and T. Honda, "Pattern dynamics and competition in a photorefractive feedback system," *J. Opt. Soc. Amer.,* **B15**, 2057, 1998.

21. P. P. Banerjee, H-L Yu, D. A. Gregory and N. Kukhtarev, "Phase conjugation, edge detection and image broadcasting using two-beam coupling configuration in photorefractive $KNbO_3$:Fe," *Opt. and Laser Tech.,* **28**, 89, 1996.

22. T. Honda and H. Matsumoto, "Buildup of spontaneous hexagonal patterens in photorefarctive BaTiO3 with a feedback mirror," *Opt. Lett.,* **20**, 1755, 1995.

23. Y. Uesu, A. Ueno, M. Kobayashi and S. Odoulov, "Hexagon formation in photorefractive crystals as mirrorless coherent oscillation," *J. Opt. Soc. Amer.,* **B15**, 2065, 1998.

24. R. Mersereau, "The processing of hexagonally sampled two-dimensional signals," *Proc. IEEE,* **67**, 930, 1979.

25. N. C. Roberts, A. G. Kirk and T. J. Hall, "Binary phase gratings for hexagonal array generation," *Opt. Comm.,* **94**, 501, 1992.

26. N. Noginova, N. Kukhtarev, T. Kukhtareva, M. Noginov, H. J. Caulfield, P. Venkateswarlu, D. Parker and P. P. Banerjee, "Photoinduced electric current in Fe doped $KNbO_3$," *J. Opt. Soc. Amer.* **B14**, 1390, 1997.

27. P. Bernasconi, I. Biaggio, M. Zgonik and P. Gunter, "Anisotropy of electron and hole drift mobility in $KNbO_3$ and $BaTiO_3$," *Phys. Rev. Lett.,* **78**, 106, 1997.

28. W. J. Firth in *Self-Organization in Optical Systems and Applications in Information Technology,* M. A. Vorontsov and W. B. Miller (eds.), Springer, Berlin, 1995.

29. C. Medrano, E. Voit, P. Amrhein and P. Gunter, "Optimization of photorefractive properties of $KNbO_3$ crystals," *J. Appl. Phys.,* **64**, 4668, 1988.

30. E. Voit, M. Z. Zha, P. Amrhein and P. Gunter, "Reduced $KNbO_3$ crystals for fast nonlinear optics," *Appl. Phys. Lett.,* **51**, 2079, 1987.

31. C. Medrano, M. Zgonik, S. Berents, P. Bernasconi and P. Gunter, "Self-pumped and incoherent phase conjugation in Fe doped $KNbO_3$," *J. Opt. Soc. Amer.,* **B11**, 1718, 1994.

32. C. Medrano, M. Ingold and P. Gunter, "Self-pumped optical phase conjugation and light oscillation in Fe doped $KNbO_3$," *Opt. Comm.,* **77**, 411, 1990.

33. G. Pauliat and P. Gunter, "Coherent light oscillations with photorefractive $KNbO_3$ crystals," *Opt. Comm.,* **66**, 329, 1988.

34. M. Zgonik, K. Nakagawa and P. Gunter, "Electrooptic and dielectric properties of photorefractive $BaTiO_3$ and $KNbO_3$," *J. Opt. Soc. Amer.,* **B12**, 1416, 1995.

35. C. Medrano, M. Zgonik,, N. Sonderer, C. Beyler, S. Krucker, J. Seglins, H. Wuest and P. Gunter, "Photorefractive effect in Cu and Ni doped KNbO3 in the visible and near infra-red," *J. Appl. Phys.,* **76**, 5640, 1994.

36. Q. W. Song, C. P. Zhang and P. J. Talbot, "Anisotropic light induced scattering and position dispersion in a $KNbO_3$:Fe crystal," *Opt. Comm.,* **98**, 269, 1993.

37. J. W. Goodman, *Introduction to Fourier Optics,* 2nd ed. McGraw-Hill, New York, 1996.

38. F. Papoff, G. D. Alessandro, G. L. Oppo and W. J. Firth, "Local and global effects of boundaries on optical pattern formation in Kerr media," *Phys. Rev.,* **A48**, 634, 1993.

39. T. Honda and P. P. Banerjee, "Threshold for spontaneous pattern formation in reflection grating dominated photorefractive media with mirror feedback," *Opt. Lett.,* **21**, 779, 1996.

40. N. Bogodaev, Y. Kuzminov, N. Kukhtarev and N. Polozkov, "Photoinduced adaptive mirror and optical generation in photorefractive strontium niobate crystals," *Sov. Tech. Phys. Lett.,* **12**, 608, 1987.

41. M. Saffman, D. Montgomery, A. A. Zozulya, K. Kuroda and D. Z. Anderson, "Transverse instability of counterpropagating waves in photorefractive media," *Phys. Rev.,* **A48**, 3209, 1993.

42. Z. Chen and N. B. Abraham, "Pattern dynamics in a bidirectional photorefractive ring resonator," *Appl. Phys.,* B S183, 1995.

43. J. Malos, M. Vaupel, K. Staliunas and C. O. Weiss, "Dynamical structures of a photorefractive oscillator," *Phys. Rev.,* **A53**, 3559, 1996.

44. B. Sturman and A. Chernykh, "Mechanism of transverse instability of counterpropagation in photorefractive media," *J. Opt. Soc. Amer.,* **B12**, 1384, 1995.

45. M. Saffman, A. A. Zozulya and D. Z. Anderson, "Transverse instability of energy-exchanging counterpropagating waves in photorefractive media," *J. Opt. Soc. Amer.,* **B11**, 1409, 1994.

46. P. P. Banerjee, J. O. Dimmock and N. V. Kukhtarev, "Self-organization in photorefractive crystals in the presence of transmission and reflection gratings," *SPIE Proc.,* **2849**, 79, 1996.

47. P. P. Banerjee, J. O. Dimmock, F. L. Madarasz and N. V. Kukhtarev, "Steady state analysis of self-organization of light into a scattering ring due to induced reflection gratings in photorefractive materials," *J. Opt. Soc. Amer. B* (submitted for publication).

48. J. O. Dimmock, P. P. Banerjee, and N. V. Kukhtarev, "Analysis of the self-organization of light due to induced reflection gratings in potassium niobate," *Proceedings of the SPIE, Photorefractive Fiber and Crystal Devices: Materials, Optical Properties and Applications III,* **3137**, 75, 1997.

49. P. Yeh, *Introduction to Photorefractive Nonlinear Optics*, Wiley, New York 1993.

50. N. V. Kukhtarev, P. Buchhave, S. F. Lyuksyutov, T. Kukhtareva, K. Sayano, F. Zhao and P. P. Banerjee, "Self-enhancement of dynamic gratings in photogalvanic crystals," *Phys. Rev. A,* **58**, 4051, 1998.

51. P. P. Banerjee, J. O. Dimmock, F. L. Madarasz and N. V. Kukhtarev, "Effect of the phase of the coupling constant on the self-organization of light in potassium niobate," *SPIE Proc. Annual Meeting,* San Diego, CA, 1998.

52. J. O. Dimmock, "Symmetry considerations in the formation and rotation of hexagonal scattering patterns in photorefractive $KNbO_3$," *SPIE Conference on Photorefractive Fiber and Crystal Devices: Materials, Optical Properties and Applications II,* Denver, **2849**, 1996.

53. O. Sandfuchs, J. Leonardy, F. Kaiser and M. R. Belic, "Transverse instabilities in photorefractive counterpropagating two-wave mixing," *Opt. Lett.,* **22**, 498, 1997.

54. W. J. Firth, "Spatial instabilities in a Kerr medium with a single feedback mirror," *J. Mod. Opt.,* **37**, 151, 1990.
55. G. D'Alessandro and W. J. Firth, "Hexagonal spatial patterns for a Kerr slice with a feedback mirror," *Phys. Rev.,* **A46**, 537–548, 1992.
56. N. B. Abraham and W. J. Firth, "Overview of transverse effects in nonlinear optical systems," *J. Opt. Soc. Amer.,* **B7**, 951, 1990.
57. S. Mallick, B. Imbert, H. Ducollet, J. P. Herriau and J. P. Huignard, *J. Appl. Phys.,* **63**, 5660, 1998.
58. D. Jones and L. Solymar, *Opt. Lett.,* **14**, 743, 1989.
59. B. I. Sturman, A. Bledowski, J. Otten and K. H. Ringhofer, *J. Opt. Soc. Amer.,* **B9**, 672, 1992.
60. N. Kukhtarev, T. Kukhtareva, A. Knyaz'kov and H. J. Caulfield, *Optik,* **97**, 7, 1994.

Chapter 4

Liquid Crystal Photorefractive Optics: Dynamic and Storage Holographic Grating Formation, Wave Mixing, and Beam/Image Processing

Iam Choon Khoo

Electrical Engineering Department, Pennsylvania State University, University Park, Pennsylvania

Summary

In this chapter, the basic mechanisms of photo-induced space-charge field formation, director axis reorientation, and refractive index changes in fullerene C_{60}- and dye-doped nematic liquid crystals (DDNLC) films are presented. These effects consist of transient as well as persistent components, and can be modulated by application of small ac and/or dc fields. Experimental observations of dynamic and high-resolution storage holo-

PHOTOREFRACTIVE OPTICS
Materials, Properties and Applications

ISBN 0-12-774810-5/$30.00

graphic grating formation, two beam coupling with gain of nearly 3000 cm^{-1}, optical limiting action at nanowatt continuous-wave (cw) laser power, and incoherent-coherent image conversion at microwatt/cm^2 light intensity level are discussed, along with some quantitative analyses of the underlying mechanisms.

4.1 Introduction

Liquid crystals are widely used in electrooptic devices because of their large broadband birefringence, dielectric anisotropy, and easy susceptibility to dc or low-frequency electric fields. The birefringence $\Delta n = n_1 - n_2 \sim 0.3$, where n_1 and n_2 are the refractive indices for fields polarized along, and perpendicular to, the director axis, respectively, spans the visible to infrared [0.4–12 μm] spectral regime [1]. The static dielectric anisotropy, $\Delta\varepsilon = \varepsilon_1 - \varepsilon_2$ is on the order of 5–10. In the liquid crystalline phase, particularly nematics, the so-called director axis of the liquid crystal that is equivalent to the c-axis of conventional photorefractive crystal, can be reoriented by the applied dc or ac fields of magnitude on the order of 1 V/μm. In conventional liquid crystal display devices, this amounts to the requirement of battery size voltages, thus making liquid crystals highly compatible with the electronic and semiconductor technology. In the context of optical image processing technology where spatial light modulators (SLM) are omnipresent, liquid crystal SLMs (LCSLM) are among the most cost-effective and widely used [1, 2].

Because of their easy susceptibility to optical fields, liquid crystals are also highly nonlinear optical materials. The nonlinear mechanisms range from electronic hyperpolarizabilities that respond in subpicosecond time scale, through thermal and density effects characterized by response times in nanoseconds to microseconds, to director axis reorientation effects with response times in milliseconds and longer time scales [1]. By far the laser-induced director axis reorientation effect gives the largest refractive index change for a given optical intensity, the so-called nonlinear index coefficient $n_2 = \Delta n / I$, where Δn is the refractive index change, and I the optical intensity. Recent studies have revealed several interesting effects, including "photorefractive-like" refractive index change in dye- or fullerene C60-doped nematic liquid crystal films [3–7]. Observation of perhaps the largest nonlinear index change mechanism in aligned methyl red-doped nematic liquid crystal film [8] has also been reported.

There is one important difference between liquid crystals and the usual photorefractive materials discussed in other chapters of this book. The refractive index change mechanism associated with the reorientation of the so-called director axis of the liquid crystal, which is equivalent to the crystalline or *c*-axis of other photorefractive crystals, is *quadratically* dependent on the applied fields, and thus may be termed a Kerr-like response. On the other hand, in all other photorefractive materials, the refractive index change is due to the Pockel effect or linear electrooptical response of the material. It is perhaps important to note here that in some recent studies of photorefractive polymeric systems [9, 10], similar crystalline axis reorientation processes were found to contribute significantly to the refractive index change, rather than the originally intended linear electrooptical effect.

In this chapter, we review the basic mechanisms of space charge field formation, director axis reorientation, the refractive index change, and experimental observations of several interesting dynamic and storage wave mixing effects. Recently obtained results on optical limiting action at nanowatt cw laser power, incoherent-coherent image conversion at μW / cm^2 light intensity level, and high-resolution holographic storage grating formation in an extraordinarily nonlinear material, methyl red–doped nematic films, are also discussed.

4.2 Nematic films under applied dc bias field

4.2.1 Space-charge field formation and refractive index change

Figure 4.1a and b depicts two typical aligned nematic liquid crystal cells, and the optical wave mixing interaction configuration. In homeotropically aligned cells, the director axis is perpendicular to the cell wall, whereas in the planar sample, the director axis lies in the plane of the cell wall. Figure 4.2 illustrates schematically three basic processes responsible for the refractive index change experienced by an extraordinarily polarized laser in these films, namely:

1. Optical charge generation → ions drift/diffusion → charge separation and space charge field formation → director axis reorientation

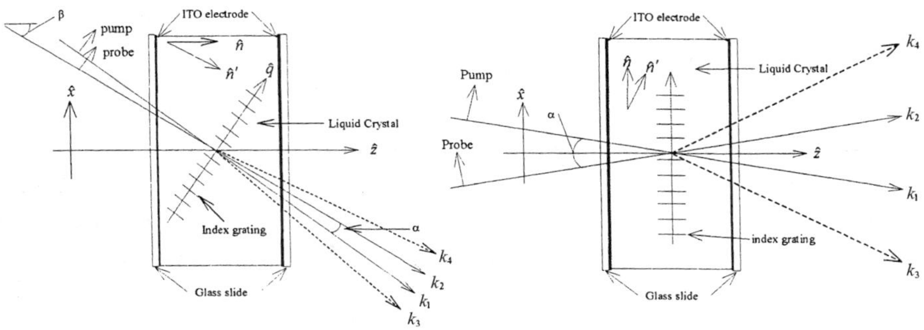

Figure 4.1: Optical wave mixing in aligned nematic liquid crystal film. (a) Homeotropic sample. (b) Planar sample. Incident laser beams E_1 and E_2 propagating along k_1 and k_2, respectively. n denotes original director axis direction, n' the reoriented director axis.

→ refractive index change (with $\pi/2$ phase shift relative to the optical intensity function)

2. Optical charge generation → ionic conduction plus director axis reorientation → space charge field formation through dielectric and conductivity anisotropies → further director axis reorientation → refractive index change
3. Optical charge generation →material flows → velocity gradient and shear stresses → director axis reorientation → refractive index change

Process (1) is analogous to the usual photorefractive crystals. For an incident optical intensity of the form $I_{\text{op}} = I_o\,(1 + m\cos(q\xi))$, a dc space charge field is created [3, 5]:

$$E_{\text{ph}} = q\nu m k_b T\,[(\sigma - \sigma_d)/(2_e\,\sigma)]\sin(q\xi) \tag{4.1}$$

where m is the intensity modulation ratio, k_b is the Boltzmann constant, σ is the illuminated conductivity, σ_d is the dark-state conductivity; $\nu = (D_- - D_+)/(D_+ + D_-)$, where D_+ and D_- are the diffusion constants for the positively and negatively charged ions, respectively; and m is the optical intensity modulation factor. The grating wave vector $q = 2\pi/\lambda_q$, where λ_q is the grating spacing. ξ is the coordinate along the direction of the grating wave vector.

Notice that E_{ph} is $\pi/2$ phase shifted from the incident optical intensity function. In [5], we showed explicitly how this gives rise to a corres-

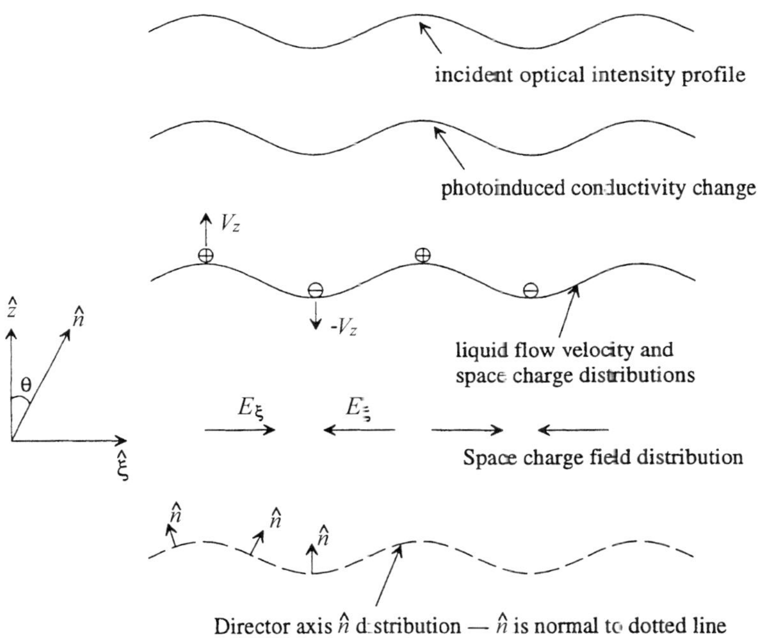

Figure 4.2: Schematic depiction of the photorefractive effect in nematic liquid crystals, where the refractive index change is due to a Kerr-like reorientation of the crystal's director axis.

ponding phase-shifted director axis reorientational function. This phase shift is responsible for the first reported [3] observation of two-beam coupling effect, energy transfer from one optical beam to another, and its dependence on the direction of the applied dc field. In Section 4.43, we discuss some of the recent studies where beam coupling gains of nearly 3000 cm^{-1} were observed [7].

Processes (2) and (3) are first pointed out in our original publication [3]. The space charge fields arise from the conductivity and dielectric anisotropies and are given by:

$$E_{\Delta\sigma} = E_z[\sigma_1 - \sigma_2)\sin\theta\cos\theta]/[\sigma_1\sin^2\theta + \sigma_2\cos^2\theta] \tag{4.2a}$$

$$E_{\Delta\varepsilon} = E_z[(\varepsilon_1 - \varepsilon_2)\sin\theta\cos\theta]/[\varepsilon_1\sin^2\theta + \varepsilon_2\cos^2\theta] \tag{4.2b}$$

where $(\sigma_1 - \sigma_2)$ is the conductivity anisotropy and $(\varepsilon_1 - \varepsilon_2)$ is the dielectric anisotropy. As these space charge fields were first investigated by Helfrich [11], we have referred to these as the Helfrich effect.

These space charge fields (Eqs. 2a and 2b) are proportional to the applied dc field E_z. For small induced reorientation angle θ, the magnitude of these space charge fields is an increasing function of θ. Once the reorientation is established, we can see from Eqs. (2a and 2b) that these space charge fields can be maintained by the dc field alone, i.e., without participation from the incident optical field at all. This behavior was indeed observed in our earlier studies [3, 5].

We have investigated a variety of photo-charge producing dopants, including laser dyes such as rhodamine-6G (R6G), methyl red, methyl orange, etc., dichroic dyes, and fullerenes C60, C70, and their derivatives. Since the incident photonic energy $h\nu$ is less than the ionization potentials of the dopant and impurities molecules in the liquid crystals, the productions of the photo-charges are attributed to excited state photochemical processes. In the case of R6G dye or other laser dye molecules, studies [12] have shown that the ions are generated by dissociation, whereas in C60 molecules the formation of charge transfer complexes [13] is the likely cause.

Detailed dynamic equations describing the director axis reorientation caused by these space charge fields, torques and flows, etc. are given in [5]. In the transient domain, when the reorientation angle is small, flows and the Helfrich field effects may be ignored, the reorientation angle θ_o induced in conjunction with an applied dc field for the configuration as depicted in Fig. 4.1a is given by

$$\theta_o = \frac{\Delta\varepsilon E_z E_{\rm ph}^{(0)} \cos\beta}{K[\pi^2/d^2 + q^2]} \tag{4.3}$$

Since $E_{\rm ph}$ varies as q, therefore $\theta_o \sim q/[\pi^2/d^2 + q^2]$ and it reaches a maximum at a grating spacing $\Lambda_{\rm max} \sim 2d$. On the other hand, in the long-time case where the reorientation angle is large, and flows and Carr-Helfrich effects are appreciable, the reorientation angle accounting for all the space charge fields, flows induced reorientation, etc. is given by:

$$\theta_o = \frac{\Delta\varepsilon E_z E_{\rm ph}^{(0)} \cos\beta}{\begin{array}{c} K[\pi^2/d^2 + q^2] + \Delta\varepsilon E_z^2/4\pi \\ {[1 + (\Delta\varepsilon/\varepsilon_1 + \Delta\sigma/\sigma_2)\cos\beta + \alpha_3\Delta\sigma\, \varepsilon_2/(\eta_2\sigma_1\Delta\varepsilon)]} \end{array}} \tag{4.4}$$

Using typical values [1] for K, $\Delta\varepsilon$, ε_1, $\Delta\sigma$, σ_2, α_3, η_2, etc., the second term in the denominator is estimated to be on the order of $K\pi^2/d^2$. In other words, we have

$$\theta_o \sim q/K[\pi^2/[(2d)^2 + q^2]. \tag{4.5}$$

Notice that in this case, corresponding to long-term illumination of the sample by the writing beams, the maximum orientation will occur if the grating spacing $\Lambda_{max} \sim 0.717\ d$. Both relationships (Eqs. 4.3 and 4.5) have been verified in our studies of transient and storage gratings in C40-doped nematic liquid crystal films [5] (Fig. 4.3).

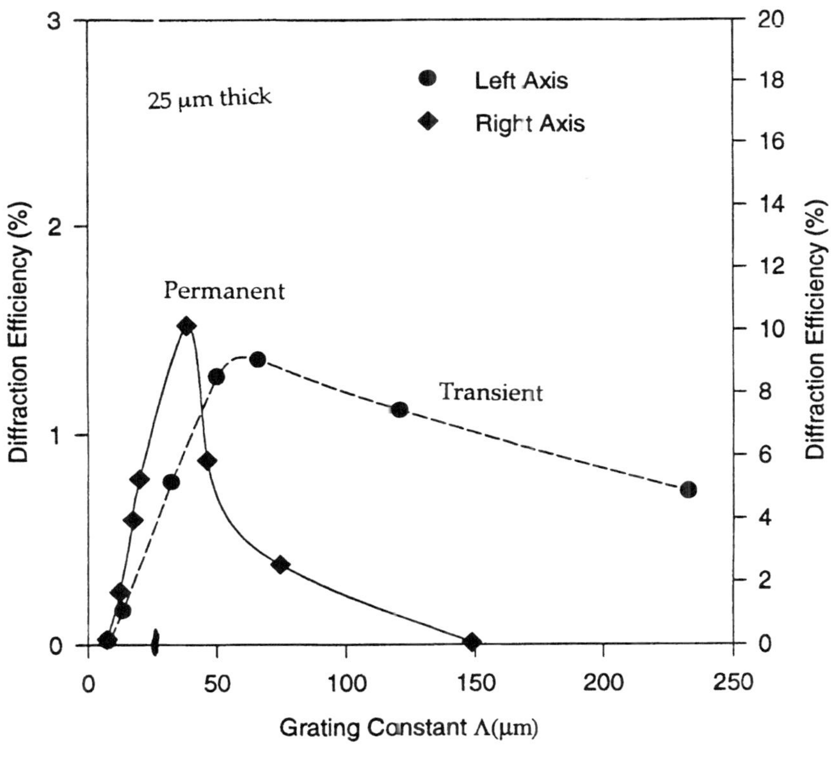

Figure 4.3: Dependence of the first-order diffraction efficiency on the grating spacing for a 25-μm C60-doped film. (a) Transient grating. (b) Permanent grating. Input beam powers: 10 and 7.6 mW. Beam diameter: 3 mm.

For an extraordinarily polarized beam, e.g., pump beam 1 in Fig. 4.1a, the refractive index change due to the director axis reorientation is given by:

$$\Delta n = n_e(\beta + \theta_o) - n_e(\beta) \sim (n_1 - n_2)(n_1/n_2)(\sin 2\beta)\theta_o. \quad (4.6)$$

Notice that Δn, because of its dependence on β, vanishes for $\beta = 0$ or 90°. This is the reason for the use of the tilted interaction geometry in the case of homeotropic cell (Fig. 4.1a). For planar aligned cells, we can either tilt the cell or create an initial reorientation of the director axis (from the plane of the cell wall) with the application of a dc field (Section 4.3.2).

4.3 Optical wave mixing effects in C60 doped films

As shown in Figs. 4.1a and 4.1b, the linearly polarized mutually coherent pump and probe beams are overlapped on the liquid crystal films at a small wave mixing angle α.The liquid crystal used is pentylcyanobiphenyl, (5CB), with traces of fullerene C60 ($\sim$ 0.05% by weight) dissolved in it. Pure undoped 5CB films were also tested. The cell windows are coated with transparent electrode Indium Tin Oxide (ITO) for applying external dc and ac fields. The laser used is the 4880 Å or the 5145 Å line of an argon-ion laser. The pump/probe beam ratio is varied by an attenuator in the path of the signal beam, while the pump power is varied by a pair of polarizers. The beam diameter at the sample is 3 mm.

4.3.1 Self-diffraction in homeotropically and planar aligned film

For the homeotropic aligned sample, the film is tilted so that the beam propagation direction makes an angle β with the director axis. The occurrence of wave mixing effects and their dependence on the applied dc voltages and incident beam directions, etc. in a homeotropically aligned sample are similar to those reported before [5]. In particular, visible amplification of the probe beam and the generation of side diffractions are observed for a dc applied voltage of 1.5 V.

Figures 4.4a and b depict, respectively, the observed first-order self-diffraction from an undoped film and a C60-doped film. Because of beam

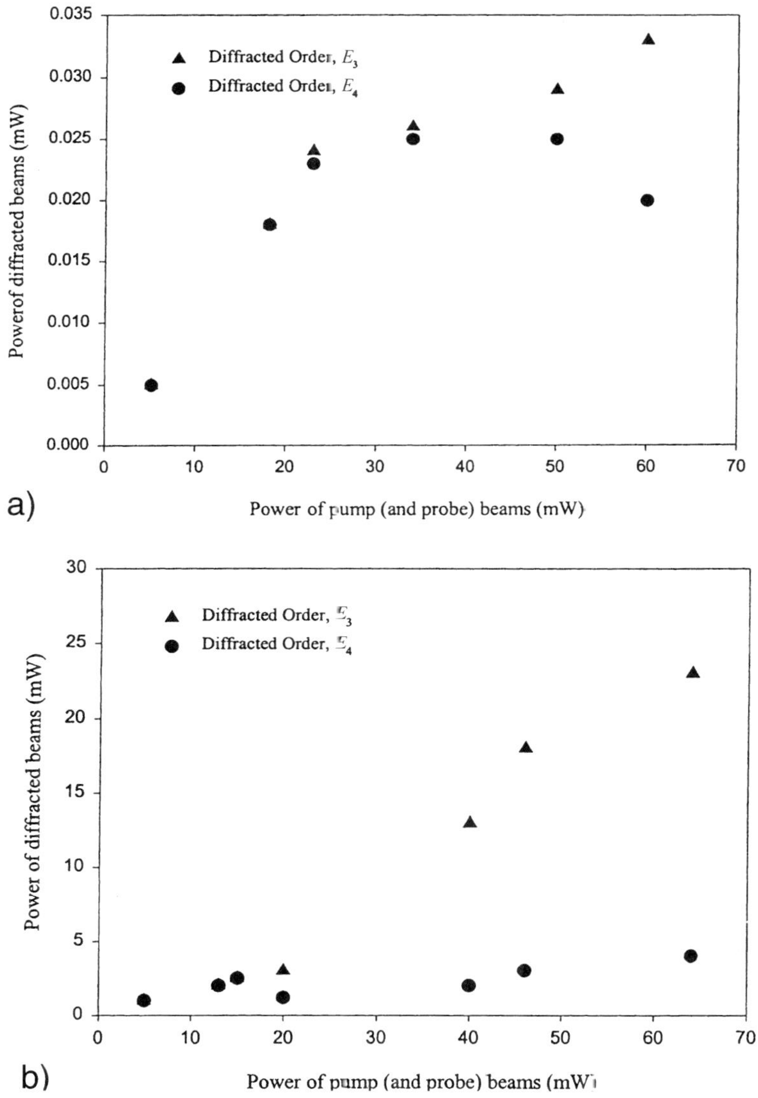

Figure 4.4: (a) ±1 order side-diffractions (E_3 and E_4) as a function of the input beam power from a undoped nematic (5CB) film. dc voltage: 3 V; film thickness: 25 μm. (b) ±1 order side-diffractions as a function of the input beam power from a C60-doped nematic (5CB) film. dc voltage: 3 V; film thickness: 25 μm.

coupling effect, one of the input beams is amplified at the expense of the other, and the self-diffractions will manifest the power asymmetry. The asymmetry is highly pronounced in the case of C60-doped film because of the much higher beam coupling effect.

For the planar case, the input beams are symmetrically (with regard to the normal to the cell wall) incident on the film. An ac field is applied across the cell walls to pretilt the director axis so that it makes a nonzero angle with the optical electric field. Although one can also tilt the sample to obtain similar effects as in homeotropically aligned cells, the use of an ac + dc bias fields provides an interaction geometry [normal incidence] not possible in homeotropic sample. This feature is important for potential application of the photorefractive nematic liquid crystal in adaptive optics and image processing devices [2].

Figure 4.5 depicts a typical dependence of the first-order self-diffraction on the applied ac + dc fields. Below the Freedericksz transition voltage, there is no director axis reorientation (i.e., $\beta = 90°$), and there is no diffraction. The diffraction rises and reaches a maximum for applied ac voltage above the Freedericksz threshold, whereby the interaction angle [90-β], (Fig. 4.1a), between the incident laser electric field and the director axis is nonzero. At higher ac voltage, the director axis will be strongly aligned along the ac field direction, and the diffraction signal will begin to decrease toward a vanishingly small value.

If the applied dc voltages are small, and/or the optical illumination time is short, these diffraction effects exhibit reversibility, i.e., the diffraction drops to a vanishingly small value when the ac field is decreased. However, if the dc bias voltage is high, the diffraction persists even when the applied ac field is reduced to zero, i.e., a permanent grating is formed, as shown in Fig. 4.5. Such permanent grating can be "switched" on/off by an ac field at millisecond speed [3].

4.3.2 Beam amplification—theory and experiments

Because of the thinness of the liquid crystal films, multiorder side diffractions are generated by the two incident beams. Such multiwave mixing processes have been quantitatively analyzed by Khoo and Liu [14] in their study of purely optically induced director axis reorientation in nematic films. Following their multiwave mixing theory, modified accordingly for a phase-shifted index grating in the present case, the equation for the

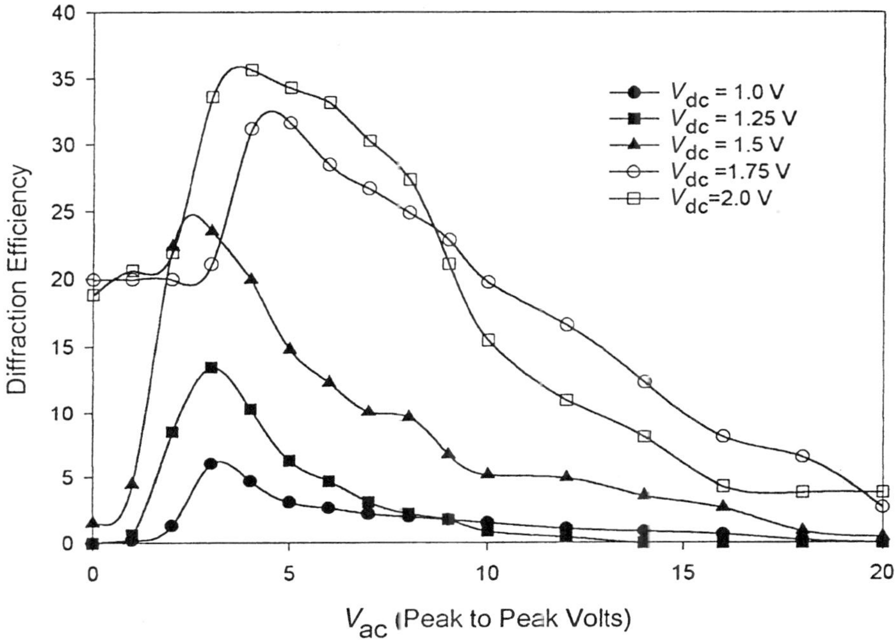

Figure 4.5: Dependence of the first-order diffraction on the applied ac field strength for a 25-μm-thick planar aligned C60-doped nematic (5CB) film. Film thickness: 25 μm. Input beam power: 20 mW each ; λ = 514.5 nm.

optical electric fields for the pump E_1, probe E_2, and diffraction E_3 on the pump side are given by:

$$dE_1/dz = -g[(E_1E_2{}^*)E_2 + (E_1E_3{}^*)E_3] + \ldots \quad (4.7)$$
$$\ldots \text{higher-order wave mixing terms}]$$

$$dE_2/dz = g[(E_2E_1{}^*)E_1 + (E_1E_3{}^*)E_1 \exp(i\Delta kz) + \ldots \quad (4.8)$$
$$\ldots \text{higher-order wave mixing terms}]$$

$$dE_3/dz = g[(E_3E_1{}^*)E_1 + (E_1E_2{}^*)E_1 \exp(i\Delta kz) + \ldots \quad (4.9)$$
$$\ldots \text{higher order wave mixing terms}]$$

and similar equations for higher-order side diffractions E_4, E_5, . . . etc.

In these equations, g is the real coupling constant (because of the $\pi/2$ phase shift) and is proportional to the induced refractive index change

Δn. The sign of g depends on the direction of the applied dc field [5]. The factor $\Delta k = |2k_1 - k_2 - k_3|$ is the wave vector mismatch, the k_i's being the magnitude of the respective optical wave vectors along the z-direction.

At a large pump probe beam ratio, i.e., a very weak probe beam, the diffracted beam E_3 is much weaker than E_2, and higher-order diffractions (E_4, E_5, etc.) are negligibly small. This justifies the consideration of only three interacting waves.

The two terms on the right-hand side of these equations describe the wave mixing contribution from the gratings formed among the two incident beams E_1, E_2, and the first-order diffracted beam E_3 on the pump side. Note that $E_2E_1{}^*E_1$ is independent of the phase (and therefore any phase aberration) of the pump beam since it is proportional to $E_1{}^*E_1$. On the other hand, the latter term $E_1E_3{}^*E_1 \exp(i\Delta kz)$ will impart phase aberration on the amplified probe beam, since it is proportional to E_1E_1. To achieve aberration-free amplification, this term should be suppressed, either by using a very weak probe, or using thick film so that the wave mixing process is in the Bragg regime ($\lambda d >> \lambda_g^2$; d = sample thickness, λ_g = grating spacing and E_3 is not generated). This is indeed demonstrated in [7] using a weak probe beam approach.

Similar to our previous report [5], there are transient and storage features in these wave mixing mediated beam amplification processes. If the exposure time is short, the effect is transient. Figure 4.6 shows the transient probe gain (after 2 s of application of the dc field) as a function of the dc voltage obtained in a 25-μm-thick C60-doped homeotropically aligned nematic film. The pump/probe beam ratio used is 239. For a 10-μm-thick sample, a probe gain value of ~18 was obtained [7]. *Gain* is defined as the ratio of the transmitted probe with applied fields on to the transmitted probe power with fields off. This corresponds to an exponential gain constant of about 2890 cm^{-1}.

4.3.3 Storage grating capability

If the applied dc field and the two incident beams are left on for a sufficiently long time , a permanent grating will be "written" onto the sample. For an applied dc voltage of 1.5 V the formation time is about 1 min. At higher dc bias voltage, e.g., 3 V, the formation time is shortened to a few seconds. Figure 4.7 depicts schematically the dynamic evolution of the grating from the transient to storage nature. Polarization microscope examinations of the permanent gratings show that they result from alter-

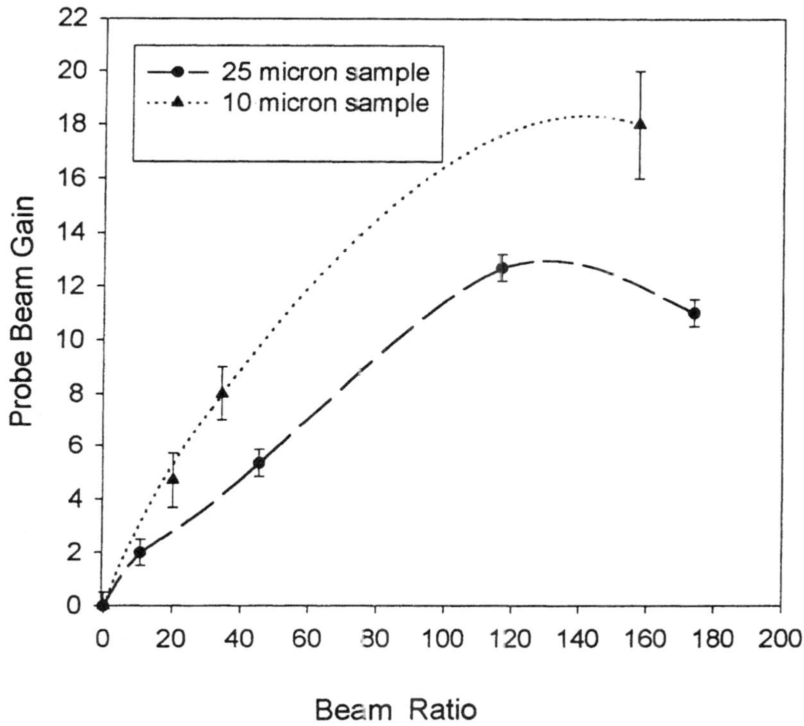

Figure 4.6: Transient probe beam amplification in nematic liquid crystal films as a function of pump/probe beam power ratio. Laser wavelength: 488 nm; wave mixing angle: 1°. β = 22.5°. Pump beam power: 30 mW.

nate "lines" of reoriented and unperturbed liquid crystal molecules. In the perturbed region, the director axis shows an increasing degree of randomness for higher-input laser power [5].

These gratings can be modulated by an ac field applied across the cell windows. An ac field of sufficient strength, typically about a few times the Freedericksz field strength [5], will align the liquid crystal molecules in the direction of the field, i.e., back to the original homeotropic alignment, and thus "erases" the grating. Typically, the switching-off response time is dependent on the ac field strength, and can be as fast as 1 ms or less [5]. When the ac field is switched off, the grating will recover in 10 ms or so, depending on the film thickness, in accordance with the usual nematic dynamics (Fig. 4.8). Our studies have shown that in C60-doped

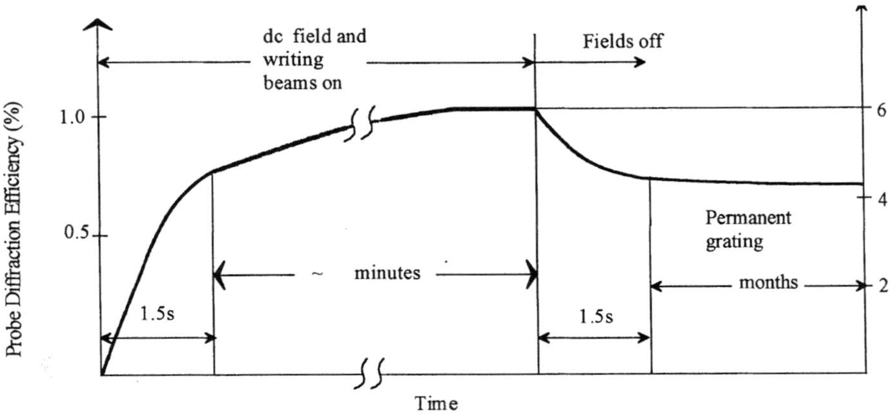

Figure 4.7: Dynamical evolution of the grating diffraction for prolonged low power (~ a few mW / cm^2) witing beam illumination. Insert shows the transient grating formation and decay for short illumination time. Higher optical intensity gives shorter response times.

film, the permanent gratings can be robust, undergoing many switched off / on cycles without diminished diffraction efficiency.

Since the director axis of the liquid crystal is dictated by the surface alignment, we believe that these "permanent" gratings are due to "permanent" realignment of the liquid crystal molecules at the surfaces, due to the adsorption of the excited charged dopant ions. In the R6G dye-doped sample, the written grating lasts for tens of minutes, whereas grating written in a C60-doped sample persists indefinitely [3, 5]. Our recent study [8] in methyl red–doped nematic films, (see Section 4.4), also shows similar dynamics to those depicted in Fig. 4.7, and may originate from the same processes.

In earlier comparative studies, we found that C60 and its derivatives provide the largest refractive index change per absorption. However, our recent studies in methyl red–doped nematic film have totally revised this finding. From [5], the n_2 value for a 25-μm-thick C60-doped film is on the order of 2.3×10^{-3} cm^2 / W ($\Delta n \sim 1.4 \times 10^{-3}$; $2I_o \sim 0.6$ W / cm^2). The absorption constant α is ~ 5 cm^{-1}, so that $n_2 / \alpha \sim 0.46 \times 10^{-3}$ cm^3 / W for C60-doped film. On the other hand, as detailed in the Section 4.4, the methyl red–doped film of the same thickness gives $n_2 / \alpha \sim 6/194 \sim 3.1 \times 10^{-2}$ cm^3 / W, which is about two orders of magnitude larger than C60-doped film.

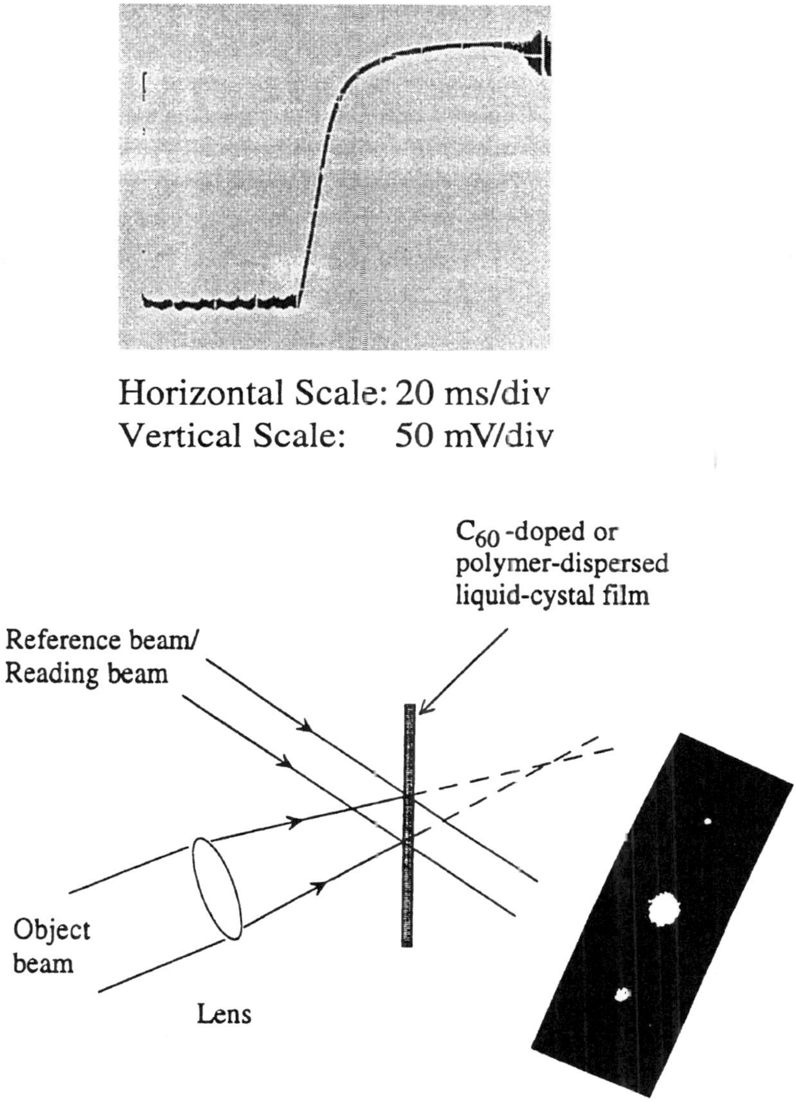

Figure 4.8: Switching dynamics of permanent holographic grating. Photograph of the diffractions and zero-order beam from the permanent phase hologram of a positive lens recorded in a C60-doped nematic film. The diffraction in the phase-matched direction shows the focusing action of the positive lens.

4.4 Methyl red–doped nematic liquid crystal films

Recent studies [8] in methyl red–doped aligned nematic liquid crystal films have revealed that the nonlinear index change effect can be orders of magnitude larger than in the R6G- or fullerene C60-doped films. More importantly, the effect in methyl red–doped films does not require a dc bias voltage at all, and may be further enhanced by an ac bias field. This feature is important as it avoids the possibility of inducing instabilities and dynamic scattering by dc bias voltages.

4.4.1 Optical wave mixing and transient grating diffraction

The insert in Fig. 4.9 illustrates the interaction geometry. The liquid crystal used is 5CB (pentylcyanobiphenyl, from EM Chemicals) with traces of the laser dye methyl red (from Aldrich Chemicals) dissolved in it at a concentration of about 0.5%. Homeotropic alignment is achieved by treating the ITO-coated cell windows with the surfactant hexadecyl-trimethyl -ammonium bromide (HTAB). The absorption constants of these doped films have been measured to be 194 cm^{-1} at the 488 nm argon laser line. The transmission is 85 and 60% for film thicknesses of 6 and 25 μm, respectively. A linearly polarized argon laser is divided into two equal-power writing beams and expanded to a beam diameter of 1.5 cm. These beams, propagating in the y-z plane, are overlapped on the liquid crystal film at an angle α, as shown in Fig. 4.9. This imparts a sinusoidal optical intensity function $I = 2I_o(1 + \cos qy)$ along the y-axis, with I_o the intensity of one writing beam. A linearly polarized 5-mW He-Ne laser is used to probe the grating.

In general, visible grating diffractions are generated with total input power as low as 100 μW, corresponding to an optical intensity $2I_o$ of 40 μW / cm^2. Maximum first-order diffraction efficiency of ~ 30% is obtained with a total writing beam intensity of ~1 mW / cm^2.

Thermal grating effects accompanying photoabsorption are ruled out by the polarization dependence and response times of the diffraction effect. If the two pump's and the probe's polarizations are in the x-direction, there is no diffraction from either the pump or the probe beams. If the probe polarization is oriented along the y-axis, i.e., along the grating wavevector direction, the diffraction is maximal. When both pump beam

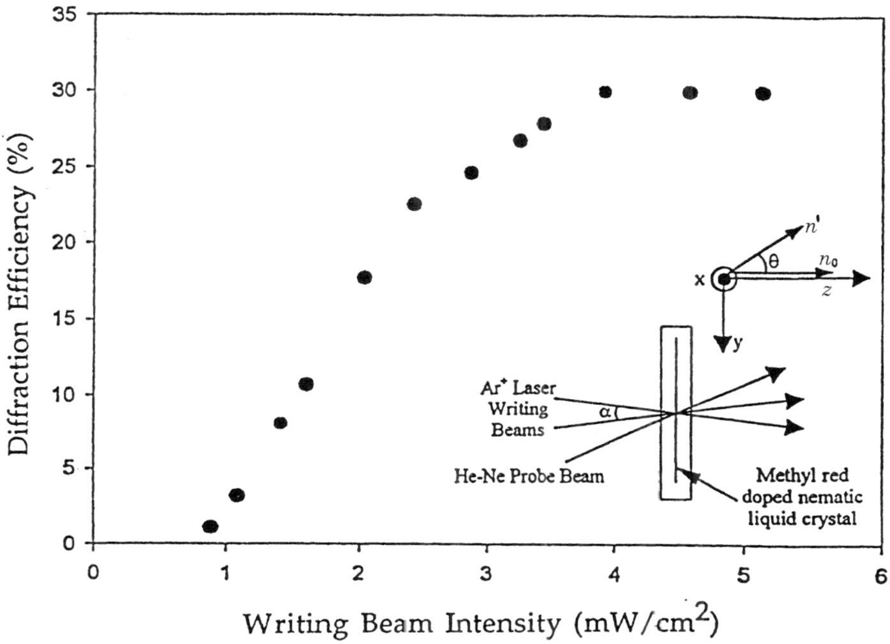

Figure 4.9: Dependence of the probe diffraction efficiency on the total writing beam intensity. Sample thicknesses = 6 μm; α = 0.04 radian; probe beam angle β = 0.1 radian. The insert depicts schematically the optical wave mixing interaction geometry.

polarizations are along the y-direction, self-diffraction from these beams is observed, while an x-polarized probe beam yields no diffraction. The refractive index change is therefore determined to be due to the reorientation of the director axis in the y-z plane, along the grating wavevector direction defined by the two writing beams.

Thermal effect is further ruled out by the observed dynamics of the effect. The writing times for these gratings have been found to be in the 10^2 ms–1 s range. When the writing beams are turned off, we recorded a decay time of 500 ms for a sample thickness of 25 μm, and a grating spacing of about 50 μm. This is typical of orientational rather than thermal effect as the latter is characterized by time constants in the submillisecond regime for thermal diffusion lengths on the order of 25–50 μm [1].

We attribute such extremely nonlinear optical response to nematic director axis reorientation by the optically induced space charge fields, similar to the effect discussed in the previous section. The unusually large response is possibly due to several factors, including the higher photocharge producing abilities of methyl red, and the large difference in the positive and negative photoions' diffusion constants, see Eq. (4.1). It is also possible that the reorientation is caused by the alignment action of the excited dye molecules. Studies of permanent grating formation in azo dyes such as methyl red–doped nematic films [14, 15] have shown that the excited dye molecules tend to lie in the plane of the cell surfaces, and in the process, reorient the liquid crystal's director axis. Flows induced director axis reorientation may also be a contributing mechanism. To what extent these various mechanisms contribute to the observed liquid crystal director axis reorientation remains a subject of active current investigation.

An estimate of the refractive index change coefficient n_2 can be obtained from Fig. 4.8, which shows the dependence of the probe diffraction efficiency from a 6-μm film for the case where the pump beams are polarized in the x-direction, and the probe polarization is along the y-axis. As reported in [8], we get a refractive index change coefficient $n_2 = 6\ \mathrm{cm}^2/\mathrm{W}$. Similar values of n_2 are obtained for the 25-μm-thick samples, and also at other argon laser lines. This value of n_2 is, to our knowledge, the largest among nonlinear optical materials known to date.

Another interesting and useful feature of the effect is that it can be enhanced by an applied ac field. Figure 4.10a shows the first-order He-Ne probe diffraction power as a function of the ac frequency for various ac voltages, from a 25-μm-thick film illuminated by two coherent argon laser beams. At such low writing beam power of 2 μW, there is no diffraction when the ac voltage is off. For ac frequencies in the range of 10^2–10^4 Hz, the diffraction increases as the applied ac voltage is raised; for frequency above 10^4 Hz, the diffraction vanishes. The exact mechanism responsible for this enhanced response by the applied ac field remains to be ascertained. We believe it is due to the difference in the ionic mobilities and charge redistribution under ac fields of different frequencies, analogous to those processes occurring in other photovoltaic materials [16, 17].

An estimate of the effective refractive index change coefficient n_2 can be obtained from Fig. 4.10a. For this case, the grating spacing $\lambda_g = 30\ \mu\mathrm{m}$ (crossing angle $\sim 1°$) and the diffraction is in the Raman-Nath regime. The first-order diffraction efficiency η is thus given by $\eta = J_1^2$

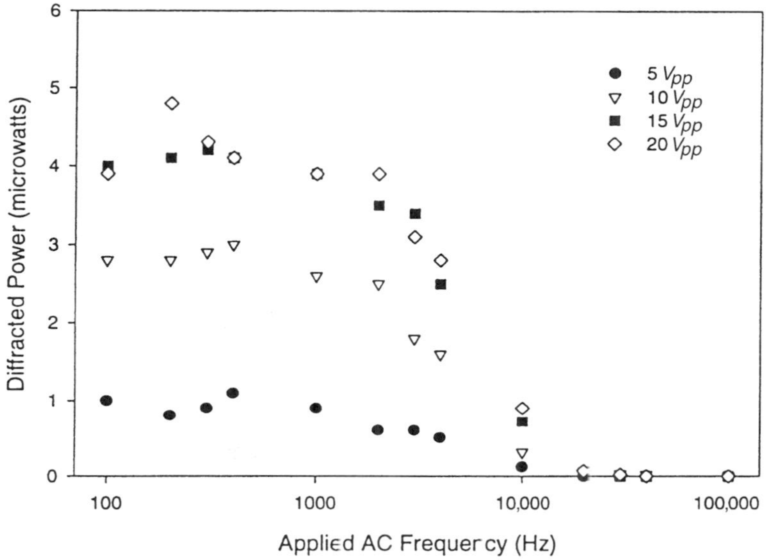

a)

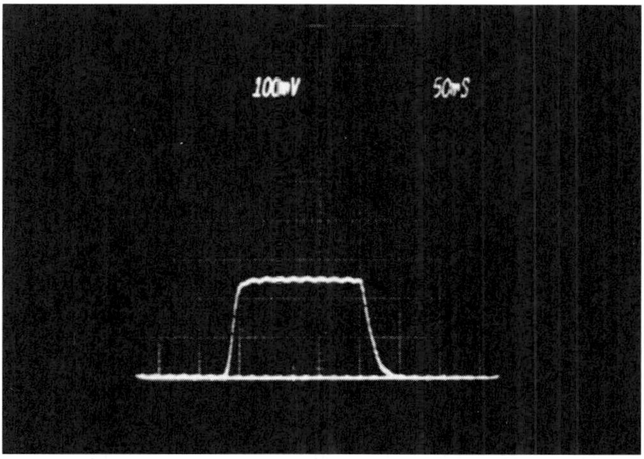

b)

Figure 4.10: (a) Observed diffracted power versus applied ac frequency for different ac voltages. Argon (λ = 488 nm) writing beam power: 2 μW each; beam diameter: 3 mm. He-Ne probe beam power: 5 mW. All beams copolarized and incident on the liquid crystal film as extraordinary rays with β = 0.4 radian and α = 0.015 radian; sample thicknesses = 25 μm. (b) Oscilloscope trace of the dynamic response of the diffracted power versus under dual-frequency ac field of V_{pp} = 20 V. Diffraction on: f = 300 Hz. Diffraction off: f = 30 kHz. Time scale: 50 ms / div.

(phase shift) ~ $(\pi \Delta n d / \lambda)^2$. From Fig. 4.2, we note that a diffraction efficiency $\eta = 10^{-3}$ is obtained for the case where ac voltage = 20 V_{pp} case. Inserting all the values for d, λ, and η in (4.1), we obtain $\Delta n \sim 2 \times 10^{-4}$. Since the intensity of the writing beam is ~ 44 μW / cm², this gives the magnitude of $n_2 = 6$ cm² / W which is similar to those reported in reference for a thinner (6-μm-thick) sample. Without the ac field, the measured n_2 value for a 25-μm-thick film is on the order of 2 cm² / W.

The ac frequency dependence as depicted in Fig. 4.10a could be used for dynamic grating switching with a dual-frequency technique. Figure 4.10b shows the on-off switching dynamics of the diffraction from the film as the applied ac field frequency is switched back and forth between 300 Hz and 30 kHz, at an ac V_{pp} = 20 V. In general, the on time is on the order of 12 ms, and the off time is about 17 ms. Faster response is possible for higher V_{pp}, or using nematics of lower viscosity and higher dielectric anisotropy.

4.4.2 Optically induced dc voltages

Just as the sinusoidal optical intensity grating produces sinusoidal space charge fields along the grating wave vector, the variation of the optical intensity along the z-direction establishes a dc space- harge field along z. This component is manifested in the form of a dc voltage drop across the entrance and exit ITO cell electrodes. Because of the higher solubility of methyl red and larger photocharge production, this component assumes a detectable magnitude. We have tested nematic films doped with various dopants, such as the dichroic dye D2, and the laser dyes R6G, C60, and methyl red; only methyl red–doped samples produce detectable photovoltages [8].

Figure 4.11 shows that the photovoltages across a 25-μm-thick methyl red–doped homeotropic cell for two different orientations ($\beta = 0$ and $\beta = 22°$). The polarity of the observed voltage is reversed if the incident direction of the light is reversed. The photovoltages rise rapidly as the input intensity is increased from a few μW / cm² to 1 mW / cm², and reaches a "saturated" value intensity above 2 mW / cm². This dependence correlates very well with the side diffraction's variation with the optical intensity, further corroborating our assertion that the observed director axis reorientation effects are due to the optically induced space charge fields.

Such photovoltaic effects have also been observed in similar nematic liquid crystalline films [18, 19]. In particular, Sato [18] reported obser-

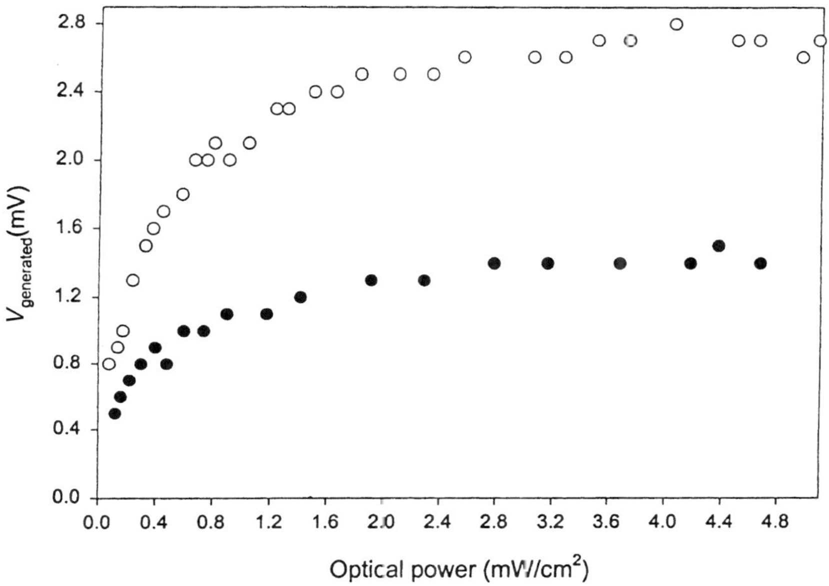

Figure 4.11: Observed photovoltage as a function of the illuminating beam intensity in a 25-μm-thick nematic film for two different incident laser direction. (Circles: $\beta = 0$; dots: $\beta = 22°$.)

vation of photoinduced voltage pulses in methyl red–doped MBBA (P-methoxybenzylidene-p-n-butylaniline) films sandwiched between SnO_2-coated Pyrex glass plates, and also in 5CB films. They attributed the photovoltaic effect to photocharge generation in the methyl red–doped nematic film, and subsequent diffusion and redistribution of the ions with different mobilities; exchanges of ions between the HTAB-coated ITO electrodes and the photocharges producing doped nematic film could also be involved.

For a particular cell thickness, the space charge field formation speed is dependent on the incident optical intensity. Figure 4.12 shows the induced voltage across the cell windows and the buildup time as a function of the beam intensity in a 6-μm-thick sample. The buildup time of these photovoltages decreases from a few seconds to 100 ms for an incident intensity of 3 mW / cm^2. At a higher intensity (not shown) of 10 mW / cm^2, the buildup time drops to below 10 ms. These dynamics are also reflected

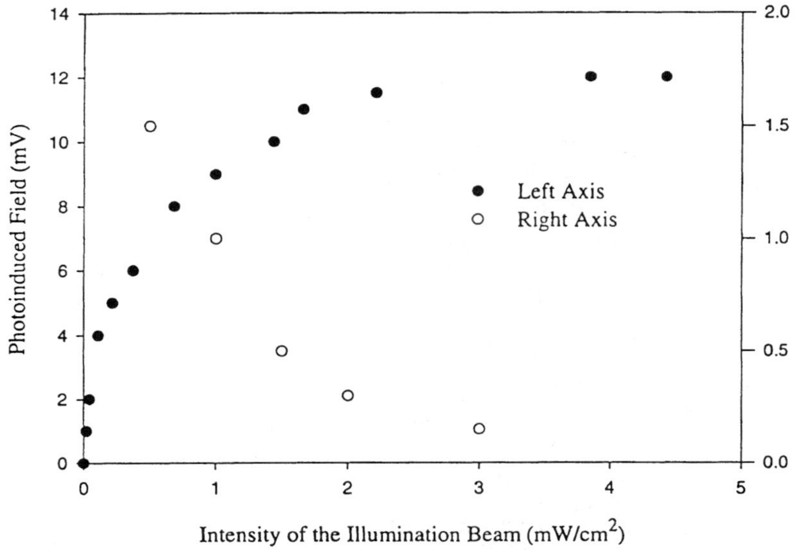

Figure 4.12: Observed dc photovoltage across the cell windows, and the buildup times as a function of the illuminating beam intensity in a 6-μm-thick sample.

in the observed response times of the nonlinear optical processes arising from the refractive index change.

4.4.3 Self-defocusing and limiting at nanowatt cw laser power

Needless to say, such extraordinarily large nonlinearity enables the performance of several all-optical switching, limiting, and image modulation and sensing processes at unprecedented low-threshold power. We describe here preliminary studies of optical limiting [20] and incoherent-coherent image conversion.

Figure 4.13 shows a typical optical limiting setup using external self-defocusing effect. A linearly polarized laser beam is focused by a 15-cm focal length input lens to a spot diameter of 0.1 mm onto a 25-μm-thick liquid crystal film placed just behind the focal plane of the input lens. The nematic film is tilted so that the incident beam makes an angle of 45° with the normal to the cell; this enhances the nonlinear refractive

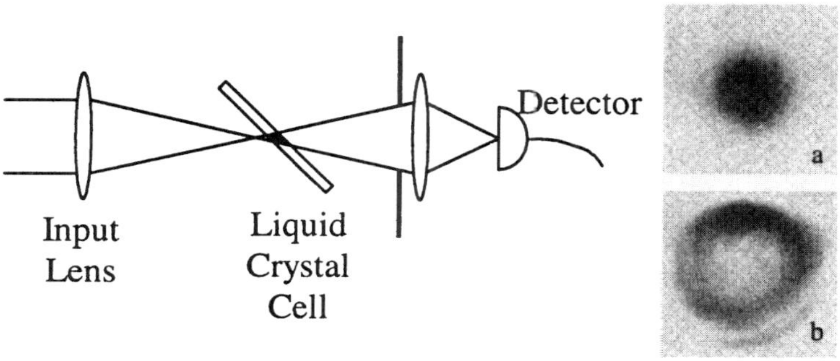

Figure 4.13: Experimental setup for optical limiting action using external self-defocusing effect. The insert is a photograph of the transmitted laser beam showing self-defocusing effect.

index change experienced by the extraordinary incident ray, (see Eq. 6). An aperture of 5-mm diameter is placed at 40 cm behind the sample to monitor the central region of the transmitted beam.

Since the incident beam size is much larger than the sample thickness, the space charge field is predominantly along the beam propagation direction z, i.e., directed from the input plane to the exit plane of the sample (Fig. 4.14). As a result of the reorientation, the director axis makes a larger angle with the optical electric field of the extraordinary input beam, and thus the refractive index change is negative. Experimentally, it is observed that the central region of the transmitted beam becomes progressively darkened, the beam divergence increases dramatically, and interference rings begin to appear, as depicted by the two photographs in Fig. 4.13. This defocusing effect with the placement of the film just behind the focal plane of the input focusing lens indicates that the sign of the nonlinear index coefficient n_2 is negative [1, 21], consistent with the director axis reorientation direction as illustrated in Fig. 4.14.

The first ring appears when the maximum (on-axis) nonlinear phase shift $\Delta\phi = n_2 I \, 2\pi d / \lambda$ imparted on the input laser field is on the order of $1.5\,\pi$ [1, 21]. In the present case, the first ring appears at an input power of around 100 nW. The intensity at the liquid crystals, accounting for the interface reflection loss and the 100-μm spot size is about 0.5 mW / cm^2. Accordingly, we have $n_2 = 2$ cm^2 / W, which is consistent with the value obtained from the grating diffraction experiments.

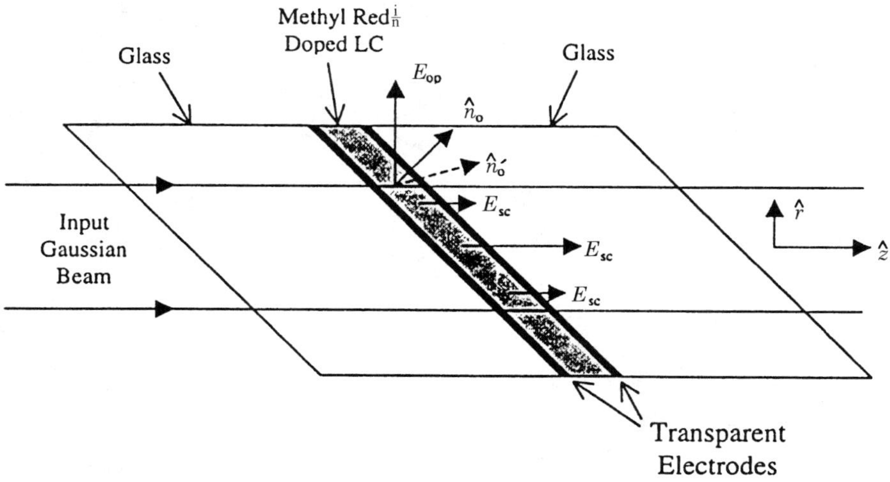

Figure 4.14: Schematic depiction of the interaction of a Gaussian laser beam with a homeotropically aligned nematic liquid crystal film.

Because of the defocusing effect, the central portion of the transmitted beam exhibits a typical optical limiting behavior (Fig. 4.15). The threshold of 70 nW, at which the transmission begins to deviate from linearity, is the lowest of all known nonresonant nonlinear materials. At an input laser power of 0.3 μW, the response time is 1.5 s, i.e., < 0.3 μJ is transmitted through the limiter. The dynamics of beam defocusing effect follow closely the photovoltage generation as described in the preceding discussion. At higher input power/intensity, the response time is shortened, and the net effect is that << 1 μJ of the input laser energy will reach the detector. This is well below the sensor/eye damage level for a long-pulse cw laser. Methyl red–doped nematic films thus rank as the lowest threshold optical limiting material among all the candidates currently under investigation [22]

4.4.4. Image processing—incoherent to coherent image conversion, adaptive optics

Using the setup similar to that employed by Hong et al.[23] (Fig. 4.16), we have demonstrated incoherent to coherent image, as well as wavelength conversion. The incoherent image-bearing optical beam, at a wavelength

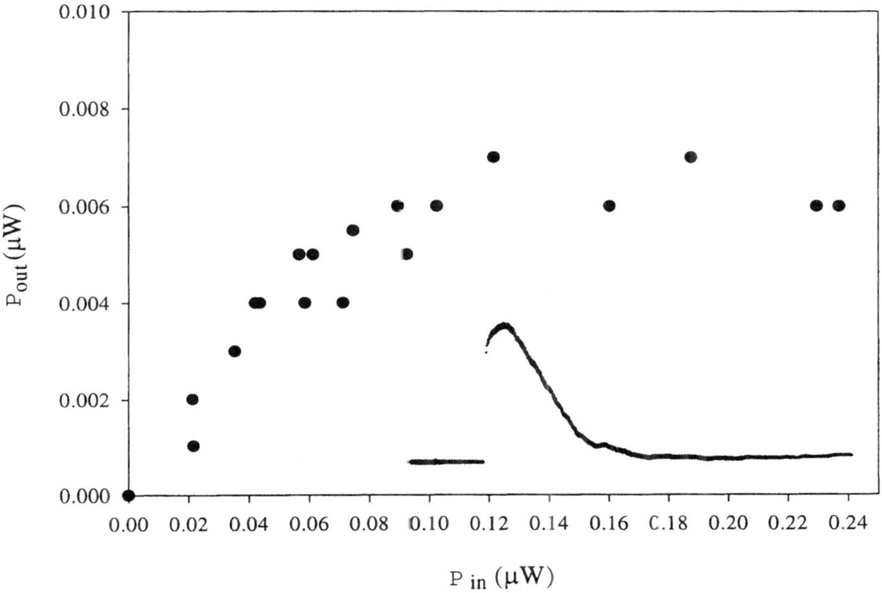

Figure 4.15: Plot of detected on-axis output power versus input laser power. Insert is an oscilloscope trace of the detected output for an input step-on cw laser showing how the later portion of the on-axis power is "switched" off by the self-defocusing effect.

of 488 nm, creates a spatial phase shift on the nematic film, which is sensed by a coherent He-Ne laser to generate a coherent image. Visible coherent images can be created with input incoherent beam intensities as low as 90 $\mu W / cm^2$.

We have also employed similar optical intensity levels in demonstrating optical phase conjugation effect using a wave-mixing configuration similar to that used by Gruneisen and Wilkes in [2]. In their work, a commercial liquid crystal Optically Addressed Spatial Light Modulators (OASLM) [1, 2] was used. Dye-doped nematic liquid crystal (DDNLC) films, however, will be much more inexpensive to fabricate as they do not require the processing of a photoconductive film and do not require a power supply. Moreover, the resolution capability of these commercial spatial light modulators is limited to around 30 lines / mm at best. On the other hand, the resolution capability of the nematic film can be 200 lp/mm or higher, as demonstrated in the study of diffraction efficiency

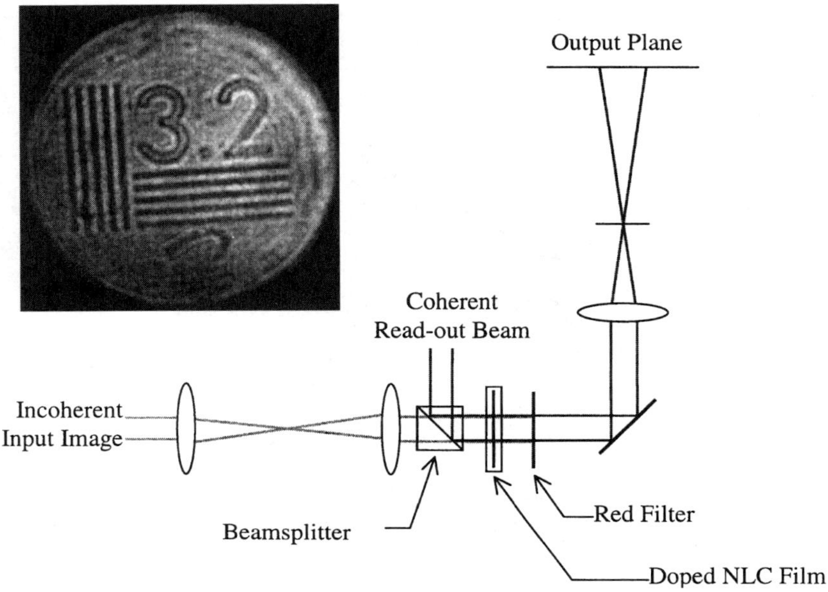

Figure 4.16: Experimental setup for incoherent to coherent image conversion. Insert is a photograph of the reconstructed coherent image. A similar setup could also be used for wavelength conversion because of the broadband birefringence of nematic liquid crystals (NLC).

dependence on grating spacing [8, 24, 25]. Clearly, the development of DDLC as a new-generation spatial light modulator is an attractive endeavor.

4.4.5 Storage holographic grating formation

As in the case of R6G and fullerene C60-doped films, these methyl red–doped nematic films also exhibit storage effect when subjected to prolonged optical illumination. The storage effect becomes pronounced if a low-frequency ac field is applied (see Section 4.4.1). When the field is turned off, a persistent component is generated, and lasts indefinitely. A plausible argument is that the excited dye molecules are adsorbed on the cell surfaces, and modify the surface alignment, causing persistent reorientation of the liquid crystals, in a manner similar to the initial alignment of the liquid crystal molecules by the surfactant molecules [1].

This light-induced reorientation has also been exploited to record permanent holographic gratings in a liquid crystal cell [24, 25] with spatial resolution higher than 1000 lines/mm. The experiments were carried out using a 50-μm-thick nematic cell. The liquid crystal used is 5CB, doped with methyl red at a concentration of 1% by weight. The inner surface of one of the glass windows (the control surface) is coated by an isotropic nonrubbed layer of poly(vynil)-cinnammate, while the inner surface of the other glass window (the reference surface) is coated by a polymide layer and rubbed to get strong planar anchoring. This surface determines the initial planar alignment of the cell. The optical information can be written by a low-power (< 1 mW) unfocused He-Cd laser ($\lambda = 0.422$ mm) and linearly polarized along a direction at 45° with respect to the initial planar orientation.

Reorientation at the control surface induces a twisted structure along the sample which can be tested by using a probe beam from a linearly polarized He-Ne laser. The probe beam polarization rotates following the induced helical order, satisfying the adiabatic condition. Thus if the probe beam impinges on the reference surface with the polarization parallel to the given planar orientation, the exit polarization will correspond to the director orientation at the control surface. If the sample is placed between crossed polarizers to have complete extinction in the initial state, the director rotation at the control surface will produce an increase in the transmitted signal. This effect can be exploited to write images on the sample [24, 25]. An important feature of this system is the low writing intensity: energy density of the order of 0.1 J / cm^2 is required to store permanent optical information. Together with its high resolution capability, these observations demonstrate that DDNLC films are one of the most sensitive ones known at present to record permanent images.

4.5 Conclusion

We have presented a comprehensive review of the optical, electrooptic, and nonlinear optical properties of nematic liquid crystals in which photocharger producing agents are introduced. These films yield large optically induced refractive index changes under low optical power and enable the realization of several optical switching and modulating processes at unprecedented low-threshold power and intensity, and writing energy fluency. Because of the broadband birefringence and transparency of the film throughout the entire visible infrared spectrum—and many dyes exist that will cover this spectrum—these specially doped nematic liquid crystal films are promising candidates for a new generation of broadband optical modulators and limiters, and other adaptive optics, transient, and storage holographic wave mixing devices.

Acknowledgment

This work was supported by Army Research Office and the Air Force Research Laboratory–Kirtland Air Force Base.

References

1. I. C. Khoo, *Liquid Crystals: Physical Properties and Nonlinear Optical Phenomena*, Wiley, New York, 1995. See also I. C. Khoo and S. T. Wu, *Optics and Nonlinear Optics of Liquid Crystals*, World Scientific Publishing, NJ, 1993.
2. M. T. Gruneisen and J. M. Wilkes, "Compensated imaging by real-time holography with optically addresses spatial light modulators," in G. Burdge and S. C. Esener (eds.), *OSA TOPS,* **14**, *Spatial Light Modulators*. See also M. A. Kramer, C. J. Wetterer and T. Martinez, Appl. Opt. **30**, 3319–3323, 1991.
3. I. C. Khoo, H. Li, Y. Liang, "Observation of orientational photorefractive effects in nematic liquid crystal," *Opt. Lett.,* **19**, 1723–1725, 1994.
4. E. V. Rudenko, A. V. Sukhov, "Optically induced spatial charge separation in a nematic and the resultant orientational nonlinearity," *JETP,* **78** (6), 875–882, 1994.
5. I. C. Khoo, "Orientational photorefractive effects in nematic liquid crystal films," *IEEE J. Quant. Elect.,* **32**, 525–534, 1996. See also I. C. Khoo, "Optical-dc-field induced space charge fields and photorefractive-like holographic grating formation in nematic liquid crystals," *Mol. Cryst. Liq. Cryst.,* **282**, 53–66, 1996.
6. G. P. Wiederrecht, B. A. Yoon and M. R. Wasielewski, "High photorefractive gain in nematic liquid crystals doped with electron donor and acceptor molecules," *Science,* **270**, 1794–1797, 1995.
7. I. C. Khoo, Brett Guenther, M. Wood, P. Chen, and Min-Yi Shih, "Coherent beam amplification with photorefractive liquid crystal," *Opt. Lett.,* **22**, 1229–1231, 1997.

8. I. C. Khoo, S. Slussarenko, B. D. Guenther and W. V. Wood, “Optically induced space charge fields, DC voltage, and extraordinarily large nonlinearity in dye-doped nematic liquid crystals.” *Opt. Lett.,* **23**, 253–255, 1998; see also I. C. Khoo, Brett D. Guenther, and S. Slussarenko, “Photo-induced space charge fields, photo-voltaic, photorefractivity, and optical wave mixing in nematic liquid crystals,” *Mol. Cryst. Liq Cryst.,* **321**, 419–438 1998.
9. A. Golemme, B. L. Volodin, B. Kippelen and N. Peyghambarian, “Photorefractive polymer-dispersed liquid crystals,” *Opt. Lett.,* **22**, 1226–1228, 1997.
10. H. Ono and N. Kawatsuki, “Orientational photorefractive effects observed in polymer-dispersed liquid crystals,” *Opt. Lett.,* **22**, 1144–1146, 1997.
11. W. Helfrich, “Conduction induced alignment of nematic liquid crystals: Basic model and stability considerations,” *Phys. Rev.,* **51**, 4092–4105, 1969.
12. F. P. Shaefer (ed.), *Dye Lasers*, Springer Verlag, Heidelberg, 1974.
13. See, e.g., Y. Wang, “Photoconductivity of fullerene doped polymers,” *Nature,* **356**, 585–587, 1992.
14. I. C. Khoo and T. H. Liu, “Theory and experiments on multiwave mixing mediated probe beam amplification,” *Phys. Rev.,* **A39**, 4036–4044, 1989.
15. A. G. Chen and D. J. Brady, “Surface stabilized holographyin an azo-dye-doped liquid crystal,” *Opt. Lett.,* **17**, 1231–1233, 1992.
16. W. H. Howland and S. J. Fonash, “On semiconductor surface evaluation using the effective surface recombination speed for Schottky-coupled photovoltage measurement,” *J. Electrochem. Soc.,* **143**, 1958–1962, 1996.
17. J. Lagowski, A. M. Kontkiewicz L. Jastrzebski and P. Edelman, “Method for the measurement of long minority carrier diffusion lengths exceeding wafer thickness,” *Appl. Phys. Lett.,* **63**, 2902–2904, 1993.
18. S. Sato, “Photovoltaic effects in MBBA cells containing organic dyes,” *Jpn. J. Appl. Phys.,* **20**, 1989–1990, 1981.
19. L. K. Vistin, P.-A. V. Kazlauskas and S. Paeda, “Photoelectric effect in liquid crystals,” *Sov. Phys. Dokl.,* **29**, 207–209 1984.
20. I. C. Khoo, M. V. Wood, B. D. Guenther, Min-Yi Shih, P. H. Chen, Zhaogen Chen and Xumu Zhang, “Liquid crystal film and nonlinear optical liquid cored fiber array for ps-cw frequency agile laser optical limiting application,” *Optics Express,* **2**(12), 1998.
21. I. C. Khoo, J. Y. Hou, T. H. Liu, P. Y. Yan, R. R. Michael and G. M. Finn, “Transverse self-phase modulation and bistability in the transmission of a laser beam through a nonlinear thin film,” *J. Opt. Soc. Am.,* **B4**, 886–891, 1987.
22. See, for example, all the materials featured in R. Crane *et al.* (eds.), “Materials for optical limiting,” *Material Research Society Proceedings,* **374**, 1995. See

also L. Tutt and T. Boggess, " A review of optical limiting mechanisms and devices using organics, fullerenes, semiconductors and other materials," *Prog. Quant. Elect.,* **17**, 299–338, 1993.

23. J. H. Hong, Frederick Vachss, S. Campbell and Pochi Yeh, "Photovoltaic spatial light modulator" *J. Appl. Phys.,* **69**, 2835–2840, 1991.
24. F. Simoni, O. Francescangeli, Y. Reznikov and S. Slussarenko, "Dye-doped liquid crystals as high-resolution recording media," *Opt. Lett.,* **22**, 549–551, 1997.
25. S. Slussarenko, O. Francescangeli, F. Simoni and Y. Reznikov, "High resolution polarization gratings in liquid crystals," *Al. Phys. Lett.,* **71**, 3613–3615, 1997.

Chapter 5

Spectral and Spatial Diffraction in a Nonlinear Photorefractive Hologram

Feng Zhao and Hanying Zhou

Jet Propulsion Laboratory, California Institute of Technology, Pasadena, California

Nonlinear photorefractive (PR) crystals are one of the most attractive materials for recording volume holographic gratings for a variety of applications such as holographic data storage, reconfigurable interconnection, narrow-band filters, optical correlators, and optical neural networks [1–3]. Their use for these applications is based on the refractive index change through electrooptic effect as a result of the space-charge field formed upon an illumination with spatial variation in intensity [4]. This recording mechanism gives rise volume PR hologram recording some unique features that are not shared by those using conventional materials such as photographic film. First and foremost, the refractive index grating recorded in a PR hologram usually has a spatial phase shift relative to the interference pattern formed by recording beams. This produces a phenomenon called energy coupling between the two beams, which is most efficient when the phase shift is $\pi/2$. Second, the PR grating recording is a dynamic process. That is, simultaneously with a hologram's recording,

PHOTOREFRACTIVE OPTICS
Materials, Properties and Applications

ISBN 0-12-774810-5/$30.00

erasure and diffraction of the incident beam from the hologram recorded previously will occur. As a result of this self-diffraction, the two recording waves interfere not only with one another, but also with the diffracted waves within the crystal volume. Finally, the diffraction from a PR hologram is anisotropic as a direct consequence of the birefringent properties of photorefractive crystals and the anisotropy of the linear electrooptic effect, which is responsible for transformation of the space-charge electric field into a phase relief. The anisotropic diffraction can give rise to such phenomena as spatial variation in diffraction efficiency, light diffraction with the polarization plane rotated, and a suppression of wavelength selectivity of volume holograms.

These important features of PR crystals affect the hologram recording process and hence the diffraction properties in different ways. For example, the beam coupling during recording will result in a nonuniform refractive index modulation ratio over the crystal thickness, which affects the spectral diffraction characteristics of a PR hologram during readout. The erasure effect during the multiplexing process will cause the diffraction efficiency to decrease rapidly to an unacceptable level for detection, and without equalization it results in nonuniform diffractions among multiplexed holograms. The anisotropy nature of photorefractive crystals, on the other hand, will affect the spatial diffraction characteristics of a PR hologram. For a page-oriented hologram memory system, this means deteriorated image fidelity.

In this chapter, we first review, in Section 5.1, the spectral diffraction properties of a reflective-type volume hologram by considering photorefractive beam coupling and recording erasure dynamics. In Section 5.2, we examine the spatial diffraction properties of a PR hologram as affected by crystal refractive index anisotropy. The crystal anisotropy effects on the fidelity of the hologram image and on the multiplexing scheme are also discussed. A discussion of the combined (intrasignal) beam coupling and the crystal anisotropy effect in PR $LiNbO_3$ crystals is given in Section 5.3, and our conclusions are given in Section 5.4.

5.1 Nonlinear beam coupling and erasure dynamics on hologram diffraction spectral characteristics

As mentioned in the preceding, the holographic recording in a PR crystal is characterized by nonlinear beam coupling and erasure dynamics during multiplexing. A thorough analysis of two-wave mixing to treat the nonlinear beam coupling in PR materials has been given by Yeh [5]. To find out the diffraction properties of a PR hologram, however, we need to go further than just considering the recording stage. Both the erasure effect during the subsequent multiplexing stage and the beam coupling effect during the final readout stage should be included to complete the analysis. To this end, a coupled-recording-wave approach [6–8] is used in our analysis, which basically assumes the following. (1) During a hologram recording, the writing beams undergo PR two-wave-mixing, producing a nonuniform refractive index modulation ratio which varies along the thickness of the PR crystal. (2) Each successive hologram recording causes an incoherent erasure to this hologram, i.e., it causes a uniform decaying in the index modulation depth but no effects on the recorded grating structure. (3) For the final stage of reconstruction, the readout beam and its diffracted beam from the hologram undergo another two-wave-mixing process. For simplicity we assume that the hologram is subjected to no further erasure at this stage (which can be achieved by fixing the hologram after recording or by using a weak readout beam).

5.1.1 Coupled-recording-wave approach for PR reflection holograms

Referring to the reflection holographic recording geometry shown in Fig. 5.1, two coherent plane waves with complex amplitudes: $R = (I_R)^{1/2} \exp(i\phi_R)$ and $S = (I_S)^{1/2}\exp(i\phi_S)$ symmetrically enter the PR crystal from

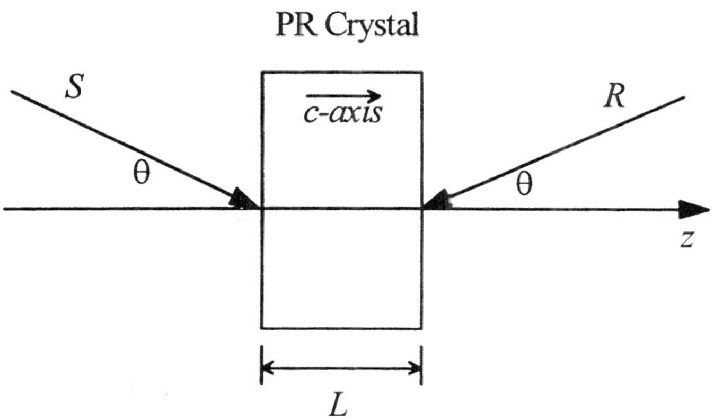

Figure 5.1: Configuration for reflection PR hologram recording.

opposite faces. The interference pattern formed by the two beams results in a refractive index modulation as

$$n = n_o + m(z)\,\Delta n_r \cos[2Kz\sin\theta + \phi_R(z) - \phi_S(z) + \phi_g]. \quad (5.1a)$$

Here Δn_r is the amplitude of the index perturbation, whose buildup characteristics can be approximately described as [1]: $\Delta n_r = \Delta n_s[1 - \exp(-t_r/\tau_r)]$, with Δn_S denoting its saturation value, and t_r and τ_r the recording time and the recording time constant, respectively. ϕ_g is the grating phase shift (assumed to be constant) relative to the intensity interference pattern, $\phi_R(z) - \phi_S(z)$ is the fringe curvature, $m(z)$ is the fringe pattern contrast (refractive-index modulation ratio) as defined by

$$m(z) = \frac{2\sqrt{I_S(z)I_R(z)}}{I_S(z) + I_R(z)}. \quad (5.1b)$$

Substituting Eq. (5.1) and the total complex light field within the crystal into the Maxwell wave equation and then following the standard slowly varying (field) approximation [2], the following nonlinear coupled equations for the recording process can be derived [9]:

$$S'(z,t_r) - i\Gamma_r m(z)\exp[-i(\phi_R(z) - \phi_S(z) + \phi_g)]R(z,t_r) = 0, \quad (5.2a)$$

$$R'(z,t_r) + i\Gamma_r m(z)\exp[i(\phi_R(z) - \phi_S(z) + \phi_g)]S(z,t_r) = 0, \quad (5.2b)$$

where $\Gamma_r = \pi\Delta n_r/\lambda\cos\theta$. From these equations, we found the recorded grating structure characteristics, i.e., the refractive index modulation ratio *m(z)* and fringe curvature $\phi_R(z) - \phi_S(z)$, as [9]

$$m(z) = [1 + m_1\exp(2\Gamma_r z\sin\phi_g)]^{-1/2}, \quad (5.3a)$$

$$\phi_R(z) - \phi_S(z) = \Gamma_r z\cos\phi_g, \quad (5.3b)$$

with

$$m_1 = \frac{[m_o^2 - \exp(-2\Gamma_r L\sin\phi_g)]^2}{4m_o(m_o + 1)[m_o + \exp(-2\Gamma_r L\sin\phi_g)]}, \quad (5.3c)$$

where $m_o = I_R(L)/I_S(0)$ is the incident beam ratio and L the crystal thickness. The arbitrary constant contained in Eq. (3b) has been set to zero without loss of generality. Eq. (5.3a) indicates that the modulation ratio will be crystal thickness dependent as long as $\phi_g \neq 0$. In Fig. 5.2

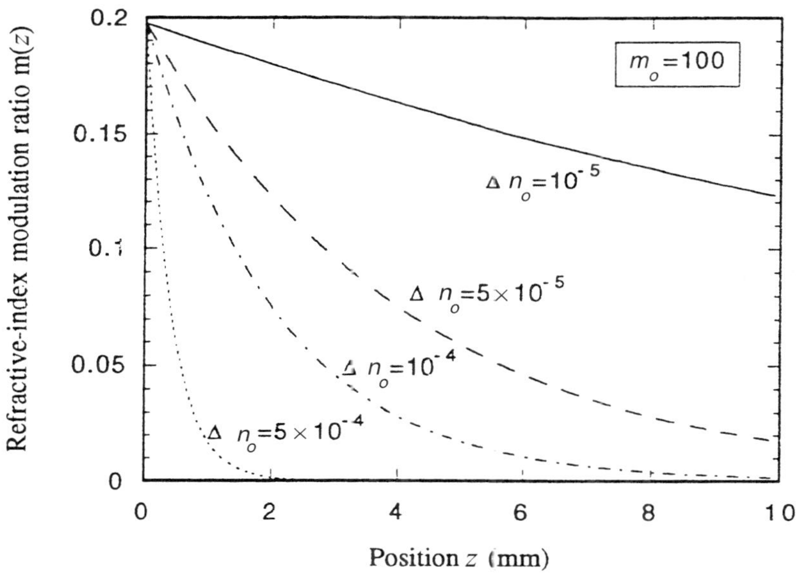

Figure 5.2: Refractive index modulation ratio $m(z)$ as a function of position z for various index perturbations Δn_o. (Source: Ref. [11]. Reprinted with permission.)

the modulation ratio $m(z)$ is plotted as a function of position z for various index perturbations, which clearly shows that the refractive-index modulation ratio could be highly nonuniform over the crystal thickness for large index perturbation. It is this nonuniformity in refractive index modulation that marks the difference between Kogelnik's coupled equations [10] for describing diffraction upon readout and the coupled-equations developed in our analysis (see below).

When successive holograms are multiplexed, the grating structure is assumed to remain the same while its amplitude decays from its recorded value to: $\Delta n_o = \Delta n_r \exp(-t_e/\tau_e)$, with τ_e denoting the erasure time constant and t_e the erasure time, which is the sum of all successive recording times. After finishing the multiplexing process, the hologram can be fixed and then reconstructed with a readout beam. A new set of coupled equations for beams $r(z)$ and $s(z)$, which originate from a single readout beam $r(z)$ and with a possible Bragg mismatch ϑ, is found as [9, 11]

$$s'(z) + f^*(z)r(z) - j\vartheta s(z) = 0 \tag{5.4a}$$

$$r'(z) + f(z)s(z) = 0\,. \tag{5.4b}$$

where

$$f(z) = \frac{j\Gamma_e \exp(j\Gamma_r z \cos\phi_g + j\phi_g)}{\sqrt{1 + m_1 \exp(2\Gamma_r z \sin\phi_g)}}, \tag{5.4c}$$

$$\Gamma_e = \frac{\pi \Delta n_o}{(\lambda_o + \Delta\lambda)\cos(\theta_o + \Delta\theta)}. \tag{5.4d}$$

with $\Delta\theta$ and $\Delta\lambda$ denoting the angle and wavelength deviations, respectively, and $\vartheta = \Delta\theta K \sin\theta - \Delta\lambda K^2/4\pi n$ [10]. Unfortunately, Eqs. (5.4) do not have an analytical solution in general, due to the z-dependent exponential factor in the expression of $f(z)$. In the following, we will treat on-Bragg and off-Bragg incident cases separately.

5.1.2 Spectral diffraction characteristics

5.1.2.1 On-Bragg readout ($\vartheta = 0$)

For Bragg incidence ($\vartheta = 0$), and when $\phi_g = \pi/2$, which happens to be the usual case for crystals such as $Sr_{1-x}Ba_xNb_2O_b$ (SBN) and $BaTiO_3$ (and having the maximum beam coupling effect), we can find the diffraction efficiency as, [9]

$$\eta = s(L)s^*(L) = \tanh^2\left\{\Gamma_e L - \frac{\Gamma_e}{\Gamma_r}\ln\left[\frac{1 + \sqrt{1 + m_1 \exp(2\Gamma_r L)}}{1 + \sqrt{1 + m_1}}\right]\right\} \tag{5.5}$$

For single recording ($\Gamma_r = \Gamma_e$) the diffraction efficiency as a function of recording time was plotted in Fig. 5.3 for beam ratio $m_o = 10$ and $m_o = 100$ (dashed curves), along with the calculations according to Kogelnik's formula [10] (solid curves). From the figure we see that even for sufficiently long recording time, the diffraction efficiency of a PR hologram does not reach to the maximum of 100% due to beam coupling effect (the PR crystal is assumed lossless). And as the recording beam ratio increases, so does the coupling effect, which in turn results in a significant drop in the achievable diffraction efficiency. In Fig. 5.4, the maximum achievable diffraction efficiency is plotted as a function of beam ratio m_o.

While the diffraction efficiency is decreased even for a single recording due to the beam coupling effect, it will be further decreased when many holograms are multiplexed due to the erasure effect. To find out the overall diffraction efficiency after multiple recording, we first performed a numerical analysis to determine the exposure time for each hologram

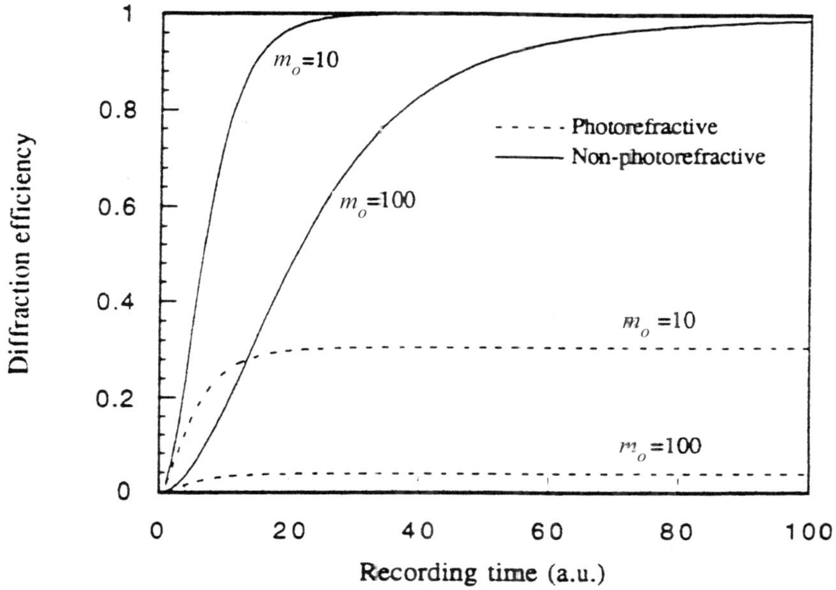

Figure 5.3: Diffraction efficiency versus recording time (single recording). (Source: Ref. [9]. Reprinted with permission.)

recording for a given (total) number of multiplexed holograms. This was done by using the sum of the recording times of all successive holograms as the erasure time for this exposure and then equating the diffraction efficiencies of Eq. (5.5) for the nth and $(n-1)$th holograms. Then the equalized diffraction efficiency can be calculated from Eq. (5.5). In Fig. 5.5 we plotted the (equalized) diffraction efficiency as the total number of holograms N for $m_o = 1$, 10, and 100. The parameter choices are: $\tau =$ 120 s, $\tau_e = 210$ s, $\phi_g = \pi/2$, $L = 2$ mm, $\lambda = 514.5$ nm, $\Delta n_s = 10^{-3}$. It is clear that due to the erasure effect, the overall diffraction efficiency is rapidly decreased as the total multiplexing number increases. For the example given above, the diffraction efficiency will drop to 0.01% (for m_o=1) when the multiplexing number N reaches 2000.

5.1.2.2 Off-Bragg readout ($\vartheta \neq 0$): Bragg selectivity

For the off-Bragg case, numerical analysis using the fourth-order Runge-Kutta method was performed. It was found that for a small Bragg devia-

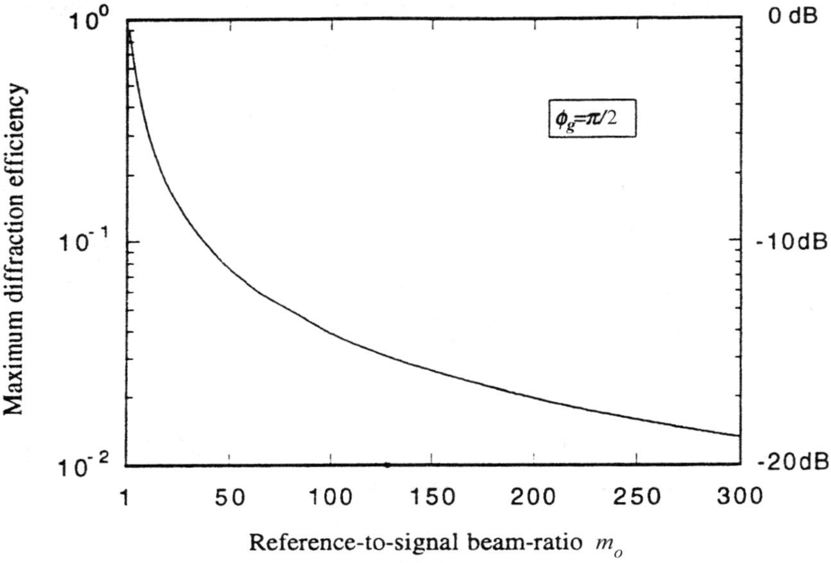

Figure 5.4: Maximum diffraction efficiency as a function of recording beam ratio (single recording). (Source: Ref. [9]. Reprinted with permission.)

tion, the difference is minor between the normalized Bragg selectivity for a PR hologram and that for a non-PR hologram. However, for large Bragg deviations, the difference can be large. For this reason, our emphasis here was given to the behavior at large off-Bragg deviations. In Figs. 5.6a and b we plotted the calculations of the normalized diffraction efficiency as a function of wavelength deviation with $\theta = 0°$, $\phi_g = 90°$, $\lambda = 614.7$ nm, and $L = 2$ mm. For comparison, calculations based on Kogelnik's formula are also plotted in the corresponding figures (dashed line, envelope only).

We see that the diffraction efficiency of a PR hologram for a large Bragg detuning can be either lower (Fig. 5.6a) or higher (Fig. 5.6b) than the calculation based on Kogelnik's formula (for non-PR holograms) depending on the recording parameters (e.g., beam ratio, index perturbation, etc.). This implicates that the nonlinear beam coupling might have an impact on Bragg selectivity (particularly at large Bragg deviation) of a PR hologram. In fact, the "apodization phenomenon"of the spectral response of a PR hologram [12] has been observed, which can be partially attributed to the beam coupling effect discussed here.

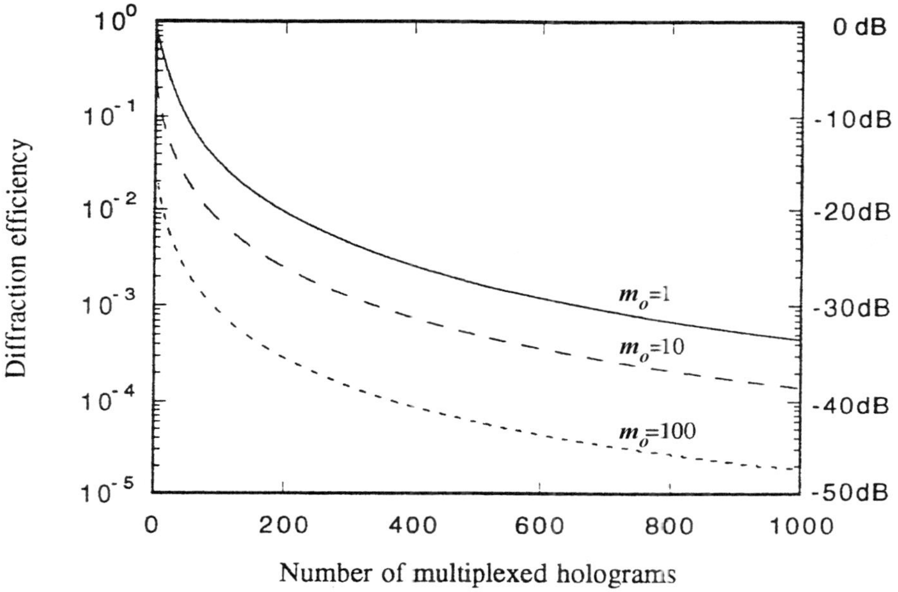

Figure 5.5: Diffraction efficiency (equalized) versus total number of multiplexed holograms. (Source: Ref. [9]. Reprinted with permission.)

5.2 Refractive-index anisotropy on hologram spatial diffraction properties

In the preceding discussion we notice that the spectral diffraction characteristics of a PR hologram is affected by the PR nonlinear beam coupling and erasure effect. In this section, we will examine the spatial diffraction characteristics of a PR hologram as affected by the refractive-index anisotropy. The spatial aspect of the diffraction is important because in a page-oriented holographic data storage system, it can affect the hologram image fidelity as well as multiplexing choice. For simplicity, our analysis is limited to a one-dimensional case which, although simple, is sufficient to illustrate the spatial variation in diffraction efficiency. The photorefractive beam coupling effect during recording will be neglected in the following analysis, and the diffraction is evaluated for on-Bragg condition.

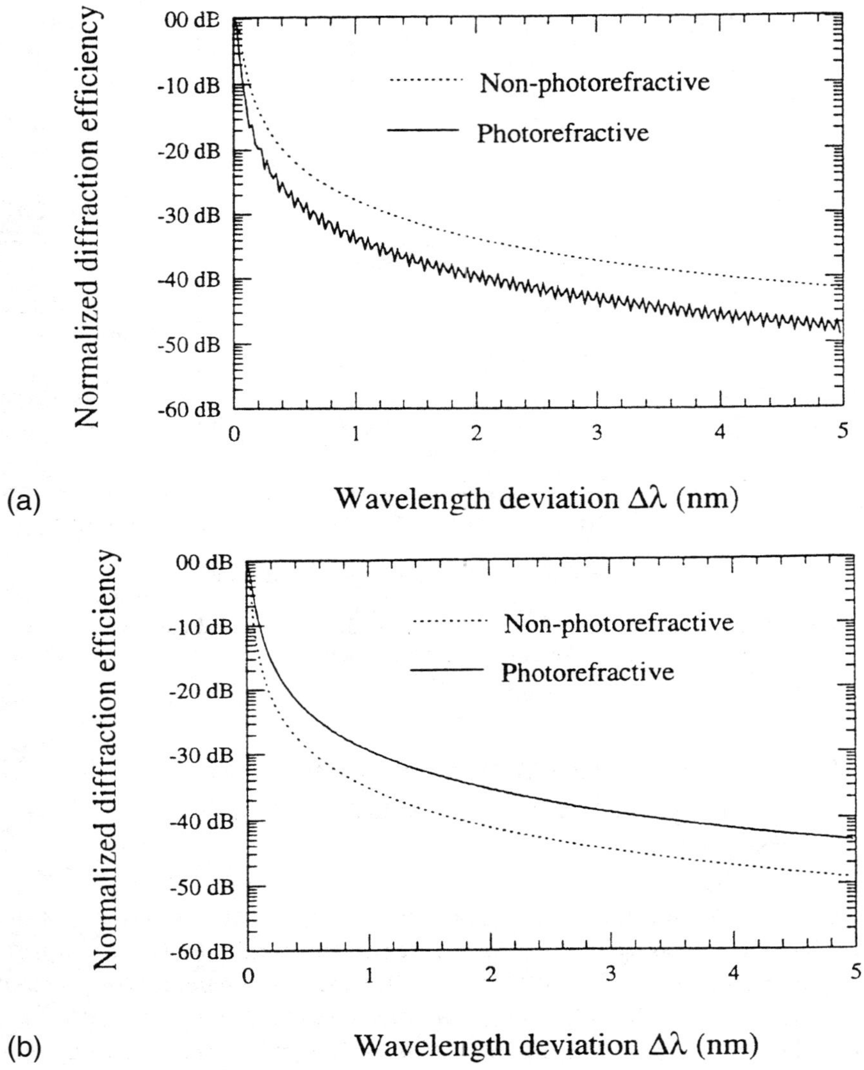

Figure 5.6: Diffraction efficiencies as a function of wavelength deviation. (a) $\Delta n_o = 3 \times 10^{-4}$, $m_o = 1$. (b) $\Delta n_o = 5 \times 10^{-4}$, and $m_o = 100$. (Source: Ref. [11]. Reprinted with permission.)

5.2.1 Spatial diffraction properties

Figures 5.7a and b show PR holographic recording configurations for transmission and reflection types, respectively. Two plane waves, S (signal) and R (reference), with incident angles of θ_1 and θ_2 (both measured inside the PR crystal) respectively, form a refractive-index grating vector $\boldsymbol{K}$ with magnitude $K = 4\pi n \sin\alpha/\lambda$, where α is half the angle between the two beams, and β is the angle of the crystal c-axis relative to the x-axis for transmission type or to the z-axis for reflection type, as shown in the figures. The signal wave S can be regarded as from one specific pixel position in a Fourier-transform hologram, or as one specific spatial frequency content in an image-plane hologram. A page of data will contain many such plane signal waves. For simplicity, we assume both the signal and the reference beams are of infinite transverse width.

Since the beam-coupling effect is neglected, Kogelnik's formula can be used to calculate the peak diffraction efficiency upon readout [10]

$$\eta \approx \left(\frac{\pi \Delta n L}{\lambda\sqrt{c_S c_R}}\right)^2, \tag{5.6}$$

for both types of recording configurations under small index perturbation approximation. Here L is the crystal thickness, λ is the free space recording wavelength, $c_s = \cos\theta_1$ and $c_R = \cos\theta_2$ are the slant factors of the signal and the reference waves, and Δn is the index perturbation caused by the interference of these two waves. The amount of the index perturbation, Δn, is determined, along with other parameters, by the effective electrooptic coefficient of the PR crystal, which, under the short-writing time condition, can be written as [1]

$$\Delta n = \frac{1}{2} n^3 r_{\text{eff}} m E_{\text{sc}} \mathbf{e}_s \bullet \mathbf{e}_R^* \frac{t}{\tau_r} \tag{5.7}$$

where n is the average (unperturbed) refractive-index of the crystal, m is the refractive-index modulation depth, t and τ_r are the recording time and the recording time constant, respectively. $\boldsymbol{e}_S$ and $\boldsymbol{e}_R$ are unit vectors of the polarization status of the two writing beams. For extraordinarily polarized beams (which is the usual case for efficient utilizing of the PR effect), we have $\boldsymbol{e}_S \cdot \boldsymbol{e}_R^* = \cos(2\alpha)$. E_{sc} and r_{eff} are the magnitude of the

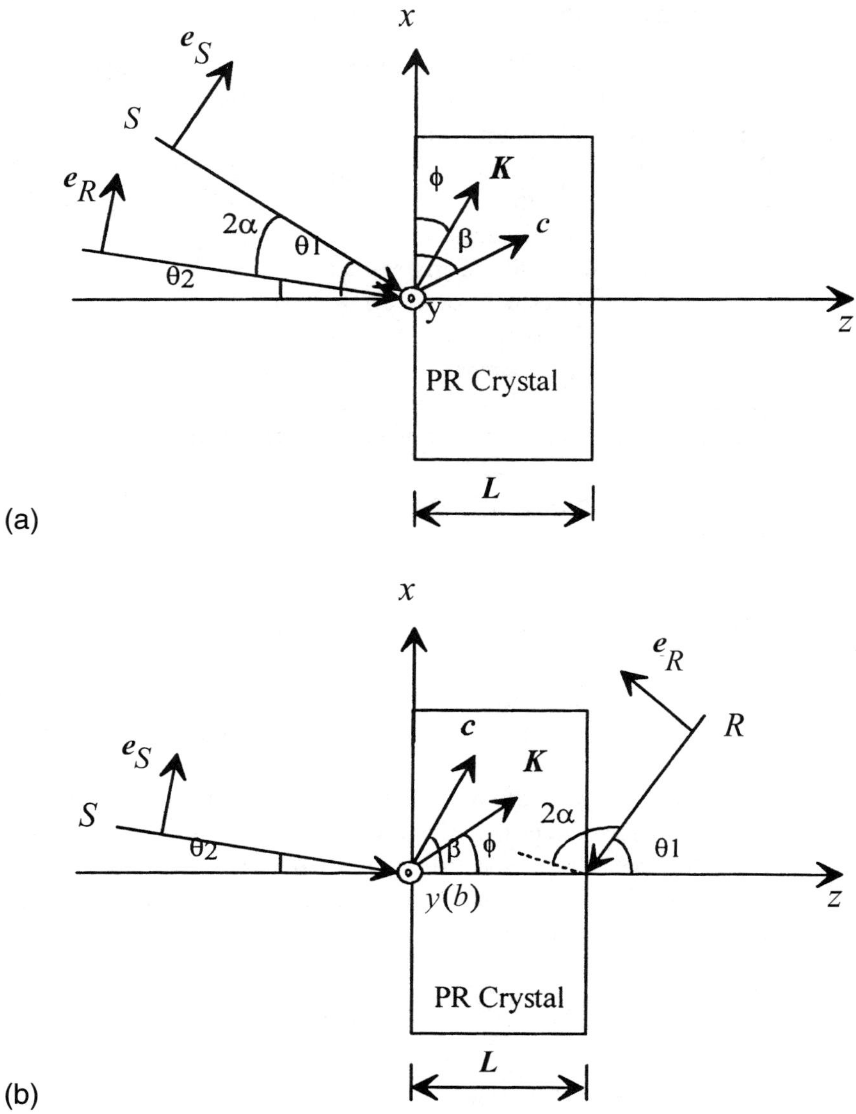

Figure 5.7: Photorefractive holographic recording configuration: (a) transmission type; (b) reflection type. (Source: Ref. [13]. Reprinted with permission.)

(steady-state) space-charge field and the effective electrooptic coefficient of the PR crystal. r_{eff} can be found as [13]

$$r_{\mathrm{eff}} = \left\{ \frac{1}{2} n_o^4 r_{13} [\cos 2\alpha - \cos(2\beta - 2\phi)] + 4 n_e^2 n_o^2 r_{51} \sin^2(\beta - \phi) + \frac{1}{2} n_e^4 r_{33} [\cos 2\alpha + \cos(2\beta - 2\phi)] \right\} \cos(\beta - \phi), \tag{5.8}$$

for both types of recording configuration. Here, as defined in Figs. 5.7a and b, for the transmission type hologram: $\phi = (\theta_2 + \theta_1)/2$ is the angle between the grating vector $\boldsymbol{K}$ and the x-axis; for the reflection type: $\phi = (\theta_2 - \theta_1)/2$ is the angle of grating vector $\boldsymbol{K}$ relative to the z-axis. β is the angle of the crystal c-axis relative to the x-axis for transmission type or to the z-axis for reflection type. n_o and n_e are the respective ordinary and extraordinary refractive indices. r_{ij} are the components of the electrooptic coefficient tensor of the materials: for $LiNbO_3$ ($n_o = 2.286$, $n_e = 2.20$): $r_{51} = 33$, $r_{13} = 9.6$, $r_{33} = 31$; for SBN ($n_o = 2.3117$, $n_e = 2.2987$): $r_{51} = 42$, $r_{13} = 67$, $r_{33} = 1340$, all are in units of 10^{-12} m/V [14]. In the absence of an external applied field, the space-charge field E_{sc} is found to be [13]

$$E_{\mathrm{sc}} = \left[\frac{E_{\mathrm{ph}}^2 \cos^2(\beta - \phi) + A^2 \sin^2 \alpha}{(1 + \varepsilon A \sin^2 \alpha / B)^2 + (\varepsilon E_{\mathrm{ph}} \cos(\beta - \phi) \sin \alpha / B)^2} \right]^{1/2}, \tag{5.9}$$

where E_{ph} is the intrinsic photovoltaic field, which is negligibly small for SBN materials but can be large for $LiNbO_3$ crystals. We shall use a moderate photovoltaic field of $E_{\mathrm{ph}} = 15$ kV/cm for $LiNbO_3$ [1] and $E_{\mathrm{ph}} \sim 0$ for SBN. Both A and B are quantities independent of material parameters, whose typical values are $A \approx 15$ kV/cm for room temperature and $B \approx 80$ kV/cm. $\varepsilon = \varepsilon_{\parallel} \cos^2(\beta - \phi) + \varepsilon_{\perp} \sin^2(\beta - \phi)$ is the effective dc dielectric constant [15] in the direction of grating vector $\boldsymbol{K}$, where $\varepsilon_{\parallel}$ and $\varepsilon_{\perp}$ are the respective parallel and perpendicular (to the c-axis) dielectric tensor elements of the PR materials with $\varepsilon_{\parallel} = 3400$ and $\varepsilon_{\perp} = 700$ for SBN, and $\varepsilon_{\parallel} = 32$ and $\varepsilon_{\perp} = 78$ for $LiNbO_3$ [14].

By retaining those terms that depend on the angles θ_1, θ_2, or β, the diffraction efficiency can then be expressed as [13]

$$\eta = c(r_{\mathrm{eff}} E_{\mathrm{sc}} \cos 2\alpha)^2 / (\cos\theta_1 \cos\theta_2), \tag{5.10}$$

where c is a constant and r_{eff} and E_{sc} are given by Eqs. (5.8) and (5.9). In Figs. 5.8a–d we plotted the diffraction efficiency (vertical axis, arbitrary units) as a function of writing angles θ_1 and θ_2 (two horizontal axes, in

(a)

(b)

(c)

(d)

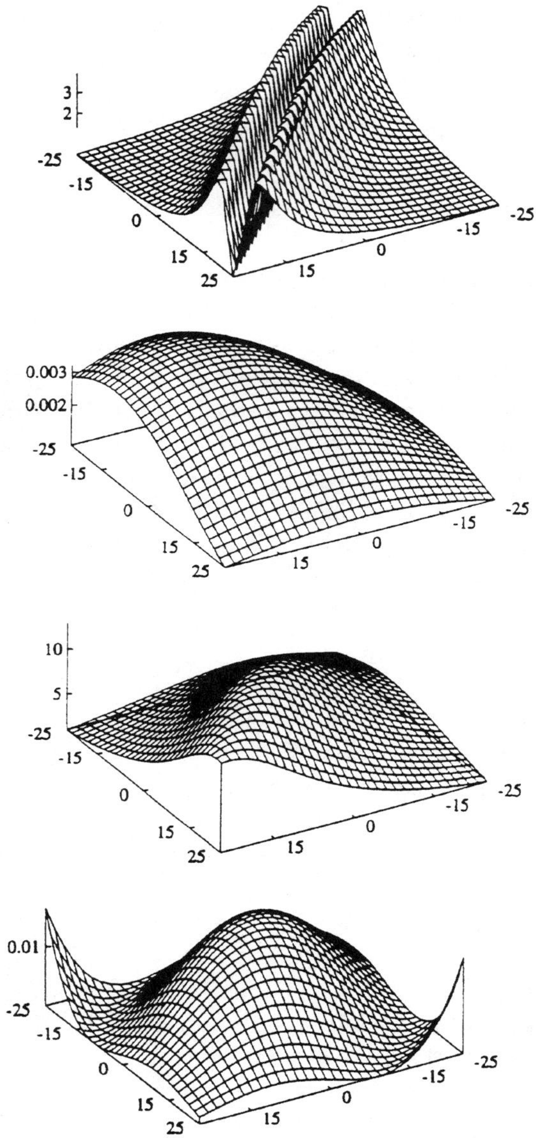

Figure 5.8: Relative diffraction efficiency versus writing angles. (a) SBN, β = 0°, transmission. (b) SBN, β = 45°, reflection. (c) $LiNbO_3$, β = 45°, transmission. (d) $LiNbO_3$, β = 0°, reflection. (Source: Ref. [13]. Reprinted with permission.)

units of degrees) for some typical recording configurations with SBN or $LiNbO_3$ crystals. Both θ_1 and θ_2 range from $-25°$ to $+25°$, which correspond to the external writing angles of $-90°$ to $+90°$ (as $n \approx 2.3$). For each fixed reference beam angle θ_2, the variation in diffraction efficiency as a function of signal beam angle θ_1 can be viewed as the intensity nonuniformity within a page of data, either from pixel to pixel as in a Fourier transform hologram, or from spatial frequency content to spatial frequency content as in an image-plane hologram. On the other hand, for each fixed signal beam angle θ_1, the variation of diffraction efficiency as a function of reference beam angle θ_2 can be viewed as the nonuniformity in diffraction from page to page in an angularly multiplexed hologram.

Clearly the diffraction efficiency varies over writing angles and is very much recording geometry and materials dependent. For example, in Fig. 5.8a where a transmission configuration is used with an SBN crystal and a crystal orientation of $\beta = 0°$, the diffraction efficiency is relatively high and uniform only within the narrow vicinity of $\alpha = (\theta_1 - \theta_2)/2 = 2°$ for grating slant angle $\phi = (\theta_2 + \theta_1)/2$ from $-25°$ to $25°$. On the other hand, for $LiNbO_3$ of transmission configuration and $\beta = 45°$ (Fig. 5.8b), the diffraction efficiency is relatively high and flat over beam angle $-10° \leq \alpha \leq 10°$ for slant angle ϕ within the range of -5–$25°$. In all cases, the variations of diffraction efficiency are different and complicated, which would complicate the recording procedure if one attempts to compensate for the nonuniformity.

5.2.2 Effect on reconstructed hologram image fidelity and on multiplexing scheme

Since an object beam is always of a certain space-bandwidth product or certain spatial frequency width, the anisotropy of diffraction efficiency will affect the fidelity of the reconstructed hologram image. By referring to Eq. (5.11), we can calculate the reconstructed hologram image intensity distribution (or spatial frequency response) with a fixed reference beam angle θ_2. Some typical results are shown in Figs. 5.9b–f (with Fig.5.9a being the input intensity distribution), where the horizontal axis represents either the pixel position in a Fourier transform hologram or the spatial frequency content in an image-plane hologram.

For a Fourier transform hologram, the variation in the diffraction efficiency would be directly related to the appearance of a reconstructed hologram image, i.e., high diffraction efficiency means bright. For an

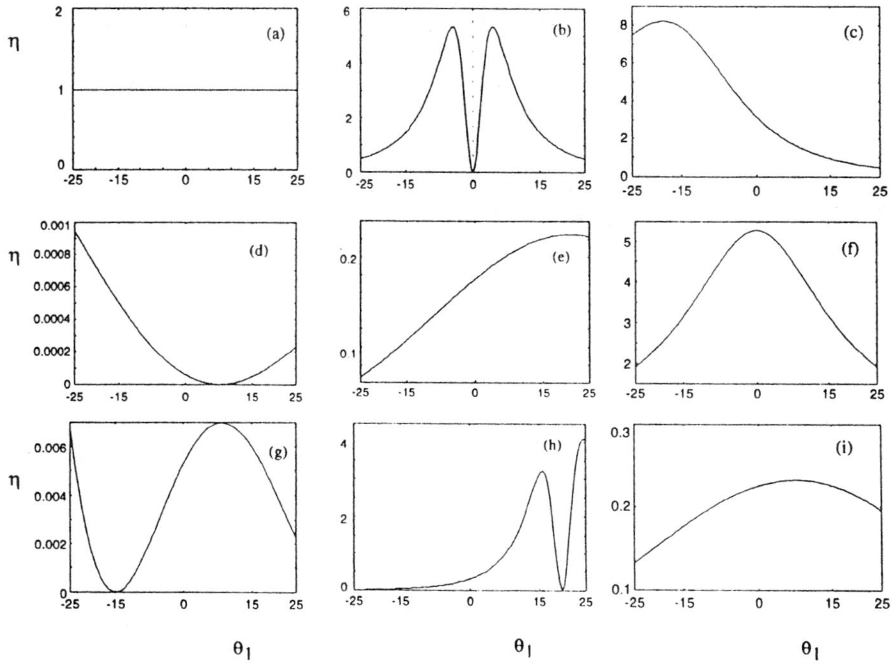

Figure 5.9: Input (a) and reconstructed (b–f)) image intensity distributions over the writing angle. (Source: Ref. [13]. Reprinted with permission.)

image-plane hologram, on the other hand, the variation in diffraction reflects the nonlinearity in the spatial frequency response of the PR hologram, which also leads to a distorted appearance of the reconstructed hologram image. The appearance of the hologram image in this latter case, however, is hard to predict because of the wide variety of spatial frequency responses, which vary from case to case as can be seen in Fig. 5.9. In either case, the fidelity of the reconstructed image will be strongly dependent on the recording geometry (including both the crystal orientation and the write-in angles). For example, in some good cases such as Fig. 5.9f, the variation in the diffraction efficiency is relatively small. In some other cases, such as Figs. 5.9b or d, the variation is much larger and more irregular. This huge variation will certainly deteriorate the fidelity of reconstructed images. In fact, we can relate the nonuniformity in diffraction efficiency roughly to the bit error rate (BER) of the hologram data by

$$\mathrm{BER} \approx \frac{1}{2}[1 - erf(x)] \approx \frac{\exp(-x^2/2)}{x\sqrt{2\pi}}, \tag{5.11}$$

where $x = (\eta_a - \eta_t)/\sigma_\eta$, with η_a, η_t, and σ_η denoting the (relative) average diffraction strength, the threshold diffraction, and the standard deviation of diffraction, respectively. Take Fig. 5.9b as an example, where an SBN crystal is used for a transmission-type hologram with its c-axis parallel to the x-axis (i.e., $\beta = 0°$). Assume an object beam has a bandwidth (in terms of angle) of $2\Delta\theta_1 \sim 6°$, which is centered at $\theta_1 \sim 4°$ with a reference beam angle set at $\theta_2 = 0°$. The deviation of the nonuniformity in diffraction efficiency can be found to be $\sigma_\eta = 0.93$ for an average diffraction of $\eta_a = 5.0$. Then, from Eq. (5.12), the BER for this particular page of data is found to be about 5×10^{-3}, assuming the threshold value is such that $\eta_t = \eta_a/2$. This high BER would impose a constraint on the object bandwidth unless a compensation procedure is implemented during the recording. It would also impose an additional constraint on the dynamic range-limited storage capacity of the PR hologram, since the minimum diffraction efficiency within a page must meet the sensitivity of a detector, while the maximum diffraction efficiency determines the dynamic range–limited storage capacity.

For high-density PR hologram storage, either an angular-multiplexing [16] or a wavelength multiplexing [17] scheme can be employed. However, since the diffraction efficiency is very much angle dependent, the variation of diffraction efficiencies among the multiplexed pages of data would inevitably occur in an angularly multiplexed hologram, as the angular multiplexing attempts to use all the accessible angle range (of the optical setup and recording medium). Therefore a (spatially) complicated exposure schedule is necessary to compensate for the variation in diffraction efficiency. On the other hand, the wavelength multiplexing scheme uses in principle only a fixed pair of writing angles. It is therefore possible to minimize the hologram image distortion within a page and among pages by using the optimal crystal orientation and writing angles. For example, in Fig. 5.9c, instead of centering an object beam at $\theta_1 = 0°$, we can choose to center it at $\theta_1 = 15°$. By doing so, the average diffraction efficiency increases from $\eta_a = 3.2$ to $\eta_a = 7.8$ (for a bandwidth of $2\Delta\theta_1 \sim 6°$), while the deviation is decreased from $\sigma_\eta = 0.42$ to $\sigma_\eta = 0.30$. If the threshold diffraction is again such that $\eta_t = \eta_a/2$, then the corresponding BER increases from about 10^{-5} to 10^{-40} — there is virtually no effect on hologram image fidelity when we properly choose the recording geometry. Clearly, by using the wavelength multiplexing scheme, we can avoid an otherwise (spatially) complicated exposure schedule. In this sense, wavelength multiplexing would be a more favorable choice over the angular one.

5.3 Anisotropic intrasignal coupling

In Section 5.1 we have considered the coupling between a plane wave of a signal beam and a plane wave of a reference beam. Since a general signal beam must consist of several plane waves [e.g., a page-encoded two-dimensional (2-D) image] to carry spatial information, intrasignal coupling will inevitably occur during a photorefractive hologram recording. The intrasignal coupling will also affect the spatial diffraction properties of a PR hologram and hence the reconstructed image fidelity by interchanging energies between plane waves components that compose the signal. In this section we will analyze the intrasignal coupling in a PR $LiNbO_3$ crystal. However, calculation of a 2-D intrasignal coupling with an arbitrary crystal orientation would be complicated, as the crystal (and most other PR crystals) is anisotropic in terms of the refractive index grating. To simplify this problem, we will consider a one-dimensional signal for some typical recording configurations under small-angle (signal field) approximation. To limit our analysis to intrasignal coupling only, coupling between each signal beam and reference beam is not included (which is the topic of Section 5.1).

Referring to Fig. 5.10, assume a number of (coherent) plane waves S_m, each with an angle of θ_m (measured inside the crystal), incident on a thick photorefractive $LiNbO_3$ crystal. These plane waves can be regarded as having been generated by an input pixel array in a Fourier transform holographic system or the signal itself in a spatial mode convertor and will undergo a multipair two-wave mixing process (intrasignal coupling) as they propagate through the crystal. For our analysis we choose two crystal orientations, the a-axis orientation (called "case A") and the b-axis orientation (called "case B"). In both cases, the c-axis of the crystal is parallel to the z-axis and the signal is of ordinary polarization for simplicity. Such a configuration is commonly used in a reflection holo-

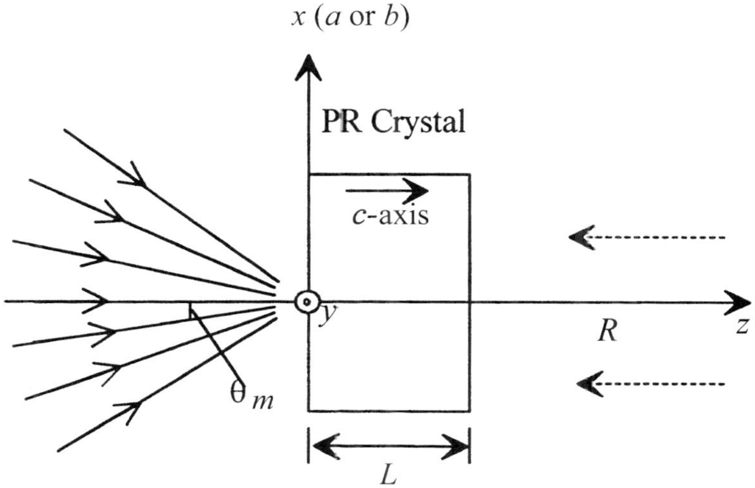

Figure 5.10: Geometry for considering intrasignal coupling. (Source: D. Zhao, et al., "Anisotropic intrasignal coupling in photoreflective LiNbO3," *Microwave and Opt. Tech. Lett.*, © 1994 John Wiley & Sons, Inc. Reprinted by permission of John Wiley & Sons, Inc.)

graphic system or in a mode conversion system for this crystal, where both the mechanism of hologram recording via the linear photovoltaic effect (since $E_{ph} \parallel c$-axis) and the electrooptic coefficient r_{33}, which is the largest in $LiNbO_3$, are efficiently utilized. The interference pattern formed by those waves will result in a refractive index perturbation in a PR crystal, which takes the form [18]

$$n_1 = \frac{1}{2I}\sum_{m}\sum_{n\neq m} -\frac{\beta}{2} r_{\text{eff}} n_o^3 \Delta E_{mn} A_m A_n^* \exp\{ik[(s_m - s_n)x + (c_m - c_n)z]\} + \text{c.c.} \tag{5.12a}$$

where $x = x$ or y, c.c. denotes the complex conjugate, A_m the amplitude of the mth plane wave, $s_m = \sin\theta_m$, $c_m = \cos\theta_m$ and $k = 2\pi n_o/\lambda$. n_o is the (ordinary) refractive index of the crystal, and λ the wavelength of the signal beam. $I = \Sigma_m A_m A_m{}^* = \Sigma_m I_m$, β is a ratio of index perturbation relative to its saturation value, E_{mn} is the magnitude of the space-charge field induced by the mth and the nth plane waves, and r_{eff} is the effective electrooptic coefficient of the crystal. For $LiNbO_3$ this coefficient can be found as

$$(r_{\text{eff}})_a = -r_{13} \sin \frac{\theta_m + \theta_n}{2}, \tag{5.12b}$$

$$(r_{\text{eff}})_b = -r_{22} \cos \frac{\theta_m + \theta_n}{2} \tag{5.12c}$$

where subscripts a and b denote cases A and B, respectively. Equation (5.12c) is derived under small-angle assumption, i.e., $-6° \leq \theta_m, \theta_n \leq 6°$ For the geometry shown in Fig. 5.10 the photovoltaic effect is negligible since the gratings formed by intrasignals are almost perpendicular to the crystal c-axis. Therefore the magnitude of the space-charge field E_{mn}, when no external field is applied, is simply

$$E_{mn} = i \frac{4\pi n_o}{\lambda} \frac{k_B T}{e} \sin \frac{\theta_m - \theta_n}{2}. \tag{5.12d}$$

where k_B is the Boltzmann constant, T is the absolute temperature, and e is the electron charge.

By extending the basic two-wave mixing theory to the multipair two-wave-mixing process [19], the coupled-wave equations for the intrasignal coupling can be derived as [18]

$$c_m \frac{1}{I_m} \frac{dI_m}{dz} = 2\gamma_a c_m - \frac{2}{I} \gamma_a \sum_n c_n I_n, \tag{5.13a}$$

$$\gamma_a = \frac{\beta}{2} r_{13} \left(\frac{2\pi n_o}{\lambda} \right)^2 n_o^4 \frac{k_B T}{e} \tag{5.13b}$$

for case A and

$$c_m \frac{1}{I_m} \frac{dI_m}{dz} = \frac{2}{I} \gamma_b \sum_n s_n I_n - 2\gamma_b s_m, \tag{5.14a}$$

$$\gamma_b = \frac{\beta}{2} r_{22} \left(\frac{2\pi n_o}{\lambda} \right)^2 n_o^4 \frac{k_B T}{e} \tag{5.14b}$$

for case B. The solutions of Eqs. (5.13) and (5.14), under small-angle approximation, can be found as [18]

$$I_m(z) = I_m(0) \exp[-2\gamma_a z (c_a / c_m - 1)], \tag{5.15a}$$

$$c_a = \sum_n I_n(0) \Big/ \sum_n I_n(0) / c_n \tag{5.15b}$$

for case A and

$$I_m(z) = f(z)I_m(0)\exp(-\,2\gamma_b z\tan\theta_m), \tag{5.16a}$$

$$f(z) = \left[\sum_n I_n(0)\exp(2\gamma_b z\tan\theta_n)\right]^{-1/c_m} \tag{5.16b}$$

for case B, where $I_m(0)$ is the intensity of the mth plane wave at its incidence upon the crystal.

From Eqs. (5.15) and (5.16) we notice that in both cases, the relative intensity distributions of the plane waves will be changed (distorted) after they are propagated through the PR crystal. However, this change has different characteristics in two cases as shown in Figs. 5.11 and 5.12, in which the output beam intensities are plotted as functions of crystal thickness z and of the (external) beam field angle, respectively. In both figures, we have used, for simplicity, seven plane waves with a uniform input intensity. The adjacent angular separation is 4° (in the air) and $\gamma_a = 4.75\ \text{cm}^{-1}$ for case A and $\gamma_b = 1.88\ \text{cm}^{-1}$ for case B (the relative values of γ_a and γ_b correspond to the ratio of the electrooptic coefficients r_{13} and r_{22}). It is clear that for the a-axis-oriented crystal, the intrasignal coupling is symmetrical with respect to the c-axis of the crystal, while it is asymmetrical for the b-axis-oriented crystal. Quantitatively, the relative intensity change in the b-axis-oriented crystal ($\sim 80\%$) is much larger than in the a-axis-oriented crystal ($< 5\%$). This can be explained by the fact that the effective electrooptic coefficient in case B is directly proportional to the cosine of the field angle, while for case A the effective electrooptic coefficient is directly proportional to the sine of the field angle (Eqs. 5.15 and 5.16). Since the field angle is assumed to be small, the effective electrooptic coefficient for case A can be considered negligible in practice. Thus, for one-dimensional applications where intrasignal coupling is unwanted, case A is preferable due to less intrasignal coupling. For usual 2-D applications, however, the intrasignal coupling is generally anisotropic, with relatively prominent case B coupling characteristics along the b-axis direction to negligible case A coupling characteristics in the perpendicular a-axis direction.

5.4 Conclusions

We have discussed both spectral and spatial diffraction properties of volume PR holograms by considering the nonlinear PR beam coupling effect,

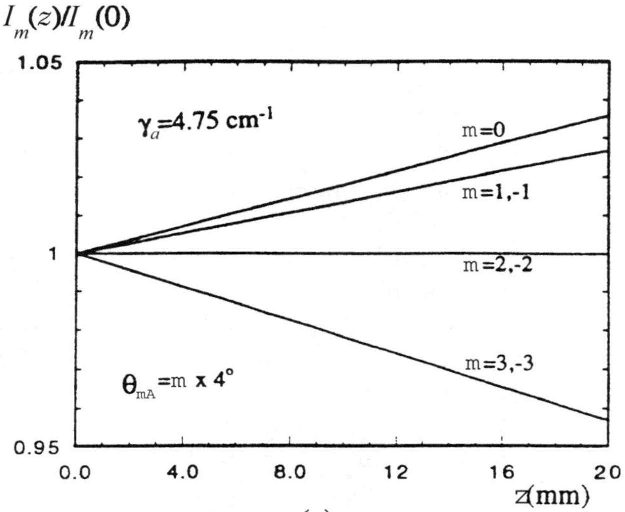

(a)

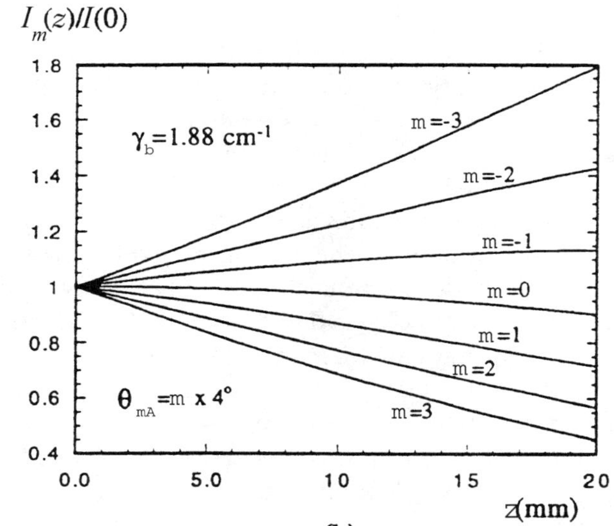

(b)

Figure 5.11: Intrasignal energy transfer as a function of crystal thickness. (a) a-axis orientation. (b) b-axis orientation. (Source: D. Zhao, et al., "Anisotropic intrasignal coupling in photoreflective LiNbO3," *Microwave and Opt. Tech. Lett.*, © 1994 John Wiley & Sons, Inc. Reprinted by permission of John Wiley & Sons, Inc.)

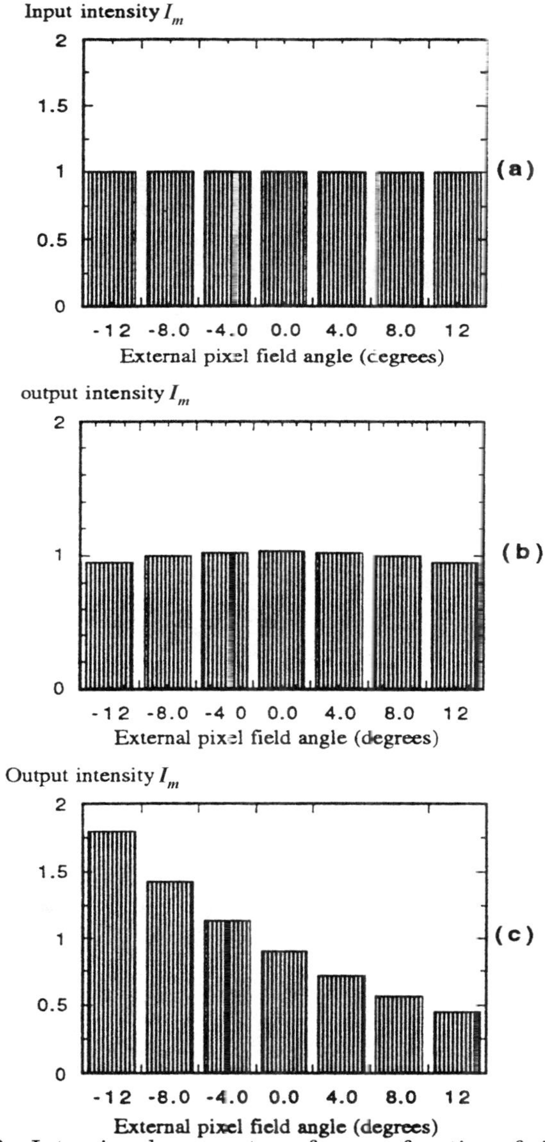

Figure 5.12: Intrasignal energy transfer as a function of signal beam field angle. (a) Input. (b) and (c) Output for a-axis orientation and b-axis orientation, respectively. (Source: D. Zhao, et al., "Anisotropic intrasignal coupling in photorefletive LiNbO3," *Microwave and Opt. Tech. Lett.*, © 1994 John Wiley & Sons, Inc. Reprinted by permission of John Wiley & Sons, Inc.)

the erasure effect, the refractive index grating anisotropy, and the combined feature of (intrasignal) beam coupling and crystal anisotropy. We have shown that the nonlinear beam coupling during a PR hologram recording will cause the maximum diffraction efficiency decreased even for a single holographic recording in a lossless PR medium, and that the erasure effect will further rapidly decrease the overall diffraction efficiency when many holograms are multiplexed. The beam coupling effect also has an impact on the Bragg selectivity of a PR hologram (especially at large Bragg deviations) and is believed to be partially responsible for the "apodization phenomenon"of the spectral response of a PR hologram observed by Rakuljic et al. [3, 12]. On the other hand, the anisotropy nature of photorefractive crystals will mainly affect the spatial diffraction characteristics of a PR hologram. For a page-oriented hologram memory system, this means deteriorated image fidelity. Its effects on the multiplexing scheme were also discussed. Because of the anisotropic nature in diffraction efficiency, the wavelength multiplexing scheme should be more advantageous to the angular one. Finally, the combined intrasignal coupling and crystal anisotropy effect are examined for PR $LiNbO_3$ crystal. The intrasignal coupling will also affect the spatial diffraction properties of a PR hologram and hence the reconstructed image fidelity by interchanging signal energy between the plane wave components that compose the signal.

Acknowledgment

The authors would like to thank Dr. F. T. S. Yu for the help and encouragement during this work.

References

1. P. Günter and J.-P. Huignard (eds.), *Photorefractive Materials and Their Applications,* Vols I and II, Springer-Verlag, New York, 1988.
2. F. T. S. Yu and S. Jutamulia, in *Optical Signal Processing, Computing, and Neural Network*, Wiley, New York, 1992, pp. 249–286.
3. G. A. Rakuljic and V. Leyva, "Volume holographic narrowband optical filter," *Opt. Lett.,* **18**, 459–461, 1993.

4. N. V. Kukhtarev, V. B. Markov, S. G. Odulov, M. S. Soskin, and V. L. Vinetskii, "Holographic storage in electro-optic crystal," *Ferroelectrics,* **22**, 949–964, 1979.
5. P. Yeh, "Two-wave mixing in nonlinear media," *IEEE J. Quantum Electron.,* **QE–23**, 484–519, 1989.
6. D. L. Staebler and J. J. Amodei, "Coupled-wave analysis of holographic storage in $LiNbO_3$," *J. Appl. Phys.,* **43**, 1042–1049, 1972.
7. D. W. Vahey, "A nonlinear coupled-wave theory of holographic storage in ferroelectric materials," *J. Appl. Phys.,* **46**, 3510–3515, 1975.
8. E. S. Maniloff and K. M. Johnson, " Maximized photorefractive holographic storage," *J. Appl. Phys.,* **70**, 4702–4707, 1991.
9. H. Zhou, F. Zhao, and F. T. S. Yu, "Effects of recording-erasure dynamics on storage capacity of a wavelength-multiplexed reflection-type photorefractive hologram," *Appl. Opt.,* **33**, 4339–4344, 1994.
10. H. Kogelnik, "Coupled wave theory for thick holograms," *Bell Sys. Tech. J.,* **48**, 2909–2947, 1969.
11. H. Zhou, F. Zhao, and F. T. S. Yu, 'Diffraction properties of reflection photorefractive volume hologram," *Appl. Opt.,* **33**, 4345–4352, 1994, *Appl. Opt,* **34**, 2593–2594, 1995.
12. G. A. Rakuljic, V. Leyva, and A. Yariv, " Optical data storage by using orthogonal wavelength-multiplexed volume holograms," *Opt. Lett.,* **17**, 1471–1473, 1992.
13. H. Zhou, F. Zhao, and F. T. S. Yu, "Angle-dependent diffraction in a thick photorefractive hologram, " *Appl. Opt.,* **34**, 1303–1309, 1995.
14. A. Yariv and P. Yeh, in *Optical Waves in Crystal,* Wiley, New York, 1984, pp. 232–233.
15. Y. Fainman, E. Klancnik, and S. H. Lee, "Optimal coherent image amplification by two-wave coupling in photorefractive $BaTiO_3$," *Opt. Eng.,* **25**, 228–234, 1986.
16. F. Mok, "Angle-multiplexed storage of 5000 holograms in lithium niobate," *Opt. Lett.,* **18**, 915–917, 1993.
17. F. Zhao, H. Zhou, S. Yin, and F. T. S. Yu, "Wavelength-multiplexed holographic storage by using the minimum wavelength channel separation in a photorefractive crystal fiber," *Opt. Comm.,* **109**, 59–63, 1993.
18. D. Zhao, H. Zhou, F. Zhao, and F. T. S. Yu, "Anisotropic intrasignal coupling in photorefractive LiNbO3," *Microwave and Opt. Tech. Lett.,* **7**, 483–486, 1994.
19. B. Fischer and M. Segev, "Photorefractive waveguide and nonlinear mode coupling effects," *Appl. Phys. Lett.,* **54**, 684–686, 1989.

Chapter 6

Holographic Memory Systems Using Photorefractive Materials

Andrei L. Mikaelian

Institute of Optical Neural Technologies, Russian Academy of Sciences, Moscow, Russia

Abstract

The chapter discusses the applications of holographic memory systems and the requirements for their parameters with regard to the progress in computer science and communications.

The methods of data recording in series or overlapping holograms are considered. Various factors limiting storage density are considered. The development of the first high-speed random access holographic data storage system (HDSS) is described. The characteristics of that system serve as a basis for evaluation of the limiting parameters of holographic data storage. The process of holographic data recording/reading is modeled, and the possibility of using the superresolution approach for increasing storage density is shown.

PHOTOREFRACTIVE OPTICS
Materials, Properties and Applications

ISBN 0-12-774810-5/$30.00

The parameters of rewritable HDSSs using photorefractive materials are considered. The problem of nondestructive reconstruction of holograms is discussed.

HDSSs using one-dimensional (1-D) holograms and their possible applications are considered.

A three-diminsional (3-D) multilayer holographic carrier is investigated. It is shown that due to interference noise the storage density of volume carriers can hardly be higher than 10^{10} bit/cm^3.

6.1 Introduction

The development of data processing and transmission systems and the problem of data storage have always been closely connected. Now there is an understanding that the lack of adequate memory systems can retard notably the advance of new information technologies. Current winchesters and compact disks (CDs) can store gigabytes of data with storage density as high as 10^{10} bit/cm^2.* Yet the bit rate (several megabits per second) and mean access times (milliseconds for hard disks and tens of milliseconds for CDs) often cannot provide efficient computations and suggest increasing.

Another aspect is constantly swelling software — modern programs gorge on hard disk (HD) space.

Different methods can be offered to increase the capacity of HDs and CDs. The use of diode lasers with shorter wavelength in CD drives (e.g., blue light with $\lambda \sim 0.4$ μm) is the most evident trick of increasing storage density. Another, more sophisticated, way is the application of near-field optics, which can give pit sizes 10 times smaller as in current DVD disks and the storage density 10 to 100 times higher. Three-dimensional storage can also be mentioned here.

All these methods employ successive (bit-by-bit) data write/read techniques which cannot provide peak rates. The use of parallelism on the hardware level (e.g., disk packs) gives a certain increase in data rates, though this approach has its own limitations.

Attempts to fully realize parallel reading and writing have inspired a renewed interest in holographic data storage. This can provide parallel

---

*In particular, four-layer, double-side digital video disks (DVDs) can now carry 17 GB of information, which corresponds to the density mentioned in the preceding discussion. The use of diamond coating and the reduced head–disk surface gap promise the same storage capacity for hard disks in the near future.

data reading at about several Gbit/s. Small-sized holographic memory systems could be used even in personal computers. Future systems are likely to be of the read-only type and have the following characteristics: 50 GB of storage capacity, access time of 0.1–1 ms, and a reading rate of no less than 100 Mbit/s. It is natural that the future will require them to become rewritable and provide better performance parameters.

The intensive introduction of fiber-optic communication lines may also call for holographic memory. The performance of current single-fiber lines can provide data transmission at 2.5 Gbit/s. The latest research in the field promises terabit rates. The constantly growing network of fiber-optic lines indicates that in time it will be available to each user. As a result, the problem of prompt retrieval of information from the Internet will become still more important. Most reference sources do not need changing often and require large-capacity memory and high data reading rates. What prediction can be made about such data storage systems for the next 10–20 years? Most likely they will have terabits of storage capacity, read-out rates of tens of Gbit/s, and access times of milliseconds to tens of milliseconds (depending on the specific use). As mentioned in [1, 2], holographic random-access memory can provide these characteristics.

Another promising field of application is neural networks. Holographic memory has a unique power of associative reading, which provides a means for automating pattern recognition and retrieval. For example, let a holographic data storage system (HDSS) contain holograms of reference patterns recorded with its own reference beam. When unknown data is fed into the holographic neural net, the intensities of restored beams will be proportional to the amplitude of correlation between the input information and reference patterns. By using a detector array in the correlation plane and an appropriate restriction on intensity, we can select the maximal signal and reconstruct the necessary pattern. A few successive repetitions of the process allow us to supress noises and perform associative pattern retrieval.

We see that the rapid development of global information systems, the growing requirements to data processing and transmission rates, and the idea of holographic neural networks have regenerated interest in holographic memory and resulted in the resumption of research in this field in the early '90s.

The researches on holographic random access data storage systems (HRADSS) and related optoelectronic components and materials were started in Russia in the mid-'60s [1–8]. The HRADSS was made by the middle '70s. Those first samples took 1 μs to find and read out 10^4 bits of data (this corresponds to a data rate of about 10 Gbit/s). The systems were tested as part of a computing system and showed high reliability [17-18]. Similar research efforts were launched in the United States [13–15] and a bit later in Germany [16]. However they did not get much support and stopped by the late '70s.

This chapter discusses the principles of HRADSS construction, their characteristics, and the possibility of attaining large storage capacity, high data rates, and high reliability. Particular attention is given to the problem of readout noise. Photorefractive materials are considered as promising storage media allowing multiple data recording and reading. I also think that readers will be interested to know more about the first holographic memory system that we developed in cooperation with some research and industrial organizations.

6.2 Data storage density of two-dimensional holograms

The storage density is the most important parameter of any holographic memory system. The greatest possible density is limited by diffraction effects and can be written as

$$\rho \cong \pi / \lambda^2 \qquad (6.1)$$

If we take $\lambda = 630$ nm, we get $\rho \cong 7.5 \times 10^8$ bit/cm^2. This value does not differ from the storage density of systems based on the "bit-by-bit" recording method and is noticeably less than the density of systems using near-field optics. Yet the possibility of parallel recording and reading of large bodies of information allows holographic memory to provide high data rates. Of course, because of various factors (noises, optics limitations, etc.), the actual value of storage density is much less than the theoretical one.

The input information for a HDSS is binary numbers where light spots represent units and dark spots zeros. Two-dimensional controllable transparencies are used for such conversion. Usually a transparency carries 10^4–10^6 bits of data and measures no greater than 2 × 2 cm. The transparencies are recorded either with overlapping by using different reference beams or as an array of spatially separated microholograms.

The overlapping method utilizes the whole area or volume of the media. The other way involves less noise. The combination of the two methods gives the best result. In this case, the data carrier will be divided into small regions each of which contains tens of overlapping microholograms. By way of example, Fig. 6.1 shows a photograph of a flat holographic media that carries 32×32 microholograms recorded with a 1.5-mm spacing.

It should be noted that the storage density and, therefore, the storage capacity are the same for both recording methods. However, this is true only in theory — when noise factors are not taken into consideration. Noise can have various sources: inhomogeneity of media and optics, unwanted reflections, etc. Among the noise sources that cannot be eliminated is the discrete structure of light-sensitive materials (photoemulsion grains, magnetic domains, photochrome molecules, etc.).

Figure 6.1: Array of spatially separated microholograms.

6.3 The effect of noise on storage density

Let us consider a medium that has N light-sensitive elements. The recording of a transparency with M bits of information takes n such elements (they are located at the maxima of the interference pattern). In reading, these elements produce fields that add up to give the light intensity proportional to n^2. For each bit in the reconstructed pattern there is n^2/N of the whole intensity.

On the other hand, the granularity of materials brings about the scattering of light with a random phase distribution. The intensity of the scattered light is proportional to N. The ratio between the signal and background light intensities is

$$\frac{P_S}{P_N} = \frac{n^2}{NM} = \frac{N}{L^2M} = \frac{N}{CL} \tag{6.2}$$

where L is the number of overlaps of holograms and $C = LM$ is the total storage capacity.

It is clear that the intensity of reconstructed patterns falls as $1/L$. Consequently, to reduce noise, the whole amount of information should be recorded in the hologram at once.

From Eq. (6.2) it also follows that the storage density of flat holograms is

$$\rho = \frac{N/S_h}{P_S/P_N}\frac{1}{L} \tag{6.3}$$

and depends on the number of overlaps. Here, S_h is the hologram area. The correctness of this formula was supported in the experiments where 600 holograms were overlapped using different reference beams [12].

6.4 The role of optics in the realization of high storage density

Optical systems determine the storage capacity of HRADSS to a large extent. Figure 6.2 illustrates this. The microholograms are recorded with a converging light wave which goes through the transparency. The image of the transparency is reconstructed in the plane a distance F away from

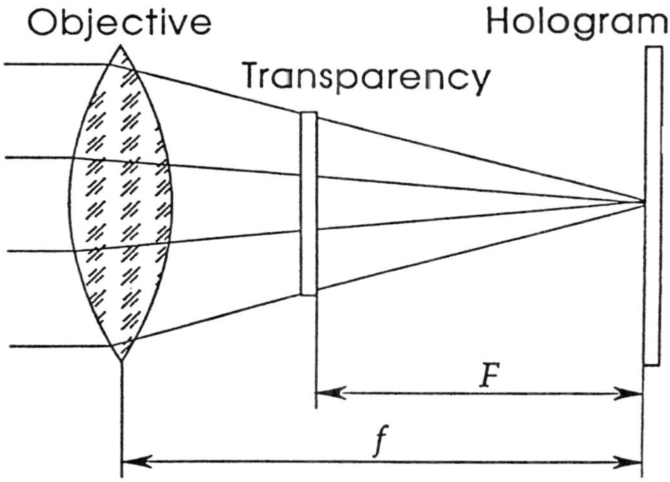

Figure 6.2: Holographic recording with a focused beam.

the hologram, which is supposed to be a square with each side measuring d_h. It is obvious that we cannot resolve elements finer than

$$\delta = 2\lambda F / d_h ,$$

i.e., the number of bits in the reconstruction cannot be greater than

$$M_{bit} = \left(\frac{D_{tr}}{\delta}\right)^2 = \frac{D_{tr}}{F}\frac{d_h^2}{4\lambda^2} \tag{6.4}$$

where D_{tr} is the transparency size. The storage density of the microhologram can thus be written as

$$\rho \cong \frac{1}{4\lambda^2}\left(\frac{D_{tr}}{F}\right)^2 = \frac{1}{8\lambda^2}\left(\frac{D}{F}\right)^2 \tag{6.5}$$

where D is the diameter of the objective in which the transparency is confined. For $D/F = 1$ and $\lambda = 0.63$ µm we get

$$\rho \cong 3 \cdot 10^7 \,\text{bit}/\text{cm}^2 \tag{6.6}$$

This value is considerably less than the limiting density defined by Eq. (6.1). We can thus increase the density by recording several transparencies in each microhologram using different reference beams.

Now we can conclude that if a flat light-sensitive medium (or a volume multilayer carrier) is used, the best holographic memory design is an array of microholograms, each of which contains several transparencies recorded at different angles. This design allows the data capacity of about 10^{10} bits, data rates of about 10 Gbit/s, and access times of a few microseconds [3, 26]. The typical array in this case will have 64 × 64 or 128 × 128 microholograms, each hologram occupying an area of about 1 mm^2 and carrying 10^5–10^6 bits of information (the last figure means five to seven overlaps) [20, 26]. These values are taken from the experiments with the first holographic memory systems, which are described.

6.5 Holographic random access data storage system

As we mentioned in the Introduction, studies in the field of holographic memory were started in Russia in the middle '60s [1–12]. The theory and limiting characteristics of holographic data storage were elaborated by that time. Various holographic memory systems and new components were developed and tested. In the late '60s all these served as a basis for setting the task of developing a holographic data storage system. The goal was a random access, read-only system capable of finding and reading a 10^4-bit data file in a time as short as 1 μs (which corresponds to a data rate of 10 Gbit/s). The following readout of a 36- or 512-bit digit from the file was meant to take the same time. Since the system was intended as a peripheral unit of mobile computing complexes, it was necessary to have a low error rate (10^{-11}–10^{-12}), small dimensions, and high resistance to vibration, temperature variations, etc. It was also proposed to investigate the possibility of increasing storage capacity to 10^{10}–10^{11} bits and decreasing the access time to 0.1 μs.

It was obvious that to provide a compact and efficient design, only low-power lasers (a few milliwatts of output power) had to be used. Hence, there was the problem of detection of weak signals. We solved it by developing a combined photoelectric detector which could reliably detect 10^{-14}–10^{-15} J per single data bit [5].

Because of fairly slow performance of acoustooptical deflectors, an important decision was made to develop and use electrooptical deflectors [11]. There was close cooperation with a number of research organizations. They were to grow ultrapure lithium niobate crystals for electrooptic deflectors, develop an optoelectronic image converter-amplifier for selecting digits from microholograms, and fabricate special fiber and optical elements and other important components.

Our team was also engaged in developing a holographic system for recording a 32 × 32 array of microholograms, each hologram storing 10^4 bits of data.

In the middle '70s, the research work resulted in making the first samples of HRADSS. One of the samples was tested as a read-only memory unit of a computing system and showed reliable performance [17, 18].

Figure 6.3 explains the HDSS operation. Figure 6.4 shows the photograph of the system without the upper cover. The device measures 156 × 60 × 44 cm^3. Its bottom part contains a power supply unit. The memory component is a cassette incorporating a 5 × 5-cm holographic plate with a 32 × 32 microhologram array (Fig. 6.1). The cassette mount provides quick (for 1 min) and accurate (to ±10 μm) installation. Each microhologram 1.35 mm in diameter has 256 36-digit numbers (32 data bits and 4 parity control bits). The information carrier is a glass substrate covered with a special, high-resolution low-noise holographic emulsion [7]. A special bleaching technique was developed to ensure stable optical properties.

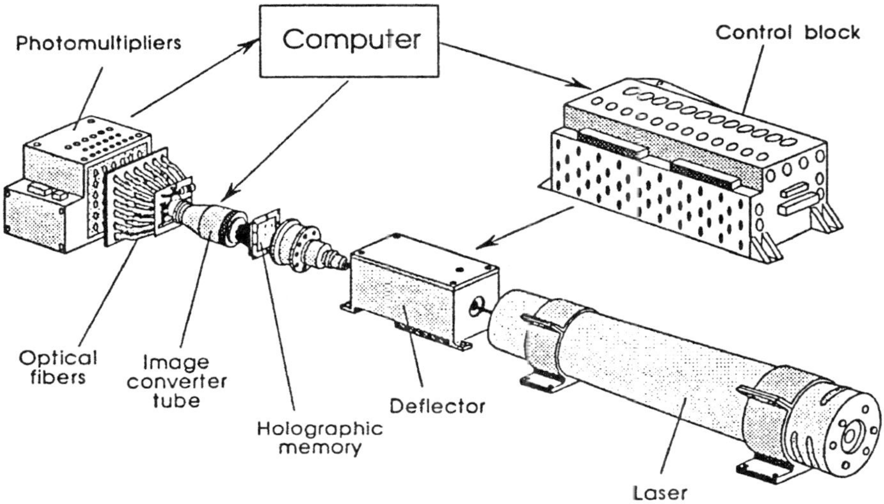

Figure 6.3: Schematics of the holographic read-out system.

Figure 6.4: Photograph of the experimental HRADSS.

This gave a 1000-h period of continuous reading from a single hologram. The diffraction efficiency of each microhologram was 8–10%, and the contrast was 20%.

An electrooptic deflector was used to direct the beam of a 8-mW He-Ne laser to any of 1024 holograms. The deflector used a 0-Z cut of a lithium niobate crystal [19–21] and provided a 1-μs switching time and 20–30% transmittance. A photograph of the deflector is shown in Fig. 6.5.

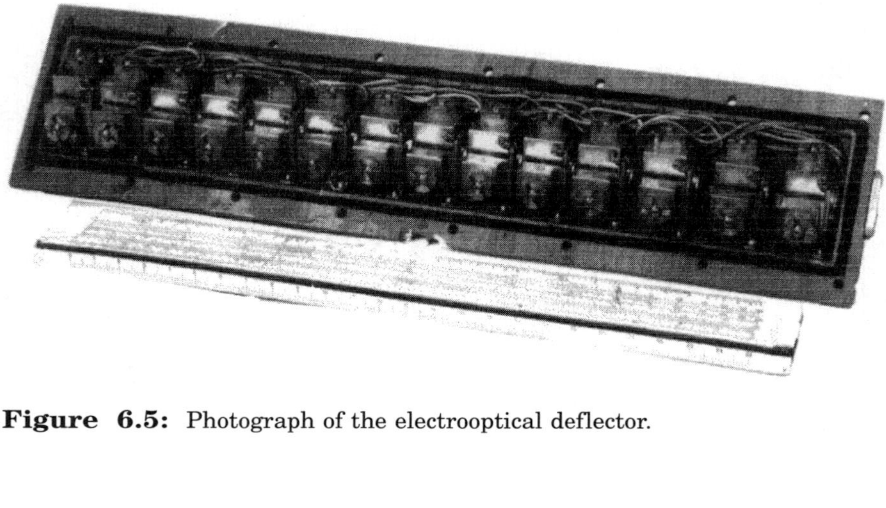

Figure 6.5: Photograph of the electrooptical deflector.

Figure 6.6 shows a beam pattern that was used for addressing all 1024 locations with a 1.6-μs time step. The signal-to-noise ratio in the pattern is over 20 dB. Figure 6.7 shows a photograph of a 64 × 64 beam pattern. The array of 1.35-mm read-out beams generated by the system consisting of the deflector and the lens measured 48 × 48 mm. The separation between the beams was 1.5 mm, the same as the intermicrohologram spacing. The nonparallelism of the beams over the whole array was not greater than ±7 in.

All the holograms give real reconstructions in the same plane, which is located 140 mm from the plate. The reconstruction measures 19.2 × 19.2 mm and consists of units (light spot 0.12 mm in diameter) and zeros spaced 0.4 mm apart. It is fed onto the photocathod of an image converter. After a specific pattern is picked, its image is amplified and goes through a fiber-optic collector to the photodetector array whose output electronics form a 36-digit number. This electronics has integrating and cutoff circuits. The cut-off level for each channel was set to the optimal value to reduce error probability.

32 x 32 = 1024
Transmittance (24—32)%

Figure 6.6: Photograph of the 32 × 32 deflector beam pattern.

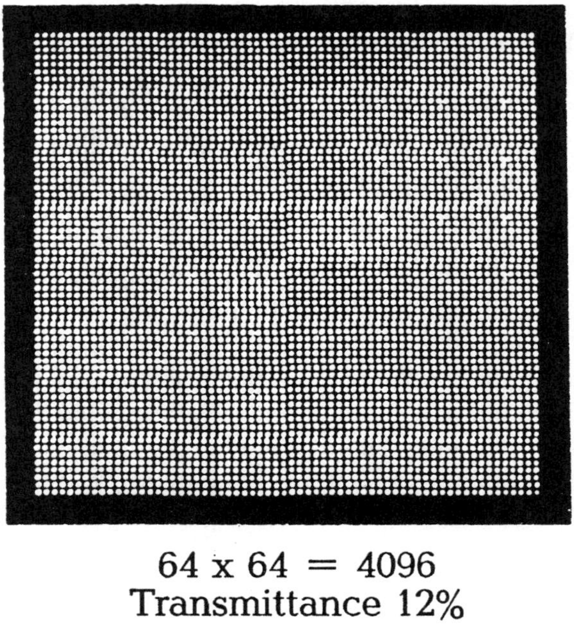

Figure 6.7: Photograph of the 64 × 64 deflector beam pattern.

Additional systems were developed to prevent an undesirable decrease in the ratio between the smallest unit-spot and greatest zero-spot intensities. A laser beam stabilization system kept the angle of incidence of the beam to within ±3.6 in. and positioned the beam on a microhologram with a precision of ±10 μm. A special system automatically adjusted the sensitivity of each detector channel in a time that makes only 0.005% of the working cycle. An accumulation time control system alleviated the effects of beam intensity instability, the variations of the deflector transfer factor and microhologram diffraction efficiency. These and some other systems allowed the departure of signal amplitude from the optimal value to not exceed ±20%.

Correcting codes were used to prevent fixed errors (e.g., dust particles in the photoemulsion, the imperfection of optics, etc.) [22]. The analysis showed that the reading of multidigit numbers gives rise to no more than one fixed error. For this reason the simple Hamming method was employed (this method can also detect double errors). The recheck of erroneous addresses was also provided in the system. The error probability was calculated to be $1.6 \cdot 10^{-15}$ in this case. The system was tested for failures in different operation modes for 7 hours. The absence of failures suggested that the error probability during the read-out process is lower than 10^{-10}.

In summary, we can say that a new type of memory — holographic random access data storage system — was created. The system enabled large bodies of information to be read at a high rate (up to 10 Gbit/s).

It should be noted that during the research we developed a number of new optoelectronic devices whose performance in certain aspects has no equal even today. Among them are the lithium niobade deflector, the multichannel photoelectric detector, and magnetooptical yttrium ferrite transparencies [23, 24] for data recording. For the first time a new concept of using unstable resonators for mode selection was offered [25] and applied for making a single-mode He-Ne laser with a Gaussian beam. An original technique to suppress piezoresonances in lithium niobate crystals, and thus to decrease the switching time of the deflector to 0.8 μs was developed.

Along with the work on the special-purpose system we continued an active search for methods to increase storage capacity of the random access memory. One idea was to use volume holographic materials and the overlapping recording technique. The other was to increase the number of spatial light modulator (SLM) elements and, therefore, the dimension of the beam array.

The experimental evidence showed that it was possible to increase the dimension of the output beam array of the electrooptic deflector to 128×128 [20, 21] and make the SLM carry as many as 10^5 bits [26]. The use of volume holographic carrier and overlapping recording could allow as many as 100 patterns to be stored in a single microhologram [12]. Thus, it was possible to raise the capacity of the holographic storage to 10^{10}–10^{11} bits with retention of the access time within a microsecond range [3, 10]. Further increase of storage capacity to 10^{12}–10^{13} bits involves the use of stack memory design and electromechanical drives. Experimental memory systems like that were tested and showed an access time of 2–3 s [35].

The access time of HDSSs is determined by the performance of an electrooptic deflector. Our investigations [20, 26] showed that if the access time is less than 1 μs, lithium niobate begins to suffer considerable losses. We proposed the use of a semiconductor laser array instead of the deflector. The research lent credence to this approach [27]. To imitate the work of a large laser array, an assembly of nine lasers was incorporated in the experimental HDSS. Addressing a particular microhologram was performed by applying the excitation current to a corresponding laser. The assembly was combined with a lens array. Signals were detected with a Photodetector Array (PDA) whose sensitivity was 10^{-12} J.

In the 20 years past since the development of our HDSS great progress has been made in optical engineering. New laser systems and high-performance optoelectronic devices have been developed. This can dramatically improve the characteristics of holographic memory systems. The creation of reliable high-capacity, high-speed HDSSs begins to take on a new meaning with the rapid development of fiber-optic means of communications (which today can operate with gigabit data flows). When incorporated into global information networks, read-only fast-access HDSSs can be effectively employed as large data archives.

6.6 Suppression of interference noise by optimizing spatial spectra of two-dimensional holograms

The system we described above had an access time of about 1 μs and a peak bit rate of 10 Gbit/s. At the same time the storage density of the 5 × 5-cm holographic memory carrier was no greater than 10^6 bit/cm^2. As was mentioned in the preceding, this fact is connected with the restrictions of the optical system and the noises that arise during the recording and writing of holograms.

Holographic recording and reading of data transparencies are accompanied by noise caused by the interference of signal beams. This kind of noise together with others lowers the contrast of reconstructed patterns (i.e., the ratio of the smallest intensity of "unit" spots to the greatest intensity of "zero" spots) at the detector input.

We found a way to suppress the interference noise. Further studies showed that it is possible to surpass the density limit determined by diffraction effects and realize the superresolution recording and reading of data transparencies. Before explaining how to do this, let us consider the nature of interference noise.

Figure 6.8 shows the schematic of the recording/reading of two-dimensional Fourier holograms, which is taken as a basis for modeling. A spatial

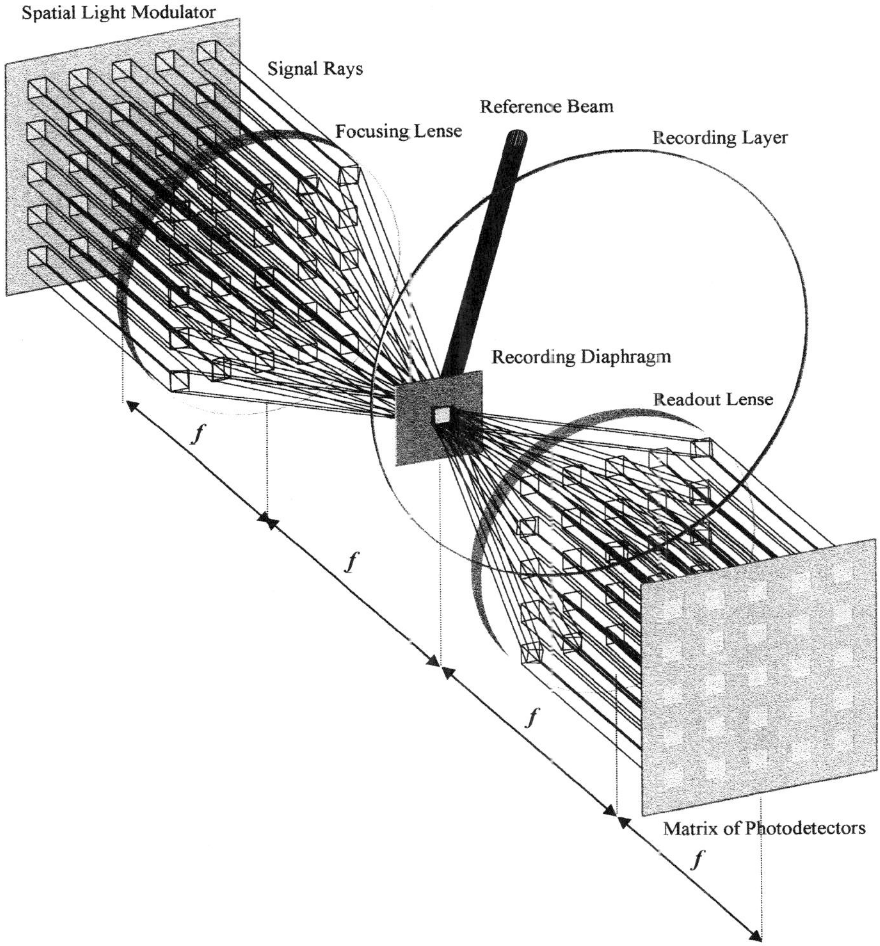

Figure 6.8: Schematic drawing of the recording/reading of 2D Fourier holograms.

light modulator (SLM) with square cells of binary transparencies is used to modulate the light in the signal beam. Transparent cells represent units, opaque cells stand for zeros. The lens focuses the light that passes through the SLM onto the holographic carrier. The diaphragm in the focal plane is used to control the spatial spectrum. Another lens focuses the reconstructions onto the 512×512 detector array whose square cells are identical to those of the SLM. Providing that the light intensity in each SLM cell has a uniform distribution (Fig. 6.9a), for the field in the diaphragm plane we have (Fig. 6.9b):

$$E(x, y) \sim \frac{\sin x}{x}\frac{\sin y}{y} \tag{6.7}$$

Figure 6.9c gives the field distribution immediately behind the diaphragm in the carrier plane and represents the cutting of all diffraction orders other than zero orders. The intensity profile on a detector cell (Fig. 6.9d) shows "tails" whose shape is different from $\sin x / x$. Therefore, it is possible to vignette the angular spectrum of the image so that the zero intensities of the noise tails of neigboring channels will be roughly at the detector cell centers and their contribution to the interference noise will fall to the minimum. This will give better "1"$_{min}$ / "0"$_{max}$ contrast values at the detector input and increase the reliability of data reading.

It is clear that contrast also depends on the distribution of zero and unit spots in the transparency. The worst case is when "1"$_{min}$ corresponds to the transparency with a unit spot surrounded by zeros and "0"$_{max}$ corresponds to the dark cell in the SLM central channel surrounded by bright unit spots.

Of course, the vignetting of the spatial spectrum changes the storage density. This fact calls for optimization: the aim is to obtain the maximum storage capacity for a specific contrast of the reconstruction.

Fast Fourier transform algorithms were used for modelling. The numerical aperture of the objective was 0.85 (which corresponds to $D / F \cong 3$), and the wavelength was 630 nm. The SLM and PDA had 512 $\times$ 512 geometrically identical cells.

In reading, the signal distribution at the photodetector input can be written as

$$E_{PDA} \sim FT^{-1}\{W(x, y, c)\} * FT\{E_{MOD}((x, y)\} \tag{6.8}$$

where $W(x, y, c)$ describes the vignetting of the spatial spectrum, and $E_{MOD}(x, y)$ is the signal distribution in the SLM cell.

In terms of spectral analysis, *W(x,y,c)* is the window function that performs the implicit weighing of input data. This function mostly determines the drop of the amplitude and the period of signal tails at the detector input. Since $W(x, y, c)$ is usually complex, the spatial spectra of 2-D Fourier holograms can be optimized in the focal plane either by placing a diaphragm or by varying the phase distribution of the field. We consider

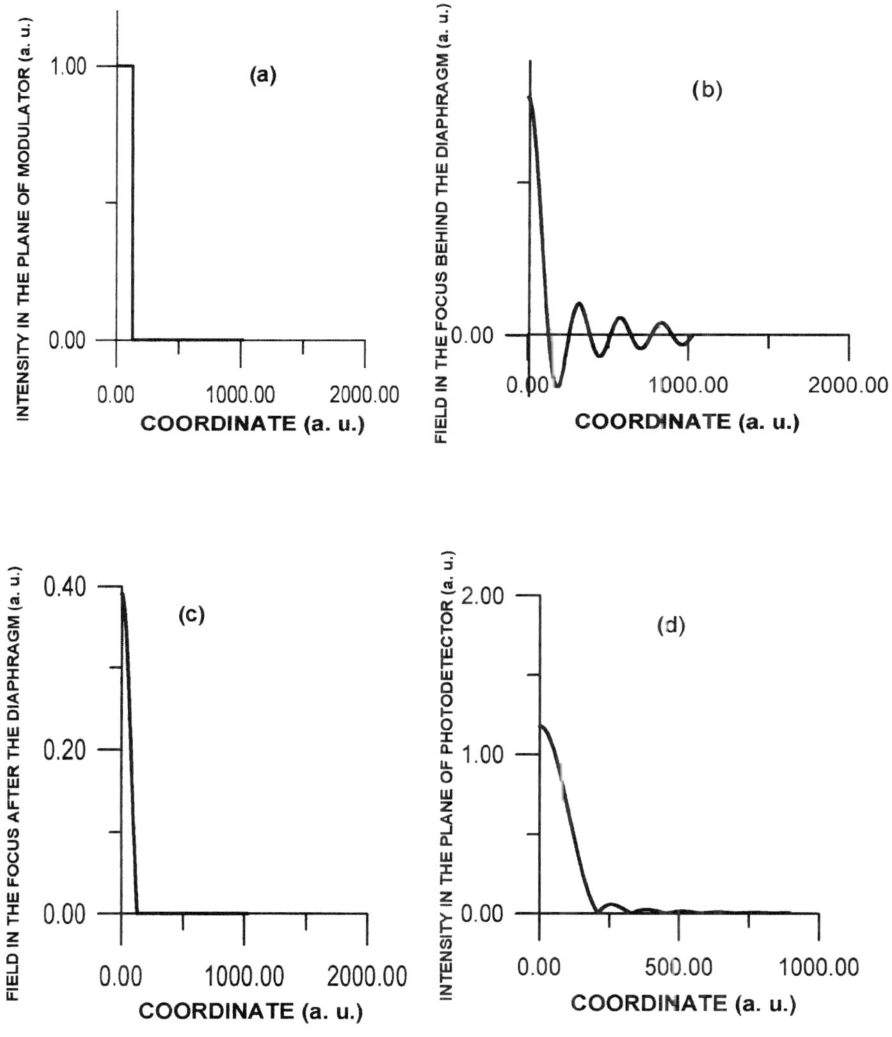

Figure 6.9: (a) Intensity distribution in the SLM plane. (b) Intensity distribution in the diaphragm plane. (c) Intensity distribution immediately behind the diaphragm. (d) Intensity distribution in the detector array plane.

the simplest case when $W(x, y, c)$ is not zero within a given rectangle and zero outside it. Parameter c determines the degree of vignetting and is the ratio of the size of the diaphragm opening to the spectrum dimension measured as the distance between the first minima ($c = 1$ when the diaphragm edges coincide with the first minima). The aim is to investigate how the contrast (the "1"$_{min}$/"0"$_{max}$ ratio) depends on parameter c and to determine the optimal value c_{opt}. Since c_{opt} and the storage density of 2-D Fourier holograms are interconnected, it is necessary to find the best relation between the density and hologram size for a given "1"$_{min}$/"0"$_{max}$ ratio.

6.7 Superresolution approach for increasing storage density

When the microhologram size is cut at the first diffraction minima and the square transparency fits into the objective aperture of size A, the storage density can be written as

$$\rho = \frac{A^2}{1 - A^2}\frac{1}{8\lambda^2} = \frac{1}{32\lambda^2}\left(\frac{D}{F}\right)^2. \tag{6.9}$$

This formula differs from Eq. (6.5) in that the intercell distance is taken equal to the cell size. When $D/F = 3$, $\lambda = 0.63\ \mu\text{m}$, $\rho \cong 7.5 \cdot 10^7$ bit/cm^2. When the spatial spectrum of the transparency is cut according to c_{opt}, the density grows considerably:

$$\rho_{max} = \rho/(c_{opt})^2 = 1.2 \cdot 10^9\,\text{bit/cm}^2 \tag{6.10}$$

This is an example of superresolution we mentioned in the preceding discussion.

Figure 6.10 shows the result of mathematical calculations. The curve gives the relation between the storage density and the contrast of the transparency reconstruction. The curve exhibits the maximum which corresponds to $c = c_{opt} \sim 0.24$–0.25. As was mentioned in the preceding discussion, $c = c_{opt}$ determines the optimal vignetting of the spatial spectrum of the transparency, in which case the minima of diffraction tails of neighboring channels roughly coincide with the centers of PDA cells.

Figure 6.11 shows the effect of PDA cell size. Curve 3 corresponds to the case when the detector cell is much smaller than the modulator cell.

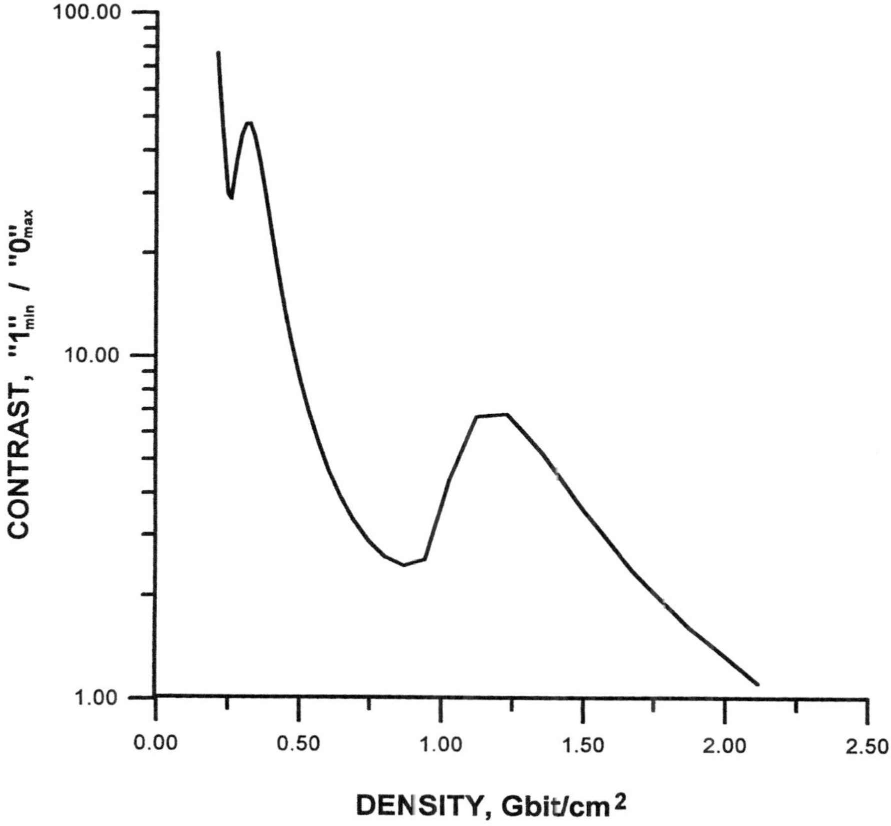

Figure 6.10: The calculated relationship between storage density and reconstruction contrast (“1” min/“0” max ratio).

Curve 2 is for the detector cell which is two times smaller than the modulator cell. The broken line reflects the case of these cells being identical in size. The figure demonstrates that decreasing the PDA cell size can lead to the reconstruction contrast growing to 11. Yet smaller cells require the use of a more powerful laser for the detector to reliably detect smaller amounts of falling light.

Figure 6.12 shows how the failure probability depends on the signal-to-noise ratio at the detector output for different values of contrast. The signal-to-noise ratio for a specific photodetector is determined by the

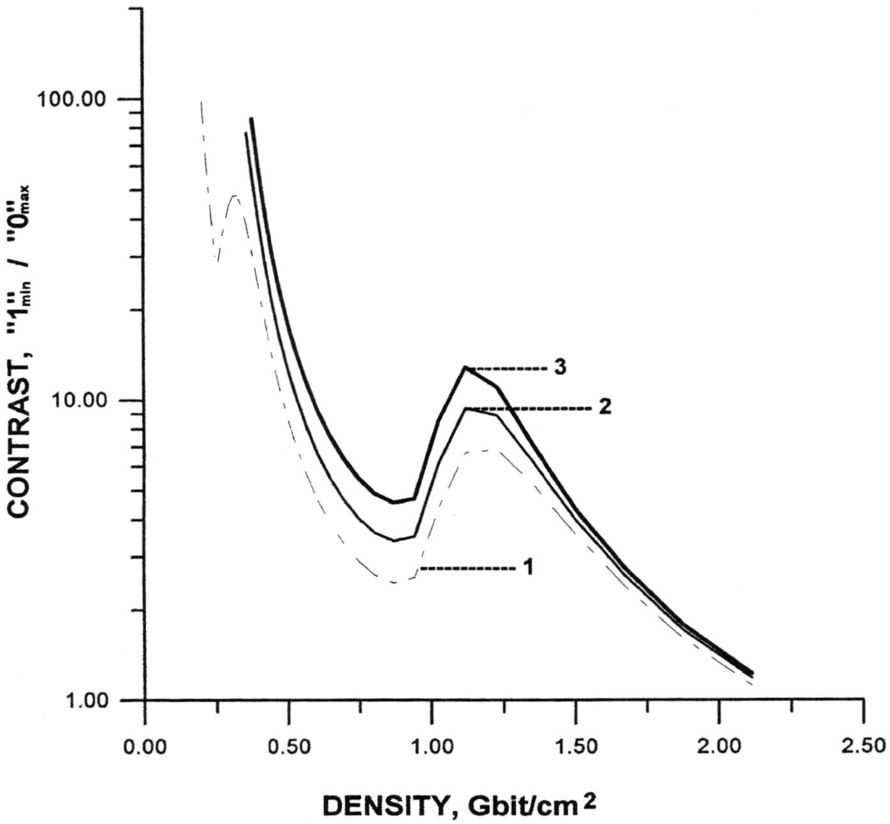

Figure 6.11: The effect of the deflector cell size on the density-contrast dependency.

intensity of "unit" spots; i.e., by the laser power and the loss in the readout channel. For example, if the response time (RC time) of a photomultiplier is about 1 μs and the "unit"-spot intensity is 10^{-10} W, the signal-to-noise ratio at the detector output will be about 10.

It might be well to point out that the superresolution technique allowing storage density higher than the diffraction limit is just the effective way of data handling, which is based on the a priori knowledge of input data (array of cells of binary intensity).

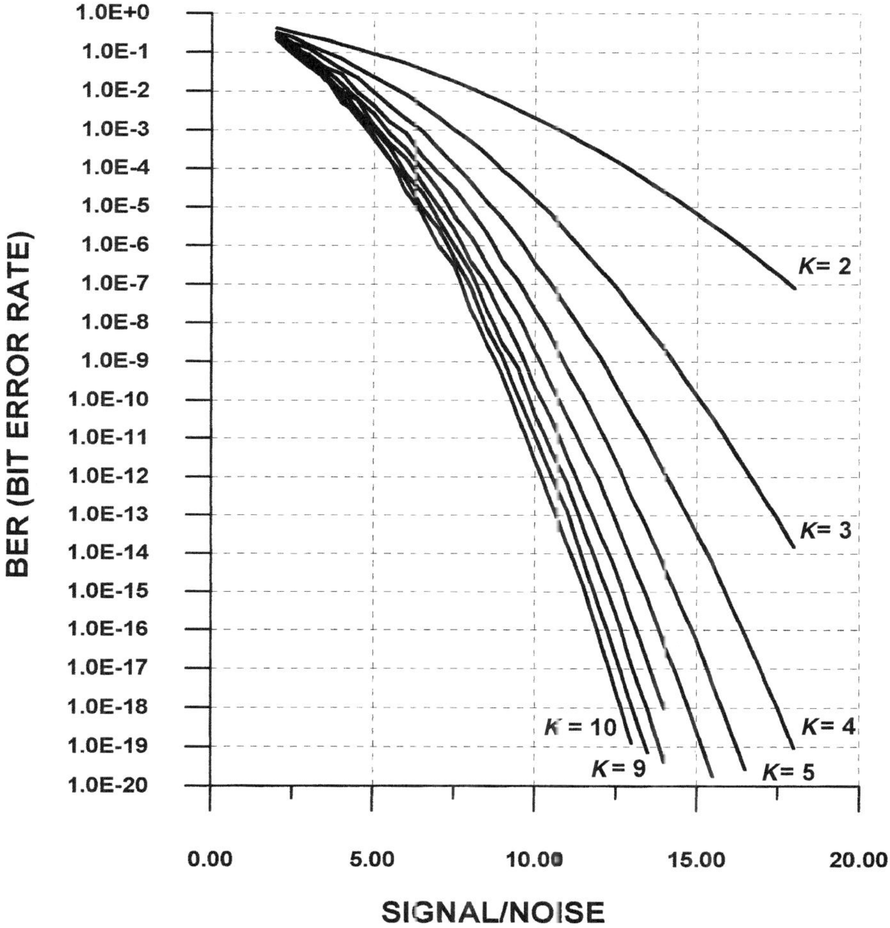

Figure 6.12: Relations between bit error rate and signal-to-noise ratio for various values of contrast.

6.8 Photorefractive materials for rewritable holograms

The search and investigation of new materials allowing the production of rewritable holograms is an important task in the development of holographic memory systems. The first such holographic medium we investigated in 1967 was photochromic film several tens of microns thick [4]. The carrier enabled us to record holograms which could store about 10^7 bit/cm^2 with a low noise level (a 5-mW He-Ne laser was used in the experiments). Yet the diffraction efficiency of those holograms was low because the material permitted the recording of only amplitude holograms.

Some time later we first tried photopolymer materials for recording phase holograms [9]. The photopolymer contained a sensitizer shifting the spectral sensitivity to 0.63 μm area. The thickness of the polymer film was 15 μm. The experiments showed that recording and erasing holograms was possible at low sensitivity (about 0.1–0.01 J/cm^2). We could perform two to three recording cycles.* The possibility of using magnetic films was also considered [8].

Another group of materials we used for multiple recording was photorefractive crystals (lithium niobate, barium strontium, and other crystals with appropriate additives). The experiments with these materials were first reported in the papers by Amodei, Phillips, and Staebler [28, 29] and Russian researchers [30].

The principal difficulty in using $LiNbO_3$:Fe crystals is known to concern their relatively low sensitivity. To raise the sensitivity, we studied the conditions of crystal growth and tried various dopants. Fe-doped lithium niobate with Fe content of 0.02–0.03 proved the best in this respect. The sensitivity at 0.44 μm was 1 J/cm^2 and at 0.63 μm it amounted to several tens of J/cm^2. The sensitivity in the course of data recording and the maximum diffraction efficiency may be different for the same energy level. It depends on the intensity of the recording light (light power density). Figure 6.13 presents curves that characterize the relationship between the diffraction efficiency and the duration of the recording cycle for various output powers of a He-Ne laser ($\lambda = 0.63$). The sensitivity is seen to grow with the power. The curves show also that the growing power brings about an increase in the maximum diffraction efficiency.

An external electric field applied along the *c*-axis of the crystal also has an effect on the recording process. Figure 6.14 shows how the diffraction efficiency varies with the light energy density for various electric-field

*It should be noted that substantial improvement of photopolymer performance has been recently made in the United States (e.g., Du Pont Corporation). This fact gives much hope for the successful use of photopolymer carriers in HDSSs.

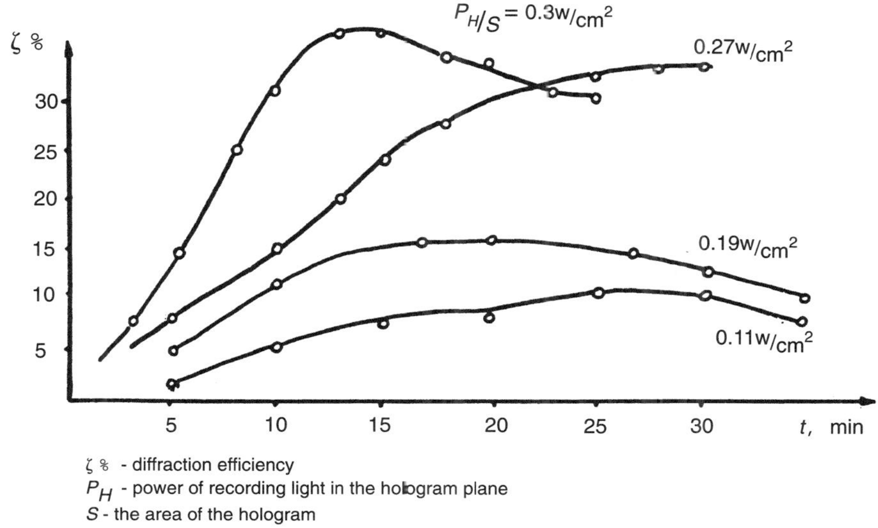

Figure 6.13: Relationship between diffraction efficiency and the recording cycle duration for various He-Ne laser output powers.

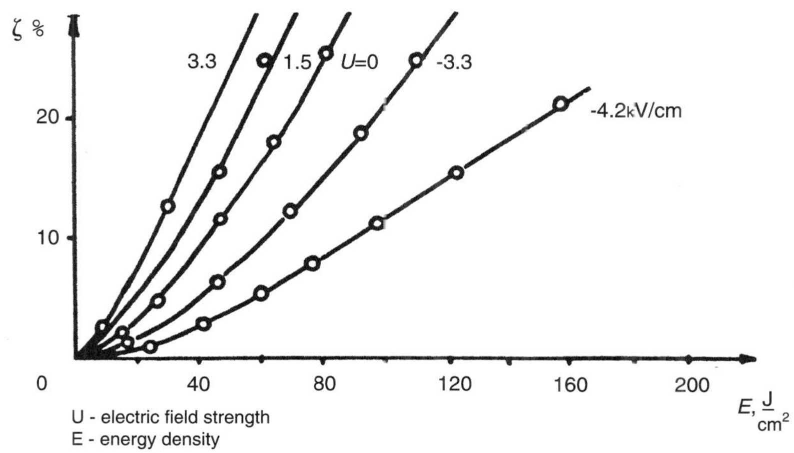

Figure 6.14: Relationship between diffraction effiency and light energy density for various electric-field strengths.

strengths. Two things should be pointed out here. First, the polarization of the applied electric field has a certain effect. Second, this effect depends on the power density of the recording light. No significant effect of the external electric field on the recording process was observed at high light power densities (above 0.2 W/cm^2). It may be attributed to strong (and nonuniform) internal electric fields induced in the crystal by high-intensity light beams. Therefore, the use of either an external electric field or more intensive recording beams is needed to get higher sensitivity and diffraction efficiency.

Holograms recorded in $LiNbO_3$ crystals can be erased either thermally (by heating it to a temperature of 150°C) or optically. No doubt, the latter is preferable because it is faster (as compared with the recording process) and can provide selective erasure. The sensitivity to optical erasure is about one order of magnitude lower than to optical recording. Sufficiently powerful, short-wavelength laser sources (e.g., He-Cd lasers) may be used for erasing. The needed light power density is about 20 J/cm^2.

The storage time for lithium niobate data carriers depends mostly on the type of the crystal used and on the storage conditions [28–30]. The storage time is short if the concentration of Fe^{3+} ions, which work as charge traps, is low. A higher Fe^{2+} ion concentration results in increased absorption and sensitivity. The samples we used in the experiments had the Fe-ion concentration of about 0.02 mass % and could store data for 2–3 months at room temperature. The diffraction efficiency of the samples dropped from 25 to 20% for this period. Natural illumination did not have much effect on the storage time. When cooled to 0°C, the crystal could store information for about a year.

The storage time can be increased substantially by applying the following stabilization technique. The crystal with a hologram is heated to about 75°C for 15–20 min and then cooled down to room temperature for 20–30 min. The heating temperature should be chosen so as not to damage the hologram. Holograms stabilized in this way can be kept for about a year without a decrease in diffraction efficiency.

Repeated readings erase the recorded data. The shorter the wavelength, the faster the erasure. The use of a 0.63-μm laser is therefore preferable for data reading. Due to weak electrooptic effects in the crystal the use of still longer wavelengths of the reading light will cause the diffraction efficiency to fall.

6.9 Holographic memory systems using photorefractive crystals

Figure 6.15 shows the schematic drawing of a rerecordable HDSS [33]. A 12-mW He-Ne laser was used for recording and reading and a 0.44-μm He-Cd laser for erasing (average power of the lasers was 100 mW). The low sensitivity of lithium niobate results in the recording, reading, and erasing of one hologram taking a few seconds. For this reason, it was possible to use an electromechanical device for moving a specific data carrier into the recording channel. The images of 10^4-bit data transparencies were recorded in a 25 × 25 × 1-mm $LiNbO_3$:Fe plate as an array of holograms each of which is about 1 mm in diameter (Fig. 6.16). The horizontal hologram spacing was 1 mm (i.e., holograms were recorded immediately one after another), and the vertical spacing was 1.5 mm. This difference in spacing is due to the fact that the reference and signal beams lie in the horizontal plane and it is in this plane that the hologram exhibits its maximum sensitivity. The signal-reference beam intensity

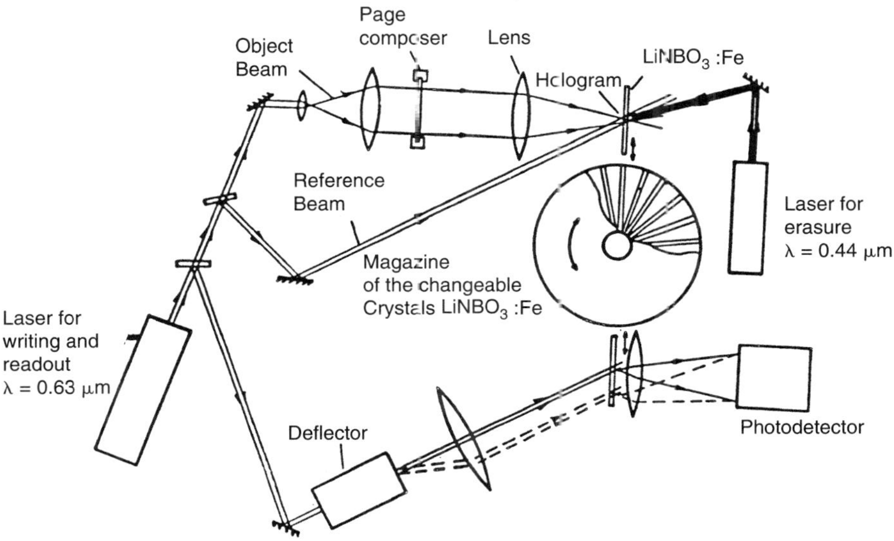

Figure 6.15: Schematic drawing of the rewritable HRADSS using photorefractive crystals.

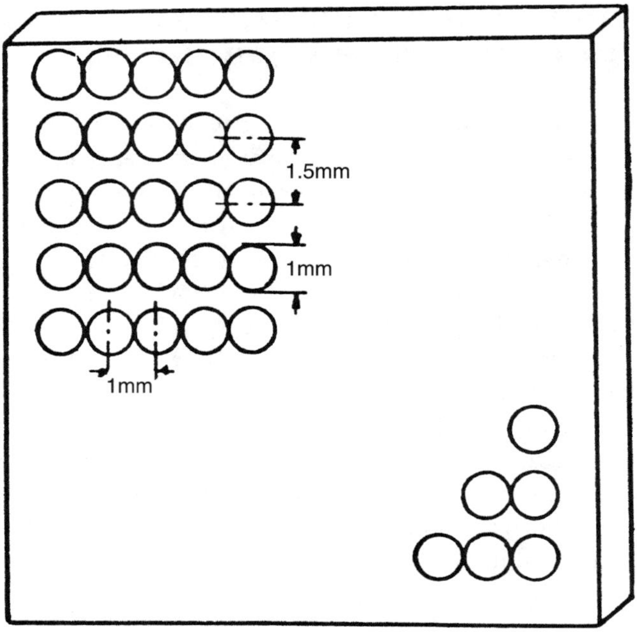

Figure 6.16: Array of microholograms in the photorefractive crystal slab.

ratio was about 1:30, the reference beam intensity being about 2 mW. To obtain the necessary diffraction efficiency of 1.5–2%, the holograms had to be exposed for 2–3 min.

The recording of holograms at the low diffraction efficiency has several advantages. First, the process is fairly fast (with a 50-mW He-Ne laser it takes less than 20 s), which is important for recording large bodies of information. Second, if no hologram stabilization is used, the diffraction efficiency falls exponentially with time [28]. The low value of initial diffraction efficiency in this case means longer storage time. Figure 6.17 shows the diffraction efficiency as a function of storage (a) and reading time (b).

The thermal erasing of holograms in $LiNbO_3$ crystals does not change the holographic properties of the material. Experiments showed that the quality of the reconstructed image after 200 erasings and recordings remains as good as that in the first cycle.

The optical erasing was done with a 0.44-μm He-Cd laser. With a light power density of 10 mW/mm^2 the duration of the erasing cycle was

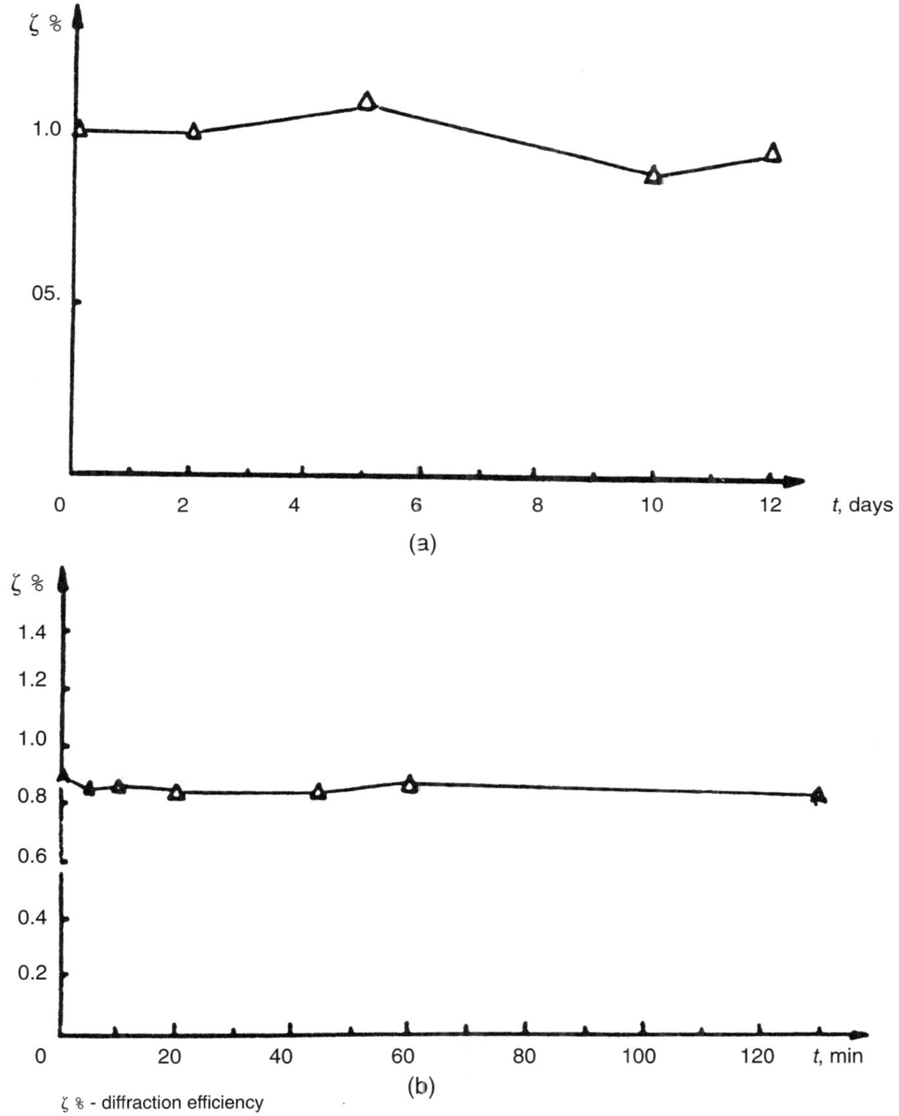

Figure 6.17: Diffraction efficiency as a function of storage (a) and reading (b) time.

found to be about 1 min (it corresponds to 60 J/cm^2). Ten successive recordings and erasings did not change the quality of holographic reconstruction noticeably. With more cycles the quality began to deteriorate because of the noise caused by the nonuniform beam intensity profile.

A magnetic page composer that could generate a binary pattern was used as the object in holographic recording. The composer is a 60-μm orthoferrite platelet deposited on a glass substrate. Both sides of the substrate carry a net-shaped system of conducting stripes. When pulses of current are applied to the appropriate pairs of perpendicular stripes, the magnetic properties of the page composer cells (the intersections of the stripes) change. When properly polarized light passes the page composer, it can be modulated to produce a specific two-dimensional binary intensity pattern.

The magnetic material that made possible the given page composer design has good magnetooptical characteristics. The Faraday rotation/losses ratio of the material is about 15° per dB [24, 33]. A fairly high coercitivity of domain walls of the material provides a hysteresis singularity and switching threshold, which is defined by the coercitivity field H_c. This permits the binary pattern to be generated and stored for almost unlimited time.

The existing state of a composer cell can be changed only when the applied field exceeds the domain wall coercitivity; that is, when the total magnetic field of the given pair of perpendicular stripes is greater than the coercitivity field. The field strength should be 10–50 oe and can be set at a desired value by polishing the orthoferrite with appropriate pressure. As compared with bubble-domain films, the orthoferrite material features higher coercitivity and domain wall mobility (Fig. 6.18). The operation speed of the orthoferrite page composer is about 10^8 bit/s, which allows a 10^4-bit array pattern to be generated in 100 μs. An example of such a pattern is shown in Fig. 6.19. With antireflection coating the transmittance of the transparency reaches 20–25%, the contrast being 300. A photograph of the experimental page composer is shown in Fig. 6.20. Its dimensions are 20 × 20 mm, and the cell spacing is 200 μm.

Experiments proved the possibility of using Fe-doped $LiNbO_3$ crystals for making rerecordable high-density random-access HDSSs. The use of a 50-mW red-light recording laser and 50-mW blue-light erasing laser enables one to achieve a total write-erase time of 30–40 s. Holograms can be stored for some months and read out for some hours without noticeable deterioration of the quality of recorded data.

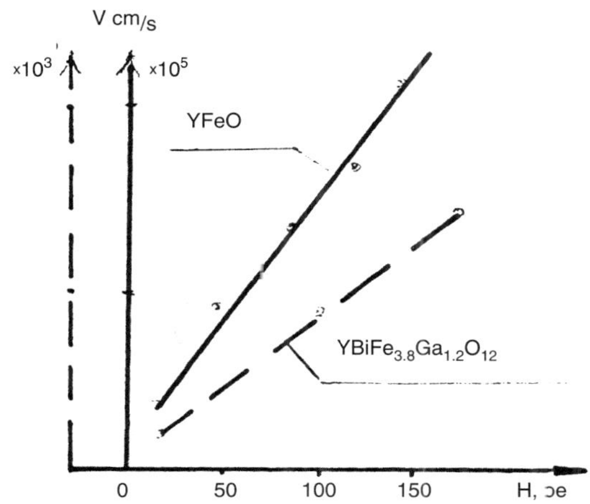

Figure 6.18: Domain wall mobility as a function of magnetic-field strength.

We also experimented with $BaSrNbO_3$:Ce which has the merit of higher sensitivity [34].

Of course, the overlapping recording of holograms (by using different reference waves) can dramatically increase the HDSS storage capacity. However, this raises the problem of erasing previously recorded holograms.

6.10 Nondestructive reading of 3-D holograms recorded in photorefractive crystals

The use of niobate lithium crystals for holographic storage involves the problem of nondestructive reconstruction of recorded information. The thermal stabilization of holograms does not protect them completely from the destructive action of reading light. Multiple readings deteriorate the quality of reconstruction, lowering the signal-to-noise ratio and leading to loss of information [30].

We found that the wavelength of reading light has little effect on photorefractive properties of the crystal [31]. In particular, we recorded

Figure 6.19: Photograph of the output pattern of the orthoferrite page composer.

holograms in a 1-mm-thick crystal with 0.63-μm light and reconstructed them at 0.63 μm, 0.69 μm (ruby laser), and 1.06 μm [yttrium-aluminum-garnet (YAG) laser]. The diffraction efficiency was 55%, 50%, and 50%, respectively, i.e., it was almost the same. This fact decided us to apply infrared semiconductor lasers for nondestructive reconstruction of holograms recorded in niobate lithium crystals. It is remarkable that even fairly high power densities do not deteriorate the quality of recorded holograms. Particularly, a 1-h exposure of the crystal to 1.06-μm light at 30 W/cm^2 power did not give any decrease of diffraction efficiency [31].

Figure 6.20: Photograph of the orthoferrite page composer.

It should be pointed out that in niobate lithium crystals the holograms of 2-D transparencies are 3-D. Such holograms are known to have high selectivity. When light of a longer wavelength is used for reading at the Bragg angle, only part of the recorded hologram is reconstructed. To reconstruct other portions of the transparency, the turning of the reading beam (or the crystal) within a certain angular range is necessary.

Figure 6.21 demonstrates the reconstruction of a transparency whose hologram was recorded in a crystal 1 mm thick with 0.44-μm light. The leftmost photograph shows the reconstruction at the same wavelength; the rightmost photograph, at 0.63 μm.

Thus, the use of longer wavelengths and the turning reading beam provides a method of nondestructive reconstruction of holograms from niobate lithium crystals.

Another method of nondestructive reading is based on two-photon holographic recording. The method also uses longer-wavelength light for reconstruction and looks promising.

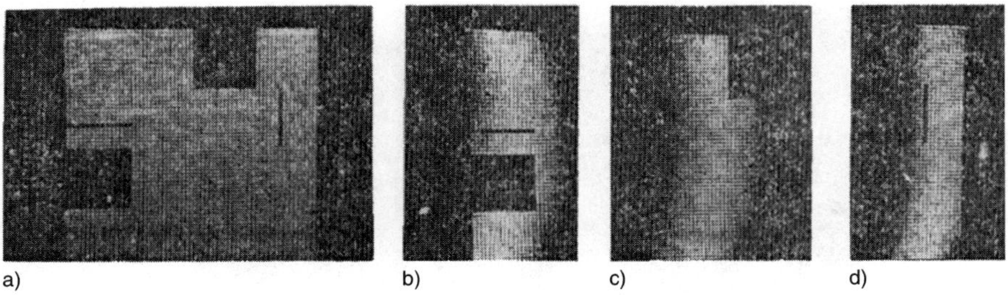

Figure 6.21: (a) Reconstruction at the recording wavelength (440 nm); (b), (c), (d) Reconstruction at 630 nm at different angles.

6.11 Application of reflection holograms

We tried to use reflection holograms in 5-mm $LiNbO_3$ crystals for data recording and reading. The angle between the reference and signal beams was about 170^0. Figure 6.22 demonstrates how diffraction efficiency of the hologram depends on the angle of the reading beam. The curve shows that it is possible to superpose holograms using tunable reading and

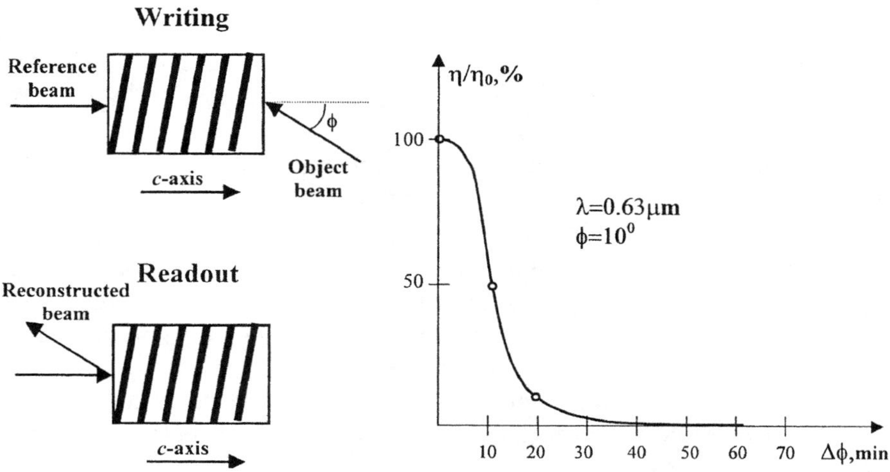

Figure 6.22: Relation between diffraction efficiency and the angle of the readout beam ($\Delta\phi$ is measured in arc minutes).

recording lasers [32, 33]. If we evaluate how diffraction efficiency varies with wavelength, we find that the change of wavelength within 100 Å allows the recording/reading of ten holograms.

Thick photorefractive crystals promise a simpler technique for making HDSSs if we use the superposition of holograms by varying the wavelength rather than the angle of reference beam. Experiments with refraction holograms also showed that a 2-h readout process almost did not change their diffraction efficiency [32, 33].

It is interesting to note some properties of reflection holograms caused by narrow interference fringes. Figure 6.23 shows the time dependences of diffraction efficiency during recording and reading. Two plots correspond to two orientations of the c-axis of the crystal relative to the reading beam. With the appropriate orientation of the c-axis, 2-h exposure to the reading light is seen to cause no appreciable drop of diffraction efficiency. This is due to self-amplification effects.

The diffraction efficiency keeps if the angle between the reading and reference beams lies within from 10′ to 30′ (minutes of arc). It means that the fringe spacing changes during reading and recording. This may be due to piezoeffects or a change of the index of refraction which are, in turn, caused by a charge accumulating in the crystal and giving rise to internal electric fields.

We see that despite certain drawbacks, $LiNbO_3$ crystals can be used for making rerecordable HDSSs. Of course, the successful application of photorefractive crystals needs further investigations of their properties and clear understanding of the mechanism of holographic recording.

6.12 Holographic memory systems using one-dimensional holograms

A moving carrier of 1-D holograms and the parallel data recording technique can be used for making high-performance memory systems. The design where the reference and signal beams are perpendicular to the carrier is interesting because in this case only the contrast of interference patterns changes. We used the design for developing a disk holographic memory system [23, 36–39].

We utilized the parallel recording of data on wide tracks, which provide the readout rate tens of times higher than that achieved in the

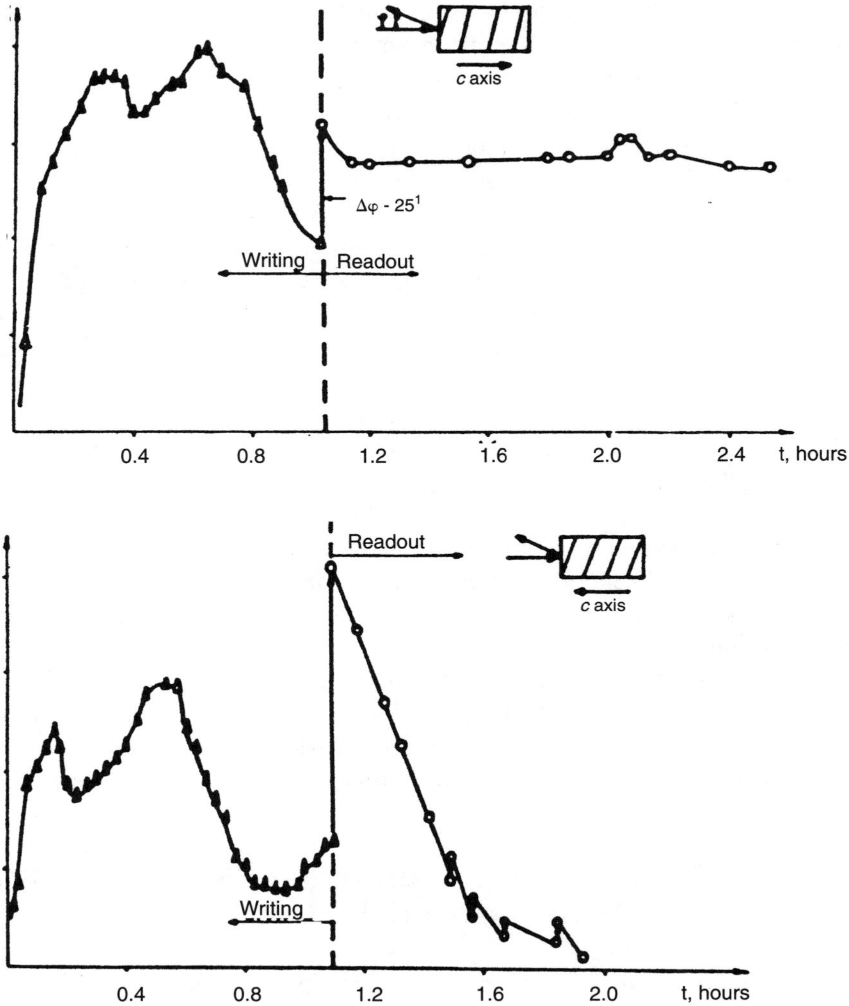

Figure 6.23: Time dependencies of diffraction efficiency in holographic recording/writing for two orientations of the c-axis of the crystal.

point-to-point method. The design also makes it possible to employ thick holograms and the 1-D hologram superposition technique. Since the results of the research were published, I only mention briefly the main characteristics and capabilities of holographic disk memory.

The experimental system (Fig 6.24) has a 34-channel electrooptic modulator for inputting electric signals. The modulator uses a $LiNbO_3$ plate 13.5 mm long and 300 μm thick. The channel spacing is 650 μm. The switching voltage is 80 V.

Thirty-four signals are recorded concurrently on a disk covered with a 5-μm film of silver halide emulsion. Each disk track is 50 μm wide and carries 1-D holograms. A 1-mW He-Ne laser is used for reconstructing holograms. The spacing between elements of the 1-D detector array is 250 μm. The diffraction efficiency of holograms is about 12%. The disk speed is 150 rpm, which provided a data rate of 16 Mbit/s. Tracks are scanned with a 50-μm step and accuracy of 2 μm. Figure 6.25 shows a photograph of the system and the output data.

The superposition of 1-D holograms involves such complex phenomena as the transformation of 1-D holograms into 3-D ones. Thicker photo-

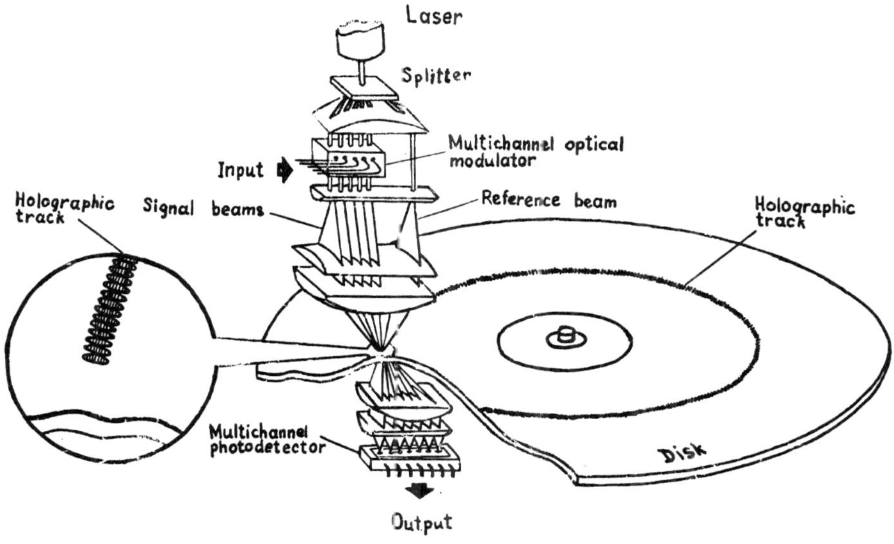

Figure 6.24: Schematic drawing of the 1D holographic disk memory system.

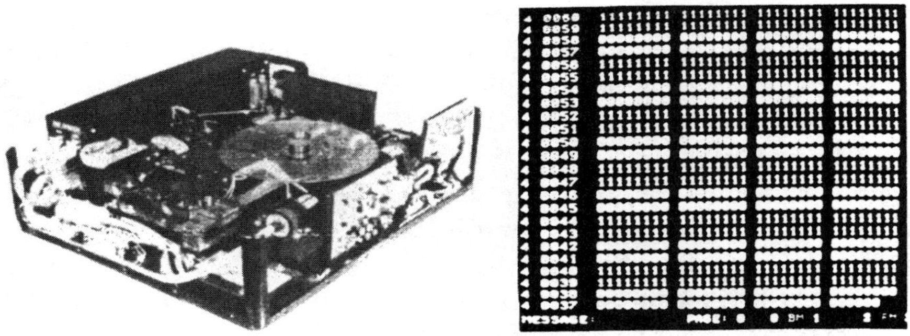

* Capacity	**1 - 1.5 Gbyte**	* Number of readout channels	**32**
* Track capacity	**1 Mbyte**	* Transfer rate	**20 – 40 Mbit/sec**
		* in the future	**200 Mbit/sec**
* Storage life	**> 20 years**	* Acces time (average)	**250 msec**

Figure 6.25: Photographs of the disk memory system (left) and the computer display with output data (right).

layers allow more overlapping holograms to be recorded. However, the beam defocusing in the perpendicular plane gives rise to errors and reduces the data storage density. The parameters of data recording that gives the best system performance were analyzed theoretically and tested experimentally [23, 39].

We also experimented with a $LiNbO_3$ plate placed on the disk surface. The best thickness of the plate was found to be 125 μm (other system parameters being the same). This permits 100 superpositions and about 10^9 bit/cm^2 of storage density. However, due to the partial erasure of previously recorded holograms, only eight superpositions could be done.

The principal advantage of using the 1-D hologram technique in holographic memory systems is the possibility to reach bit rates as high as hundreds of Mbit/s. In particular, the system is unrivaled for the fast reading of large data files when relatively long access times (about tens of milliseconds) are tolerable. Another important application is the real-time recording of continuous data flows.

The system may find more areas of application with development of necessary optoelectronic components. However, read/write holographic memory systems are now hardly feasible because of the absence of suitable reversible materials.

6.13 Three-dimensional multilayer holographic memory

The use of volume holographic materials and appropriate recording techniques (e.g., hologram superposition) can increase the data storage density appreciably. In some papers the volume density is claimed to be of the order of $1/\lambda^3$ (by analogy with the 2-D density, which is $\sim 1/\lambda^2$). Actually, this is not so because the reconstruction of any 3-D hologram is accompanied by noise. The noise is mostly stray interference, which is caused in the long run by overlapping holograms not being perfectly orthogonal [23]. Various methods are used to suppress interference noise (e.g., spatial and frequency coding of the reference beams). Yet the full suppression is impossible.

The analysis of volume holographic media and the evaluations of the maximum storage density have been recently made in many papers. We think that given an acceptable noise level (corresponding to a reconstruction contrast of about 5–10), holographic memory systems with storage density of over 10^{10} bit/cm^3 are hardly feasible. This conclusion is substantiated by a study of a 3-D multilayer data carrier [40], the results of which are presented in the following discussion.

Consider a holographic carrier consisting of a few thin layers in which transmission phase Fourier holograms are recorded. The holograms are reconstructed in transmission by focusing the beam successively on each layer. This simple 3-D memory design helps to understand the character of the noise that attends the reconstruction of holograms. In the particular case of a two-layer carrier (Fig. 6.26), we see that the signal rays reconstructed from the second layer go through the first layer and reconstruct the rays, part of which falls on the detector and produces the noise background. Besides, the part of the reference beam that passes the second layer reconstructs holograms in the first layer. Since the holograms recorded in the layers have low diffraction efficiency, the intensity of such noise rays is comparable with that of the signal rays from the second layer. The noise rays can be cut off by using a diaphragm (see Fig. 6.27 where this trick is explained for a three-layer design).

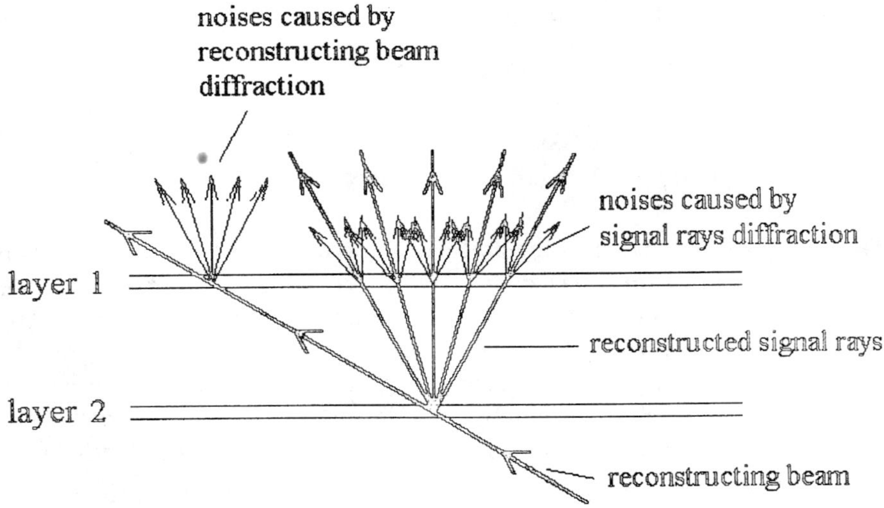

Figure 6.26: The nature of noise in reading two-layer holographic carrier.

The geometric considerations can help to estimate the size of the diaphragm. To avoid the cutting of signal beams during reconstruction from the third layer, the radius of the diaphragm opening must not be smaller than

$$r_{\min} = (h_1 + h_2)\, tg\alpha\,. \qquad (6.11)$$

These notations are clear from Fig. 6.27.

On the other hand, the diaphragm should block the stray rays from the first and second layers; i.e., its radius should not be greater than the minimum of the two values

$$r_{\max 1} = h_1(tg\beta - tg\alpha)\,, \qquad (6.12)$$

$$r_{\max 2} = h_2 tg\beta - (h_1 + h_2)\, tg\alpha\,. \qquad (6.13)$$

This means that the following condition must be met

$$r_{\max 1} \geq r_{\min}\,,\; r_{\max 2} \geq r_{\min} \qquad (6.14)$$

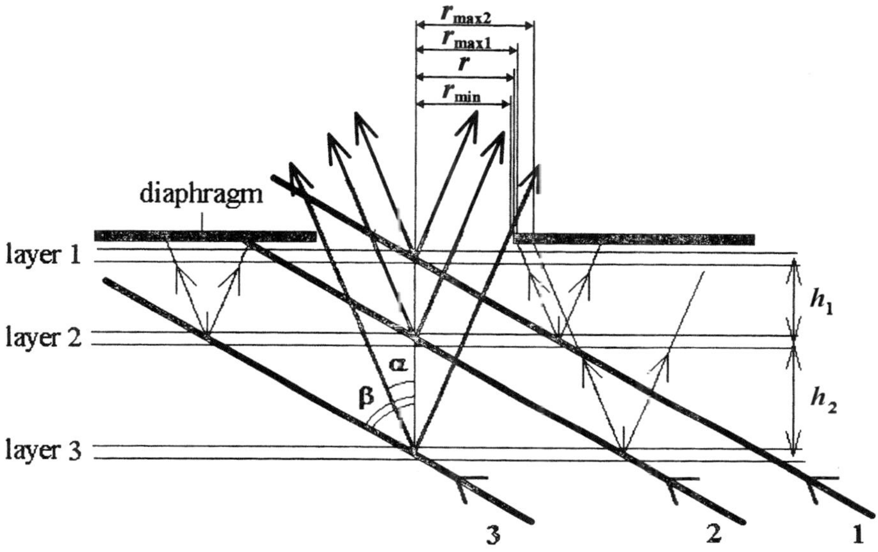

1, 2, 3 - the path of the reconstructing beam and the boundary signal rays in the information read-out from the 1st, 2nd and 3rd layers, respectively

Figure 6.27: Schematic drawing explaining the role of the diaphragm.

In the limiting case of $tg\beta = (\sqrt{2}+2)tg\alpha$ $h_2 = h_1\sqrt{2}$ and Eq. (6.14) changes into $r_{min} = r_{max}1 = r_{max}2$. Formula (11) in this case gives a single value for *r:*

$$r = h_1(1 + \sqrt{2})tg\alpha\,. \tag{6.15}$$

The above considerations also apply to a two-layer carrier design.

Let us evaluate parameter *r*. In particular, if $\alpha = 28°$, β must be about 62° to satisfy the first of the inequalities (Eq. 6.14). The smallest spacing between the layers h_{1min} is geometrically determined to be:

$$h_{1min} = L/(tg\beta - tg\alpha)\,, \tag{6.16}$$

where L is the hologram size (in the case of 1-D holograms, the size is measured across the interference fringes). If $L = 50$ μm, $h_{1min} \approx 37$ μm and $h_{2min} \approx 52$ μm. The universal diaphragm for the three-layer design will measure 95 × 95 μm.

For a carrier consisting of more layers, the use of the universal diaphragm for reconstruction from all layers requires more complicated optics. A variable-diameter diaphragm appears to be more justifiable in this case. If the maximal diffraction angle is constant and equal to $\sqrt{2} + 2$ for all layers, the spacing between the third and fourth layers can be calculated to be 74 μm and the spacing between the fourth and fifth layers to be 135 μm. The diaphragm for reading from the fourth layer will be roughly 174 × 174 μm, and from the fifth layer, 317 × 317 μm.

6.14 Interference noises in three-dimensional data carriers and volume storage density

Let us now look more closely at the noise caused by diffraction of signal rays on the information layers located on their way to the objective. Holograms recorded in these layers can be regarded as phase gratings resulting from the interference of the object beams with the reference beam and with each other. Let us call the first kind of gratings signal gratings, and the second kind intermodulation gratings. Signal gratings have more narrow fringes (higher spatial frequency) because during recording the angle between the reference and object beams is quite large. Since the layers do have thickness, these gratings are angle selective and do not produce rays that reach the detector. For this reason, only the diffraction of signal beams on intermodulation gratings makes a major contribution to the noise in the reading channel.

In our study we started from the assumption that holograms in the layers have low diffraction efficiency and phases in the modulator channels are distributed randomly. We computed the contrast (the "1"$_{min}$/"0"$_{max}$ ratio in a reconstruction) as a function of different parameters [40]. Figure 6.28 exemplifies how the reconstruction contrast falls with the number of layers. It is seen that with diffraction efficiency of no greater than 5%, a fivefold increase in storage capacity brings about just a small growth of noise (a few percentage points).

Figure 6.29 shows how the storage density grows with the number of information layers. The quality of reconstruction (i.e., the "1"$_{min}$/"0"$_{max}$

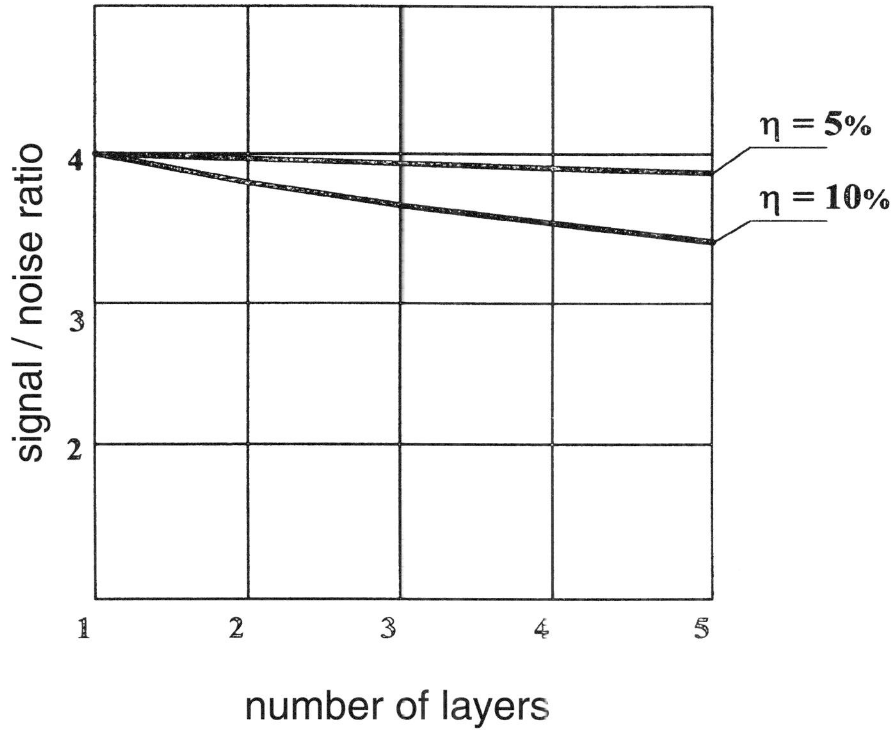

Figure 6.28: Relationship between the signal-to-noise ratio and the number of holographic layers.

ratio) here is assumed to keep constant. The parameters of the single-layer carrier were given above (Fig. 6.11). It follows from the figure that with diffraction efficiency of no greater than 1% it is possible to have over 10^{10} bit/cm^2 of storage density and fairly good reconstruction contrast.

As may be seen from Fig. 6.30, the thickness of the multilayer carrier is about 1 cm. Note that the computations were made for a 512 × 512-bit input array and $\lambda = 0.63$ μm; i.e., for the same conditions as with the single-layer carrier (see preceding discussion).

It is clear that even if the superresolution technique is used for recording in each layer, the requirement of reliable data reading (necessary reconstruction contrast is 6–8) would not allow the multilayer holographic carrier to have a storage density higher than 10^{10} bit/cm^3.

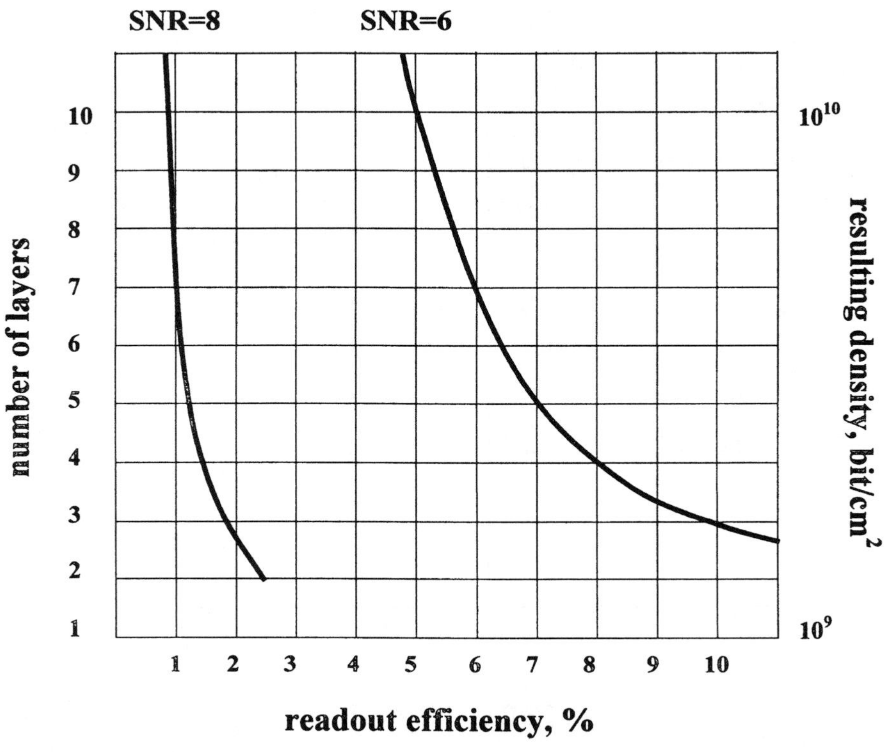

Figure 6.29: Relationship between the storage density (number of layers) and diffraction efficiency for two signal-to-noise ratios of 6 and 8.

The results of the study give us the basis to think that the use of volume holographic carriers will hardly allow further increase in storage density. Moreover, it is clear that volume media will suffer from interference noise to a greater extent than multilayer structures.

6.15 Conclusion

It should be pointed out that the development of random-access HDSSs will certainly require the use of hologram superposition, in one form or another. It is due to the fact that SLMs used for inputting information

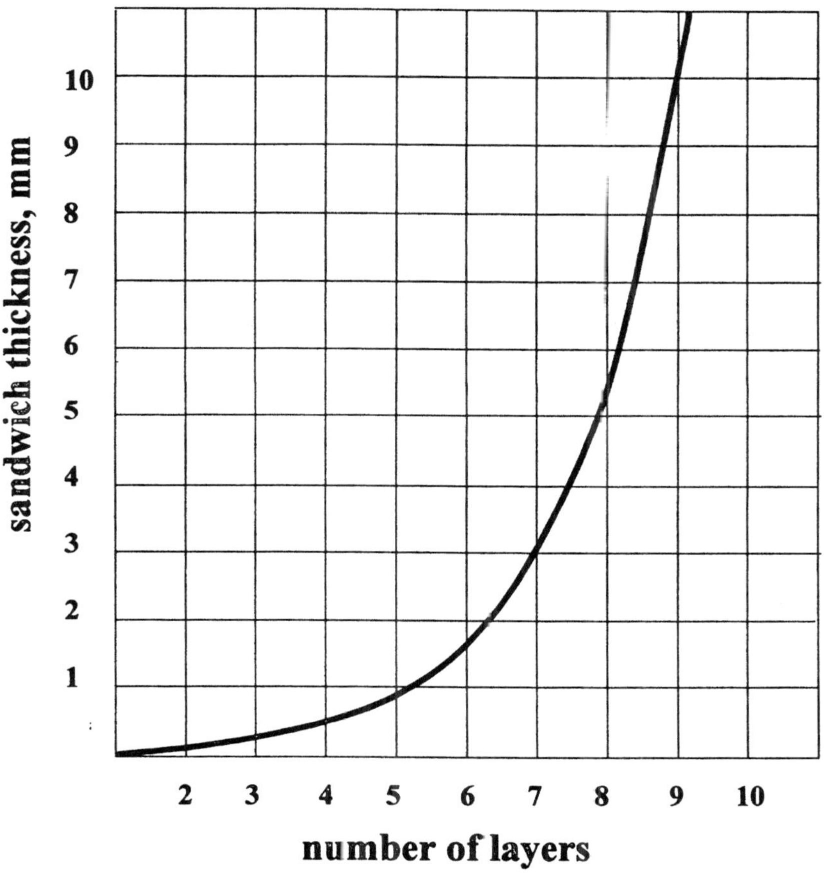

Figure 6.30: Thickness of the sandwich carrier versus the number of layers.

can hardly carry more than 10^6 bits. Therefore, even in the case of flat thick-layer media (when the portions of the carrier with recorded microholograms are spaced apart), a certain number of hologram overlaps (on the order of several tens) will be necessary to reach high storage density (10^8–10^{10} bit/cm^2). The search of effective data recording methods using hologram superposition is thus an important problem on the way to the development of high-density HDSSs.

Acknowledgment

In conclusion, I also would like to thank the Russian Foundation for Basic Research for the support of most of our recent studies in the field (grant number 96-01-00114).

References

1. A. L. Mikaelian, V. L. Bobrinev, A. P. Axenchikov, E. H. Gulanian, *Popov Society Conference on Quantum Electronics Proc.,* 1967.
2. A. L. Mikaelian, *Holography,* Znanie, Moscow, 1968.
3. A. L. Mikaelian, V. L. Bobrinev, S. M. Naumov, L. Z. Sokolova, *Radiotechnika i Electronika,* No. 1, 115–124, 1969; *IEEE J. Quantum Electronics,* **QE-6** (4), 193–198, 1970; *Revue d'Optique,* **1**(2), 1970.
4. A. L. Mikaelian, V. L. Bobrinev, A. P. Axenchikov, E. H. Gulanian, V. V. Shatun, *IEEE J. Quantum Electronics,* **QE-4**, 757–762, 1968, *Dokladi Akadem. Nauk SSSR,* **181**, 1105, 1968.
5. A. L. Mikaelian and V. L. Bobrinev, *Optoelectronics,* **2**, 198–199, 1970.
6. A. L. Mikaelian *et al., Holographic Memory System*, USSR patent 368816, May 28, 1970.
7. A. L. Mikaelian and V. L. Bobrinev, *Pisma JETP,* **4**(5), 271–275, 1966.
8. A. L. Mikaelian, L. M. Klukin, N. M. Pomerantsev, V. A. Fabrikov, *Voprosi Radioelektroniki,* No. 2, 3–11, 1967.
9. A. L. Mikaelian, V. L. Bobrinev, N. G. Nakhodkin, E. H. Gulanian, *et al., Kvantovaia Elektronika,* No. 1, 145–146, 1971; *Revue d'Optique,* **1**(2), 1970.
10. A. L. Mikaelian, V. L. Bobrinev, A. A. Akselrod, M. M. Koblova, *et al., Kvantovaia Elektronika,* No. 1, 79–84, 1971.
11. A. L. Mikaelian, M. M. Koblova, E. A. Zasovin, *Radiotechnika i Electronika,* No. 8, 1768–1770, 1970.
12. A. L. Mikaelian, V. L. Bobrinev, L. Z. Sokolova, *Dokladi Akad. Nauk,* **191**(4), 799–801, 1970.
13. L. K. Andersson, S. Broido, G. T. Lamacchia, L. H. Lin, *IEEE Conference on Laser Engineering and Applications,* 1967.
14. F. M. Smith and C. H. Gallaher, *Bell Syst. Techn. J.,* No. 46, VII–VIII, 1967.
15. L. K. Andersson, *Bell Lab. Rec.,* No. 46, 319–325, 1968.
16. H. Kiemle, *Appl. Opt.,* **13**(4), 803–807, 1974.

17. A. A. Akselrod, V. I. Bobrinev, V. G. Voronin, *et al.*, *Kvantovaia Elektronika,* No. 5, 995–998, 1978.

18. A. L. Mikaelian, V. L. Bobrinev, Yu. S. Vinokurov, *et al.*, in V. A. Barachevsky (ed.), *Random Access Holographic Data Storage System in Proc.* Nauka, Leningrad, 1987.

19. A. L. Mikaelian, M. M. Koblova, E. A. Zasovin, B. S. Kiseliov, Electrooptical detector, USSR patent 445368, Sept. 14, 1974; USSR patent 577919, Dec. 25, 1975.

20. A. L. Mikaelian, M. M. Koblova, E. A. Zasovin, B. S. Kiseliov, *Kvantovaia Elektronika,* No. 12, 2550–2557, 1978.

21. A. L. Mikaelian, *Laser Focus,* 37–38, 1973.

22. Yu. S. Vinokurov and A. A. Akselrod, *Voprosi Radioelektroniki,* No. 5, 104–111, 1981.

23. A. L. Mikaelian, *Optical Methods for Information Technologies,* Nauka, Moscow, 1990; English version by Allerton Press, NY, 1994.

24. A. M. Balbashov, A. A. Komlev, A. L. Mikaelian, *et al.*, *Kvantovaia Elektronika,* **4**(9), 1933–1945, 1977; Magnetooptical SLM for holographic memory systems, USSR patent 683540, Dec. 10, 1977.

25. A. L. Mikaelian, A. V. Korovizin, L. V. Naumova, *Pisma JETP,* **2**(1), 37–40, 1965.

26. A. L. Mikaelian, *Radiotekhnika,* **36**(11), 3–24, 1981.

27. V. L. Bobrinev, V. S. Vorobiev, Y. H. Kagan, M. A. Maiorchuk, A. L. Mikaelian, N. B. Nikontov, *Avtometria, Nauka,* No. 5, 52–56, 1977.

28. J. J. Amodei, W. Phillips, D. L. Staebler, *Appl. Opt.,* **11**(2), 390–396, 1972.

29. J. J. Amodei, W. Phillips, D. L. Staebler, *RCA Revue,* No. 3, 94–102, 1972.

30. V. L. Bobrinev, Z. G. Vasilieva, E. H. Gulanian, A. L. Mikaelian, *Pisma JETP,* **18**(4), 267–270, 1973.

31. E. H. Gulanian, I. R. Dorosh, A. L. Mikaelian, *et al.*, *Kvantovaia Elektronika,* **6**(5), 1097–1100, 1976.

32. A. L. Mikaelian, E. H. Gulanian, E. I. Dmitrieva, I. R. Dorosh, *Kvantovaia Elektronika,* **5**(2), 1978.

33. A. L. Mikaelian, *Optical Information Processing,* Vol. 2, Plenum, New York, 217–233, 1976; "Erasable holographic memory," *Laser Focus,* Nov. 1973.

34. V. V. Voronov, E. H. Gulanian, A. L. Mikaelian, *et al.*, *Kvantovaia Elektronika,* **6**(9) 1993–1999, 1979.

35. A. L. Mikaelian and V. I. Bobrinev, *Radiotekhnika,* No. 5, 7–10, 1974.

36. E. H. Gulanian and N. A. Kostrov, *Voprosi Radioelektroniki,* No. 6, 81–86, 1973.
37. A. L. Mikaelian, A. F. Vanin, E. H. Gulanian, S. A. Prokopenko, *Kvantovaia Elektronika,* **17**(5), 1074–1085, 1987.
38. A. L. Mikaelian, E. H. Gulanian, Y. S. Vinokurov, *et al., Int. J. Optical Computing,* **1**(1), 93–100, 1990.
39. A. L. Mikaelian, E. H. Gulanian, B. S. Kretlov, *et al., Proc. SPIE,* No. 1621, 194–204, 1991.
40. D. M. Abakumov, N. A. Ashurbekov, E. H. Gulanian, A. L. Mikaelian, *Int. J. Optical Memory and Neural Networks,* **5**(4), 295–305, 1996.

Chapter 7

Cross Talk in Volume Holographic Memory

Xianmin Yi, Pochi Yeh,

Department of Electrical Engineering, University of California, Santa Barbara, California

Claire Gu

Department of Electrical Engineering, University of California, Santa Cruz, California

Volume holographic memory offers many desirable features, including compactness, high storage density, and fast parallel access [1–4]. The information storage capacity of such a system is ultimately limited by the geometric factor $O(V/\lambda^3)$, where V is the volume of the holographic medium and λ is the wavelength of light [1]. There have been reports of the storage of 5000 high-resolution holograms in $LiNbO_3$:Fe via angle multiplexing [5]. Various multiplexing schemes have been proposed and studied. These include angle multiplexing [5], wavelength multiplexing [6, 7], phase code multiplexing [8], peristrophic multiplexing [9], shift multiplexing [10], sparse-wavelength angle multiplexing [11], and spectral-hole angle multiplexing [12]. Cross talk noise due to diffraction is a fundamental limitation to storage capacity. Cross talk and other system limitations together reduce the storage capacity of current holographic

PHOTOREFRACTIVE OPTICS
Materials, Properties and Applications

ISBN 0-12-774810-5/$30.00

memory systems from the ultimate storage capacity $O(V/\lambda^3)$ by nearly two orders of magnitude. In Section7.1 of this chapter, we consider the cross talk noise and the cross talk–limited storage capacity in Fourier-plane holographic memories with angle- and wavelength-multiplexing schemes and image-plane holographic memory with angle-multiplexing scheme.

Cross-talk noise may also arise from the expansion or shrinkage of the storage medium during and after the hologram recording process. For example, the storage medium will always expand or shrink when the temperature changes. Thermal expansion coefficients for various storage media are different. While being relatively small for photorefractive crystals, they are much larger for organic polymers. Another example is the shrinkage of organic polymers during polymerization. When photoinduced polymerization is utilized as the recording mechanism, the organic polymer needs to be completely polymerized after the holograms have been recorded via photoinduced polymerization. That will induce a postrecording shrinkage of the storage media. The expansion or shrinkage of the storage medium may not only increase the cross talk noise but also cause other limitations to the storage system. In fact, it is one of the dominant limitations to volume holographic memories using organic polymers as storage media at the present time. Its effects, often broadly referred to as grating detuning, will be the subject of Section 7.2 of this chapter.

7.1 Cross talk

In this section, we consider the cross talk noise and the cross talk–limited storage capacity in angle-multiplexed Fourier plane holographic memory, wavelength-multiplexed Fourier plane holographic memory, and angle-multiplexed image plane holographic memory.

7.1.1 Angle-multiplexed Fourier plane holographic memory

Fourier-transform holograms are angle multiplexed in a volume holographic medium with the setup shown in Fig. 7.1. Plane reference waves with certain angle separations propagate in the yz-plane. The size of the holographic medium is $a \times b \times t$, where a and b are the transverse dimensions, and t is the thickness of the medium. The input object is

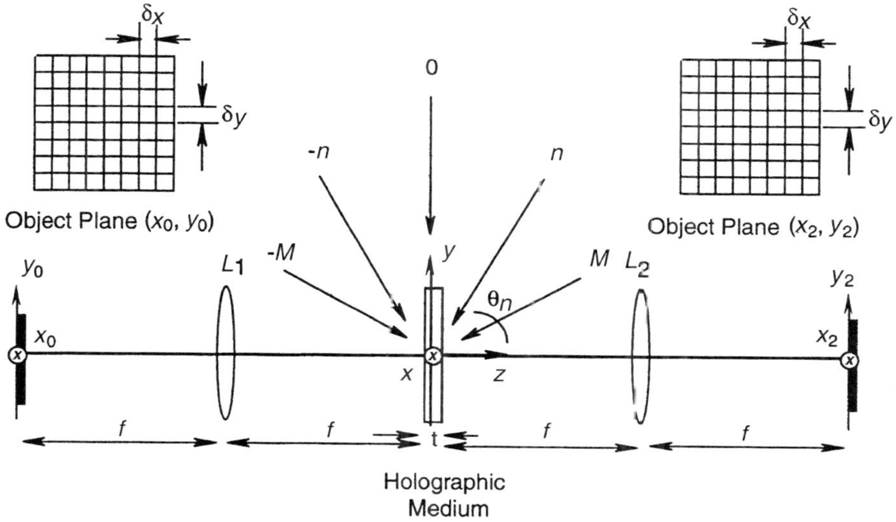

Figure 7.1: System configuration for angle-multiplexed Fourier plane holographic memory. The z-axis is chosen to be the optical axis of the imaging system. The yz-plane is the plane of incidence for the reference wave.

made up of rectangular pixels of size $\delta_x \times \delta_y$ in the x_0y_0-plane. The pixels can be modulated to be either on or off. The optical detector in the output plane (x_2, y_2) is also made up of closely packed rectangular pixels of size $\delta_x \times \delta_y$.

Assume that $N = 2M + 1$ phase (index) holograms are stored inside a volume holographic medium. We label these holograms as $m = -M$, $-(M - 1), \ldots, 0, \ldots, (M - 1), M$. With the presence of these phase holograms, the modulation of the dielectric constant of the medium can be written as

$$\Delta\varepsilon \propto \sum_{m=-M}^{M} R_m^* S_m + \text{c.c.} \tag{7.1}$$

where S_m represents the amplitude of the mth object and c.c. is the complex conjugate term. The mth reference beam R_m is given by

$$R_m \propto \exp(i\mathbf{k}_m\mathbf{r}) \tag{7.2}$$

where $\mathbf{k}_m$ is the wavevector for the mth reference beam. In addition, with standard Fourier optics analysis [18], the mth signal beam can be expressed as

$$S_m(r) \propto \exp(ikz)\int dx_o dy_o f_m(x_0, y_0)\exp\left[-i\frac{2\pi}{\lambda f}(xx_0 + yy_o)\right] \exp\left[-i\frac{\pi}{\lambda f}\frac{z}{f}(x_o^2 + y_o^2)\right] \tag{7.3}$$

where $f_m(x_0, y_0)$ is the mth input object; x, y, and z are the coordinates at the back focal plane of lens L_1; and f is the focal length. By substituting Eqs. (7.2) and (7.3) into Eq. (7.1), one can obtain $\Delta\varepsilon$ = explicitly.

During the readout, a plane wave with wavevector $\mathbf{k}_n$ is incident upon the volume holographic medium. The diffracted plane-wave component $E(\mathbf{r})$, with wavevector $\mathbf{k}_d$, can be derived from standard scalar diffraction theory [19]. The following expression assumes that the Born and paraxial approximations are valid:

$$E(\mathbf{r}) \approx \frac{k_0^2}{4\pi}\frac{\exp(ikr)}{r}\int d\mathbf{r}' \exp(-i\mathbf{K}\cdot\mathbf{r}')\,\Delta\varepsilon(\mathbf{r}') \tag{7.4}$$

where $\mathbf{K} = \mathbf{k}_d - \mathbf{k}_n$. A plane wave at the front focal plane of lens L_2 is converted to a point at the rear focal plane of lens L_2. We can thus relate the diffracted wave vector $\mathbf{k}_d$ to its coordinates at the output plane (x_2, y_2) as

$$\mathbf{k}_d = \left[\frac{2\pi}{\lambda}\frac{x_2}{f}, \frac{2\pi}{\lambda}\frac{y_2}{f}, \frac{2\pi}{\lambda}\left(1 - \frac{x_2^2}{2f^2} - \frac{y_2^2}{2f^2}\right)\right]. \tag{7.5}$$

Using Eqs. (1)–(3) and (5), we obtain the optical wave at the output plane as

$$g(x_2, y_2) \propto \sum_{m=-M}^{M}\int dx_0 dy_0 f_m(x_0, y_0)$$
$$\operatorname{sinc}\left[\frac{t}{2\pi}\left(k_{mz} - k_{nz} + \frac{\pi}{\lambda}\frac{x_2^2 - x_0^2 + y_2^2 - y_0^2}{f^2}\right)\right] \tag{7.6}$$
$$\times \operatorname{sinc}\left[\frac{a}{2\pi}\left(k_{mx} - k_{nx} + \frac{2\pi}{\lambda}\frac{x_2 + x_0}{f}\right)\right]\operatorname{sinc}\left[\frac{b}{2\pi}\left(k_{my} - k_{ny} + \frac{2\pi}{\lambda}\frac{y_2 + y_0}{f}\right)\right].$$

According to Eq. (7.6), the output consists of diffraction from all the stored images, as indicated by the summation over m. The amplitude of each image is determined by the difference between the input wave vector $\mathbf{k}_n$ and that of the corresponding reference beam used for the recording of the holograms according to the sinc functions. Suppose the reading beam is identical to one of the reference beams during recording: the term with $\mathbf{k}_m = \mathbf{k}_n$ in Eq. (7.6) corresponds to the reconstructed image; all other terms in Eq. (7.6) with $\mathbf{k}_m \neq \mathbf{k}_n$ give rise to the interpage cross talk noise. For the term with $\mathbf{k}_m = \mathbf{k}_n$ in Eq. (7.6), the sinc functions that operate on the x and y components indicate the spatial frequency cutoff due to the finite transverse size of the recording medium. These are the sources of the intrapage interpixel cross talk noise. According to Eq. (7.6), the interpage cross talk and the intrapage interpixel cross talk are coupled in the most general case. Fortunately, they can be considered separately in the cross talk analysis of a low-noise data storage system.

We first consider the interpage cross talk noise. We assume that the transverse dimensions of the medium are greater than the spatial bandwidth of the images. Then the last two sinc functions in Eq. (7.6) can be approximated by δ functions. Thus, Eq. (7.6) can be easily integrated. This leads to

$$g(x_2, y_2) \propto \sum_{m=-M}^{M} f_m\left(-x_2 - \frac{\lambda f}{2\pi}\Delta K_{mnx}, -y_2 - \frac{\lambda f}{2\pi}\Delta K_{mny}\right)$$
$$\times \operatorname{sinc}\left[\frac{t}{2\pi}\left(\Delta K_{mnz} + \frac{\Delta K_{mnx}x_2 + \Delta K_{mny}y_2}{f} + \lambda\frac{\Delta K_{mnx}^2 + \Delta K_{mny}^2}{4\pi}\right)\right] \tag{7.7}$$

where $\Delta\boldsymbol{K}_{mn} = \boldsymbol{k}_m - \boldsymbol{k}_n$.

According to Eq. (7.7), each noise pattern is a shifted-input object pattern with a reduced amplitude represented by the sinc functions. The amount of shift and amplitude reduction depend on the momentum mismatch between the reading beam ($\mathbf{k}_n$) and the corresponding reference beam for recording ($\mathbf{k}_m$). Remember that each input object is made up of randomly distributed on-pixels. If the interpage cross talk noise is large at a particular point in the x_2y_2-plane, it means that a lot of on-pixels covering that point, each from a particular hologram, add up in phase. As a result, the noise will remain large in the vicinity of that point with a range of about 1 pixel. Therefore, the noise-to-signal ratio (NSR) calculated at one point in the output plane will represent fairly well the cross talk noise of the pixel at that point when the light is detected by an area detector of the same size as a pixel.

We assume that the input objects bear random data patterns and the probability for a pixel to be on or off is exactly 1/2. The images are represented by $|f_m|^2 = 1$ for the pixels that are turned on and $|f_m|^2 = 0$ for the pixels that are turned off. In this case, the NSR at an arbitrary point in the output plane can be written as

$$\mathrm{NSR} = \frac{1}{2}\sum_{m\neq n} \mathrm{sinc}^2 \left[\frac{t}{2\pi}\left(\Delta K_{mnz} + \frac{\Delta K_{mnx}x_2 + \Delta K_{mny}y_2}{f} + \lambda\frac{\Delta K_{mnx}^2 + \Delta K_{mny}^2}{4\pi}\right)\right]. \tag{7.8}$$

According to the preceding equation, the interpage cross talk noise for the nth hologram mainly comes from the holograms whose wavevectors $\mathbf{k}_m$ are very close to $\mathbf{k}_n$. In the holographic memory system shown in Fig. 7.1, the momentum mismatch $\Delta\mathbf{K}_{mn}$ can be expressed in terms of the hologram reference beam angles θ_m and θ_n. For those holograms with $|\theta_m - \theta_n| << 1$, the momentum mismatch can be written as

$$\Delta K_{mnx} = 0 \tag{7.9a}$$

$$\Delta K_{mny} = \frac{2\pi}{\lambda}(\theta_m - \theta_n)\left[-\cos\theta_n + \frac{1}{2}\sin\theta_n(\theta_m - \theta_n)\right] \tag{7.9b}$$

$$\Delta K_{mnz} = \frac{2\pi}{\lambda}(\theta_m - \theta_n)\left[-\sin\theta_n - \frac{1}{2}\cos\theta_n(\theta_m - \theta_n)\right]. \tag{7.9c}$$

By substituting Eq. (7.9) into Eq. (7.8) and keeping only the first-order term of $\theta_m - \theta_n$ in the argument of each sinc function, the NSR can be written as

$$\mathrm{NSR} = \frac{1}{2}\sum_{m\neq n} \mathrm{sinc}^2\left[\frac{t}{\lambda}(\theta_m - \theta_n)\sin\theta_n\left(1 - \cot\theta_n\frac{y_2}{f}\right)\right]. \tag{7.10}$$

In our general analysis, we assume that the angle separation between the reference beams of the adjacent holograms is

$$\Delta\theta = \frac{s\lambda}{t\sin\theta} \tag{7.11}$$

where θ is the averaged reference beam angle of the two adjacent holograms, and s is a dimensionless hologram separation parameter. According to Eq. (7.11), s is equal to the number of sinc lobes by which the center pixel of each hologram is separated from its two adjacent holograms. Note that the angle separation between the reference beams of the adjacent holograms is a function of the averaged reference beam angle θ. We denote the hologram whose reference beam angle θ is $\pi/2$ as the 0th hologram. Then the relation between the hologram number n and the hologram reference beam angle θ_n can be obtained from Eq. (7.11) as

$$n = \frac{t \cos\theta_n}{s\lambda}. \tag{7.12}$$

Note that $|n|$ increases linearly as θ_n deviates from $\pi/2$ and the increase slows down as θ_n approaches 0 or π. The maximum value of $|n|$ is $t/(s\lambda)$.

We now define a cross talk variable as

$$p_\theta \equiv \frac{\cot\theta_n y_2}{sf}. \tag{7.13}$$

Using Eqs. (7.11)–(7.13), the NSR, Eq. (7.10), can be written as

$$\text{NSR} = \frac{1}{2}\sum_{m \neq n} \text{sinc}^2[s(m - n)(1 - sp_\theta)]. \tag{7.14}$$

According to Eq. (7.14), when the cross talk variable p_θ approaches $1/s$, the NSR will be much larger than one. This is a result of hologram degeneracy. When the hologram reference beam angle θ approaches 0 or π, $\cot\theta$ increases rapidly to infinity. This leads to a rapid increase of the cross talk noise. In the following analysis, we assume that θ is not close to 0 or π. Thus, under the approximation $\cot\theta \approx \cos\theta$, the cross talk variable p_θ can be written as, according to Eqs. (7.12)–(7.13),

$$p_\theta \approx n\frac{\lambda y_2}{tf}. \tag{7.15}$$

Note that the maximum cross talk variable is directly proportional to the number of the stored holograms times the size of the objects in the y_2 direction. Using Eq. (7.14), we can first find the maximum cross talk variable p_θ for an arbitrary s so that the NSR for a smaller p_θ will be smaller than a given NSR requirement NSR_{re}. We will denote this value of p_θ as $p_{\theta\max}(NSR_{re}, s)$. Then we can find the maximum value of $p_{\theta\max}(NSR_{re}, s)$ for the optimal s. We will denote this value of p_θ as

$p_{\theta\max}(NSR_{re})$. $p_{\theta\max}(NSR_{re})$ is the upper bound of the product of the number of the stored holograms and the normalized size $\lambda y_2/tf$ of the objects in the y_2 direction for a given NSR requirement.

According to Eq. (7.14), NSR is a function of both the hologram separation s and the cross talk variable p_θ. We denote it as NSR $= C_\theta(s, p_\theta)$. Using Eq. (7.14), we obtain the following recurrence relationship,

$$C_\theta(s, p_\theta) = C_\theta(1, s^2 p_\theta - s + 1)\,. \tag{7.16}$$

In Fig. 7.2, we show the NSR as a function of the cross talk variable p_θ for hologram separation $s = 1$, i.e., NSR $= C_\theta(1, p_\theta)$. For the outermost reading beam, $n = M$, there is only one adjacent reference beam, $n = M - 1$. As a result, the cross talk noise is smaller than that for the reading

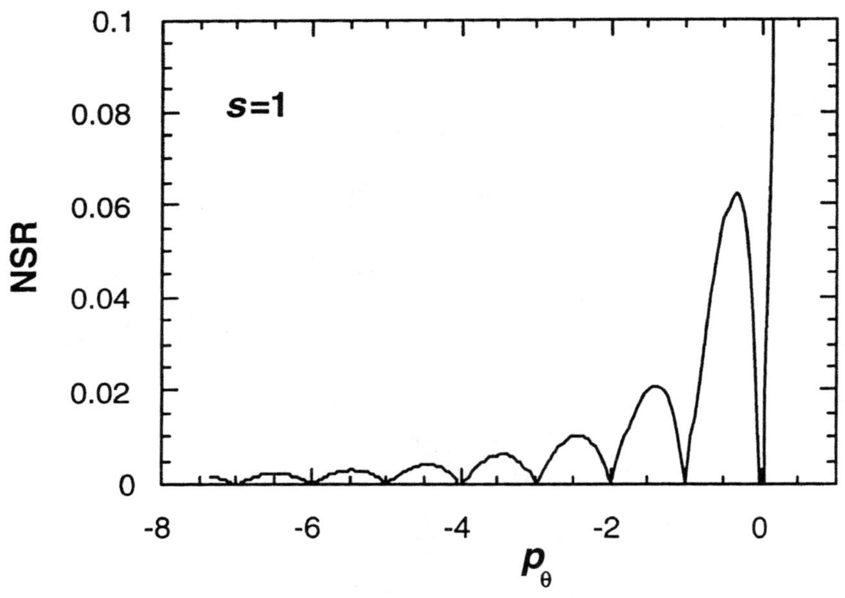

Figure 7.2: The noise-to-signal ratio (NSR) of the interpage cross talk noise as a function of the cross talk variable p_θ for hologram separation s = 1, i.e., NSR $= C_\theta(1, p_\theta) \approx 2\sum_{n=1}^{\infty} \text{sinc}^2[n(1 - p_\theta)]$. The height of the main lobe is much greater than 1. The first side lobe peaks at around $p_\theta = -1/2$ with a height of about 0.06. For a typical M of several hundreds, the side lobes are virtually independent of M. Note that 2M + 1 is the number of recorded holograms.

beams located slightly away from the edge (with slightly smaller n). In our simulation, we have chosen n to be at least 30 less than M. Using Eq. (7.16) we can easily obtain from Fig. 7.2 the plot of the NSR as a function of the cross talk variable p_θ for an arbitrary hologram separation s, i.e., NSR $= C_\theta(s, p_\theta)$. What we need to do is to move the p_θ axis to the left by $s-1$ units and then reduce the scale of the p_θ axis by a factor of s^2. Once we get the plot of NSR $= C_\theta(s, p_\theta)$, we can interpret it in two ways: NSR as a function of the position y_2/f in the output plane for a given hologram n or NSR as a function of the hologram number n at a given position y_2/f in the output plane.

Referring to Fig. 7.2, we note that NSR $= C_\theta(s, p_\theta)$ looks like part of the sinc^2( ) function. It consists of a main lobe and a series of side lobes with decreasing height. As we can see from Eq. (7.14) when we separate the center pixel ($p_\theta = 0$) of each hologram by s sinc lobes from the two adjacent holograms, the pixels at other locations ($p_\theta \neq 0$) will be separated by $(1 - sp_\theta)$ sinc lobes from the two adjacent holograms. $s(1 - sp_\theta)$ can be either larger than s or smaller than s, depending on whether p_θ is negative or positive. Hologram degeneracy occurs when $s(1 - sp_\theta)$ is much smaller than one. This leads to a main lobe which is much higher than one. The height of the first three side lobes are 0.06, 0.02, and 0.01.

Using Fig. 7.2, the slopes of the lobes of NSR $= C_\theta(s, p_\theta)$ can be estimated numerically. The slopes of the main lobe and the first side lobe near $p_\theta = 0$ are approximately 1/2. The slopes of the first side lobe and the second side lobe near $p_\theta = 1$ are approximately 1/4. The slopes of the second side lobe and the third side lobe near $p_\theta = 2$ are approximately 1/6. Now we are ready to estimate the maximum cross talk variable $p_{\theta\max}(NSR_{re})$ for a given NSR requirement.

When $0.06 < NSR_{re} < 0.2$, NSR_{re} is larger than the maximums of all the side lobes of the function NSR $= C_\theta(s, p_\theta)$. In this case, $p_{\theta\max}(NSR_{re}, s)$ is limited by the main lobe and can be estimated as

$$p_{\theta\max}(NSR_{re}, s) = \frac{1}{s} - \frac{1}{s^2} + \frac{2NSR_{re}}{s^2} \tag{7.17}$$

which is a quadratic function of $1/s$. At $s = 2 - 4NSR_{re}$, the cross talk variable reaches its maximum value,

$$p_{\theta\max}(NSR_{re}) = \frac{1}{4 - 8NSR_{re}}. \tag{7.18}$$

For example, when $NSR_{re} = 0.1$, the maximum cross talk variable $p_{\theta\max}(NSR_{re})$ will be 0.34 at $s = 1.6$. We recall $p_{\theta(\max}(NSR_{re})$ is the upper bound of the product of the number of stored holograms M and the normalized size $\lambda y_2/tf$ of the objects. So, for $y_{2\max} = 0.5f$, $\lambda = 0.5\mu m$, $t = 1cm$, a maximum cross talk parameter of 0.34 yields $M = 1.4 \times 10^4$.

When $0.02 < NSR_{re} < 0.06$, NSR_{re} is lower than the maximum of the first side lobe and is larger than the maxima of all the other side lobes of the function NSR $= C_\theta(s, p_\theta)$. In this case, the possible values of s can be either larger than 2 or equal to 1. For $s \geq 2$,

$$p_{\theta\max}(NSR_{re}, s) = \frac{1}{s} - \frac{2}{s^2} + \frac{4NSR_{re}}{s^2}. \tag{7.19}$$

At $s = 4 - 8NSR_{re}$, $p_{\theta\max}(NSR_{re}, s)$ reaches its local maximum in the region of $s \geq 2$:

$$p_{\theta\max}(NSR_{re}) = \frac{1}{8 - 16NSR_{re}}. \tag{7.20}$$

For example, when $NSR_{re} = 0.05$, the maximum cross talk variable $p_{\theta\max}(NSR_{re})$ will be 0.14 at $s = 3.6$. For $s = 1$,

$$p_{\theta\max}(NSR_{re}) = 2NSR_{re}. \tag{7.21}$$

For example, when $NSR_{re} = 0.05$, the maximum cross talk variable $p_{\theta\max}(NSR_{re})$ will be 0.1 at $s = 1$.

In this category, there are two locally optimal values of s: $s = 1$ or $s = 4 - 8NSR_{re}$. For $s = 1$, the storage capacity depends linearly on the required NSR. In addition, the cross talk noise and storage capacity are also sensitive to the exact value of s. For $s = 4-8NSR_{re}$, the storage capacity is insensitive to the required NSR. In addition, the cross talk noise and storage capacity are also insensitive to the exact value of s. As for the storage capacity, the cross talk–limited storage capacity for $s = 4-8NSR_{re}$ is larger than that for $s = 1$. The main advantage for choosing $s = 1$ is that it requires a smaller angle tuning range.

For most practical applications, the required noise-to-signal ratio falls into one of the two categories discussed in the preceding. The maximum cross talk variable $p_{\theta\max}(NSR_{re})$ can be obtained similarly for $NSR_{re} < 0.02$. In short, the interpage cross talk noise, due to the finite thickness of the holographic medium, limits the number of the stored holograms and the size of objects in the y_2 direction.

We now consider the intrapage interpixel cross talk noise. We start from the term $m = n$ in Eq. (7.6):

$$g(x_2, y_2) \propto \int dx_0 dy_0 f_m(x_0, y_0)\,\mathrm{sinc}\left[\frac{t}{2\lambda f^2}(x_2^2 - x_0^2 + y_2^2 - y_0^2)\right] \times \mathrm{sinc}\left[\frac{a}{\lambda f}(x_2 + x_0)\right]\mathrm{sinc}\left[\frac{b}{\lambda f}(y_2 + y_0)\right]. \tag{7.22}$$

The last two sinc functions in Eq. (7.22) represent the spatial frequency cutoff due to the finite transverse size of the holographic medium. The first sinc function describes the effect of the thickness of the holographic medium on the spatial frequency cutoff. Equation (7.22) is a general expression for studying the intrapage interpixel cross talk noise. When the thickness of the holographic medium is small compared with its transverse size, the first sinc function in Eq. (7.22) can be replaced by one, and Eq. (7.22) can be simplified as

$$g(x_2, y_2) \propto \int dx_0 dy_0 f_m(x_0, y_0)\,\mathrm{sinc}\left[\frac{D}{\lambda f}(x_2 + x_0)\right]\mathrm{sinc}\left[\frac{D}{\lambda f}(y_2 + y_0)\right] \tag{7.23}$$

where, for simplicity, we assume $a = b = D$. As described in the beginning of this section, the mth input object bearing arbitrary data can be expressed as

$$f_m(x_0, y_0) = \sum_{l=1}^{l2} A_{ml}\,\mathrm{rect}\left(\frac{x_0 - l_x\delta}{\delta}\right)\mathrm{rect}\left(\frac{y_0 - l_y\delta}{\delta}\right) \tag{7.24}$$

where we assume square pixels with $\delta_x = \delta_y = \delta$, l is a shorthand notation for the two integers l_x and l_y that specify the address of each pixel. A_{ml} can be either one or zero depending on the on- or off-state of the corresponding pixel. Equation (7.23) is a space invariant transformation of the input object $f_m(x_0, y_0)$ into the output image $g(x_2, y_2)$. Without loss of generality, let us examine the image amplitude of the input pixel at the origin (0, 0):

$$g(x_2, y_2) \propto \int dx_0\,\mathrm{rect}\left(\frac{x_0}{\delta}\right)\mathrm{sinc}\left[\frac{D}{\lambda f}(x_2 + x_0)\right]\int dy_0\,\mathrm{rect}\left(\frac{y_0}{\delta}\right) \mathrm{sinc}\left[\frac{D}{\lambda f}(y_2 + y_0)\right]. \tag{7.25}$$

We note that the image amplitude is a product of two convolution functions of rectangular function and sinc function along the x- and y-axes, respectively.

According to Eq. (7.25), each square pixel of the object is transformed into a convolution of a square and two sinc functions by virtue of diffraction. The diffraction-limited image of each pixel thus consists of a main lobe and a series of side lobes. The physical overlap of the images of these pixels leads to the interpixel cross talk. In Fig. 7.3, we show $g(x_2, y_2)$ with $x_2 > 0$ and $y_2 > 0$ for the case $\delta = \lambda f / D$. We can see several qualitative results from Fig. 7.3. The main lobe of the convolution function occupies an area larger than the rectangular pixel, and may overlap with adjacent pixels when the pixels are closely packed. The side lobes along the same row or column as the pixel being considered is much higher than the side lobes that are not on the same row and column as the pixels being considered. As a result, we can neglect the cross talk noise induced by the pixels that are not on the same row and column as the pixel being considered. The height of the side lobes decreases as their distance from the main lobe increases.

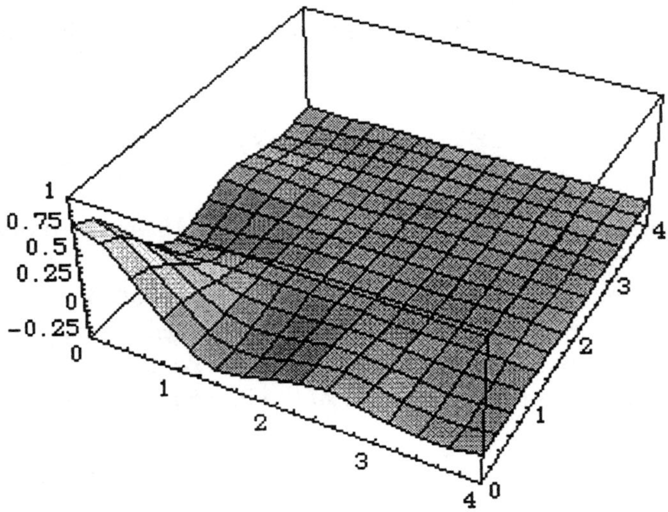

Figure 7.3: The image of a pixel for the case $\delta = \lambda f/D$. The two horizontal axes are x_2 and y_2, respectively. The units of these two axes are $\lambda f/D$. The vertical axis is $g(x_2, y_2)$.

The height of the side lobes of the convolution function is determined by the ratio between the size δ of the pixels and the width $\lambda f/D$ of the sinc function side lobes. When $\delta D/\lambda f$ is an odd integer, the height of the side lobes of the convolution function is maximum. A convolution side lobe maximum is reached when the sinc function and the rectangular function are positioned as in Fig. 7.4a. For the case of $\delta D/\lambda f = 1$, the maximum can be estimated as

$$\begin{aligned}
&\left|\int dx_0 \operatorname{sinc}\left(\frac{D}{\lambda f}x_0\right)\operatorname{rect}\left(\frac{x_0 - x}{\delta}\right)\right| \\
&\qquad \approx \frac{\lambda f}{\pi D x}\left|\int dx_0 \sin\left(\delta\frac{D}{\lambda f}x_0\right)\operatorname{rect}\left(\frac{x_0 - x}{\delta}\right)\right| \\
&\qquad \approx \left(\frac{2\lambda^2 f^2}{\pi^2 D^2}\right)\frac{1}{x}.
\end{aligned} \tag{7.26}$$

Note that the height of the side lobe maximum decreases as $1/x$ where x is the distance of the side lobe from the main lobe. When $\delta D/\lambda f$ is an even integer, the height of the side lobes of the convolution function is minimum. A convolution side lobe maximum is reached when the sinc function and the rectangular function are positioned as in Fig. 7.4b. For the case of $\delta D/\lambda f = 2$, the maximum can be estimated as

$$\begin{aligned}
&\left|\int dx_0 \operatorname{sinc}\left(\frac{D}{\lambda f}x_0\right)\operatorname{rect}\left(\frac{x_0 - x}{\delta}\right)\right| \\
&= \left|\int dx_0 \operatorname{sinc}\left(\frac{D}{\lambda f}x_0\right)\operatorname{rect}\left(\frac{x_0 - x - \frac{1}{2}\frac{\lambda f}{D}}{\frac{\delta}{2}}\right)\right. \\
&\quad \left. + \int dx_0 \operatorname{sinc}\left(\frac{D}{\lambda f}x_0\right)\operatorname{rect}\left(\frac{x_0 - x + \frac{1}{2}\frac{\lambda f}{D}}{\frac{\delta}{2}}\right)\right| \\
&\approx \frac{2\lambda^2 f^2}{\pi^2 D^3}\left(\frac{1}{x - \frac{1}{2}\frac{\lambda f}{D}} - \frac{1}{x + \frac{1}{2}\frac{\lambda f}{D}}\right) \\
&= \frac{2\lambda^3 f^3}{\pi^2 D^3}\frac{1}{x^2}.
\end{aligned} \tag{7.27}$$

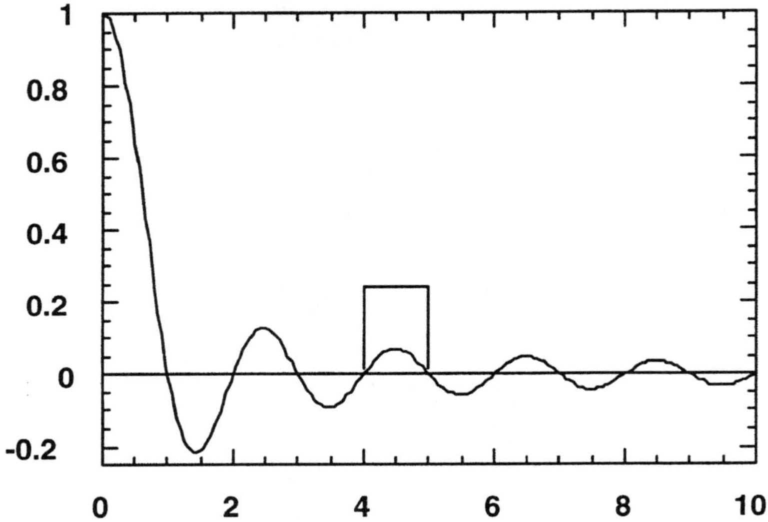

(a)

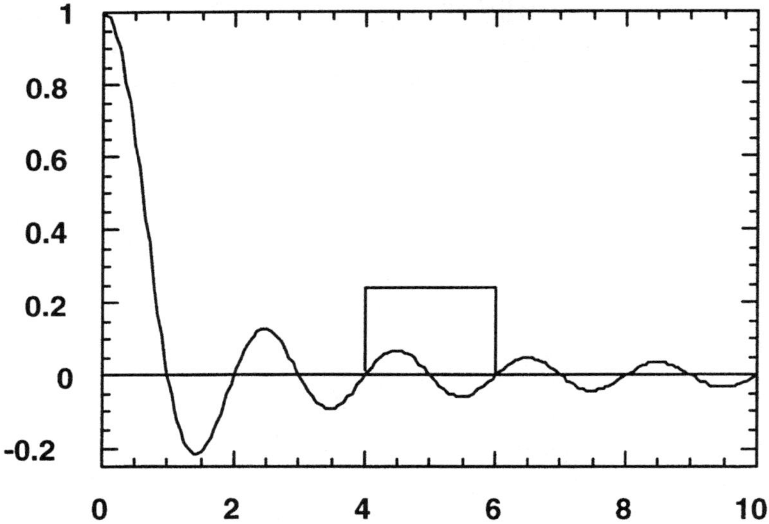

(b)

Figure 7.4: Relative position between the sinc function and the rectangular function so that a convolution side lobe maximum is reached for (a) $\delta D/\lambda f = 1$ and (b) $\delta D/\lambda f = 2$.

Note that the height of the side lobe maximum decreases as $1/x^2$, where x is the distance of the side lobe from the main lobe.

In Fig. 7.5, we show the convolution function along the x-axis for $\delta = \lambda f/D$ and $\delta = 2\lambda fD$, respectively. The case $\delta = \lambda f/D$ is a worst case (i.e., highest noise). When the pixels are closely packed, the interpixel cross talk noise will be intolerable. To reduce the interpixel cross talk noise, proper separation between adjacent pixels has to be made. The case $\delta = 2\lambda f/D$ is a near-optimal case. In this case, the pixel size is close to the diffraction limitation. The interpixel cross talk noise is negligible even if the pixels are closely packed. In this section, we choose $\delta = 2\lambda f/D$ to be the optimal pixel size in estimating the storage capacity.

A general quantitative analysis of the intrapage interpixel cross talk noise can be found in [15]. The analysis shows that the cross talk noise increases quickly as the pixel size decreases from $\delta = 2\lambda f/D$. The analysis also shows that the cross talk noise is small when the pixel size is larger than $\delta = 2\lambda f/D$ whether the ratio between the pixel size δ and the width $\lambda f/D$ of the sinc function side lobes is even or odd. One reason is that the height of the side lobes of the convolution function decreases as their distance from the main lobe increases. Another reason is that the oscillating side lobes from different pixels tend to cancel each other. However, reducing the cross talk noise by choosing a large pixel size is undesirable. For example, increasing the pixel size from $\delta = 2\lambda f/D$ to $\delta = 4\lambda f/D$ will decrease the storage capacity by a factor of 4.

Now we consider the effect of finite transverse size and the effect of finite thickness of the holographic medium together. The finite transverse size of the holographic medium induces the intrapage interpixel cross talk noise. It limits the size of the pixels. A near optimal pixel size is $2\lambda f/D$. In this case, the intrapage interpixel cross talk noise is negligible even when the pixels are closely packed. The finite thickess of the holographic medium is a main source of the interpage cross talk noise. It limits the number of the holograms and the size of the objects in the y_2 direction (in the plane of incidence). The upper bound of $p_\theta = n\lambda y_2/tf$ due to the interpage cross talk noise is $p_{\theta\max}(NSR_{re})$. The size of the objects in the x_2 direction is limited by the paraxial approximation requirement, i.e., $|x_2|/f < \alpha$, where α can be considered as the numerical aperture of the object (α is usually less than 1). Combining these three limitations, the cross talk–limited storage density can be written as

$$\rho_\theta = \frac{2\alpha p_{\theta\max}(NSR_{re})}{\lambda^3}. \tag{7.28}$$

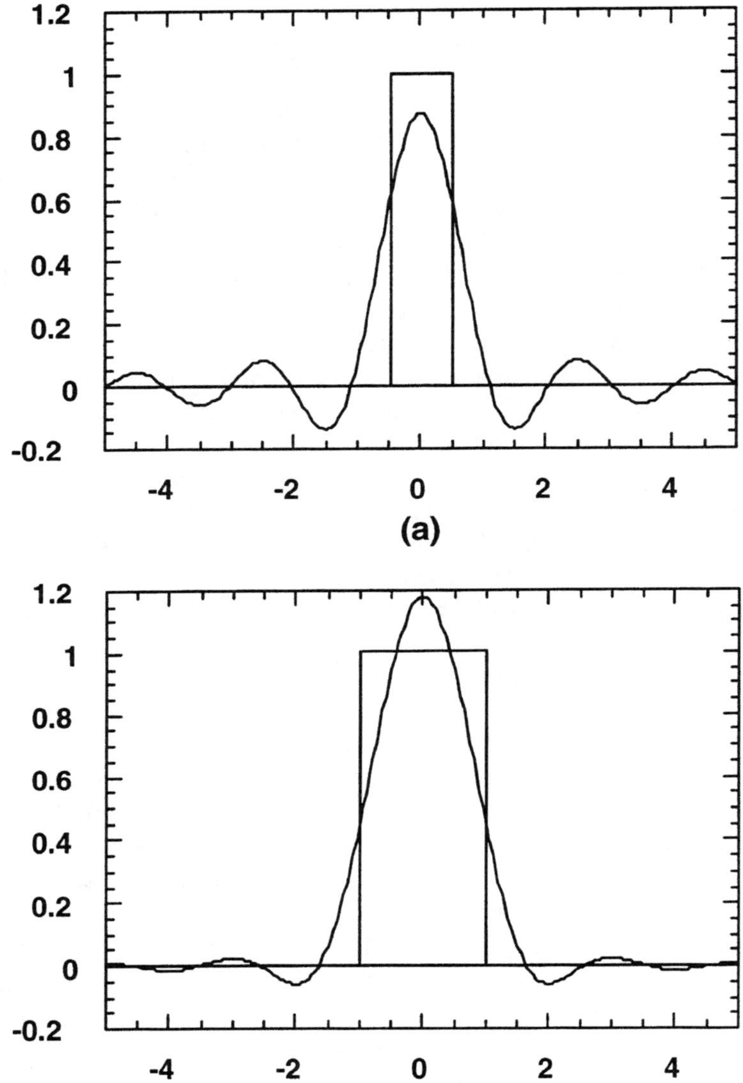

Figure 7.5: Convolution function $\int dx_0 \text{rect}(x_0/\delta)\text{sinc}[D(x_2 + x_0)/(\lambda f)]$ for the case (a) $\delta = \lambda f/D$ and (b) $\delta = 2\lambda f/D$. The rectangular function in each figure represents the pixel. The horizontal axis is the x_2-axis. The horizontal axis is in unit of $\lambda f/D$.

The cross talk–limited storage density is reduced from the ultimate storage density of $1/\lambda^3$ by a factor related to the NSR requirement.

7.1.2 Wavelength-multiplexed Fourier plane holographic memory

In this section, we consider the cross talk noise and the cross talk–limited storage capacity in wavelength-multiplexed Fourier plane holographic memory [14, 17, 20–22]. The formulation for this case is similar to that of the angle-multiplexed Fourier plane holographic memory. Thus, we omit the formulation for this case and present the main results directly.

Fourier-transform holograms are wavelength multiplexed in a volume holographic medium with the setup shown in Fig. 7.6. The counter-propagating configuration is the optimal configuration for wavelength multiplexing. A plane wave R_m of wavelength λ_m interferes with S_m, the Fourier transform of the mth object image, also of wavelength λ_m. We label the holograms $m = -M, -(M+1), \ldots, 0, \ldots, (M+1), M$. The NSR of the interpage cross talk noise can be written as

$$\mathrm{NSR} = \frac{1}{2}\sum_{m\neq n} \mathrm{sinc}^2\left[-2t\left(\frac{1}{\lambda_m} - \frac{1}{\lambda_n}\right) + \frac{t(x_2^2 + y_2^2)}{2f^2}\left(\frac{\lambda_m}{\lambda_n^2} - \frac{1}{\lambda_n}\right)\right] \tag{7.29}$$

where λ_n is the reading wavelength and is equal to one of the recording wavelengths, x_2 and y_2 are the coordinates of the output plane, t is the

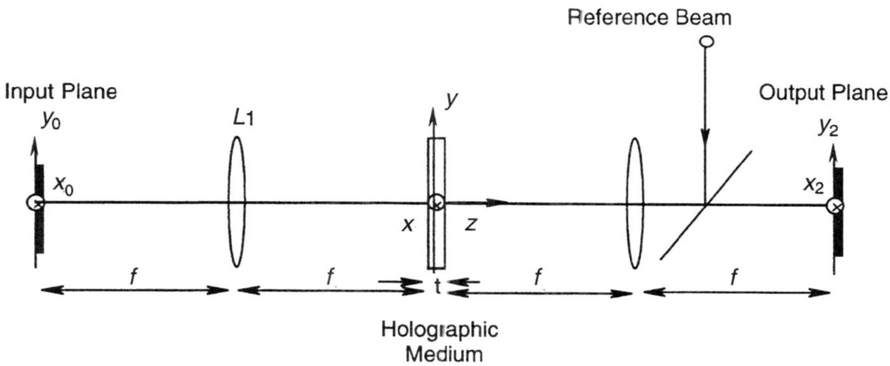

Figure 7.6: System configuration for wavelength-multiplexed Fourier plane holographic memory.

thickness of the holographic medium and f is the focal length. The frequency separation between the adjacent holograms is

$$\Delta\nu = s\frac{c}{2t} \tag{7.30}$$

where c is the speed of light in vacuum and, s is a hologram separation parameter (an arbitrary positive real number). When s is an integer, the center pixel of each hologram is placed at one of the zeros of the sinc functions of the two adjacent holograms. The individual frequencies are given as

$$v_m = \frac{c}{\lambda_m} = m\Delta\nu + \nu_o \qquad (m = -\mathrm{M}, \ldots, 0, 1, 2, \ldots, M) \tag{7.31}$$

We further define a cross talk variable as

$$p_\lambda = \frac{x_2^2 + y_2^2}{2f^2} \tag{7.32}$$

where f is the focal length of the lenses. Using Eqs. (7.30)–(7.32), Eq. (7.29) can be written as

$$\mathrm{NSR} = \frac{1}{2}\sum_{m\neq n} \mathrm{sinc}^2[s(m - n)(1 + p_\lambda)]\,. \tag{7.33}$$

We now denote $C_\lambda(s, p_\lambda) = \mathrm{NSR}$ which is a function of both s and p_λ. According to Eq. (7.33), we obtain a recurrence relationship,

$$C_\lambda(s, p_\lambda) = C_\lambda(1, sp_\lambda + s - 1)\,. \tag{7.34}$$

In Fig. 7.7, we show the noise-to-signal ratio as a function of p_λ for $s = 1$, i.e., $\mathrm{NSR} = C_\lambda(s = 1, p_\lambda)$. For the outermost reading point, $n = M$, there is only one adjacent reference point, $n = M - 1$. As a result, the cross talk noise is smaller than that of the reading points located slightly away from the edge (with smaller n). Other than this, the cross talk noise is independent of n. In our simulation, we choose n to be at least 30 less than M. According to Eq. (7.34), we can easily obtain from Fig. 7.7 the plot of the NSR as a function of p_λ for an arbitrary s, i.e., $\mathrm{NSR} = C_\lambda(s, p_\lambda)$. All we need to do is to move the p_λ axis to the right by $s - 1$ units then reduce the scale of the p_λ axis by a factor of s. Physically, $C_\lambda(s, p_\lambda)$ gives NSR as a function of position in the output plane (x_2, y_2).

Notice that $\mathrm{NSR} = C_\lambda(s, p_\lambda)$ looks like part of the $\mathrm{sinc}^2(\ )$ function. It consists of a series of side lobes with decreasing height. The height of

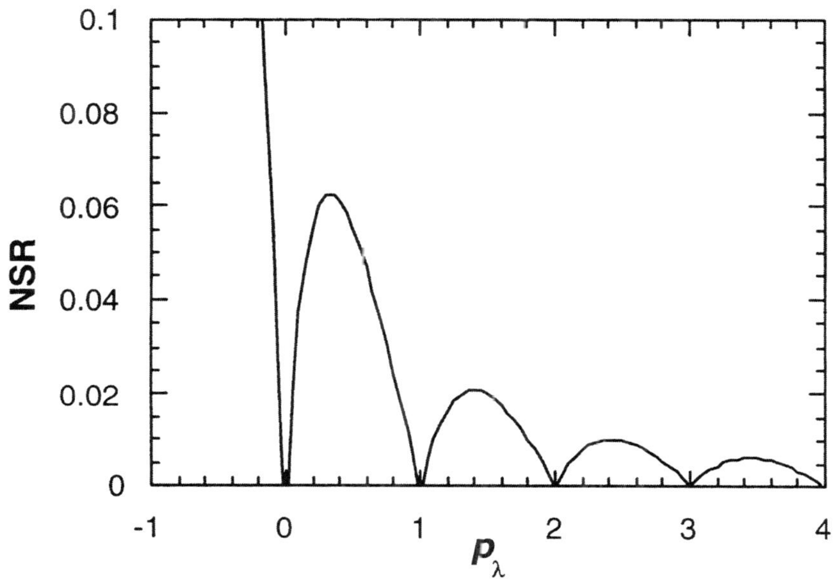

Figure 7.7: The NSR of the interpage cross talk noise as a function of $p_\lambda = (x_2^2 + y_2^2)/(2f^2)$, i.e., NSR = $C_\lambda(s, p_\lambda)$ for $s = 1$.

the first three side lobes are 0.06, 0.02, and 0.01. As we can see from Eq. (7.33), when we separate the center pixel ($p_\lambda = 0$) of each hologram by s sinc lobes from the two adjacent holograms, the pixels at other locations ($p_\lambda \neq 0$) will be separated by $s(1 + p_\lambda)$ sinc lobes from the two adjacent holograms. Since p_λ is always larger than zero, $s(1 + p_\lambda)$ will always be larger than s. In most practical systems, $s \geq 1$, the interpage cross talk noise is induced only by the side lobes of the sinc function response of the stored holograms. As a result, the maximum cross talk noise is always much less than one for $s \geq 1$.

We assume that the frequency tuning range is $(\lambda, u\lambda)$, where $u = \lambda_{\max}/\lambda_{\min}$. According to Eq. (7.30), the number of holograms that can be stored is $(1 - u^{-1})2t/(s\lambda)$. The minimum pixel size is limited by the longest wavelength. As an estimation, we assume that the minimum size of the pixels is $\delta_{\min} = 2u\lambda f/D$. Since the size of the object is not related to wavelength, the storage capacity is proportional to $u^{-2} - u^{-3}$. The storage capacity is maximum at $u = 3/2$.

When $0.06 < NSR_{re} < 0.2$ and $s \geq 1$, or when $0.02 < NSR_{re} < 0.06$ and $s \geq 2$, the interpage cross talk noise will be smaller than the required

NSR NSR_{re}. The size of the objects will be limited by the paraxial approximation requirement, i.e., $x_2/f < \alpha$ and $y_2/f < \alpha$. In this case, the storage density is

$$\rho_\lambda = \frac{8\alpha^2}{27s\lambda^3}. \tag{7.35}$$

Now we consider the case of $NSR_{re} < 0.06$ and $s = 1$. According to Fig. 7.7, $C_\lambda(1, p_\lambda) \approx p_\lambda/2$. The NSR can be written as

$$\mathrm{NSR} = \frac{x_2^2 + y_2^2}{8f^2} \qquad \begin{pmatrix} 0.06 < NSR_{re} < 0.2, & s \geq 1 \\ 0.02 < NSR_{re} < 0.06, & s \geq 2 \end{pmatrix}. \tag{7.36}$$

Since the maximum cross talk noise is about 0.06, the maximum size of the input object is limited by the cross talk noise. Thus, the storage density becomes

$$\rho_\lambda = \frac{32NSR_{re}}{27\lambda^3} \qquad (NSR_{re} < 0.06,\ s = 1). \tag{7.37}$$

The cross talk–limited storage density is reduced from the ultimate storage density of $1/\lambda^3$ by a factor related to the NSR requirement.

7.1.3 Angle-multiplexed image plane holographic memory

In this section, we calculate the NSR that is due to cross talk between image plane holograms purely from geometrical constraints.

Image plane holograms are angle-multiplexed in a volume holographic medium with the setup shown in Fig. 7.8. Two 4-f imaging systems are used to image the input object onto the medium for recording and then to the output plane for readout. The z-axis is chosen to be the optical axis of the imaging system. The yz-plane is the plane of incidence. Plane reference waves with certain angle separations propagate in the yz-plane. The input object with dimensions $a_x \times a_y$ in the (x_1, y_1) plane is made up of closely spaced rectangular pixels of size $\delta_x \times \delta_y$. The amplitude transmittance of each pixel can be modulated to be 1 and -1. This assumption is made only for simplicity. The more realistic case that the amplitude transmittance of each pixel is modulated to be 1 and 0 can be treated similarly. To obtain the maximum storage density, the transverse size of the holographic medium is taken to be equal to the size of the objects. The optical detector at the output plane (x_4, y_4) is also made up of closely

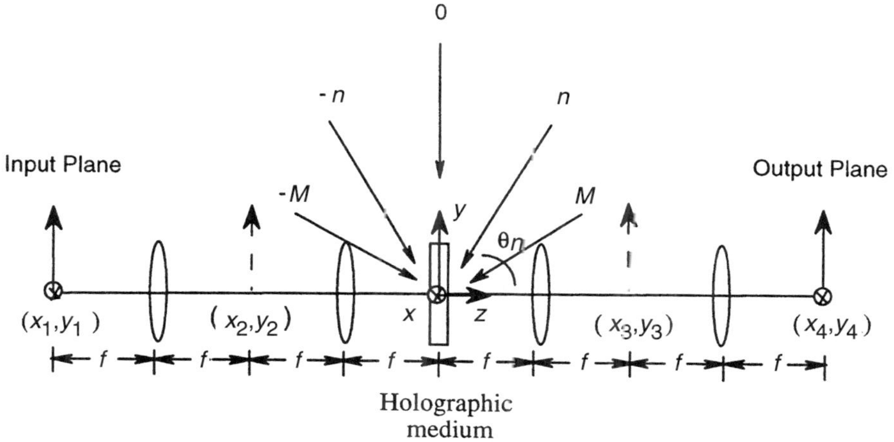

Figure 7.8: Recording and readout geometry for image plane holographic storage.

packed rectangular pixels of size $\delta_x \times \delta_y$. $N = 2M + 1$ phase (index) holograms are stored inside a volume holographic medium. We label these holograms as $m = -M, -(M - 1), \ldots, 0, \ldots, (M - 1), M$.

In image plane holographic memory, the input object is generated with a liquid crystal display (LCD), and the output image is detected with a charge coupled device (CCD) detector array. To ensure relatively simple pixel matching, the positions of the LCD pixels are fixed during the recording of multiple holograms, and the positions of the CCD pixels are also fixed during the readout process. The NSR of the cross talk noise in the output plane can be easily calculated if we assume that the position of the LCD display with respect to the optical axis is random, with a uniform distribution of probability over the size of a pixel $\delta_x \times \delta_y$, while the position of the CCD detector is fixed. Before doing so, we first prove that the NSR obtained under the assumption of a random input object position is equal to the averaged NSR over a detector pixel in the more realistic case that both the position of the input object and the position of the output detector are fixed.

Referring to Fig. 7.8, the optical wave amplitude at the intermediate plane (x_2, y_2) is the Fourier transform of the optical wave amplitude at the input plane (x_1, y_1), and the optical wave amplitude at the output plane (x_4, y_4) is the Fourier transform of the optical wave amplitude at the intermediate plane (x_3, y_3). Let $g_m(x_1, y_1)$ be the mth input object and

$G_m(x_2, y_2)$ be the optical wave amplitude at the intermediate plane (x_2, y_2) during the recording of the mth hologram, then we have

$$G_m(x_2, y_2) = \frac{1}{\lambda f}\iint g_m(x_1, y_1)\exp\left(-i2\pi\frac{x_2x_1}{\lambda f} - i2\pi\frac{y_2y_1}{\lambda f}\right)dx_1dy_1. \tag{7.38}$$

Let $E(x_3, y_3)$ and $e(x_4, y_4)$ be the optical wave amplitudes at the intermediate plane (x_3, y_3) and the output plane (x_4, y_4), respectively, during the readout of the nth hologram, then we have

$$e(x_4, y_4) = \frac{1}{\lambda f}\iint E(x_3, y_3)\exp\left(-i2\pi\frac{x_4x_3}{\lambda f} - i2\pi\frac{y_4y_3}{\lambda f}\right)dx_3dy_3. \tag{7.39}$$

In Eqs. (7.38) and (7.39), λ is the wavelength and f is the focal length of the lenses.

Again referring to Fig. 7.8, if the (x_2, y_2)-plane is taken as the input plane and the (x_3, y_3)-plane as the output plane, the image plane holographic storage system can be viewed as a Fourier plane holographic storage system. Since the size of the input objects is at most the same as the transverse size of the holographic medium, all the optical waves passing through the (x, y)-plane are effectively recorded. For mathematical simplicity, we can take the transverse size of the holographic medium to be infinitely large when considering the cross talk noise. Using the previously derived results on the cross talk noise in angle-multiplexed Fourier plane holograms, we can express the optical wave amplitude in the intermediate plane (x_3, y_3) as

$$\begin{aligned} E(x_3, y_3) \propto \int_{m=-M}^{M} & G_m\left(-x_3 - \frac{\lambda f}{2\pi}\Delta K_{mnx}, -y_3 - \frac{\lambda f}{2\pi}\Delta K_{mny}\right) \\ & \times \operatorname{sinc}\left[\frac{t}{2\pi}\left(\Delta K_{mnz} + \frac{\Delta K_{mnx}x_3 + \Delta K_{mny}y_3}{f} + \lambda\frac{\Delta K_{mnx}^2 + \Delta K_{mny}^2}{4\pi}\right)\right] \end{aligned} \tag{7.40}$$

where $\Delta\boldsymbol{K}_{mn} = \boldsymbol{k}_m - \boldsymbol{k}_n$ is the difference between the mth reference wave vector $\boldsymbol{k}_m$ and the reading beam wave vector $\boldsymbol{k}_n$. ΔK_{mna} is the component of $\Delta\boldsymbol{K}_{mn}$ in the α direction. Since all the reference beams lie in the yz-plane, we have $\Delta K_{mnx} = 0$.

Using Eqs. (7.38)–(7.40), we can express the optical wave amplitude of at the output plane as

$$e(x_4, y_4) = \sum_{m=-M}^{M}\frac{1}{A}\exp\left(i\frac{2\pi y_4 B}{A}\right)\int_{-A/2+y_4}^{A/2+y_4} g_m(x_4, y)\exp\left[i2\pi\left(-\frac{B}{A} + \frac{\Delta K_{miy}}{2\pi}\right)y\right]dy \tag{7.41}$$

where A and B are shorthand notations defined as

$$A = \frac{\lambda t}{2\pi}\Delta K_{mny} \tag{7.42}$$

$$B = \frac{t}{2\pi}\left(\Delta K_{mnz} + \frac{\lambda}{4\pi}\Delta K_{mny}^2\right). \tag{7.43}$$

Suppose the reading beam is identical to one of the reference beams during recording: the term with $\mathbf{k}_m = \mathbf{k}_n$ in Eqs. (7.40) and (7.41) corresponds to the reconstructed signal; all other terms in Eqs. (7.40) and (7.41) with $\mathbf{k}_m \neq \mathbf{k}_n$ give rise to the cross talk noise. Equation (7.41) is normalized so that the amplitude of the reconstructed signal at the output plane is $g_n(x_4, y_4)$. For further reference, we denote the amplitude of the cross talk noise in the intermediate plane (x_3, y_3) with $N(x_3, y_3)$ and the amplitude of the cross talk noise in the output plane (x_4, y_4) with $n(x_4, y_4)$.

We now consider the case that both the position of the input object and the position of the output detector are fixed. According to Eq. (7.41), the amplitude of the cross talk noise at the output plane (x_4, y_4) for a given y_4 is a constant of x_4 within a pixel, while the amplitude of the cross talk noise at the output plane (x_4, y_4) for a given x_4 is a varying function of y_4 within a pixel. The statistical properties of the cross talk noise at different pixels are identical as long as the pixels are not close to the edges of the output images. Using Eq. (7.41), we can write the autocorrelation function of the cross talk noise amplitude as

$$\langle n(x_4 + \Delta x_4, y_4 + \Delta y_4) n^*(x_4, y_4)\rangle = \sum_{\substack{m=-M \\ m\neq n}}^{M} \frac{1}{A^2}\exp\left(i2\pi B\frac{\Delta y_4}{A}\right)\int_{-A/2}^{A/2} dy' \int_{-A/2}^{A/2} dy''$$
$$\times \langle g_m(x_4 + \Delta x_4, y' + y_4 + \Delta y_4) g_m^*(x_4, y'' + y_4)\rangle \exp\left[i2\pi\left(-\frac{B}{A} + \frac{\Delta K_{miy}}{2\pi}\right)(y' - y'' + \Delta y_4)\right] \tag{7.44}$$

where $\langle\ \rangle$ denotes the ensemble average over random data patterns. In arriving at Eq. (7.44), we have used the relation

$$\langle g_{m'}(x_4 + \Delta x_4, y' + y_4 + \Delta y_4) g_{m''}^*(x_4, y'' + y_4)\rangle = 0 \tag{7.45}$$

for $m' \neq m''$. This relation can be easily justified under the condition that the data patterns for different holograms are independent of each other. We further define

$$\begin{aligned}&\overline{\langle n(x_4 + \Delta x_4, y_4 + \Delta y_4)n^*(x_4, y_4)\rangle} \\ &\quad = \frac{1}{\delta_x\delta_y}\int\int_{\delta x \times \delta y} \langle n(x_4 + \Delta x_4, y_4 + \Delta y_4)n^*(x_4, y_4)\rangle dx_4 dy_4\end{aligned} \tag{7.46}$$

$$\begin{aligned}&\overline{\langle g_m(x_4 + \Delta x_4, y' + y_4 + \Delta y_4)g_m^*(x_4, y'' + y_4)\rangle} \\ &\quad = \frac{1}{\delta_x\delta_y}\int\int_{\delta x \times \delta y} \langle g_m(x_4 + \Delta x_4, y' + y_4 + \Delta y_4)g_m^*(x_4, y'' + y_4)\rangle dx_4 dy_4 .\end{aligned} \tag{7.47}$$

Using Eqs. (7.44), (7.46), and (7.47), we obtain

$$\begin{aligned}\overline{\langle n(x_4 + \Delta x_4, y_4 + \Delta y_4)n^*(x_4, y_4)\rangle} &= \sum_{\substack{m=-M \\ m\neq n}}^{M} \frac{1}{A^2}\exp\left(i2\pi B\frac{\Delta y_4}{A}\right)\int_{-A/2}^{A/2} dy' \int_{-A/2}^{A/2} dy' \\ &\times \overline{\langle g_m(x_4 + \Delta x_4, y' + y_4 + \Delta y_4)g_m^*(x_4, y'' + y_4)\rangle} \\ &\exp\left[i2\pi\left(-\frac{B}{A} + \frac{\Delta K_{miy}}{2\pi}\right)(y' - y'' + \Delta y_4)\right].\end{aligned} \tag{7.48}$$

The integrations in Eqs. (7.46) and (7.47) are over the size of a pixel in the output plane. $\overline{\langle g_m(x_4 + \Delta x_4, y' + y_4 + \Delta y_4)g_m^*(x_4, y'' + y_4)\rangle}$ is independent of the exact location of the integration area as long as the integration area is not close to the edges of the input objects. Therefore, according to Eq. (7.48), $\overline{\langle n(x_4 + \Delta x_4, y_4 + \Delta y_4)n^*(x_4, y_4)\rangle}$ is also independent of the exact location of the integration area.

The case that the position of the input object with respect to the optical axis is random, with a uniform distribution of probability over the size of a pixel $\delta_x \times \delta_y$, while the position of the output detector is fixed can be treated similarly. In fact, we can interpret the integrations in Eqs. (7.46) and (7.47) as taking the ensemble average over the random input object position. Under this interpretation, Eq. (7.48) gives the autocorrelation function of the cross talk noise amplitude in the output plane as the ensemble average both over the random data pattern and over the random input object position. Therefore, we have proved that the autocorrelation function of the cross talk noise amplitude in the output plane in the case that the position of the input object is random over the size of a pixel δ_x

$\times\ \delta_y$ is equal to the averaged autocorrelation function over a detector pixel in the case that the position of the input object is fixed. In both cases, the position of the output detector is assumed to be fixed.

With the above proof, we now directly calculate the NSR of the cross talk noise in the output plane under the assumption of a random input object position. Since Fourier transform is a linear transformation, the autocorrelation function of the optical amplitude in the image plane can be obtained from the autocorrelation function of the optical amplitude in the Fourier plane and vice versa. In our analysis of the cross talk noise, we first obtain the autocorrelation function of an input object in the input plane and the autocorrelation function of its spectrum which is equivalent to the optical amplitude in the (x_2, y_2) plane. Then we obtain the autocorrelation function of the noise amplitude in the (x_3, y_3) plane with the help of previous results on the cross talk noise in angle-multiplexed Fourier plane holograms. Finally, we obtain the autocorrelation function of the noise amplitude in the output plane.

Assuming the pixels of each object are randomly modulated, we can obtain the autocorrelation function of an input object by taking the average both over space and over random data patterns,

$$\begin{aligned}&\langle g_m(x_1, y_1) g_m^*(x_1', y_1')\rangle \\ &= \operatorname{rect}\left(\frac{x_1 + x_1'}{2a_x}\right)\Lambda\left(\frac{x_1 - x_1'}{\delta_x}\right)\operatorname{rect}\left(\frac{y_1 + y_1'}{2a_y}\right)\Lambda\left(\frac{y_1 - y_1'}{\delta_y}\right)\end{aligned} \tag{7.49}$$

where rect(x) is the rectangular function and $\Lambda(x)$ is the triangular function. In the case that the position of the input object is fixed, the input object at two pointss within the same pixel is fully correlated while the input object at two point in two different pixels is totally uncorrelated. In this case, the autocorrelation function of the input object depends on the exact pixel location. Under the assumption of random input object location, the correlation of the input object at two points within the input object is independent of the exact location of the pixels. The correlation depends only on the separation between the two points and the correlation length is about the size of a pixel. This is described by the two triangular functions in Eq. (7.49). In deriving Eq. (7.49), we have also assumed that no light can pass through the input plane outside the input object. Under this assumption, the correlation at two points with one inside the input object and the other outside the input object is zero. This is described by the two rectangular functions in Eq. (7.49). This description is valid when the size of the input objects is much larger than the size of the pixels. In typical holographic storage systems, the number of pixels per row and the number of pixels per column are of the order of 1000. Therefore, the condition is well satisfied.

The Fourier transform, which relates $g_m(x_1, y_1)$ and $G_m(x_2, y_2)$, is a linear transform. Using Eqs. (7.38) and (7.49), we can obtain the autocorrelation function of the optical wave amplitude at the intermediate plane (x_2, y_2) during the recording of the mth hologram

$$\begin{aligned}\langle G_m(x_2, y_2)G_m^*(x_2', y_2')\rangle &= a_x\Delta_x a_y\delta_y \times \mathrm{sinc}\left[\frac{(x_2 - x_2')a_x}{\lambda f}\right]\\ \mathrm{sinc}^2\left[\frac{(x_2 + x_2')\delta_x}{2\lambda f}\right]&\mathrm{sinc}\left[\frac{(y_2 - y_2')a_y}{\lambda f}\right]\mathrm{sinc}^2\left[\frac{(y_2 + y_2')\delta_y}{2\lambda f}\right].\end{aligned} \tag{7.50}$$

Note that there is no cross-correlation between different objects with random data patterns and between their Fourier spectrums. According to Eq. (7.50), the spread of the spectrum is inversely proportional to the size of the pixels, and the correlation length of the spectrum is inversely proportional to the size of the input objects.

The optical wave amplitude in the intermediate plane (x_3, y_3) is related to the optical wave amplitude at the intermediate plane (x_2, y_2) by Eq. (7.63). Equation (7.40) is a linear transformation from $G_m(x_2, y_2)$ to $E(x_3, y_3)$. Using Eqs. (7.50) and (7.63), we obtain

$$\begin{aligned}\langle N(x_3, y_3)N^*(x_3', y_3')\rangle &\propto \sum_{\substack{m=-M\\ m\neq n}}^{M} a_x\delta_x a_y\delta_y\\ &\times \mathrm{sinc}\left(\frac{x_3 - x_3'}{\lambda f}a_x\right)\mathrm{sinc}^2\left[\frac{(x_3 + x_3')\delta_x}{2\lambda f}\right]\\ &\mathrm{sinc}\left(\frac{y_3 - y_3'}{\lambda f}a_x\right)\mathrm{sinc}^2\left[\left(\frac{y_3 + y_3'}{2\lambda f} + \frac{\Delta K_{miy}}{2\pi}\right)\delta_y\right]\\ &\times \mathrm{sinc}\left[\frac{t}{2\pi}\left(\Delta K_{miz} + \Delta K_{mny}\frac{y_3}{f} + \frac{\lambda}{4\pi}\Delta K_{mny}^2\right)\right]\\ &\mathrm{sinc}\left[\frac{t}{2\pi}\left(\Delta K_{miz} + \Delta K_{mny}\frac{y_3'}{f} + \frac{\lambda}{4\pi}\Delta K_{mny}^2\right)\right]\end{aligned} \tag{7.51}$$

where $\langle N(x_3, y_3)N^*(x_3', y_3')\rangle$ is the autocorrelation function of the noise amplitude in the (x_3, y_3) plane.

Remember that the image plane holographic storage system can be viewed as a Fourier plane holographic storage system if the (x_2, y_2) plane is taken as the input plane and the (x_3, y_3) plane as the output plane. Therefore, the angles of the reference beams in the case of image plane holographic storage are arranged in the same way as in the case of Fourier plane holographic storage. Eqs. (7.9), (7.11) and (7.12), regarding the angles of the reference beams and the momentum mismatch $\Delta\mathbf{K}_{mn}$, apply in this section as well.

In order to proceed from Eq. (7.51), we need to make two approximations. The first one is $y_3 \approx y_3' \approx (y_3 + y_3')/2$ in the last two sinc functions in Eq. (7.51). This requires $t\Delta K_{mny}(y_3 - y_3')/(4\pi f) << 1$. According to the third sinc function in Eq. (7.51), $y_3 - y_3'$ is of the order of $\lambda f/a_y$. According to Eqs. (7.9), (7.11) and (7.12), $\Delta K_{mny} = 2\pi(m - n)s\,\mathrm{ctan}(\theta_n)/t$. $\mathrm{ctan}(\theta_n) < 1$, since the reference beam angles are arranged around $\pi/2$ to reduce the cross talk noise. $|m - n|s < 100$, since the cross talk noise comes mainly from adjacent holograms. Typically, $\lambda = 0.5\ \mu\mathrm{m}$, $t = 1$ cm, $a_y = 1$ cm. Then we have $t\Delta K_{mny}(y_3 - y_3')/(4\pi f) < 2.5 \times 10^{-3}$. The second approximation is to neglect the term $\Delta K_{miy}\delta_y/(4\pi)$ in the fourth sinc function in Eq. (7.51). This requires $\Delta K_{miy}\delta_y/(4\pi) << 1$. As an estimation, the number of pixels per column in the y direction is of the order of 1000. Then we have $\Delta K_{miy}\delta_y/(4\pi) < 5 \times 10^{-2}$. Therefore, these two approximations are valid under typical conditions. With these two approximations, Eq. (7.51) can be simplified as

$$\langle N(x_3, y_3)N^*(x_3', y_3')\rangle \propto \sum_{\substack{m=-M \\ m\neq n}}^{M} \mathrm{sinc}^2\left[\frac{t}{2\pi}\left(\Delta K_{miz} + \Delta K_{mny}\frac{y_3 + y_3'}{2f} + \frac{\lambda}{4\pi}\Delta K_{mny}^2\right)\right]$$
$$\times\,\mathrm{sinc}\left(\frac{x_3 - x_3'}{\lambda f}a_x\right)\mathrm{sinc}^2\left[\frac{(x_3 + x_3')\delta_x}{2\lambda f}\right]\mathrm{sinc}\left(\frac{y_3 - y_3'}{\lambda f}a_y\right)\mathrm{sinc}^2\left[\left(\frac{y_3 + y_3'}{2\lambda f}\right)\delta_y\right]. \tag{7.52}$$

Equation (7.52) is ready for another Fourier transform. Using Eqs. (7.39) and (7.52), we obtain the autocorrelation function of the noise amplitude in the output plane,

$$\langle n(x_4, y_4)n^*(x_4', y_4')\rangle = \delta_y \mathrm{rect}\left(\frac{x_4 + x_4'}{2a_x}\right)\Lambda\left(\frac{x_4 - x_4'}{\delta_x}\right)\mathrm{rect}\left(\frac{y_4 + y_4'}{2a_y}\right)$$
$$\times \int dY_3 \mathrm{sinc}^2(Y_3\delta_y)\exp[-i2\pi Y_3(y_4 - y_4')] \sum_{\substack{m=-M \\ m\neq n}}^{M} \tag{7.53}$$
$$\mathrm{sinc}^2\left[\frac{t}{2\pi}\left(\Delta K_{miz} + \Delta K_{mny}\lambda Y_3 + \frac{\lambda}{4\pi}\Delta K_{mny}^2\right)\right]$$

where $Y_3 = (y_3 + y_3')/(2\lambda f)$ is simply an integration argument. Equation (7.53) is normalized so that the signal intensity is one. The two rectangular functions in Eq. (7.53) show that the cross talk noise in the output plane is uniform within the output image and is zero outside the output image. Since we are interested only in the cross talk noise within the output image, we can replace the two rectangular functions in Eq. (7.53) with one. The triangular function in Eq. (7.53) is the result of the random input object position assumption. The random input object position assumption is used as an approximation to study the case that the input object position is fixed. When the input object position is fixed, according to Eq. (7.41), the cross talk noise in the output plane for a given y-coordinate is independent of the x-coordinate within any one pixel. Since the pixels of the optical detector at the output plane are matched with the pixels of the output image, we can replace x' with x in Eq. (7.53). These simplifications lead to

$$\langle n(x_4, y_4) n^*(x_4, y_4') \rangle = \delta_y \int dY_3 \operatorname{sinc}^2(Y_3 \delta_y) \tag{7.54}$$

$$\exp[-i2\pi Y_3(y_4 - y_4')] \sum_{\substack{m=-M \\ m \neq n}}^{M} \operatorname{sinc}^2\left[\frac{t}{2\pi}\left(\Delta K_{miz} + \Delta K_{mny}\lambda Y_3 + \frac{\lambda}{4\pi}\Delta K_{mny}^2\right)\right].$$

Equation (7.54) gives the averaged autocorrelation function of the noise amplitude in the output plane along the y-coordinate only. When $y_4 = y_4'$, Eq. (7.54) gives the averaged amplitude variance of the noise at an arbitrary point in the output plane. Since Eq. (7.54) has been normalized so that the signal intensity is one, the amplitude variance of the noise is equal to the NSR of the noise. The range of the integration in Eq. (7.54) requires special attention. Here, we choose the range of the integration to be from $-1/\delta_y$ to $1/\delta_y$. In other words, we include only the main lobe of the function $\operatorname{sinc}^2(Y_3\delta_y)$ in the range of the integration. Physically, this is equivalent to inserting a low-pass filter $\operatorname{rect}[y_2\delta_y/(2\lambda f)]$ at the intermediate plane (x_2, y_2) during the hologram recording process. According to the analysis in Section 7.1, the interpixel cross talk noise introduced by such a low-pass filter is negligible. In general, all lenses have a finite numerical aperture, which acts as a low-pass filter. According to Eq. (7.54), the noise due to the main lobe of the function $\operatorname{sinc}^2(Y_3\delta_y)$ not only has a large variance at an arbitrary point but also has a correlation length about the size of 1 pixel. The variance of such kind of noise at an arbitrary point can be used to describe the noise over 1 pixel. On the other hand, the noise due to all the side lobes of the function $\operatorname{sinc}^2(Y_3\delta_y)$

has a correlation length much smaller than the size of one pixel. Even if the noise due to all the side lobes has a variance comparable to (or larger than) that due to the main lobe, it will be negligible in typical storage systems where the size of the detector pixel is close to the size of the object pixel. Therefore, in the case of an area detector pixel, the NSR of the cross talk noise can be written

$$\text{NSR} = \delta_y \int_{-1/\delta_y}^{1/\delta_y} dY_3 \text{sinc}^2(Y_3\delta_y) \sum_{\substack{m=-M \\ m\neq n}}^{M} \text{sinc}^2\left[\frac{t}{2\pi}\left(\Delta K_{miz} + \Delta K_{mny}\lambda Y_3 + \frac{\lambda}{4\pi}\Delta K_{mny}^2\right)\right]. \tag{7.55}$$

Remember $Y_3 = (y_3 + y_3')/(2\lambda f)$. The function $\text{sinc}^2(Y_3\delta_y)$ in Eq. (7.55) can be interpreted as the signal intensity distribution in the intermediate plane (x_3, y_3). The summation in Eq. (7.55) can be interpreted as the NSR in the intermediate plane (x_3, y_3). According to Eq. (7.55), the NSR at the output image plane (x_4, y_4) is simply the weighted average of the NSR in the intermediate Fourier plane (x_3, y_3), and the weighting function is the signal intensity at the intermediate Fourier plane (x_3, y_3). Note that the NSR in the Fourier plane is a varying function of the y-coordinate while the NSR in the image plane is independent of the y-coordinate. Since the nonuniformly distributed noise in the Fourier plane is uniformly distributed in the image plane, the NSR in the image plane is expected to be smaller than the worst NSR in the Fourier plane.

The summation in Eq. (7.55) has been studied in detail in Section 7.1. Using the notations defined in that section, we can rewrite Eq. (7.55) as

$$\text{NSR} = 2\int_{-1}^{1} d\xi \text{sinc}^2(\xi) C_\theta\left(s, \frac{n\lambda^2}{t\delta_y}\xi\right). \tag{7.56}$$

Note that the factor 2 comes from the assumption that the pixels are modulated to be 1 and -1 in this section instead of 1 and 0 as in Section 7.1. When the pixels are assumed to be turned on and off and the pixels being turned on are assumed to have random phase, the NSR will be reduced by half. We define the cross talk variable for angle-multiplexed image plane holograms as

$$p_\theta' = \frac{n\lambda^2}{t\delta_y}. \tag{7.57}$$

According to Eq. (7.56), the NSR is a function of the hologram separation parameter s and the cross talk variable p'_θ we denoted as NSR = $C'_\theta(s, p'_\theta)$.

Generally speaking, the NSR increases with the cross talk variable for a given hologram separation. According to the definition of the cross talk variable, the cross talk noise limits the minimum size of the pixels in the y-direction and the maximum number of the stored holograms. We denote the maximum cross talk variable for a given NSR requirement and a given hologram separation as $p'_{\theta\max}(s, NSR_{re})$. We denote the maximum cross talk variable for a given NSR requirement and an arbitrary hologram separation as $p'_{\theta\max}(NSR_{re})$. The size of the pixels in the x-direction is limited by the finite numerical aperture of the lenses. If α' is the numerical aperture of the lenses, the minimum pixel size in the x-direction will be $\delta_x = \lambda/\alpha'$. As has been mentioned in the preceding discussion, the size of image is limited by the transverse size of the holographic medium. Then we can obtain the storage density as

$$\rho_\theta = \frac{2\alpha' p'_{\theta\max}(NSR_{re})}{\lambda^3}. \tag{7.58}$$

The cross talk–limited storage density is reduced from the ultimate storage density of $1/\lambda^3$ by a factor related to the NSR requirement. It is interesting to note the similarity between Eqs. (7.58) and Eq. (7.28). Although the definitions of α' in Eq. (7.58) and α in Eq. (7.28) are different, both of them are the result of the finite numerical aperture of the lenses and their values are close to each other.

We first consider the case that the hologram separation parameter s is equal to one. In this case, we have $C_\theta(s = 1, p'_\theta) = 0.5p_\theta$. Carrying out the integration in Eq. (7.56), we obtain

$$\text{NSR} = 0.4p'_\theta \tag{7.59}$$

$$p'_{\theta\max}(s = 1, NSR_{re}) = 2.5NSR_{re}. \tag{7.60}$$

Comparing Eqs. (7.60) and (7.21) and taking into account the factor 2 mentioned in the paragraph after Eq. (7.56), we can see that the cross talk–limited storage density of image plane holographic memory is about twice that of the Fourier plane holographic memory when the hologram separation parameter s is equal to one. As has been mentioned, image plane holographic memory can be viewed in some sense as Fourier plane holographic memory. Therefore, the small difference in the cross talk–limited storage density between these two cases is not surprising.

Analytical expression for the maximum cross talk variable $p'_{\theta\max}$ (NSR_{re}) is not available in general. For a given hologram separation, we can use Eq. (7.56) to generate a plot of the NSR versus the cross talk variable p'_{θ}. From that plot, the maximum cross talk variable $p'_{\theta\max}(s, NSR_{re})$ can be immediately found out. The qualitative properties revealed in the special case of hologram separation $s = 1$ remain valid for arbitrary hologram separation s.

In this section, we have analyzed the cross talk noise and the cross talk–limited storage capacity for angle-multiplexed Fourier plane holographic memory and wavelength-multiplexed Fourier plane holographic memory. A complete analysis is provided for the case of angle-multiplexed Fourier plane holographic memory. Previous works have been limited to the case that the angle separation between the adjacent holograms is an integral number of sinc lobes and that the reference beam angle is close to $\pi/2$. In our analysis, we rederived a general expression for the interpage cross talk noise so that it applies to a wider range of the reference beam angles. We also considered the general case when the angle separation between the reference beams of the adjacent holograms is arbitrary. In general, the interpage cross talk noise, due to the finite thickness of the holographic medium, limits the number of the holograms and size of the object. The intrapage interpixel cross talk noise, due to the finite transverse size of the holographic medium, limits the size of the pixels. In addition, the cross talk–limited storage density is reduced from the ultimate geometric limit by a factor related to the NSR requirement. The cross talk noise of wavelength-multiplexed Fourier plane holographic memory is also analyzed. The behavior of the interpage cross talk noise in this case is different from that of the angle-multiplexed Fourier plane holographic memory. When the hologram spacings in terms of the hologram separation parameter s are chosen to be the same for these two cases, the worst NSR for angle multiplexing will be larger than that for wavelength multiplexing. On the other hand, the minimum pixel size in wavelength-multiplexed Fourier plane holographic memory has to be larger than that in angle-multiplexed Fourier plane holographic memory. As a result, the cross talk–limited storage densities for these two cases are close to each other. We have also presented a statistical theory of the cross talk noise in angle-multiplexed image plane holograms and obtained the cross talk–limited storage density. We find that the cross talk noise limits the size of the pixels and the number of recorded holograms and that the cross talk–limited storage density is reduced from the ultimate density of $1/\lambda^3$ by a factor related to the SNR. We also find that the cross talk noise in image plane holograms is lower than that in Fourier plane holograms. Overall, in Fourier plane holograms the cross talk noise increases as the object size increases, whereas the noise in image plane holograms depends on the pixel size. In either case, the cross talk noise increases as object beam bandwidths in the recording plane become higher.

7.2 Grating detuning

In this section, we consider the effects of the expansion or shrinkage of the storage medium in Fourier plane holographic memories with plane and Gaussian reference waves. Results concerning the Bragg mismatch, pixel displacement, and impulse broadening are obtained and discussed.

Since the expansion or shrinkage of organic polymers can be up to an order of magnitude larger than that of other storage media such as photorefractive crystals, we will model the storage medium according to the properties of organic polymers. The qualitative results remain valid for other storage media. When they are used as storage media, organic polymers are usually made into thin films sandwiched between two layers of supporting substrates. The thickness of the thin film ranges from 100 μm to 1 mm. The thin film can expand or shrink freely in the direction perpendicular to the thin film. When the thin film is adhered to the substrates and the expansion or shrinkage properties of the thin film and the substrate are different, the expansion or shrinkage of the thin film in the direction parallel to the thin film will be affected by the substrate. As a result, the expansion or shrinkage of the thin film in the direction perpendicular to the thin film will be different from that in the direction parallel to the thin film. When the thin film is not adhered to the substrates or the expansion or shrinkage properties of the thin film and the substrate are the same, the expansion or shrinkage of the thin film in the direction perpendicular to the thin film will be the same as that in the direction parallel to the thin film. The anisotropic expansion or shrinkage of the thin film will also result in an optical anisotropy. The anisotropic refractive index of the thin film will be uniaxial with the optical axis perpendicular to the thin film. An ordinarily polarized optical wave propagating through the thin film will still experience an isotropic refractive index.

We assume that the storage media is a thin media with the transverse size much larger than the thickness. The (x, y, z)-coordinate system is chosen as shown in Fig. 7.9 with the origin at the center of the storage media. The storage media at the origin is assumed to remain at the origin during the expansion or shrinkage of the storage medium. The expansion or shrinkage of the storage medium is always assumed to be isotropic along the xy-plane in the following analysis. Consider two points on the storage medium. Let their distances along the xy-plane before and after the expansion or shrinkage of the storage medium be L_h and L'_h, respectively, then the expansion coefficient along the xy-plane is defined as $\rho_h = L_h/L'_h$. Let their distances along the z-axis before and after the expansion or shrinkage of the storage medium be L_v and L'_v, respectively, then the expansion coefficient along the z-axis is defined as $\rho_v = L_v/L'_v$. When $\rho_v = \rho_h$, the expansion or shrinkage of the storage medium is isotropic. When $\rho_v \neq \rho_h$, the expansion or shrinkage of the storage medium is anisotropic. A change of the storage medium density is often accompanied with a change of the refractive index. The refractive index of the storage medium is assumed to be isotropic in the following analysis. Let the refractive indices of the storage medium before and after the expansion

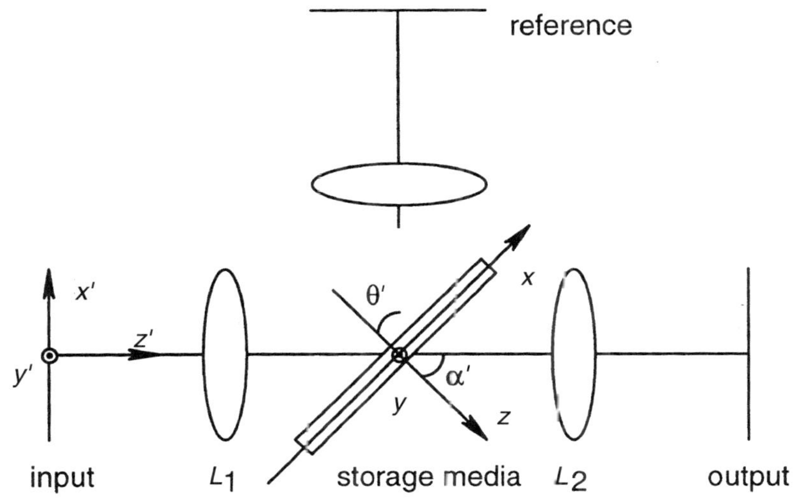

Figure 7.9: System configuration of a Fourier plane holographic memory.

or shrinkage of the storage medium be n and n', then the index tuning coefficient is defined as $\rho_n = n'/n$. The effects of the expansion or shrinkage of the storage medium may be partially compensated by using different recording wavelength and readout wavelength. Let the wavelengths of the recording wave and the readout wave be λ and λ', respectively, then the wavelength tuning coefficient is defined as $\rho_\lambda = \lambda/\lambda'$. Furthermore, notice that $|\rho - 1| << 1$, where ρ can be any one of the coefficients defined above. Therefore, only effects to the first-order term of $\rho - 1$ need to be considered.

In the following analysis, we will study only the Fourier plane holographic memory. Again referring to Fig. 7.9, a 4-f imaging system is utilized to transform the input object onto the storage medium during the recording and then to reconstruct the output image during readout, where f is the focal length of the Fourier-transforming lenses. Holographic recording and readout are essentially linear processes with respect to the signal wave amplitude. Therefore, it is sufficient to study only the impulse response of the holographic storage system. During the recording, a point source at (x', y') on the input plane will be transformed into a single plane wave by the front lens L_1. The wavevector of the plane wave is denoted as $\boldsymbol{k}_s$, $(\boldsymbol{k}'_s)$ when it is inside (outside) the storage medium. The front and back surfaces of the storage medium are assumed to be optically flat and parallel to each other. As a result, the plane wave passes through the storage medium without changing its direction. It is then refocused into a point image at $(-x', -y')$ on the output plane. The plane wave and the recording reference wave record a hologram in the storage medium. When the readout reference wave is directed upon the recorded hologram, a diffracted wave is generated. The diffracted wave may consist of one or more plane wave components. The wavevector of a plane wave component of the diffracted wave is denoted as $\boldsymbol{k}_d$ $(\boldsymbol{k}'_d)$ when it is inside (outside) the storage medium. The plane wave component will be refocused by the rear lens L_2 into a point image at $(-x' - \Delta x', -y' - \Delta y')$ on the output plane. If the diffracted wave consists of a single plane-wave component, then the impulse response of the holographic storage system is simply a point image on the output plane. If the diffracted wave consists of multiple plane wave components, then the impulse response of the holographic storage system has a finite spatial spread on the output plane. It is our task to obtain the displacement $\boldsymbol{d}$the relative diffraction efficiency η, and the spatial spread of the storage system impulse response as a function of the point source position (x', y') on the input plane.

Consider the case that the diffracted wave consists of a single plane wave component. Then the impulse response of holographic storage system is simply a point image on the output plane. The displacement $\boldsymbol{d} = \Delta x'\hat{x}' + \Delta y'\hat{y}'$ is a function of the point source position (x', y'), i.e., $\boldsymbol{d} = \boldsymbol{d}(x', y')$. In general, the difference between two values of the displacement $\boldsymbol{d}$ due to two point sources within the same pixel on the input plane will be much smaller than the size of the pixel. So the presence of the displacement $\boldsymbol{d}$ will not distort the shape of individual pixels. However, the difference between two values of the displacement $\boldsymbol{d}$ due to two point sources separated far away on the input plane can be comparable to or larger than the size of the pixel. Therefore, $\boldsymbol{d}$ is often referred to as pixel displacement. The effect of the pixel displacement $\boldsymbol{d}$ on the storage system depends on the detection scheme. In the oversampling detection scheme, the number of pixels of the output detector array is much larger than that of the input spatial light modulator. Each image pixel is detected by multiple detector pixels. The displacement of the image pixel can be subsequently calculated and corrected through digital signal processing. In this detection scheme, pixel displacement is not a problem. The drawback of this detection scheme is that a large detector pixel density is required and that complicated digital signal processing will slow down the output. In the pixel-matching detection scheme, the number of pixels of the output detector array is the same as that of the input spatial light modulator. Each input pixel is to be reconstructed precisely onto the corresponding detector pixel. This leads to a simple detector array and fast output rate, but places stringent requirements on the optical system. For an ideal optical system in which the output image is identical to the input object, pixel matching can be achieved if the pixel size and position of the output detector array are the same as those of the input spatial light modulator. For a realistic optical system, the pixel displacement $\boldsymbol{d}(x', y')$, if not compensated, will lead to pixel mismatch between the output image and the detector array. Therefore, the study of pixel displacement is relevant mainly to the pixel-matching detection scheme.

Some useful relations among the quantities mentioned above can be easily obtained. The wavevector $\boldsymbol{k}_s'$ of the recording plane wave generated by a point source at (x', y') on the input plane outside the storage media is given by

$$\boldsymbol{k}_s' = \frac{2\pi}{\lambda}\left(-\frac{x'}{l_s}\hat{x}' - \frac{y'}{l_s}\hat{y}' + \frac{f}{l_s}\hat{z}'\right) \tag{7.61}$$

where $l_s = \sqrt{x'^2 + y'^2 + f^2}$. The wavevector of the readout plane wave component refocused into a point image at $(-x' - \Delta x', -y' - \Delta y')$ on the output plane outside the storage media is given by

$$\boldsymbol{k}'_d = \rho_\lambda \frac{2\pi}{\lambda}\left(-\frac{x' + \Delta x'}{l_d}\hat{x}' - \frac{y' + \Delta y'}{l_d}\hat{y}' + \frac{f}{l_d}\hat{z}'\right) \tag{7.62}$$

where $l_d = \sqrt{(x' + \Delta x')^2 + (y' + \Delta y')^2 + f^2}$. According to Eqs. (7.61) and (7.62), the pixel displacement $\boldsymbol{d}$ can then be written as

$$\boldsymbol{d} = -(\Delta u k'_{sx'} + u\Delta k'_{x'})\hat{x}' - (\Delta u k'_{sy'} + u\Delta k'_{y'})\hat{y}' \tag{7.63}$$

where $u = f/k'_{sz'}$, $\Delta u = -f\Delta k'_{z'}/k'^{\,2}_{sz'}$, $\Delta \boldsymbol{k}' = \boldsymbol{k}'_d - \tau \boldsymbol{k}'_s$ and τ is an arbitrary constant. Strictly speaking, τ should be equal to one. However, as long as the value of τ is close to one, the above relation will be accurate to the first order of $|\tau - 1|$. In the following analysis, we will choose $\tau = \rho_h$. $\Delta \boldsymbol{k}'$ can be first obtained in the (x, y, z)-coordinates, then transformed back to the (x', y', z')-coordinates. A wavevector $\boldsymbol{k}'$ can be projected onto the (x', y', z')-coordinate as $\boldsymbol{k}' = k'_{x'}\hat{x}' + k'_{y'}\hat{y}' + k'_{z'}\hat{z}'$, and onto the (x, y, z)-coordinate as $\boldsymbol{k}' = k'_x\hat{x} + k'_y\hat{y} + k'_z\hat{z}$. Then the coordinate transformation can be written as

$$(k'_x, k'_y, k'_z) = (k'_{x'}\cos\alpha' + k'_{z'}\sin\alpha', k'_{y'}, -k'_{x'}\sin\alpha' + k'_{z'}\cos\alpha') \tag{7.64}$$

where α' is the angle between the z-axis and the z'-axis. The wavevectors $\boldsymbol{k}$ and $\boldsymbol{k}'$ of the same plane wave inside and outside the storage medium are uniquely related by the boundary conditions, which can be written in the (x, y, z)-coordinate as

$$\boldsymbol{k}'_x = \boldsymbol{k}_x \quad \text{and} \quad \boldsymbol{k}'_y = \boldsymbol{k}_y. \tag{7.65}$$

Using Eqs. (7.61), (7.64), and (7.65), the wavevector $\boldsymbol{k}_s$ of the recording plane wave inside the storage medium can be written explicitly as a function of the coordinates (x', y') of the original point source.

The effects of the expansion or shrinkage of the storage medium depend on the wave front of the reference wave. In angle multiplexing and wavelength multiplexing, the reference wave consists of a single plane-wave component. In shift multiplexing and code multiplexing, the reference wave consists of multiple plane-wave components. In the following analysis, we will study the case that the reference wave is a plane wave and the case that the reference wave is a Gaussian wave separately.

7.2.1 Plane reference wave

In this section, we consider the case that the reference wave is a plane wave. Referring to Fig. 7.10, we assume that both the recording reference wave and the readout reference wave propagate in the xz-plane. The wavevector of the recording reference wave is denoted as $\boldsymbol{k}_r$ ($\boldsymbol{k}'_r$) inside (outside) the storage medium. The angle between the wavevector of the recording reference wave and the z-axis is denoted as θ (θ') inside (outside) the storage medium. Besides tuning the wavelength, we can also tune the incident angle of the readout reference beam to compensate the effects of expansion or shrinkage of the storage medium. The angle between the wavevectors of the recording reference wave and the readout reference wave is denoted as $\Delta\theta$ ($\Delta\theta'$) inside (outside) the storage medium. Two plane waves will write a sinusoidal grating. The wavevector of the grating is denoted as $\boldsymbol{k}_g$ ($\boldsymbol{k}'_g$) before (after) the expansion or shrinkage of the storage medium. We assume that the transverse size of the storage medium is infinitely large. As a result, the diffracted wave will also be a single plane wave. The Bragg-matching condition along the xy-plane is strictly required. This condition determines the wavevector of the dif-

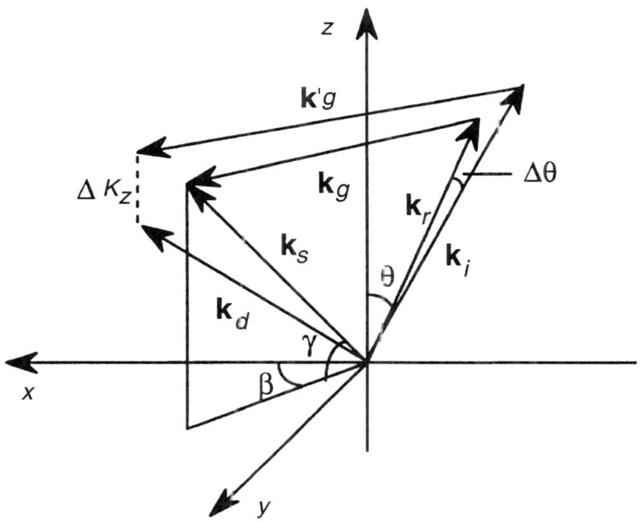

Figure 7.10: k-space representation of holographic recording and readout. Both the signal wave and the reference wave are plane waves.

fracted wave. A small Bragg-mismatch can be tolerated along the z-axis due to the finite thickness of the storage medium. This mismatch will lead to a reduced diffraction efficiency.

The qualitative discussion above will now be described mathematically. Since the optical waves are plane waves and the gratings are sinusoidal gratings, a geometrical approach will be sufficient and convenient for our analysis. The wavevector $\boldsymbol{k}_s$ of the recording signal wave inside the storage medium can be more conveniently represented by its angular coordinates β and γ as is shown in Fig. 7.9. β and γ are functions of the point source position (x', y') on the input plane, i.e., $\beta = \beta(x', y')$ and $\gamma = \gamma(x', y')$. Explicit expressions for the trigonometrical functions of β and λ can be easily obtained from Eqs. (7.1), (7.4) and (7.5). For example,

$$\tan\beta = \frac{y'}{x'\cos\alpha' - f\sin\alpha'} \tag{7.66}$$

$$\cos\gamma = \frac{\sqrt{(x'\cos\alpha' - f\sin\alpha')^2 + y'^2}}{nl_s} \tag{7.67}$$

where n is the refractive index of the medium before the recording, and $l_s = \sqrt{x'^2 + y'^2 + f^2}$. The wavevectors of the recording signal wave and the recording reference wave inside the storage medium can be written as

$$\boldsymbol{k}_r = \frac{2\pi n}{\lambda}(-\sin\theta\hat{x} + 0\hat{y} + \cos\theta\hat{z}), \tag{7.68}$$

$$\boldsymbol{k}_s = \frac{2\pi n}{\lambda}(\cos\gamma\cos\beta\hat{x} + \cos\gamma\sin\beta\hat{y} + \sin\gamma\hat{z}). \tag{7.69}$$

The wavevectors of the grating before and after the expansion or shrinkage of the storage medium can be written as

$$\boldsymbol{k}_g = \frac{2\pi n}{\lambda}[(\cos\gamma\cos\beta + \sin\theta)\hat{x} + \cos\gamma\sin\beta\hat{y} + (\sin\gamma - \cos\theta)\hat{z}], \tag{7.70}$$

$$\boldsymbol{k}'_g = \frac{2\pi n}{\lambda}[\rho_h(\cos\gamma\cos\beta + \sin\theta)\hat{x} + \rho_h\cos\gamma\sin\beta\hat{y} + \rho_v(\sin\gamma - \cos\theta)\hat{z}]. \tag{7.71}$$

The wavevector of the readout reference wave inside the storage medium can be written as

$$\boldsymbol{k}_i = \rho_n\rho_\lambda\frac{2\pi n}{\lambda}(-\sin\theta\hat{x} + 0\hat{y} + \cos\theta\hat{z}) - \Delta\theta\frac{2\pi n}{\lambda}(\cos\theta\hat{x} + 0\hat{y} + \sin\theta\hat{z}). \tag{7.72}$$

Using the Bragg-matching condition in the xy-plane, $k_{dx} = k_{ix} + k'_{gx}$, $k_{dy} = k_{iy} + k'_{gy}$, and $k_{dz} = \sqrt{(2\pi n \rho_n \rho_\lambda/\lambda)^2 - k_{dx}^2 - k_{dy}^2}$, we obtain the wavevector of the diffracted wave as

$$\begin{aligned}\boldsymbol{k}_d = & \hat{x}\frac{2\pi n}{\lambda}[\rho_h \cos\gamma\cos\beta + (\rho_h - \rho_n\rho_\lambda)\sin\theta - \Delta\theta\cos\theta] + \hat{y}\frac{2\pi n}{\lambda}\rho_h \cos\lambda\sin\beta \\ & + \hat{z}\frac{2\pi n}{\lambda}\left\{\rho_n\rho_\lambda\sin\gamma + (\rho_n\rho_\lambda - \rho_h)\frac{\cos^2\gamma}{\sin\gamma} + [(\rho_n\rho_\lambda - \rho_h)\sin\theta\right. \\ & \left. + \Delta\theta\cos\theta]\frac{\cos\gamma\cos\beta}{\sin\gamma}\right\},\end{aligned} \tag{7.73}$$

and the Bragg mismatch along the z-axis, $\Delta K_z = k_{dz} - k_{iz} - k'_{gz}$, can be written explicitly as

$$\begin{aligned}\Delta K_z = & \frac{2\pi n}{\lambda}(\rho_n\rho_\lambda - \rho_v)(\sin\gamma - \cos\theta) + \frac{2\pi n}{\lambda}(\rho_n\rho_\lambda - \rho_h)\frac{\cos\gamma}{\sin\gamma}(\cos\gamma + \cos\beta\sin\theta) \\ & + \frac{2\pi n}{\lambda}\Delta\theta\left(\sin\theta + \frac{\cos\gamma\cos\beta}{\sin\gamma}\cos\theta\right).\end{aligned} \tag{7.74}$$

The relative diffraction efficiency can be written as

$$\eta = \mathrm{sinc}^2\left(\frac{\Delta K_z t}{2\pi}\right) \tag{7.75}$$

where t is the thickness of the storage medium. Whether t is the thickness before or after the expansion or shrinkage of the storage medium makes no significant difference in Eq. (7.75). According to the boundary condition, the wavevector $\boldsymbol{k}'_d$ can be obtained as $k'_{dx} = k_{dx}$, $k'_{dy} = k_{dy}$, and $k'_{dz} = \sqrt{(2\pi\rho_\lambda/\lambda)^2 - k'^2_{dx} - k'^2_{dy}}$, and the wavevector $\boldsymbol{k}'_s$ can be written as $k'_{sx} = (2\pi n/\lambda)\cos\gamma\cos\beta$, $k'_{sy} = (2\pi n/\lambda)\cos\gamma\sin\beta$ and $k'_{sz} = \sqrt{(2\pi/\lambda)^2 - k'^2_{sx} - k'^2_{sy}}$. Then the wavevector difference $\Delta\boldsymbol{k}' = \boldsymbol{k}'_d - \rho_h\boldsymbol{k}'_s$ can be written explicitly as

$$\begin{aligned}\Delta\boldsymbol{k}' = & \hat{x}\frac{2\pi n}{\lambda}[(\rho_h - \rho_n\rho_\lambda)\sin\theta - \Delta\theta\cos\theta] \\ & + \hat{z}\frac{2\pi n}{\lambda}\frac{1}{\sqrt{\dfrac{1}{n^2} - \cos^2\gamma}}\left\{\frac{(\rho_\lambda - \rho_h)}{n^2} + \cos\gamma\cos\beta[(\rho_n\rho_\lambda - \rho_h)\sin\theta + \Delta\theta\cos\theta]\right\}.\end{aligned} \tag{7.76}$$

Equation (7.76) expresses the wavevector difference in the (x, y, z)-coordinate. The wavevector difference in the (x', y', z')-coordinate can be obtained by using the coordinate transformation given by Eq. (7.64). Then the displacement $\boldsymbol{d}$ of the reconstructed image point on the output plane can be obtained subsequently from Eq. (7.73).

Now, we have completed the general formulation for the case of the plane reference wave. Using Eqs. (7.74) and (7.75), we can obtain the relative diffraction efficiency when the signal wave is reconstructed during readout. Using Eqs. (7.63), (7.64) and (7.76), we can then obtain the displacement $\boldsymbol{d}$ of the reconstructed image point on the output plane. There will be no spatial spread of the reconstructed image point on the output plane.

As an example, we consider the case of tuning the wavelength of the reading reference wave to compensate the effects of the isotropic expansion or shrinkage of the storage medium. We choose $\rho_n\rho_\lambda = \rho_h = \rho_v$ and $\Delta\theta = 0$. According to Eqs. (7.74) and (7.75), the Bragg mismatch is fully compensated for plane waves generated by point sources at any position on the input plane. The reading reference wave and the recording reference wave are in the same direction inside the storage medium. The angle between the reading reference wave and the recording reference wave outside the storage medium is given by $\Delta\theta' = (\rho_n - 1)\tan\theta'$. Each plane wave component of the readout signal wave propagates in the same direction as the corresponding plane wave component of the recording signal wave inside the storage medium. The angular direction in the (x, y, z)-coordinate of a plane-wave component of the recording signal wave outside the storage medium is denoted as (β', γ'). Due to the change of the refractive index during the expansion or shrinkage of the storage medium, the angular direction of the corresponding plane wave component of the readout signal wave outside the storage medium will deviate from (β', γ') by $\Delta\beta' = 0$ and $\Delta\gamma' = (1 - \rho_n)\cot\gamma'$. In other words, outside the storage medium, the readout plane wave will propagate in the plane formed by the recording plane wave and the z-axis with an angular deviation $\Delta\gamma'$ from the recording plane wave. According to Eq. (7.76), the wavevector difference can be simplified as

$$\Delta\boldsymbol{k}' = \hat{z}\frac{2\pi}{\lambda}\frac{1 - \rho_n}{\sqrt{1 - n^2\cos^2\gamma}}. \tag{7.77}$$

Using Eqs. (7.63), (7.64), and (7.77), the pixel displacement can be written explicitly as

$$\boldsymbol{d} = (1 - \rho_n)\frac{x'^2 + y'^2 + f^2}{f(f\cos\alpha' + x'\sin\alpha')}[(f\sin\alpha' - x'\cos\alpha')\hat{x}' + (-y'\cos\alpha')\hat{y}'] . \tag{7.78}$$

If the refractive index does not change (i.e., $\rho_n = 1$), the pixel displacement will be zero. Since a change of the volume of the storage medium usually induces a change of its refractive index (i.e., $\rho_n \neq 1$), the pixel displacement is usually not zero. When the signal wave propagates perpendicularly to the storage medium (i.e., $\alpha' = 0$), the pixel displacement can be further simplified as

$$\boldsymbol{d} = (\rho_n - 1)\left[1 + \frac{x'^2 + y'^2}{f^2}\right](x'\hat{x}' + y'\hat{y}') \tag{7.79}$$

According to Eq. (7.79), the pixel displacement is circularly symmetric with respect to the origin, and the displacement of the pixel at the origin is zero. The linear terms of x' and y' in Eq. (7.79) represent the amplification of the input object by a factor of ρ_n. The nonlinear terms of x' and y' in Eq. (7.79) represent a distortion of the input object. When the signal wave propagates obliquely to the storage medium (i.e., $\alpha' \neq 0$), the pixel displacement is no longer circularly symmetrical with respect to the origin. In the vicinity of the origin, Eq. (7.78) can be simplified as

$$\boldsymbol{d} = (1 - \rho_n)\left(\tan\alpha' f\hat{x}' - \frac{x'\hat{x}'}{\cos^2\alpha'} - y'\hat{y}'\right) \tag{7.80}$$

According to Eqs. (7.78) and (7.80), the output image is shifted along the x-axis (described by the constant term of x' and y'), amplified with different amplification factors along the x'-direction and the y'-direction (described by the linear terms of x' and y'), and distorted (described by the nonlinear terms of x' and y'). The pixel displacement is to be compared with the size of the pixels. The size of the pixels ($\delta \times \delta$) is determined by the size of the input object ($a \times a$) and the number of pixels per page ($N \times N$). We consider the following case with a signal wave incidence angle $\alpha' = 2\pi/9$, a size of the input object $a \times a = 0.7f \times 0.7f$, a number of pixels $N \times N = 1000 \times 1000$, and a change of the refractive index of the storage medium $\rho_n = 1.01$. For later comparison, we note that, if the refractive index of the storage medium is $n = 1.5$, the signal wave will have an angular spread of $\pi/15 < \pi/2 - \gamma < \pi/5$ inside the storage medium. Using Eq. (7.78) with these parameters, we obtain the pixel displacement in terms of the pixel size for several important positions of the input object as $\boldsymbol{d}(0,0) = -12.0\delta\hat{x}'$, $\boldsymbol{d}(a/2,0) = -6.1\delta\hat{x}'$, $\boldsymbol{d}(-a/2,0) = -27.0\delta\hat{x}'$, $\boldsymbol{d}(0,\pm a/2) = -13.5\delta\hat{x}' \pm 5.6\delta\hat{y}'$, $\boldsymbol{d}(a/2,\pm a/2) = -6.7\delta\hat{x}' \pm 4.8\delta\hat{y}'$ and $\boldsymbol{d}(-a/2,\pm a/2) = -30.0\delta\hat{x}' \pm 8.8\delta\hat{y}'$. To implement the pixel-matching detection scheme, the pixel displacement has to be much smaller than the size of the pixels. This can be achieved by reducing the change of the refractive index $(\rho_n - 1)$, reducing the number of pixels per page, or adjusting the optical system to compensate the pixel displacement.

As another example, we consider the case that the recording reference wave propagates along the z-axis (i.e., $\theta = 0$). The Bragg mismatch and the wavevector difference can be simplified as

$$\Delta K_z = \frac{2\pi n}{\lambda}\left\{\frac{1-\sin\gamma}{\sin\lambda}[\rho_n\rho_\lambda - \rho_h - (\rho_h - \rho_v)\sin\gamma] + \Delta\theta\frac{\cos\gamma}{\sin\gamma}\cos\beta\right\} \quad (7.81)$$

$$\Delta\boldsymbol{k}' = -\hat{x}\frac{2\pi n}{\lambda}\Delta\theta + \hat{z}\frac{2\pi n}{\lambda}\frac{1}{\sqrt{\frac{1}{n^2}-\cos^2\gamma}}\left[\frac{(\rho_\lambda - \rho_h)}{n^2} + \Delta\theta\cos\gamma\cos\beta\right]. \quad (7.82)$$

When the signal wave and the reference wave enter the storage medium from the same side, $\sin\gamma$ is positive. When the signal wave and the reference wave enter the storage medium from opposite sides, $\sin\gamma$ is negative. Due to the refractive index of the storage medium, the range of $\sin\gamma$ is at most $\sqrt{1 - 1/n^2} < |\sin\gamma| < 1$. For instance, if $n = 1.5$ then $0.7 < |\sin\gamma| < 1$. The term $1 - \sin\gamma$ in Eq. (7.78) is much smaller when the two waves enter the storage medium from the same side. Therefore, the Bragg mismatch due to the expansion or shrinkage of the storage medium can be significantly reduced by choosing the proper system configuration. Meanwhile, according to Eq. (7.81), whether the signal wave and the reference wave enter the storage medium from the same side or from the opposite sides does not affect the pixel displacement. In the following analysis of this example, we will consider only the case that the signal wave and the reference wave enter the storage medium from the same side.

First, we consider the case that there is no angle detuning of the reading reference wave (i.e., $\Delta\theta = 0$). The Bragg mismatch and the wave-vector difference can be simplified

$$\Delta K_z = \frac{2\pi n}{\lambda}\frac{1 - \sin\gamma}{\sin\gamma}[\rho_n\rho_\lambda - \rho_h - (\rho_h - \rho_v)\sin\gamma], \tag{7.83}$$

$$\Delta \boldsymbol{k}' = \hat{z}\frac{2\pi}{\lambda}\frac{\rho_\lambda - \rho_h}{\sqrt{1 - n^2\cos^2\gamma}}. \tag{7.84}$$

Note that Eq. (7.84) is of the same form as Eq. (7.77). The general properties of the pixel displacement for this case have been studied in detail in the last example. We define a parameter $\sin\gamma_0$ as $\sin\gamma_0 = (\rho_n\rho_\lambda - \rho_h)/(\rho_h - \rho_v)$ and rewrite the Bragg mismatch as

$$\Delta K_z = \frac{2\pi n}{\lambda}(\rho_h - \rho_v)(1 - \sin\gamma)(\sin\gamma_0 - \sin\gamma)\frac{1}{\sin\gamma}. \tag{7.85}$$

When the expansion or shrinkage of the storage medium is anisotropic (i.e., $\rho_h \neq \rho_v$), the Bragg-matching condition can be satisfied by only one value of γ (i.e., $\gamma = \gamma_0$). The Bragg mismatch is a simple function of $\sin\gamma$. Given the range of $\sin\gamma$, the Bragg mismatch can be minimized over the entire range of $\sin\gamma$ by choosing an optimal value of $\sin\gamma_0$. The minimization of the Bragg mismatch also depends on the control over the parameters ρ_n, ρ_λ, ρ_h, and ρ_v that is available to us because $\rho_h - \rho_v$ in Eq. (7.85) may be a function of $\sin\gamma_0$. We will consider the Bragg mismatch in each situation separately.

One common situation is that the expansion coefficients ρ_n, ρ_h, and ρ_v of the storage medium are specified while the wavelength of the reading reference wave can be tuned. In this situation, $\rho_h - \rho_v$ is a constant. Figure 7.10 shows the Bragg mismatch as a function of γ. The simulation parameters are: $\lambda = 0.5\ \mu$m, $n = 1.5$, $t = 1$ mm, $\rho_h - \rho_v = 0.01$, and $\pi/15 < \pi/2 - \gamma < \pi/5$. The dashed curves are the Bragg mismatch as a function of γ for six different values of $\sin\gamma_0$. The values of $\sin\gamma_0$ are 0.0, 0.5, 0.7, 0.8, 0.9, and 1.0 from the bottom curve up to the top curve. The solid curve is the Bragg mismatch as a function of γ with the optimal value of $\sin\gamma_0$ equal to 0.835. With this optimal value of $\sin\gamma_0$, the maximum Bragg mismatch over the entire range of $\sin\gamma$ is found to be $(\Delta K_z t/2\pi)_{\max} = 0.22$. Given the optimal value of $\sin\gamma_0$, the wavelength tuning parameter ρ_λ is uniquely determined. It will be different from the expan-

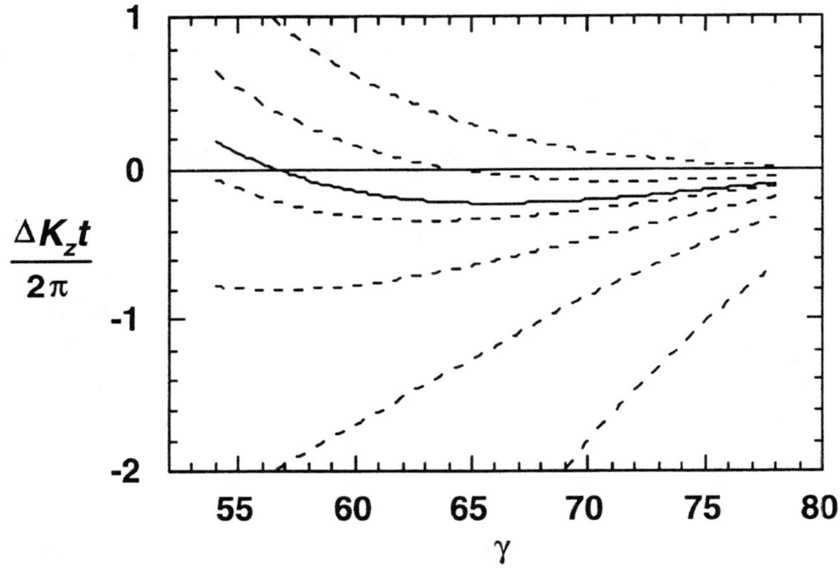

Figure 7.11: Bragg mismatch as a function of the angle γ.

sion coefficient ρ_h. Therefore, according to Eq. (7.84), the pixel displacement will not be zero. In general, when the expansion or shrinkage of the storage medium is anisotropic, the Bragg mismatch can not be totally compensated by tuning the wavelength of the reading reference wave. When the optimal wavelength tuning parameter ρ_λ is different from one, the Bragg mismatch with the optimal wavelength detuning can be significantly smaller than that without wavelength detuning. When the optimal wavelength tuning parameter ρ_λ is already one, the Bragg mismatch with the optimal wavelength detuning is the same as that without wavelength detuning. Whether the optimal wavelength tuning parameter ρ_λ is one or not depends on the relative values of the expansion coefficients ρ_n, ρ_h, and ρ_v. This suggests that the optimal wavelength tuning of the reading reference wave can be replaced by the optimal control of the relative values of the expansion coefficients ρ_n, ρ_h, and ρ_v in reducing the Bragg mismatch due to the anisotropic expansion or shrinkage of the storage medium.

The option of tuning the wavelength of the reference wave may not always be available (i.e., $\rho_\lambda = 1$). We define the volume expansion coefficient as $\rho_c = \rho_h^2\, \rho_v$. For $\rho << 1$, it can be shown that $(\rho_c - 1) = 2(\rho_h - 1) + (\rho_v - 1)$. If the expansion coefficients ρ_n and ρ_c are specified, the Bragg mismatch can be minimized by tuning the difference between the expansion coefficients ρ_h and ρ_v. In this situation, we have $\rho_h - \rho_v = (3\rho_n - \rho_c - 2)/(3 \sin \gamma_0 + 1)$ where $3\rho_n - \rho_c - 2$ is a constant. Consider the following parameters: $\lambda = 0.5\ \mu\text{m}$, $n = 1.5$, $t = 1$ mm, $(\rho_n - 1) - (\rho_c - 1)/3 = 0.01$, and $\pi/15 < \pi/2 - \gamma < \pi/5$. Then the optimal value of $\sin \gamma_0$ is 0.835, and the corresponding maximum Bragg mismatch is $(\Delta K_z t/2\pi)_{\max} = 0.19$. As a comparison, when the expansion or shrinkage of the storage medium is isotropic and the wavelength of the reading reference wave is the same as the recording reference wave, the maximum Bragg mismatch is $(\Delta K_z t/2\pi)_{\max} = 7.5$. A reduction of the maximum Bragg mismatch by a factor of more than 35 has been achieved by optimizing the anisotropic expansion or shrinkage of the storage medium. Given the optimal value of $\sin \gamma_0$, the expansion coefficient ρ_h is uniquely determined. It will be different from one. Therefore, according to Eq. (7.24), the pixel displacement will not be zero.

According to Eq. (7.84), if we choose $\rho_\lambda = \rho_h$, then $\Delta \boldsymbol{k}' = 0$. In other words, we can tune the wavelength of the reading reference wave to eliminate the pixel displacement. If the expansion coefficients ρ_n and ρ_c are specified, the Bragg mismatch can be minimized by tuning the difference between the expansion coefficients ρ_h and ρ_v . In this situation, we have $\rho_h - \rho_v = (\rho_n - 1)/\sin \gamma_0$ where $\rho_n - 1$ is a constant. Using the following parameters: $\lambda = 0.5\ \mu\text{m}$, $n = 1.5$, $t = 1$ mm, $\rho_n = 1.01$ and $\pi/15 < \pi/2 - \gamma < \pi/5$, and an optimal value of $\sin \gamma_0 = 0.835$, we obtain a maximum Bragg mismatch of $(\Delta K_z t/2\pi)_{\max} = 0.27$. As a comparison, when the expansion or shrinkage of the storage medium is isotropic and the wavelength of the reading reference wave is tuned to eliminate the pixel displacement, the maximum Bragg mismatch is $\Delta K_z t/2\pi)_{\max} = 7.5$. A reduction of the maximum Bragg mismatch by a factor of more than 25 has been achieved by optimizing the anisotropic expansion or shrinkage of the storage medium. As another comparison, when the expansion or shrinkage of the storage medium is isotropic and the wavelength of the reading reference wave is tuned to eliminate the Bragg mismatch, the maximum pixel displacement will be more than 30 times the pixel size if the number of pixels per page is 1000×1000.

Second, we consider the case when there is an angle detuning of the reading reference wave (i.e., $\Delta\theta \neq 0$). According to Eq. (7.82), the pixel displacement can be written explicitly as

$$\boldsymbol{d} = (\rho_\lambda - \rho_h)\frac{x'^2 + y'^2 + f^2}{f(f\cos\alpha' + x'\sin\alpha')}[(f\sin\alpha' - x'\cos\alpha')\hat{x}' + (-y'\cos\alpha')\hat{y}'] + \Delta\theta\frac{n\sqrt{x'^2 + y'^2 + f^2}}{f(f\cos\alpha' + x'\sin\alpha')}[(x'^2 + f^2)\hat{x}' + x'y'\hat{y}']. \tag{7.86}$$

The two terms of Eq. (7.86) cannot be canceled by tuning the relative values of $\rho_\lambda - \rho_h$ and $\Delta\theta$. Therefore, an angle detuning of the reading reference wave will lead to an additional pixel displacement. On the other hand, an angle detuning of the reading reference wave can be used to reduce the Bragg mismatch. According to Eq. (7.81), the Bragg mismatch is to be minimized over the entire range of both γ and β by choosing an optimal value of $\Delta\theta$. Since the signal wave enters the storage medium obliquely, $\cos\beta$ is close to one over the entire range of β. Therefore, we can first minimize the Bragg mismatch over the entire range of γ for $\cos\beta = 1$ by choosing an optimal value of $\Delta\theta$, then adjust $\Delta\theta$ slightly from that value so that the Bragg mismatch is minimized over the entire range of both γ and β. For instance, when the expansion or shrinkage of the storage medium is isotropic and there is no wavelength detuning of the reading reference wave, the Bragg mismatch can be written as

$$\Delta K_z = \frac{2\pi n}{\lambda}(\rho_h - \rho_v)\frac{1 + \cos\phi}{\cos\phi}\tan\frac{\phi}{2}\left(\tan\frac{\phi}{2} - \tan\frac{\phi_0}{2}\cos\beta\right) \tag{7.87}$$

where $\phi = \pi/2 - \gamma$ and $\tan\phi_0/2 = -\Delta\theta/(\rho_n - \rho_h)$. Due to the refractive index of the storage medium, the range of $\cos\phi$ is at most $\sqrt{1 - 1/n^2} < |\cos\phi| < 1$. For instance, if $n = 1.5$ then $0.7 < |\cos\phi| < 1$. Therefore, we can replace $\cos\phi$ approximately with one in Eq. (7.87). Let the maximum value of ϕ be denoted as ϕ_m. The Bragg mismatch can be minimized by choosing $\phi_0 = \phi_m$, with a corresponding angle detuning of $\Delta\theta = -(\rho_n - \rho_h)\tan\phi_m/2$. Consider an example with the following parameters: $\lambda = 0.5\ \mu\text{m}$, $n = 1.5$, $t = 1$ mm, $\rho_n - \rho_h = 0.01$, $\pi/15 < \phi < \pi/5$, $\alpha = 2\pi/9$, and $a \times a = 0.7f \times 0.7f$. We obtain an optimal value of $\Delta\theta$ of -0.00325, and a corresponding maximum Bragg mismatch of $(\Delta K_z t/2\pi)_{\max} = 1.9$. As a comparison, when there is no angle tuning of the reading reference wave, the maximum Bragg mismatch is $(\Delta K_z t/2\pi)_{\max} = 7.5$. The maximum Bragg mismatch with the optimal angle detuning is about four times smaller than that with no angle tuning. This is a small reduction compared with the reduction that can be achieved by tuning the wavelength of the reading reference wave or by tuning the anisotropic expansion coefficients of the storage medium.

In general, the Bragg-matching condition can not be satisfied by all the plane-wave components of the signal wave. Due to symmetry, it is natural to require that the signal wave generated by a point source on the x' axis of the input plane be read out with the Bragg-matching condition being satisfied. According to Eq. (7.84), this requirement translates to

$$\begin{aligned}\Delta\theta = \frac{1}{\sin(\phi_0 + \theta)}[(\rho_h - \rho_n\rho_\lambda)(\sin^2\phi_0 + \sin\phi_0\sin\theta) \\ + (\rho_v - \rho_n\rho_\lambda)(\cos^2\phi_0 + \cos\phi_0\cos\theta)]\end{aligned} \tag{7.88}$$

where $\phi_0 = \pi/2 - \gamma(x',0)$. θ can be obtained from θ' with the relation $n\sin\theta = \sin\theta'$. $\Delta\theta'$ can be obtained from $\Delta\theta$ with the relation $\Delta\theta' = (\rho_n - 1)\tan\theta' + n\Delta\theta\cos\theta/\cos\theta'$. If we choose a specific value of ϕ and one of the two parameters $\Delta\theta$ and ρ_λ, we can use Eq. (7.88) to determine the other one. Then numerical simulation can be performed.

7.2.2 Gaussian reference wave

In this section, we consider the case when the reference wave is a fundamental Gaussian wave. In a shift-multiplexed holographic storage system, a spherical wave is often used as a reference wave. An iris is placed in front of the storage medium so that only the part of the storage medium that is exposed to the signal wave is exposed to the reference wave. A spherical wave with a finite beam spot can be modeled by a fundamental Gaussian wave. Using a Gaussian beam spot can simplify the analytical study significantly.

Under the paraxial approximation, a Gaussian wave that propagates perpendicularly through a dielectric interface remains a Gaussian wave. We define an auxiliary function

$$\begin{aligned}&G(x, y, z; n, \lambda, \omega_0, z_m, \delta) = \\ &\frac{\omega_0}{\omega(z_m + z)}\exp\left\{-i[k(z_m + z) - \zeta(z_m + z)] - r^2\left[\frac{1}{\omega^2(z_m + z)} + \frac{ik}{2R(z_m + z)}\right]\right\}\end{aligned} \tag{7.89}$$

where $\omega^2(z) = \omega_0^2(1 + z^2/z_0^2)$, $R(z) = z(1 + z_0^2/z^2)$, $\zeta(z) = \tan^{-1}(z/z_0)$, $z_0 = \pi\omega_0^2 n/\lambda$, $k = 2\pi n/\lambda$, and $r^2 = (x + \delta)^2 + y^2$. It represents a Gaussian wave propagating along the z-axis of the (x, y, z)-coordinate with a wavelength λ and a minimum spot size ω_0 located at $(-\delta, 0, -z_m)$ in a medium with refractive index n. In the following analysis, we have $z_m + z >> z_0$, $z_m + z >> \delta$, $\omega(z_m + z) >> \delta$. Typical values are: $z_m + z$ and $\omega(z_m + z)$ are of the order of 1 mm, and λ, z_0, and δ are of the order of 1 μm. Under this condition, the following approximations are accurate: $\zeta(z_m + z) \approx \pi/2$, $R(z_m + z) \approx z_m + z$, $\omega(z_m + z) \approx \omega_0(z_m + z)/z_0$, $r^2/\omega^2(z_m + z) \approx (x^2 + y^2)/\omega^2(z_m + z)$, and $ikr^2/[2R(z_m + z)] \approx ik(x^2 + y^2 + 2\delta x)/[2R(z_m + z)]$.

Using this auxiliary function, the recording Gaussian reference wave $E_r(x,9y, z)$ and the readout Gaussian reference wave $E_i(x, y, z)$ inside the storage medium can be written as

$$E_r(x, y, z) = G(x, y, z; n, \lambda, \omega_{0_r}, z_r, 0) \tag{7.90}$$

$$E_i(x, y, z) = G(x, y, z; \rho_n n, \rho_\lambda \lambda, \omega_{0_i}, z_i, \delta)\,. \tag{7.91}$$

We assume that the recording Gaussian reference wave outside the storage medium has a minimum spot size ω'_{0r} at location z'_r. Under the paraxial approximation, the parameters ω_{0r} and z_r in Eq. (7.62) can be determined by $\omega_{0r} = n\omega'_{0r}$ and $z_r = n(z'_r - t/2) + t/2$. We assume that the recording Gaussian reference wave outside the storage medium has a minimum spot size ω'_{0i} at location z'_i. Under the paraxial approximation, the parameters ω_{0i} and z_i in Eq. (7.63) can be determined by $\omega_{0i} = \rho_n n\omega'_{0i}$ and $z_i = \rho_n n(z'_{\rm i} - t/2) + t/2$. In general, we can tune the wavelength and the minimum spot location of the readout Gaussian reference wave to compensate the effects of the expansion or shrinkage of the storage medium. We choose $\omega'_{0r} = \omega'_{0i}$, since tuning the minimum spot size of the readout Gaussian reference wave does not compensate the effects of the expansion or shrinkage of the storage medium.

Remember that the storage system is linear with respect to the signal wave amplitude and that we are considering the impulse response of the storage system. A point source on the input plane will generate a plane wave near the Fourier plane. Then the signal wave inside the storage medium can be written as

$$E_s(x, y, z) = \exp[-i(k_{sx}x + k_{sy}y + k_{sz}z)]\,. \tag{7.92}$$

The dielectric perturbation induced by the signal wave and the recording reference wave before the expansion or shrinkage of the storage medium can be written as

$$\Delta\varepsilon(x, y, z) = E_r^*(x, y, z)E_s(x, y, z) + \text{c.c.} \tag{7.93}$$

where c.c. refers to complex conjugate. The dielectric perturbations before and after the expansion or shrinkage of the storage medium are related by

$$\Delta\varepsilon'(x, y, z) = \Delta\varepsilon(\rho_h x, \rho_h y, \rho_v z). \tag{7.94}$$

Due to the finite spot size of the Gaussian reference wave, the diffracted wave during readout always consists of multiple plane-wave components. Under Born approximation, the amplitude of the plane-wave component of the diffracted wave can be written as

$$A_d(k_{dx}, k_{dy}, k_{dz}) \propto \int_V E_i(x, y, z)\Delta\varepsilon'(x, y, z)\exp[i(k_{dx}x + k_{dy}y + k_{dz}z)]dxdydz\,. \tag{7.95}$$

The integrations over x and y in Eq. (7.95) can be carried out analytically with the following integral equation

$$\int_{-\infty}^{+\infty} \exp(ib\omega - a\omega^2)d\omega - \sqrt{\frac{\pi}{a}}\exp\left(-\frac{b^2}{4a}\right) \tag{7.96}$$

where $\mathrm{Re}[a] > 0$ and $\mathrm{Re}[\sqrt{\pi/a}] > 0$. Substituting Eqs. (7.89)–(7.94) into Eq. (7.95) and using Eq. (7.96), we obtain

$$A_d(k_{dx}, k_{dy}, k_{dz}) \propto \int_t dz \exp[-ik_i(z_i + z) + ik_r(z_r + \rho_v z) + i(k_{dz} - \rho_v k_{sz})z]$$
$$\times \frac{\omega_{oi}}{\omega_i(z_i + z)} \frac{\omega_{or}}{\omega_r(z_r + \rho_v z)} \frac{\pi}{a(z)} \exp\left[-\frac{\left(k_{dz} - \rho_h k_{sx} - \dfrac{\delta k_i}{z_i + z}\right)^2 + (k_{dy} - \rho_h k_{sy})^2}{4a(z)}\right] \tag{7.97}$$

where $\omega_r(z_r + \rho_v z) = \omega_{0r}(z_r + \rho_v z)/z_{0r}$, $\omega_i(z_i + z) = \omega_{0i}(z_i + z)/z_{0i}$, $z_{0r} = \pi\omega_{0r}^2 n/\lambda$, $z_{0i} = \pi\omega_{0i}^2 n\rho_n\rho_\lambda/\lambda$, $a(z) = a_r(z) + ia_i(z)$, $a_r(z) = \rho_h^2/\omega_r^2(z_r + \rho_v z) + 1/\omega_i^2(z_i + z)$, and $a_i(z) = -k_r\rho_h^2/[2(z_r + \rho_v z)] + k_i/[2(z_i + z)]$.

Here, we introduce another detuning coefficient $\rho_m = z_r/z_i$. $a_i(z)$ can be minimized by choosing $\rho_m = \rho_h^2/(\rho_n\rho_\lambda)$. Then we have $a_i(z) = k_r z\rho_m(\rho_n\rho_\lambda\rho_v - \rho_h^2)/[2(z_r + \rho_v z)(z_r + \rho_m z)]$. Also remember that $|\rho - 1| << 1$ where ρ can be ρ_h, ρ_v, ρ_n, ρ_λ and ρ_m. Typically, $|\rho - 1|$ is of the order of 10^{-2}. Under this condition, the following approximations are still accurate: $\omega_{0i} \approx \omega_{0r}$, $z_{0i} \approx z_{0r}$, $\omega_r(z_r + \rho_v z) \approx \omega_r(z_r + z)$, $\omega_i(z_i + z) \approx \omega_r(z_r + z)$, $a_r(z) \approx 2/\omega_r^2(z_r + z)$, and $a_i(z) \approx kz(\rho_n\rho_\lambda - \rho_h^2/\rho_v)/[2(z_r + z)^2]$.

Over the range of integration in Eq. (7.97), we have $z_r >> z$. Typical values are: z_r is of the order of 1 cm, z is of the order of 1 mm. Under this condition, we can neglect the higher-order terms of z/z_r in Eq. (7.97) and obtain

$$A_d(k_{dx}, k_{dy}, k_{dz}) \propto \int_t dz \frac{a_r}{a(z)} \exp\left[-\frac{(k_{dx} - \rho_h k_{sx} - k_s\delta/z_r)^2 + (k_{dy} - \rho_h k_{sy})^2}{4a(z)}\right]$$
$$\exp[i(\rho_v k_r - k_i + k_{dz} - \rho_v k_{sz})z] \tag{7.98}$$

where $a(z) = a_r + ia_i(z)$, $a_r(z) = 2/\omega_r^2(z_r)$, $a_i(z) = [(\pi n/\lambda)(\rho_n\rho_\lambda - \rho_h^2/\rho_v)/z_r^2]z$.

We will continue the discussion of Eq. (7.98) under the assumption that the beam spot of the Gaussian reference beam at the storage medium is much larger than the thickness of the storage medium, i.e., $\omega_r(z_r) >> t$. This assumption is valid under typical experimental conditions. We further require that $a_i(z)$ be comparable to or smaller than a_r over the range of integration in Eq. (7.98). The physical meaning of this requirement will be apparent from an example in the following discussion. Under these two conditions, the first exponential term in Eq. (7.98) is significant only in the vicinity of $k_{dx} = \rho_h k_{sx} + k_s\delta/z_r$ and $k_{dy} = \rho_h k_{sy}$, where the second exponential term in Eq. (7.98) remains to be approximately constant. Therefore, Eq. (7.98) can be further simplified as

$$A_d(k_{dx}, k_{dy}, k_{dz}) \propto \int_t dz \frac{a_r}{a(z)} \exp\left[-\frac{(k_{dx} - \rho_h k_{sx} - k_s\delta/z_r)^2 + (k_{dy} - \rho_h k_{sy})^2}{4a(z)}\right]$$
$$\exp(i\Delta k_z z) \tag{7.99}$$

where $\delta K_z = \rho_v k_r - k_i + \sqrt{k_d^2 - (\rho_h k_{sx} + k_s\delta/z_r)^2 - (\rho_h k_{sy})^2} - \rho_v k_{sz}$ is the Bragg mismatch along the z-direction.

First, we consider the case when the expansion or shrinkage of the storage medium is isotropic, i.e., $\rho_h = \rho_v$. We choose $\rho_n\rho_\lambda = \rho_h (= \rho_v)$ and $\delta/z_r = 0$. Then, we have $a_i(z) = 0$ and $\Delta K_z = 0$. In other words, there is no quadratic phase curvature mismatch between the recording Gaussian reference wave $E_r(\rho_h x, \rho_h y, \rho_v z)$ and the readout Gaussian reference wave $E_i(x, y, z)$ and there is no phase mismatch along the z-direction. Equation (7.99) can be simplified as

$$A_d(k_{dx}, k_{dy}, k_{dz}) \propto \exp\left[-\frac{(k_{dx} - \rho_h k_{sx})^2 + (k_{dy} - \rho_h k_{sy})^2}{4a_r} \right]. \quad (7.100)$$

The angular spread of the diffracted wave is due to the finite spot size of the Gaussian reference wave at the storage medium. The angular spread of the diffracted wave limits the minimum pixel size.

Second, we consider a hypothetical case that both $a_i(z)$ and ΔK_z in Eq. (7.99) are nonzero and constant of position z. We will refer to $a_i(z)$ simply as a_i. Equation (7.99) can be simplified as

$$A_d(k_{dx}, k_{dy}, k_{dz}) \propto \exp\left[-\frac{(k_{dx} - \rho_h k_{sx} - k_s \delta / z_r)^2 + (k_{dy} - \rho_h k_{sy})^2}{4(a_r + i a_i)} \right] \operatorname{sinc}\left(\frac{\Delta K_z t}{2\pi} \right). \quad (7.101)$$

The phase mismatch along the z-direction reduces the diffraction efficiency by a factor of $\operatorname{sinc}^2[\Delta K_z t/(2\pi)]$. It is desirable that $\Delta K_z t/(2\pi) << 1$. The quadratic phase curvature mismatch between the recording Gaussian reference wave $E_r(\rho_h x, \rho_h y, \rho_v z)$ and the readout Gaussian reference wave $E_i(x, y, z)$ expands the angular spread of the diffracted wave by a factor of $1 + a_i^2/a_r^2$. Expanding the angular spread of the diffracted wave increases the minimum pixel size, therefore reduces the storage capacity. It is desirable that a_i/a_r be comparable to or less than one. $a_i/a_r = 1$ translates to a quadratic phase curvature mismatch of 2 across the beam spot of the Gaussian reference waves at the storage medium, i.e., from $r = 0$ to $r = \omega_r(z_r)$.

Third, we consider the general case in which the expansion or shrinkage of the storage medium can be anisotropic. In this case, both $a_i(z)$ and ΔK_z in Eq. (7.99) are in general nonzero. Since $a_i(z)$ is not a constant of position z, the integration in Eq. (7.99) cannot be carried out analytically. It is apparent that the behavior of Eq. (7.99) resembles the hypothetical case considered in the preceding discussion. Therefore, we will analyze and minimize $a_i(z)$ and ΔK_z instead.

As described in Eq. (7.95), the diffracted wave consists of multiple plane-wave components. The pixel displacement can be represented by the displacement of the central plane-wave component. In addition, ΔK_z is also the Bragg mismatch for the central plane-wave component. Therefore, the pixel displacement and the Bragg mismatch for the case of Gaussian reference wave are the same as those for the case of plane reference wave. In the equivalent case of plane reference wave, the recording plane reference wave propagates perpendicularly to the storage medium and the angle detuning of the reading plane reference wave corresponds to δ/z_r. This case has already been studied in detail in Section 7.2.1. The results on the Bragg mismatch and the pixel displacement can be directly applied here. We will use the results of Section 7.2.1 to study the behavior of $a_i(z)$.

For the case when there is no spatial shift of the reading reference wave (i.e., $\delta = 0$), we have

$$\frac{a_i(t/2)}{a_r} = \frac{\pi n t}{4\lambda}(\rho_h - \rho_v)(\sin\gamma_o - 1)\tan^2\phi_g \tag{7.102}$$

where ϕ_g is the half-apex angle of the Gaussian beam, i.e., $\tan\phi_g = \omega_r(z_r)/z_r$. We now consider a numerical example. Let the angular spread of the reference wave and the angular spread of the signal wave be both about 40° outside the storage medium. This is almost the maximum angular spread we can allocate so that the reference wave propagates perpendicularly to the storage medium and there is no angular overlap between the signal wave and reference wave. Then we have inside the storage medium with a refractive index $n = 1.5$: $\phi_g = \pi/15$, and $\pi/15 < \pi/2 - \gamma < \pi/5$. The other parameters are: $\lambda = 0.5$ μm, $n = 1.5$, $t = 1$ mm, and $\rho_h - \rho_v = 0.01$. These parameters are the same as the parameters we used in the analysis of the equivalent case of plane reference wave. The optimal value of $\sin\gamma_o$ to minimize the Bragg mismatch has been found to be 0.0835. Then we have $a_i(t/2)/a_r = 0.18$, which indicates a negligible increase of the diffracted wave angular spread.

For the case when the expansion or shrinkage of the storage medium is isotropic and there is no wavelength detuning of the reading reference wave, we have

$$\frac{a_i(t/2)}{a_r} = \frac{\pi n t}{4\lambda}(\rho_n - \rho_h)\tan^2\phi_g \tag{7.103}$$

Consider the parameters: $\lambda = 0.5$ μm, $n = 1.5$, $t = 1$ mm, $\rho_n - \rho_h = 0.01$, and $\phi_g = \pi/15$. Then we have $a_i(t/2)/a_r = 1.1$, which leads to a small increase of the diffracted wave angular spread.

Choosing an optimal value of $\rho_m = \rho_h^2/(\rho_n\rho_\lambda)$ is important in reducing $a_i(t/2)/a_r$ in the above two cases. If ρ_m had been chosen to be one, $a_i(t/2)/a_r$ could have been an order of magnitude larger than the corresponding optimal values.

In this part, we have studied the effects of the expansion or shrinkage of the storage medium on Fourier-plane holographic memory. The effects include Bragg mismatch, pixel displacement and impulse broadening. These effects can always be eliminated by reducing the expansion or shrinkage of the storage medium. However, a large expansion or shrinkage often occurs in organic polymers. Therefore, it is important to eliminate or minimize those effects due to the presence of the expansion or shrinkage of the storage medium. Bragg mismatch is always detrimental to holographic memories. We found that it can be significantly reduced by choosing the appropriate system configuration, carefully designing the expansion or shrinkage characteristics of the material, and tuning the wavelength and angle of the reading reference wave. Pixel displacement is detrimental when the pixel-matching detection scheme is used. It can be compensated by relocating the pixels of the detector array correspondingly. Otherwise, we found that once the Bragg mismatch is minimized in the optimization there are not enough free parameters left to further reduce the pixel displacement. We did find one situation in which the pixel displacement is eliminated by tuning the wavelength of the reference wave while the Bragg mismatch is reduced by controlling the anisotropic expansion or shrinkage characteristics of the storage medium. Impulse broadening occurs only when the reference wave consists of multiple plane-wave components. It increases the minimum pixel size. We found that it can be reduced by adjusting the wavefront of the reading reference wave.

7.3 Conclusions

In summary, we have studied two kinds of limitations of holographic memories. One is the cross talk noise, due to the finite size of the storage medium. The other is the grating detuning, due to the expansion or shrinkage of the storage medium. In general, they are all geometrical limitations, and they may all lead to cross talk among the pixels if not properly addressed. Both limitations have been studied analytically. Various configurations are compared and possible optimizations to eliminate or minimize are suggested.

References

1. P. J. van Heerden, "Theory of information storage in solids," *Applied Opt.* **2**, 393, 1963.
2. E. G. Ramberg, "Holographic information storage," *RCA Review,* **33**, 5–53, 1972.
3. D. Psaltis and F. Mok, "Holographic memories," *Sci. Am.,* **11**, 70–76, 1995.
4. J. F. Heanue, M. C. Bashaw, and L. Hesselink, "Volume holographic storage and retrieval of digital data," *Science,* **265**, 749–752, 1994.
5. F. H. Mok, "Angle-multiplexed storage of 5000 holograms in lithium niobate," *Opt. Lett.,* **18**, 915–917, 1993.
6. G. A. Rakuljic, V. Leyva and A. Yariv, "Optical data storage by using orthogonal wavelength-multiplexed volume holograms," *Opt. Lett.,* **17**, 1471–1473, 1992.
7. S. Yin, H. Zhao, F. Zhao, M. Wen, Z. Yang, J. Zhang and F. T. S. Yu, "Wavelength multiplexed holographic storage in a sensitive photorefractive crystal using a visible-light tunable diode laser," *Opt. Commun.,* **101**, 317–321, 1993.
8. C. Denz, G. Pauliat, G. Roosen and T. Tschudi, "Volume holographic multiplexing using a deterministic phase encoding technique," *Opt. Commun.,* **85**, 171–176, 1991.
9. K. Curtis, A. Pu and D. Psaltis, "Method for holographic storage using peristrophic multiplexing," *Opt. Lett.,* **19**, 993–994, 1994.
10. A. Pu, G. Barbastathis, M. Levene and D. Psaltis, "Shift multiplexed holographic 3-D disk," *Optical Computing Technical Digest,* **10**, 219–221, 1995.
11. S. Campbell, X. Yi and P. Yeh, "Hybrid sparse-wavelength angle-multiplexed optical data storage system," *Opt. Lett.,* **19**, 2161–2163, 1994.
12. X. Yi, P. Yeh and C. Gu, "Cross-talk noise and storage density of volume holographic memory with spectral hole burning materials," NLO'96.
13. C. Gu, J. Hong, I. McMichael, R. Saxena and F. H. Mok, "Cross-talk-limited storage capacity of volume holographic memory," *J. Opt. Soc. Am.,* **A 9**, 1978–1983, 1992.
14. X. Yi, P. Yeh and C. Gu, "Statistical analysis of cross-talk noise and storage capacity in volume holographic memory," *Opt. Lett.,* **19**, 1580–1582, 1994.
15. J. Hong, I. McMichael and J. Ma, "Influence of phase masks on crosstalk in holographic memory," *Opt. Lett.* (in press).

16. M. C. Bashaw, J. F. Heanue, A. Aharoni, J. F. Walkup and L. Hesselink, "Cross-talk considerations for angular and phase-encoded multiplexing in volume holography," *J. Opt. Soc. Am.*, **B 11**, 1820–1836, 1994.

17. J. F. Heanue, M. C. Bashaw and L. Hesselink, "Sparse selection of reference beams for wavelength- and angular-multiplexed volume holography," *J. Opt. Soc. Am.*, **A 12**, 1671–1676, 1994.

18. J. W. Goodman, *Introduction to Fourier Optics*, McGraw-Hill, New York, 1968.

19. J. D. Jackson, *Classical Electrodynamics,* Wiley, New York, 1975, pp. 427–432.

20. A. Yariv, "Interpage and interpixel cross-talk in orthogonal (wavelength-multiplexed holograms," *Opt. Lett.*, **18**, 652–654, 1993.

21. K. Curtis, C. Gu and D. Psaltis, "Cross talk in wavelength-multiplexed holographic memories," *Opt. Lett.*, **18**, 1001–1003, 1993.

22. F. T. S. Yu, F. Zhao, H. Zhou and S. Yin, "Cross-talk in a wavelength multiplexed reflection-type photorefractive fiber hologram," *Opt. Lett.*, **18**, 1849–1851, 1993.

23. X. Yi, S. Campbell, P. Yeh and C. Gu, "Statistical analysis of cross-talk noise and storage capacity in volume holographic memory II: image plane holograms," *Opt. Lett.*, **20**, 779–781, 1995.

24. K. Curtis and D. Psaltis, "Cross talk for angle and wavelength multiplexed image plane holograms," *Opt. Lett.*, **19**, 1774–1776, 1994.

25. J. W. Goodman, *Statistical Optics,* Wiley, New York, 1985, pp. 371–374.

Chapter 8

Imaging and Storage with Spherical-Reference Volume Holograms

George Barbastathis

Department of Mechanical Engineering, Massachusetts Institute of Technology, Cambridge, MA

David J. Brady

Department of Electrical and Computer Engineering Beckman Institute for Advanced Science and Technology, University of Illinois at Urbana-Champaign, Urbana, IL

8.1 Introduction

Volume holography was first proposed by van Heerden [1] as a method for holographic storage. Van Heerden's idea was to superimpose several pages of data as low-bandwidth modulations on volume gratings of slightly different spatial periodicities; then use the Bragg-selectivity effect, well-known from x-ray crystallography, to retrieve pages selectively: the desired page is Bragg-matched and diffracts strongly, whereas all other pages are Bragg-mismatched and silent. Soon thereafter, Leith with co-workers [12] and Kogelnik [3] calculated the angle and wavelength selec-

PHOTOREFRACTIVE OPTICS
Materials, Properties and Applications

ISBN 0-12-774810-5/$30.00

tivities of volume gratings, while the discovery of the photorefractive effect [4] provided a class of materials appropriate for storing volume holograms. Today, research interest in volume holography remains strong in the data storage area [5–7], as well as in optical neural networks [8–10] and communications [11]. To date commercial ventures based on volume holograms are for spectral filtering [2] and three-dimensional (3-D) storage devices [6, 7, 13–17]. In this chapter we review volume holographic systems and materials (Section 8.2), present the fundamentals of volume diffraction theory (Section 8.3), where these techniques are based, and discuss two-volume holographic applications, shift-multiplexed storage [18] (Section 8.4) and 3-D incoherent optical imaging [19] (Section 8.5).

Volume holograms are usually recorded by interfering two mutually coherent optical beams in a photosensitive material, as shown in Fig. 8.1a. Typically, the spatially-varying intensity pattern modulates the refractive index, although absorption holograms are also possible. One of the beams, called the "signal beam," carries the information to be stored in the hologram, usually effected as a phase or amplitude modulation on the beam profile by a spatial light modulator (SLM). The second beam is usually as simple as a plane [2] or a spherical [18] wave (although exceptions do exist [20, 21]) and provides the phase reference for the interference pattern formation (as in the case of planar holography [22, 23]), hence the name "reference beam". After exposure is complete, the hologram operates as a diffractive element: when illuminated by a "probe beam," it diffracts a

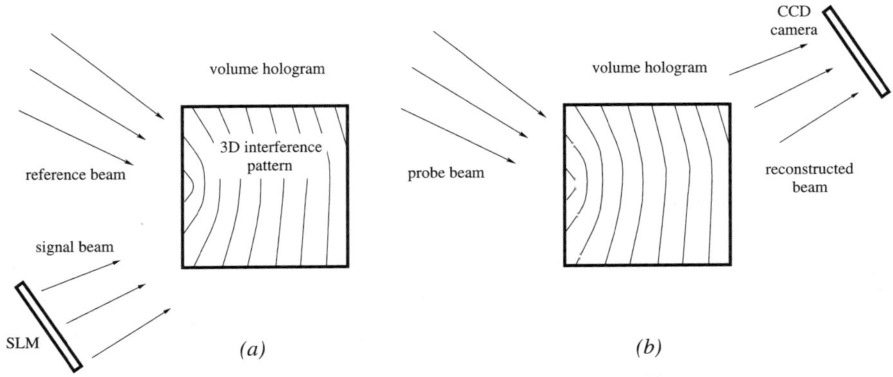

Figure 8.1: (a) Volume holographic recording geometry. (b) Reconstruction geometry.

"reconstructed beam" which is measured by an intensity detector, such as a charge-coupled device (CCD) camera (Fig. 8.1b). Interferometric detection that yields both amplitude and phase information about the diffracted beam is also possible, but seldom used in practice.

The probing operation of a volume hologram is in effect a complex transformation between the input (probe) and output (reconstructed) optical fields. The number of degrees of freedom (or rank) of the transformation equals the number of resolvable volume elements ("voxels") at wavelength λ contained within the volume V of the hologram, i.e., $N^3 \equiv V/\lambda^3$. Comparatively, surface elements, such as thin lenses, thin holograms, etc. are fundamentally reduced to N^2 degrees of freedom only (within the same transverse aperture). However, complete optical access to all N^3 degrees of freedom of a volume hologram is a tough control problem, constrained by the response of the holographic material [24]; practical solutions known to date suffer from dynamic range limitations [25–27], although promising new materials [28, 29] and recording techniques [30] appeared recently. We will review volume holographic materials and other systems issues in Section 8.2.

Two typical probe → reconstruction maps implemented by a volume hologram are: from a set of reference beams to a set of stored data pages, in the case of storage (see Figs. 8.2, and 8.6 and Sections 8.2.1 and 8.4); and from an input pattern to an output pattern or response, in the case of neural networks (such as correlators, see Figures 8.4 and 8.7, and Sections 8.2.1 and 8.4.1). In [19], we introduced yet another possible mapping, from a polychromatic extended object to an intensity pattern that can be combined with other patterns (captured, e.g., by scanning the object mechanically) to reconstruct the object in its native space; in other words, the volume hologram performs an *imaging* transformation (see Section 8.5). Our main purpose in this chapter is to provide an understanding of how these mappings result from the physical principles of volume diffraction in the particular case of spherical-reference volume holograms.

8.2 Volume holographic systems

8.2.1 Multiplexing schemes and architectures

Holograms superimposed ("multiplexed") within the same volume holographic material are indexed by the reference beam that was used when they were recorded; i.e., each hologram is recorded with a reference beam that has a unique distinguishable feature from the reference beams of the other holograms. Commonly used features that define the multiplexing schemes are:

- the *angle* of incidence of the reference beam on the material ("angle" [2, 31] and "peristrophic" [32] multiplexing schemes),
- the *color* of the recording beams, ("wavelength" multiplexing scheme [2, 33, 34]),
- the *phase* modulation on the reference beam profile ("phase-code" multiplexing scheme [20]), and
- the *position* of the reference beam relative to the material ("shift" multiplexing schemes [18, 21]).

A single hologram is selectively retrieved by illuminating the holographic memory by the corresponding reference beam that was used for the recording of that particular hologram. The reconstruction obtained this way consists primarily of the selected hologram, along with small contributions from other multiplexed holograms, which of course are undesired as cross talk [35–40].

The selection of multiplexing scheme depends on the desired system properties; e.g., the phase-code method provides very low cross talk between superimposed holograms [38, 40] but the maximum number of holograms that can be multiplexed is limited by the space-bandwidth product of available SLMs. Combinations have also been used in hybrid multiplexing schemes that provide higher capacity, e.g., angle + peristrophic [41], and angle + wavelength [42]. A detailed discussion and comparison of holographic multiplexing methods is the topic of [43].

For example, in the popular angle multiplexing scheme of Figure 8.2, the reference beam for the mth hologram is a plane wave incident at angle θ_m. The mth hologram is retrieved selectively by illuminating the exposed material with the corresponding plane wave at angle θ_m. The reconstruction consists of the contents of the mth hologram only with minimal cross talk from other holograms if the angular separation between adjacent reference beams was an integral multiple of

$$\Delta\theta = \frac{\lambda}{2L\sin\theta}. \tag{8.1}$$

This quantity is known as "angle Bragg selectivity." Since $\Delta\theta$ is propor-

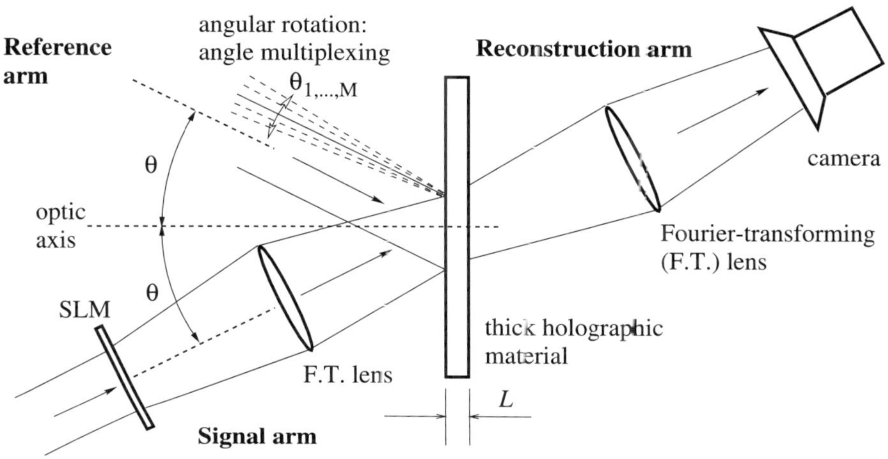

Figure 8.2: Angle-multiplexed holographic memory architecture

tional to λ/L, the selectivity improves by using shorter wavelengths or thicker materials. We will not go through a detailed calculation of the angle selectivity condition here (see instead, e.g., [2, 43]), but instead we will provide the corresponding derivation for the shift multiplexing method in Section 8.4.

It is important to note that the multiplexed holograms share the entire volume of the recording material; therefore, holographic storage is fundamentally different than layered volume storage methods, such as the digital versatile disk (DVD) and two-photon storage [44]. One might think of the process of Bragg-matching a single hologram in the presence of multiple holograms sharing the same medium as similar to tuning a receiver to a radio station; the matching angle θ_m (for the angle-multiplexing scheme) corresponds to the resonance frequency of the receiver, and the Bragg separation $\Delta\theta$ corresponds to the quality factor Q that determines the receiver bandwidth.

The arrangement of recording/probe beams with respect to the material in a holographic system is an important design decision. Three possibilities, shown in Fig. 8.3, are used extensively in practice:

- In the *transmission* geometry (Fig. 8.3a) the two recording beams are incident on the same face of the holographic material. The reconstruction appears on the opposite face as an extension of the

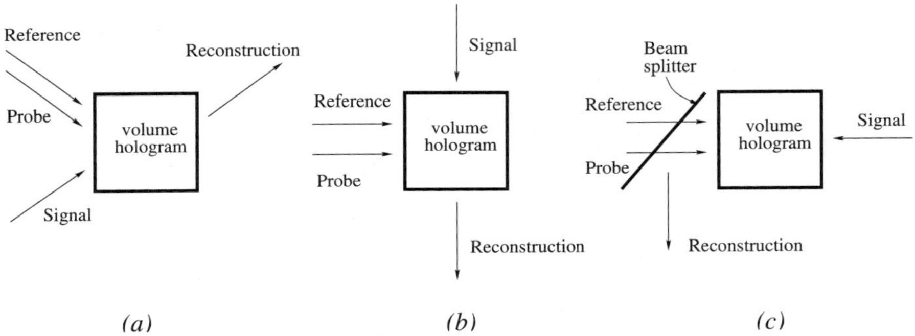

Figure 8.3: Simplified holographic recording geometries: (a) Transmission geometry. (b) 90° geometry. (c) Reflection geometry.

signal beam (i.e., as a replica of the signal beam if it were present and propagated beyond the holographic material). It is most popular in the holographic disk architecture [5, 41, 45, 46].

- The *90° geometry* (Fig. 8.3b) is similar to the transmission geometry, except the reference and signal beams are incident on two normal faces of a cubelike recording medium. It has been used extensively in high-capacity demonstrations [6, 13, 14] and compact implementations [17, 47].
- In the *reflection* geometry (Fig. 8.3c) the recording beams are incident on two opposite faces of the holographic material and (approximately) counterpropagating. In this case the reconstruction is counter propagating and on the same side of the medium as the probe beam (hence the term "reflection"). A beam-splitter separates the reconstruction from the probe. The reflection geometry has been popular in wavelength-multiplexed systems [34], because it provides the optimal-wavelength Bragg selectivity.

Interesting trade-offs between the storage capacity and noise performance of a holographic memory system are posed by the location of the holographic material with respect to the optical arrangement that images the SLM onto the CCD detector in the signal/reconstruction arms [48]. The four best-known possibilities are

- the *image-plane* geometry, where the SLM is imaged onto the holographic material and the reconstruction in turn is imaged onto the CCD;

- the *Fourier-plane* geometry, where the SLM is Fourier-transformed onto the holographic material and the reconstruction in turn is Fourier-transformed onto the CCD;
- the *Fresnel-zone* geometry, where the material is placed in the Fresnel zone of the SLM;
- and the *van der Lugt* geometry, which is similar to the Fourier-plane geometry, except the SLM is adjacent to the Fourier-transforming lens.

Another option is to use a probe that is phase-conjugated with respect to the reference beam; the reconstruction itself is then phase-conjugated, and the need for using imaging elements is eliminated [47].

Several other systems issues are important for holographic storage and have been treated extensively, e.g., page-oriented error correction [49–51] and channel modulation [52, 53], and pixel matching [54] (i.e., minimizing aberration distortion by using unit magnification in the optical system between the SLM and the detector).

Holographic memories have been used extensively for pattern recognition in the optical correlator architecture (Fig. 8.4). In that case the probe beam is not used to index a particular hologram, but rather to effect a parallel search of the entire memory [55–57]. Suppose the memory

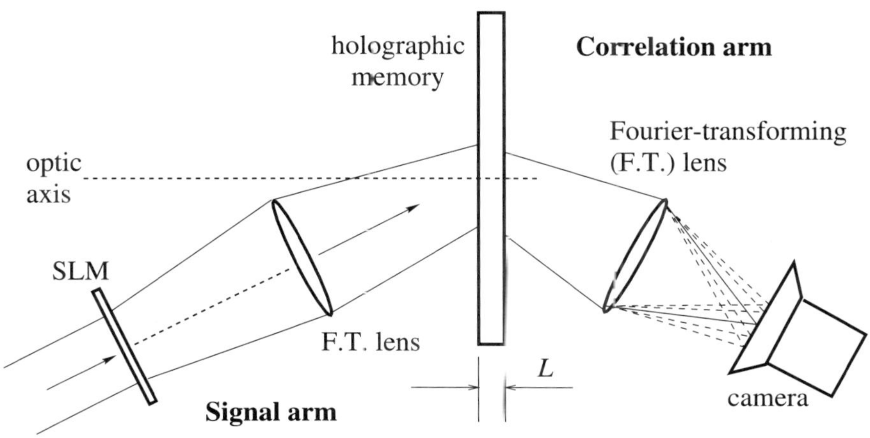

Figure 8.4: Optical correlator using an angle-multiplexed holographic memory (see Fig. 8.2).

contains M patterns f_m ($m = 1, \ldots, M$) in the Fourier-plane geometry. If the probe beam contains a new pattern g, then at the Fourier transform the reconstruction consists of the correlations $g^* f_m$ of the novel pattern with all the stored patterns at once. The parallel correlation operation is obtained at the expense of losing shift invariance in one dimension at the output plane. Several applications have used this mode of holographic memory operation successfully [58–60]. We will see in Section 8.5 that the correlation operation performed by volume holograms is useful as an imaging operation as well.

8.2.2 Volume holographic materials

The choice of holographic material is the most critical issue in holographic memory system design. A wide range of materials have been tried for holographic storage with various degrees of success [61, 62]. Here we consider photorefractive and photopolymer materials only, because so far they have been the most popular in experiments and high-capacity demonstrations.

Photorefractive crystals, such as Fe-doped $LiNbO_3$, $Sr_xBa_{1-x}NbO_3$ (SBN:x), and $BaTiO_3$ were the first materials to be used for holographic storage [63]. During recording, the refractive index change occurs via the electro-optic effect after a spatially varying space-charge field is established in the crystal from the diffusion or drift of photo-excited charges away from the illuminated regions [64–66]. The space-charge field sustains itself after removal of the recording beams, but decays because of thermal electronic excitation in the dark, or uniform photo-excitation during hologram readout. Decay occurs also as a result of superimposing more holograms in the same location of the material. As a result of the erasure of existing holograms when new holograms are recorded, the dynamic range of the material is not fully utilized, and the diffraction efficiency (defined as the portion of the reference beam power diffracted into the hologram) of $M \gg 1$ equal-strength holograms is [25, 27]

$$\eta(M) = \frac{(M/\#)^2}{M^2}. \tag{8.2}$$

The parameter $M/\#$ (pronounced "M-number") depends highly on material parameters, such as absorption coefficient, doping levels, recombination lifetimes, etc., but also on the beam profiles and intensities, and the stability of the experimental arrangement; it is, therefore, a *system* parameter [27]. Typical photorefractives have $M/1$ or less, but there are exceptions [13].

Photorefractive holograms are semipermanent, and, therefore, appropriate for optically erasable, rewritable and refreshable random access memory architectures [6, 13, 14, 47, 67, 68], or when dynamic holography is required, e.g., two-wave mixing [69, 70], phase-conjugation [71, 72], optical novelty filters [73], self-waveguiding [74, 75], etc. Photorefractives are often used also in applications that require permanent storage, because the crystal thickness can be large (several millimeters or centimeters), thus providing high capacity. For permanent storage in photorefractives, a number of techniques have been devised for getting around the erasure problem, and include thermal fixing [76–81], electrical fixing [82–86], two-lambda readout [87–91], and two-photon recording [30, 92–95]. A comprehensive review of nonvolatile photorefractive storage is given in [96].

A different class of holographic recording mechanisms is based on photochemical changes initiated by exposure to the recording beams. The most common example is photoinduced polymerization in the DuPont polymer HRF-150 [97–99], where recording occurs as refractive index modulation. Because of density changes in the exposed areas, it is permanent and does not significantly degrade over time. Despite the different recording mechanism, the diffraction efficiency as function of number of superimposed holograms still follows the rule (2). The HRF-150 has been demonstrated to have approximately $M/6$, and has been used successfully in a number of high capacity demonstrations of holographic storage [41, 46, 100].

The selection of material and multiplexing technique depends on the application. Storage in photopolymers is permanent, hence they target read-only memory (ROM) or write-once-read-many (WORM) storage applications. Unfortunately, the thickness of photopolymer films is limited by considerations of mechanical stability and optical quality. The highest capacity ever achieved in the DuPont polymer is 12 bits/μm^2 [46] using shift multiplexing with a 100-μm-thick film. This surface density is higher than the DVD-ROM by a factor of 2. Recently, samples of thickness up to 5 mm were fabricated using a poly(methyl methacylate) PMMA polymer matrix to host the photosensitive material phenanthrenequinone (PQ) [29, 101, 102]. Theoretical calculations [18, 45] show that the achievable density at 5 mm hologram thickness is as high as 200 bits/μm^2. Therefore, PQ-doped PMMA seems promising as a replacement to the DuPont HRF-

150 polymer and nonvolatile photorefractive storage for permanent high-density holographic memories.

8.3 Volume diffraction theory

In holographic materials, the recording of a volume hologram is expressed as a spatial modulation $\Delta\varepsilon(\mathbf{r})$ of the refractive index inside a finite volume $V_{\mathcal{H}}$. A generalized version of the volume diffraction geometry is given in Fig. 8.5. The probe field $E_{\mathrm{p}}(\mathbf{r})$ and index modulation $\Delta\varepsilon(\mathbf{r})$ are expressed in the xyz-coordinate system. For notational clarity, we use a different $x'y'z'$-coordinate system for the diffracted field $E_{\mathrm{d}}(\mathbf{r})$. When the hologram is illuminated by a probe field $E_{\mathrm{p}}(\mathbf{r})$, the diffracted field $E_{\mathrm{d}}(\mathbf{r}')$ is found as the solution to Maxwell's equations in the presence of the modulated

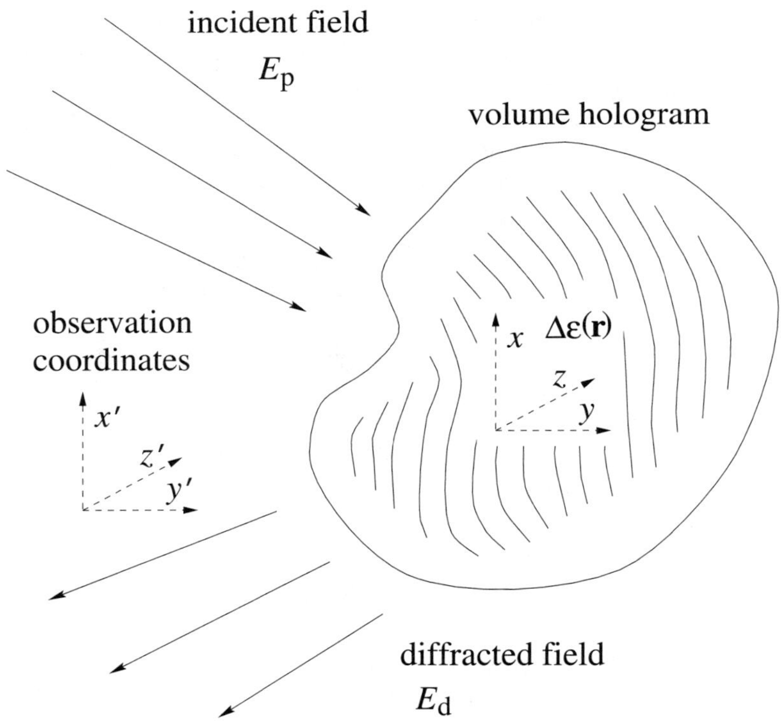

Figure 8.5: Volume diffraction geometry.

refractive index. The solution is simplified if we assume that the magnitude of the modulation is much smaller than the unmodulated refractive index ε_0:

$$|\Delta\varepsilon(\mathbf{r})| \ll \varepsilon_0, \qquad \mathbf{r} \in V_{\mathcal{H}}, \tag{8.3}$$

and apply the weak diffraction approximation (also known as "Born's approximation"). When several holograms are superimposed in the same volume, the inverse-square dynamic range law (8.2) means that individual holograms have low diffraction efficiencies, and, hence, the approximation remains always valid. The diffracted field is the 1st-order term in Born's scattering expansion [103]

$$E_{\mathrm{d}}(\mathbf{r}') = \iiint_{V_{\mathcal{H}}} E_{\mathrm{p}}(\mathbf{r})\Delta\varepsilon(\mathbf{r}) \times \frac{\exp\{ik|\mathbf{r} - \mathbf{r}'|\}}{|\mathbf{r} - \mathbf{r}'|}\, \mathrm{d}^3\mathbf{r}, \tag{8.4}$$

where $k = 2\pi/\lambda$ is the wavenumber.

Note that the last term in the integrand is actually the scalar Green's function for free space. Therefore, the diffracted field is the coherent summation of fields emitted by infinitesimal radiators distributed within the entire volume of the hologram. The amplitude of each radiator is the product of the local index modulation $\Delta\varepsilon(\mathbf{r})$ with the probe field $E_{\mathrm{p}}(\mathbf{r})$ in the same location.

Expression (8.4) works well when spherical waves are involved in the recording of the hologram, and we will use it extensively in the remainder of this chapter. It is quite general, because in principle it can be applied to all the recording and multiplexing configurations described in Section 8.2.1. For planar reference waves, however, it is computationally more efficient to represent the diffracted field and the grating in wavevector space instead [43]. If the reconstructed grating is strong enough to violate Born's approximation, it is better to use a coupled-wave approach for the calculation of the diffracted field [3].

8.4 Shift multiplexing

8.4.1 Introductory remarks

Shift multiplexing [18], or the use of spherical-reference volume holograms for data storage, was motivated by two needs. First, it was needed to have a mechanically stable access method to the multiplexed holograms.

Indeed, in a shift-multiplexed system, such as the one shown in Fig. 8.6, overlapping holograms are recorded with small relative displacements of the recording beams (the spherical beam reference and an arbitrary signal beam), hence the name "shift." The required relative translation δ between adjacent holograms is called "shift Bragg selectivity." Thus, the reference and signal are incident on location $\delta_m = (m - 1)\delta$ when recording the mth hologram. To reconstruct the same hologram, a spherical probe beam illuminates the memory at the same location δ_m. Relative translation between a light beam and the holographic material is, therefore, the only mechanical operation required for the implementation of a holographic memory. In the popular holographic disk architecture, where holograms are arranged along tracks, the required motion is conveniently provided by the disk rotation mechanism, while different tracks are accessed through radial head motion. Both motions have been well studied by the CD-ROM and DVD industries and provide enough precision without further worries

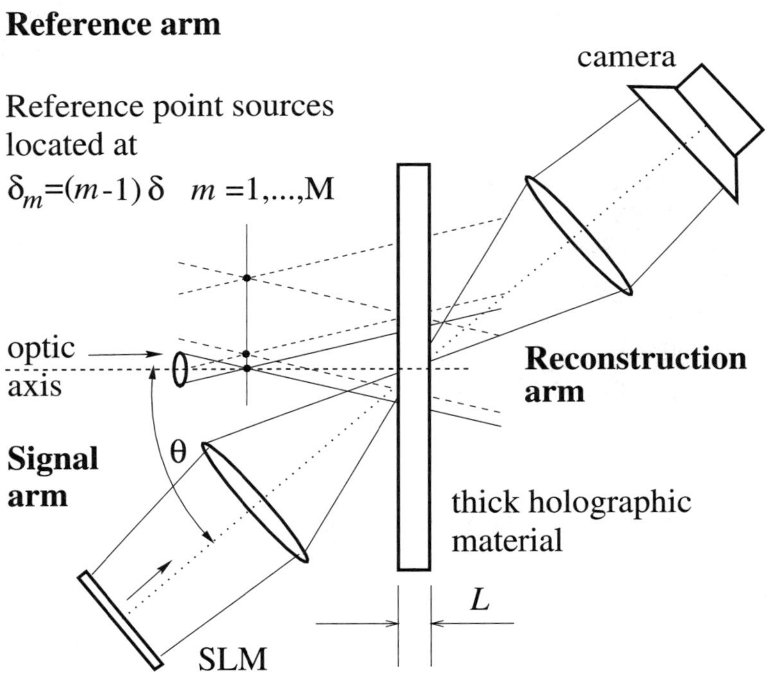

Figure 8.6: Shift-multiplexed holographic memory architecture.

for the holographic memory designer. By contrast, angle multiplexing requires both angular and spatial scanning of the probe beam, greatly complicating the mechanical design of the readout head. Other techniques that do not require mechanical access at all are also available, e.g., wavelength and phase-code multiplexing, but are limited by the state-of-the-art in tunable lasers and SLM, respectively.

The second motivation for spherical-reference holograms was for use in lensless correlator architectures, like the one shown in Fig. 8.7, as a compact, misalignment-tolerant implementation (note the inverted direction of the reference beams). The device functions as follows: when the superimposed holograms are illuminated by a test pattern, they diffract the cross-correlations of the test pattern with all stored patterns simultaneously. (In fact, vignetting effects may limit the accuracy of this operation, but this need not concern us here.)

8.4.2 Volume diffraction from spherical-reference holograms

We now consider the transmission geometry of Fig. 8.8, with a plane-wave signal beam. The reference beam used for recording is a spherical

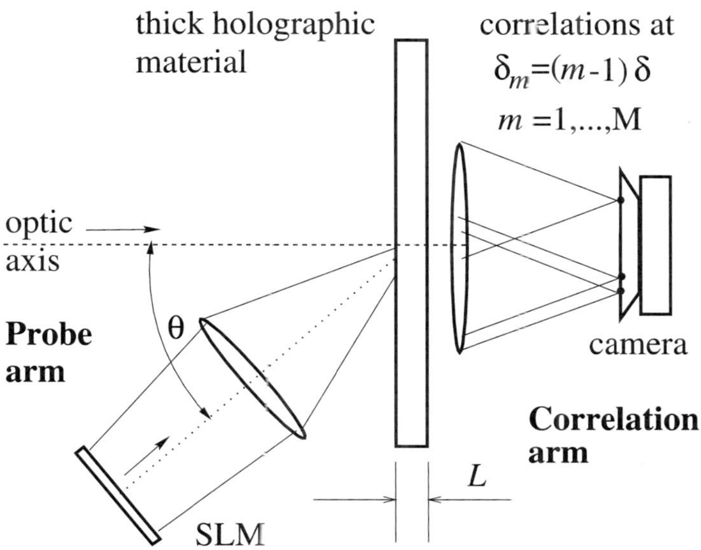

Figure 8.7: Shift-multiplexed correlator.

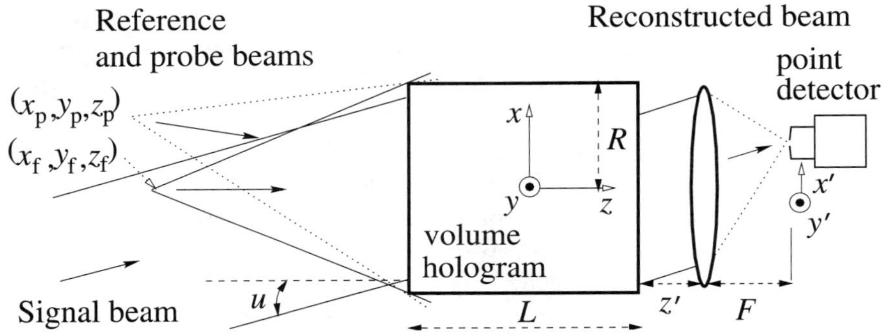

Figure 8.8: Transmission geometry with spherical wave reference and plane wave signal beams.

wave at wavelength λ produced by a point source at $\mathbf{r}_f = x_f\hat{\mathbf{x}} + y_f\hat{\mathbf{y}} + z_f\hat{\mathbf{z}}$. We express this wave in the paraxial approximation as

$$E_f(\mathbf{r}) = \exp\left\{i2\pi\frac{z - z_f}{\lambda} + i\pi\frac{(x - x_f)^2 + (y - y_f)^2}{\lambda(z - z_f)}\right\}. \tag{8.5}$$

Note that here and in the sequel we have neglected a term of the form $1/\lambda(z - z_f)$ because it varies with z much slower than the exponential term. The signal beam is a plane wave propagating at angle $u \ll 1$ with respect to the $\hat{z}$-axis. In the paraxial approximation, it is expressed as

$$E_s(\mathbf{r}) = \exp\left\{i2\pi\left(1 - \frac{u^2}{2}\right)\frac{z}{\lambda} + i2\pi u\frac{x}{\lambda}\right\}. \tag{8.6}$$

The modulation of the material refractive index resulting from exposure to beams E_f, E_s is given by

$$\Delta\varepsilon(\mathbf{r}) = |E_f(\mathbf{r}) + E_s(\mathbf{r})|^2. \tag{8.7}$$

In volume diffraction calculations we neglect all but $E_f^*(\mathbf{r})E_s(\mathbf{r})$ from the interference pattern (8.7). It is easy to show, by direct substitution in the volume diffraction equation (8.4), that the remaining three terms are Bragg-mismatched and do not diffract significantly.

The probe field is a spherical wave at the same wavelength λ emanating at $\mathbf{r}_p = x_p\hat{\mathbf{x}} + y_p\hat{\mathbf{y}} + z_p\hat{\mathbf{z}}$. The expression for the probe field is

$$E_p(\mathbf{r}) = \exp\left\{i2\pi\frac{z - z_p}{\lambda_p} + i\pi\frac{(x - x_p)^2 + (y - y_p)^2}{\lambda_p(z - z_p)}\right\}. \tag{8.8}$$

To find the diffracted field at detector coordinates $\mathbf{r}'$ (located near the focus $\mathbf{r}_s$ of the signal beam) we use the volume diffraction formula (8.4). We simplify by assuming that the holographic medium is disk-shaped with radius R in the xy-plane, and thickness L along the z-direction, and making the paraxial approximation, i.e., we assume that R is smaller than any longitudinal distance that the fields propagate. We then obtain

$$E_d(\mathbf{r}'') = \iiint E_p(\mathbf{r})\Delta\varepsilon(\mathbf{r}) \operatorname{circ}\left(\frac{\sqrt{x^2+y^2}}{R}\right) \operatorname{rect}\left(\frac{z}{L}\right) \times \exp\left\{i2\pi \frac{z''-z}{\lambda_p} + i\pi \frac{(x''-x)^2+(y''-y)^2}{\lambda_p(z''-z)}\right\} d^3\mathbf{r}. \tag{8.9}$$

The actual field on the detector is the Fourier transform of $E_d(\mathbf{r}'')$, i.e.,

$$\tilde{E}_d(\mathbf{r}') = \iint_{-\infty}^{+\infty} E_d(\mathbf{r}'') \exp\left\{-i2\pi \frac{x'x''+y'y''}{\lambda_p F}\right\} dx''dy'', \tag{8.10}$$

where F is the focal length of the Fourier-transforming lens (see Fig. 8.13), and constant phase factors have been omitted. The limits of integration in (8.10) are taken to be infinite by assuming that the effective aperture of the system is set by the transverse size R of the hologram rather than by the lens. We can then substitute (8.9) into (8.10), and perform the x'', y'' integrations right away. The result is

$$\tilde{E}_d(\mathbf{r}') = \iiint \exp\{i\pi A(z)(x^2+y^2)\} \exp\{i2\pi\,[B_x(z)x + B_y(z)y]\} \times \exp\{i\pi C(z)\} \operatorname{circ}\left(\frac{\sqrt{x^2+y^2}}{R}\right) \operatorname{rect}\left(\frac{z}{L}\right) d^3\mathbf{r}. \tag{8.11}$$

The coefficients $A(z)$, $B_x(z)$, $B_y(z)$, $C(z)$ are expressed as

$$A(z) = \frac{1}{\lambda(z-z_f)} - \frac{1}{\lambda(z-z_p)}, \tag{8.12}$$

$$B_x(z) = -\frac{x_p}{\lambda(z-z_p)} + \frac{x_f}{\lambda(z-z_f)} - \frac{x'}{\lambda F} + \frac{u}{\lambda}, \tag{8.13}$$

$$B_y(z) = -\frac{y_p}{\lambda(z-z_p)} + \frac{y_f}{\lambda(z-z_f)} - \frac{y'}{\lambda F}, \tag{8.14}$$

$$C(z) = \frac{x_p^2+y_p^2}{\lambda(z-z_p)} - \frac{x_f^2+y_f^2}{\lambda(z-z_f)} + \left(\frac{x'^2+y'^2}{\lambda F^2} - \frac{u^2}{\lambda}\right) z. \tag{8.15}$$

The integral in Eq. (8.11) is further simplified by using cylindrical coordinates. We define the parameters ρ, ϕ, $B(z)$, $\alpha(z)$ by

$$\begin{cases} x = \rho \cos \phi, & B_x(z) = B(z)\cos \alpha(z) \\ y = \rho \sin \phi, & B_y(z) = B(z)\sin\alpha(z) \end{cases} \Leftrightarrow \tag{8.16}$$

$$\begin{cases} \rho = \sqrt{x^2 + y^2}, & B(z) = \sqrt{B_x(z) + B_y(z)} \\ \tan \phi = y/x, & \tan \alpha(z) = B_y(z)/B_x(z) \end{cases} \tag{8.17}$$

(the sign of the inverse tangent is taken to conform with the quadrant of x, y, and $B_x(z)$, $B_y(z)$, respectively). Equation (8.11) then becomes:

$$\tilde{E}_{\mathrm{d}}(\mathbf{r}') = \int_{-L/2}^{L/2} \exp\{i\pi C(z)\} \int_0^R \exp\{i\pi A(z)\rho^2\} \times \int_{-\pi}^{\pi} \{-i2\pi B(z)\rho \cos(\phi - \alpha(z))\} \, \mathrm{d}\phi\rho\mathrm{d}\rho\mathrm{d}z. \tag{8.18}$$

The innermost integral results in the zero-order Bessel function of the first kind,

$$\int_{-\pi}^{\pi} \exp\{- i2\pi B(z)\phi \cos(\phi - \alpha(z))\}\mathrm{d}\phi = 2\pi J_0(2\pi B(z)\rho). \tag{8.19}$$

The next-level integral occurs in the calculation of the 3-D-PSF of a lens near focus [104], and is written as

$$\int_0^1 \exp\left\{-\frac{i}{2} u\rho^2\right\} J_0(v\rho)\rho\mathrm{d}\rho = \mathcal{L}(u, v) \tag{8.20}$$

where the real and imaginary parts of the function $\mathcal{L}(u, v)$ are expressed in terms of the *Lommel* functions. In terms of the $\mathcal{L}$ function, the diffracted field at the detector is expressed as

$$\tilde{E}_{\mathrm{d}}(\mathbf{r}') = 2\pi R^2 \int_{-L/2}^{L/2} \exp\{i\pi C(z)\} \, \mathcal{L}(2\pi A(z)R^2, 2\pi B(z)R) \, \mathrm{d}z. \tag{8.21}$$

This result has an interesting interpretation. Since $\mathcal{L}(. , .)$ describes the amplitude transmitted from a quadratic lens also, Eq. (8.21) means that the diffracted light from the volume hologram is the coherent superposition of *several* "lenses" stacked in the $\hat{\mathbf{z}}$ direction. If the probe source is at the common front focus of all these virtual "lenses," then the "lenses" are all in phase and give a strong reconstruction in the back focal point. The volume hologram is then *Bragg-matched*. If the probe moves around (or changes its color, as we will see in Section 8.5) the "lens" contributions are in general out of phase, resulting in *Bragg mismatch*, and the reconstructed amplitude drops. Exceptional combinations of probe wavelength and position where strong diffraction still occurs do exist, and we will refer to them as *degeneracies* (Sections 8.4.4 and 8.5).

For storage applications, we are interested in the behavior of the diffraction efficiency, i.e., the amplitude of the diffracted field $\tilde{E}_{\rm d}(\mathbf{r}')$, as function of the relative displacement $\mathbf{r}_{\rm p} - \mathbf{r}_{\rm f}$ of the reference and probe beams. This can be calculated by numerical integration of Eq. (8.21). In the next two sections we consider probe source translation (relative to the reference source) in two directions, $\hat{\mathbf{x}}$ and $\hat{\mathbf{y}}$. Translation of the probe in the $\hat{\mathbf{x}}$ direction (Section 8.4.3) results in Bragg-mismatch, which is useful for storage because holograms can be multiplexed in that direction. Translation in the $\hat{\mathbf{y}}$ direction (Section 8.4.3) results in translation of the reconstruction by a proportional amount without significant drop in diffraction. Therefore, it is degenerate according to the preceding definition.

8.4.3 Shift selectivity in the transmission geometry

Consider a probe beam displaced by δ_x in the $\hat{\mathbf{x}}$ direction with respect to the reference, i.e., in the plane defined by the signal beam direction and the optical axis. Specifically, we assume $\mathbf{r}_{\rm f} = (0, 0, -z_0)$ and $\mathbf{r}_{\rm p} = (\delta_x, 0, -z_0)$, with $z_0 > 0$. As the probe moves, the reconstruction moves also, to a location found by maximizing the slowly varying component of the integrand, i.e., setting the arguments to $\mathcal{L}(.\,,.)$ equal to zero. From (8.12) we see that $A(z) = 0$ always. From Equations (8.13) and (8.14) it follows that we cannot satisfy $B(z) = 0$ simultaneously for all z. Instead, we satisfy

$$B_x(0) = 0 \Leftrightarrow \frac{x'}{F} = u - \frac{\delta_x}{z_0} \tag{8.22}$$

$$B_y(0) = 0 \Leftrightarrow \frac{y'}{F} = 0 \tag{8.23}$$

If Eqs. (8.22) and (8.23) are satisfied, then, ignoring terms of order z or higher in the Taylor expansion of $\mathcal{L}(.\,,.)$, we obtain

$$\tilde{E}(\mathbf{r}') \approx 2\pi R^2 \int_{-L/2}^{L/2} \exp\{i\pi C(z)\}\, \mathrm{d}z. \tag{8.24}$$

Next we approximate $C(z) \approx C_0 + C_1 z$, where, using Eqs. (8.22) and (8.23),

$$C_0 = C(0) = -\frac{\delta_x^2}{\lambda z_0}, \qquad C_1 = \left.\frac{dC}{dz}\right|_{z=0} = -\frac{2\delta_x u}{z_0}. \tag{8.25}$$

Ignoring terms of higher order in z in the Taylor expansion of $C(z)$, we find the explicit (but approximate) expression

$$\tilde{E}_d\left(u - \frac{\delta_x}{z_0}, 0, z'\right) \approx \exp\left\{i2\pi\frac{z'}{\lambda}\left[1 - \left(u - \frac{\delta_x}{z_0}\right)^2\right] - i\pi\frac{\delta_x^2}{\lambda z_0}\right\} \times \operatorname{sinc}\left(\frac{\delta_x u L}{\lambda z_f}\right). \tag{8.26}$$

The shift Bragg selectivity is, by definition, the in-plane displacement of the probe beam required to reach the first null of the sinc function, i.e.,

$$\Delta\delta_x \approx \frac{\lambda z_f}{uL}. \tag{8.27}$$

Note the analogy of (27) and (1) if we correspond $\Delta\theta \leftrightarrow \Delta\delta_x/z_f$, the apparent relative angular motion resulting from shift δ_x.

The plot of the calculated diffraction efficiency $|\tilde{E}_d|^2$ using Eq. (8.21) with observation points given by Eqs. (22) and (23) as function of δ_x is given in Fig. 8.9 for the geometry of Fig. 8.8 with $\mathbf{r}_f = (0, 0, -10^4\lambda)$, $u = 0.2$, $R = 500\lambda$, $L = 4 \times 10^3\lambda$. From the plot we see that the location of the minima in diffraction efficiency is in good agreement with the estimate for the locations of the sinc nulls ($\Delta\delta_x = 12.5\lambda$) predicted by Eq. (8.27).

8.4.4 Volume holographic degeneracies in the transmission geometry

We now consider the case of out-of-plane probe source displacement by δ_y, i.e., along the $\hat{\mathbf{y}}$ direction. Let $\mathbf{r}_f = (0, 0, -z_0)$ and $\mathbf{r}_p = (0, \delta_y, -z_0)$, with $z_0 > 0$. The reconstruction in this case also translates in response to the translation of the probe, to a location found by setting $B(0) = 0$ (as in the previous section). The result is:

$$\frac{x'}{F} = u \qquad \frac{y'}{F} = -\frac{\delta_y}{z_0}. \tag{8.28}$$

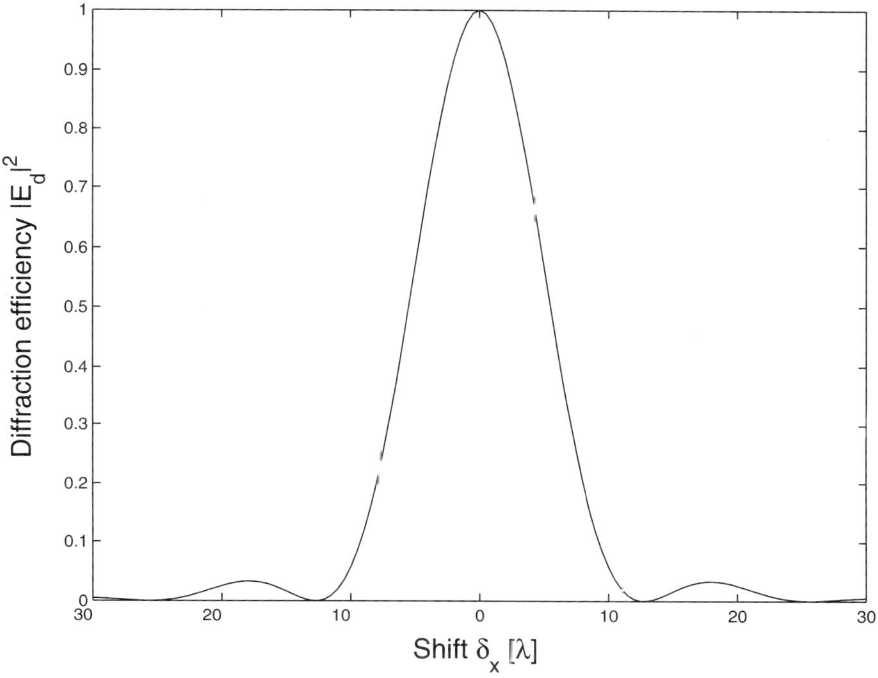

Figure 8.9: Numerical calculation of the diffraction efficiency as function of the probe displacement δ_x in the $\hat{x}$-direction.

Substituting Eq. (8.28) into Eq. (8.15) we find that $C_1 = 0$; therefore, to estimate the diffraction integral (8.21) we need the z^2 term in the Taylor expansion of $C(z)$. The result is

$$\tilde{E}_{\mathrm{d}}\left(u, -\frac{\delta_y}{z_0}, z'\right) \approx \exp\left\{i2\pi\frac{z'}{\lambda}(1-u^2) - i\pi\frac{\delta_y^2}{\lambda z_0}\right\} \times \int_{-L/2}^{L/2} \exp\left\{-i2\pi\frac{\delta_y^2 z^2}{\lambda z_0^3}\right\} \mathrm{d}z. \quad (8.29)$$

We conclude that the reconstructed diffraction efficiency decreases slower as function of probe translation out-of-plane rather than in-plane, because the sinc function in (8.26) decreases, as function of δ_x, faster than the integral in (8.29) as function of δ_y. This statement is verified by the numerical calculation of Fig. 8.10, which was done for the same parameters as Fig. 8.9 but using (8.21) with observation points given by Eq. (8.28). This slow Bragg-mismatching behavior is called *degeneracy* because the hologram is (relatively) insensitive to probe position in the $\hat{\mathbf{y}}$ direction.

8.5 Imaging with volume holograms

8.5.1 Introductory remarks

The ubiquity of digital computers and communications networks has revolutionized several technologies, and imaging is no exception. Digital imaging is not necessarily restricted to the superior image storage and processing flexibility available with digitized images; it can also be used

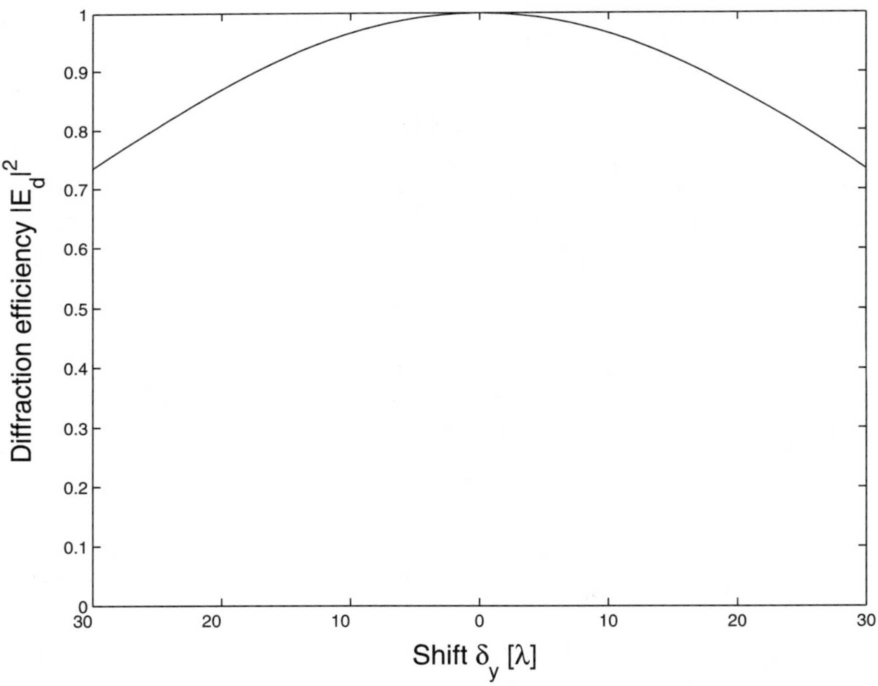

Figure 8.10: Numerical calculation of the diffraction efficiency as function of the probe displacement δ_y in the $\hat{y}$-direction.

to make more features of the natural environment accessible to advanced visual sensors by freeing the designer from the requirement of physical analogy between the intensity distribution on the sensor (e.g., a camera) and the projection of the objects to the two-dimensional grid (2-D) of camera pixels. Instead, the designer is allowed to select a representation space where information about object features of interest is maximal, then collect several projections of the objects in that space, and reconstruct the objects in their native 3-D space with minimal a priori assumptions. For example, it was recently shown that the intensity distributions collected from a simple lensless pinhole camera scanned along a line can be inverted as a Radon transform to reconstruct 3-D distributions of incoherent light sources [105]. Other promising methods of optical tomography are optical coherence tomography [106] and coherence imaging [107–109].

The layout of a general hybrid imaging system is shown in Fig. 8.11a. The required imaging (also referred to as inversion) transformations are divided in two parts, analog and digital, performed by the front-end optics and the back-plane electronics and software, respectively. The balance between analog and digital operations is determined by the location of the intensity detector [e.g., a CCD camera or complementary metal oxide semiconductor (CMOS) sensor], which makes field information available to the digital parts of the system while discarding the field phase. For example, in classical imaging systems forming 2-D projections of 3-D objects (Fig. 8.11b), there is no digital part. All transformations are analog, performed by a lens-based imaging system, and the detector is located at the focal plane.

Optimal image sensor performance requires that the designer have maximum freedom in the specification of both the analog and digital parts of the system. Whereas electronics and software can be configured almost arbitrarily to implement any computable function, achieving arbitrary response from optical elements is more difficult. For example, cost-effective lens design is usually limited to surfaces of revolution because of surface machining considerations. Similarly, planar diffractive optical elements are restricted by the minimum feature size of lithographic or other surface patterning techniques.

Volume holograms provide the most general optical response as analog elements [8, 24]. This is because volume diffraction operates on a propagating field, whereas surface elements only provide the initial condition for the field on a surface (see the degree-of-freedom comparison of volume and surface transformations in Section 8.1). In the case of imaging

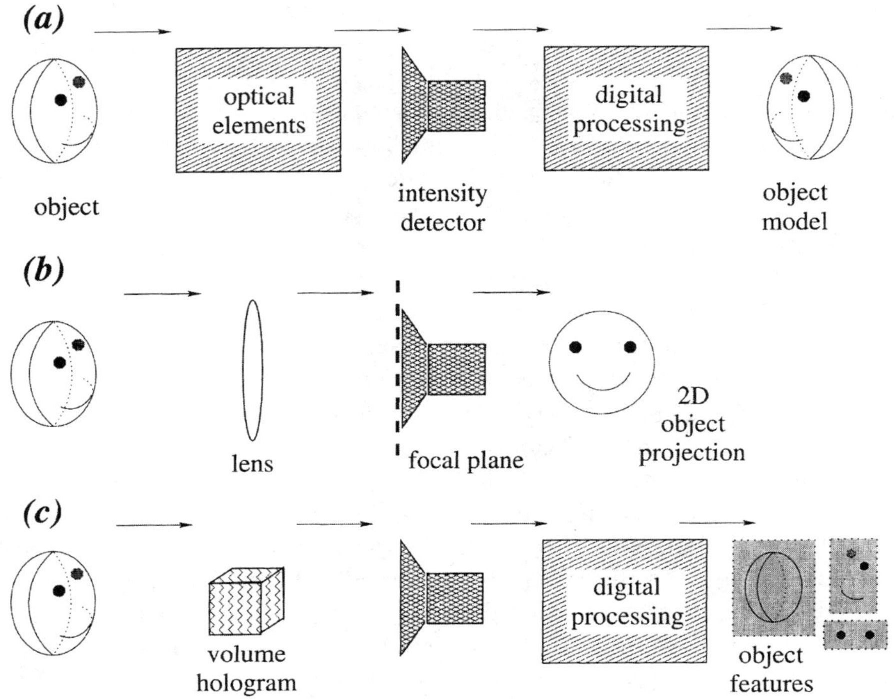

Figure 8.11: Imaging systems. (a) Hybrid imaging system with analog/optical and digital/electronic or software parts. (b) Classical imaging system with a focal plane and no digital processing. (c) Volume holographic imaging system with feature extraction.

systems, volume holographic analog transformations can potentially facilitate complicated superresolution, feature extraction, model-based imaging, recognition, etc. operations at the back plane (Fig. 8.11c). We do not attempt to solve the general design problem here, but rather demonstrate a simple tomographic imaging operation of spherical-reference volume holograms. We utilize a combination of the spatial selectivity and degeneracy properties first introduced in Section 8.4 in the context of shift multiplexing [18].

During the data retrieval and correlation operations that we discussed in Sections 8.2–8.4, the volume hologram performs a projection of the probe field onto the stored patterns. In the retrieval case, the projection is simply one of the stored patterns (with small cross talk contributions from other stored patterns). In the correlation case, we obtain instead a superposition of correlations. Now suppose that the probe field, which so far was silently taken to be deterministic, is replaced by the incoherent radiation produced by an extended polychromatic object, as in Fig. 8.12. Restricting ourselves to primary-source type (e.g., luminescent or fluorescent) objects, we aim to obtain the position in 3-D space, intensity, and color of the elementary radiators that comprise the object. In other words, we seek to *image* the extended object. The reconstructed field is a projection of the probe field onto the stored patterns, with contributions from distinct object radiators adding incoherently [19]. While there are a multitude of operations that could be performed this way, depending on the contents of the hologram, we concentrate here on tomographic operations only. Suppose that the volume hologram contains a single interference pattern formed by a reference point source and a simple signal field, e.g., a plane or spherical wave. We can immediately anticipate that, for a monochromatic object, the reconstruction primarily contains contributions from the vicinity of the original reference source location, because this location is *Bragg-matched* according to the theory of shift multiplexing (Section 8.4). The measurement of the focused diffracted field by a point detector is proportional to the object intensity at this Bragg-matched location. Other probe sources do not correlate well with the stored pattern (because they are Bragg-mismatched), and are rejected from the measurement. Using volume diffraction theory, we extend this elementary understanding to polychromatic objects in the next four sections. We show that

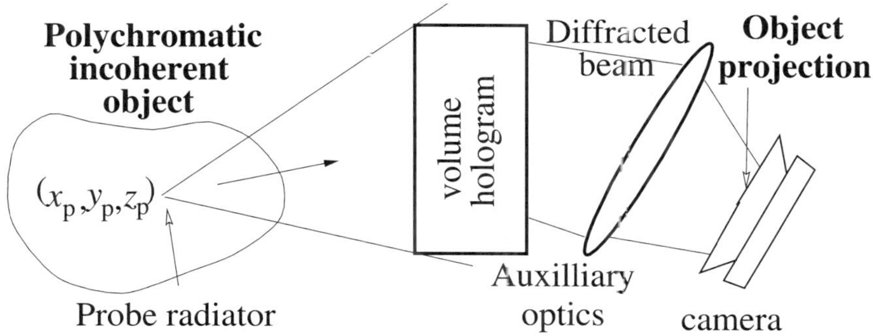

Figure 8.12: Three-dimensional imaging as volume holographic correlation.

the correlation operation establishes a correspondence between a slice of the object, the shape of which depends on the geometry, and the detector surface; i.e., it is a *tomographic* operation ("tomographic" means "registering a slice" in Greek). Thus, the spatial selectivity property exploited for storage in the shift multiplexing technique extends naturally to an important imaging operation.

8.5.2 Reflection geometry, plane-wave signal

First we consider the reflection geometry of Fig. 8.13, with a plane wave signal beam. The reference beam used for recording is a spherical wave at wavelength λ_f produced by a point source at $\mathbf{r}_f = x_f\hat{\mathbf{x}} + y_f\hat{\mathbf{y}} + z_f\hat{\mathbf{z}}$. We express this wave in the paraxial approximation as

$$E_f(\mathbf{r}) = \exp\left\{i2\pi\frac{z - z_f}{\lambda_f} + i\pi\frac{(x - x_f)^2 + (y - y_f)^2}{\lambda_f(z - z_f)}\right\}. \tag{8.30}$$

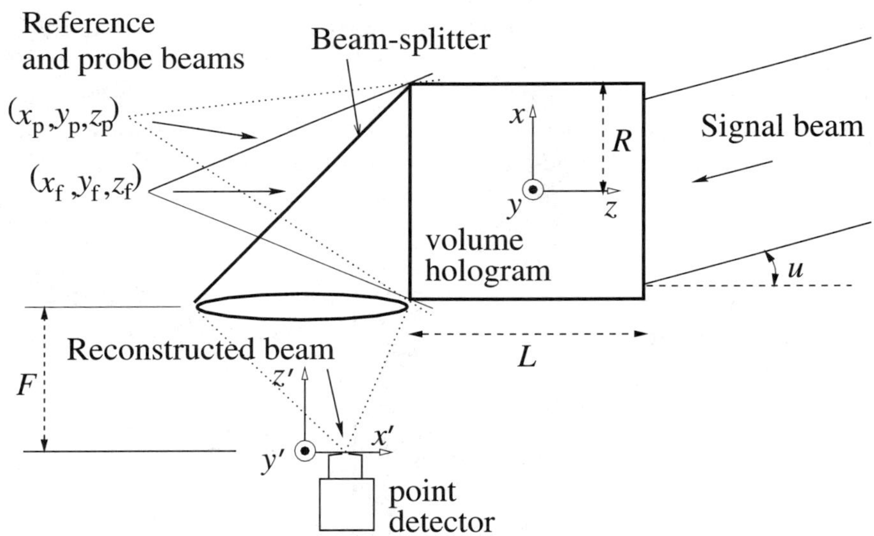

Figure 8.13: Reflection geometry with spherical wave reference and plane-wave signal beams.

The signal beam is a plane wave propagating at angle $u \ll 1$ with respect to the $\hat{\mathbf{z}}$-axis. In the paraxial approximation, it is expressed as

$$E_s(\mathbf{r}) = \exp\left\{-i2\pi\left(1 - \frac{u^2}{2}\right)\frac{z}{\lambda} + i2\pi u\,\frac{x}{\lambda}\right\}. \tag{8.31}$$

The probe field is a spherical wave at wavelength λ_p emanating at $\mathbf{r}_p = x_p\hat{\mathbf{x}} + y_p\hat{\mathbf{y}} + z_p\hat{\mathbf{z}}$. The expression for the probe field is

$$E_p(\mathbf{r}) = \exp\left\{i2\pi\,\frac{z - z_p}{\lambda_p} + i\pi\,\frac{(x - x_p)^2 + (y - y_p)^2}{\lambda_p(z - z_p)}\right\}. \tag{8.32}$$

To find the diffracted field at the detector coordinates $\mathbf{r}'$ (located near the focus $\mathbf{r}_s$ of the signal beam) we use the volume diffraction formula [Eq. (8.4)] in a disk-shaped holographic medium with radius R in the xy-plane, and thickness L along the z-direction. The procedure is similar to that followed for the shift selectivity calculation in Section 8.4, leading to Eq. (8.21) and will not be repeated here. The result is a similar integral of the form

$$\tilde{E}_d(\mathbf{r}') = 2\pi R^2 \int_{-L/2}^{L/2} \exp\{i\pi C(z)\}\ \mathcal{L}\left(2\pi A(z)R^2, 2\pi B(z)R\right) dz \tag{8.33}$$

where the coefficients $A(z)$, $B_x(z)$, $B_y(z)$, $C(z)$ are expressed as

$$A(z) = \frac{1}{\lambda_p(z - z_p)} - \frac{1}{\lambda_f(z - z_f)}, \tag{8.34}$$

$$B_x(z) = -\frac{x_p}{\lambda_p(z - z_p)} + \frac{x_f}{\lambda_f(z - z_f)} - \frac{x'}{\lambda_p F} + \frac{u}{\lambda_f}, \tag{8.35}$$

$$B_y(z) = -\frac{y_p}{\lambda_p(z - z_p)} + \frac{y_f}{\lambda_f(z - z_f)} - \frac{y'}{\lambda_p F}, \tag{8.36}$$

$$C(z) = \frac{x_p^2 + y_p^2}{\lambda_f(z - z_p)} - \frac{x_f^2 + y_f^2}{\lambda_f(z - z_f)} + \left[4\left(\frac{1}{\lambda_p} - \frac{1}{\lambda_f}\right) - \frac{x'^2 + y'^2}{\lambda_p F^2} + \frac{u^2}{\lambda_f}\right] z\,. \tag{8.37}$$

Some properties of the diffracted field are now apparent:

1. The hologram is Bragg-matched if the probe is a replica of the reference, i.e., if $\lambda_p = \lambda_f$ and $\mathbf{r}_p = \mathbf{r}_f$, and the detector is placed so that $x'/F = u$, $y'/F = 0$. In general, if the probe position and/or

wavelength are changed, Bragg mismatch occurs as in the transmission geometry case (Section 8.4).

2. In Section 8.4.4 we found that, for a probe source at the same wavelength as the reference ($\lambda_p = \lambda_f$) in the transmission geometry, degeneracy occurs if the probe source is translated out-of-plane with respect to the signal beam and the optical axis of the reference. Here we will derive the degeneracy condition for the reflection geometry, and will generalize it by allowing the probe wavelength λ_p to vary as well as the probe position $\mathbf{r}_p$. We will refer to the locus of $(\lambda_p, \mathbf{r}_p)$ that maximizes the diffraction efficiency as the *degeneracy surface* of the volume hologram.

According to the above definition, the degeneracy surface consists of the stationary points of the diffraction integral. To obtain these, we must (1) maximize the slow-varying term $\mathcal{L}(.\,,.)$, or, equivalently, set the arguments $A(z)$ and $B(z)$ equal to zero, at least to zero order in the Taylor expansion in z; and (2) minimize the variation of the exponential term, at least to first order in the Taylor expansion in z, by setting $\partial C/\partial z = 0$. For later convenience, we define the parameter $\mu = \lambda_p/\lambda_f$, and, using Eqs. (8.34)–(8.36), obtain

$$z_p = \frac{z_f}{\mu}, \qquad \frac{x_p}{z_p} = \mu\left(\frac{x_f}{z_f} - u\right) + \frac{x'}{F'}, \qquad \frac{y_p}{z_p} = \mu\frac{y_f}{z_f} + \frac{y'}{F}. \tag{8.38}$$

Substituting the last two relations of Eq. (8.38) into Eq. (8.37) results in the following quadratic equation in μ:

$$D\mu^2 + G\mu - H = 0, \quad \text{where}$$

$$D = \left(\frac{x_f}{z_f} - u\right)^2 + \frac{y_f^2}{z_f^2}, \qquad H = 4 - 2\frac{x'^2 + y'^2}{F^2}, \tag{8.39}$$

$$G = 4 - 2u\frac{x_f}{z_f} + \frac{x'^2 + y'^2}{F^2} - \left(\frac{x_f}{z_f} - u - \frac{x'}{F}\right)^2 - \left(\frac{y_f}{z_f} - \frac{y'}{F}\right)^2.$$

The degeneracy surface is obtained by setting μ equal to the root of Eq. (8.39) that is closest to 1, then substituting in Eq. (8.38). The result for a particular numerical example, using $\mathbf{r}_f = (100, -100, -2500)\ \lambda$, $u = 0$, $-0.3 \le x'/F, y'/F \le 0.3$, is shown in Fig. 8.14. In this figure we mapped diffracted color to brightness, so that the darkest grid point corresponds to $\mu = 0.899$, and the brightest to $\mu = 1.0$.

Two important properties of the degeneracy surface of Fig. 8.14 must be pointed out:

1. The degeneracy surface has a finite thickness because the reconstructed intensity from points and colors near the surface falls off smoothly according to Eq. (8.33). As the probe moves away from the degeneracy surface in the *normal* direction, the diffraction efficiency behavior is similar to the in-plane Bragg selectivity curve of Fig. 8.9.
2. The diffraction efficiency from points belonging to the degeneracy falls off as one moves from the center of the surface (i.e., the point that corresponds to $\lambda_p = \lambda_f$, $\mathbf{r}_p = \mathbf{r}_f$) toward the edges because higher-order z terms in the exponents cause weak Bragg-mismatch. The falloff of the diffraction efficiency as the probe moves *on* the degeneracy surface is similar to the out-of-plane Bragg selectivity curve of Fig. 8.10.

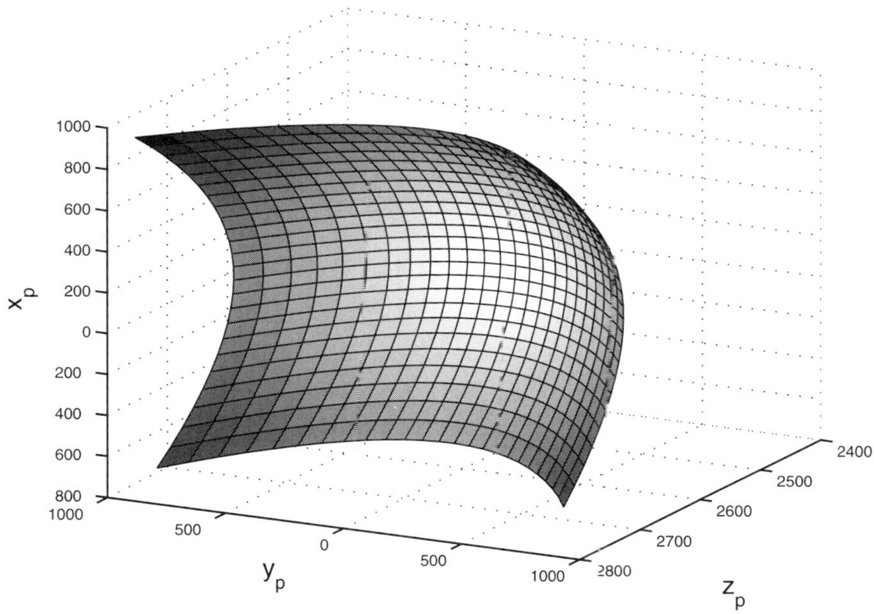

Figure 8.14: Computed degeneracy surface (space and color) of the reflection recording geometry with a plane-wave signal beam (see Fig. 8.13).

The tomographic operation of the volume hologram effected by the degeneracies is now apparent: Suppose that the hologram is illuminated by the incoherent superposition of several point sources forming an extended polychromatic object, as in Fig. 8.12. Then, the object portion that belongs to the degeneracy surface reconstructs the hologram, whereas the remaining object components are Bragg-mismatched, and isolated from the reconstructed field. The locations and wavelengths of the Bragg-matched object portion map on a regular detector grid at the output Fourier-transform plane as shown in Fig. 8.14. By scanning the hologram in two dimensions, a complete four-dimensional (4-D) (space + color) map of the object is obtained. The resolution and dynamic range of this tomographic operation follow from Eq. (8.33), but will not be given here.

8.5.3 Reflection geometry, spherical wave signal

We now consider the case of a reflection hologram recorded with a spherical wave signal beam, as shown in Fig. 8.15. The signal beam, counterpropagating with respect to the reference, comes to a focus at $\mathbf{r}_s = x_s\hat{\mathbf{x}} + y_s\hat{\mathbf{y}} + z_s\hat{\mathbf{z}}$. The expression of the electric field for this wave is

$$E_s(\mathbf{r}) = \exp\left\{i2\pi\frac{z - z_s}{\lambda_f} + i\pi\frac{(x - x_s)^2 + (y - y_s)^2}{\lambda_f(z - z_s)}\right\}. \tag{8.40}$$

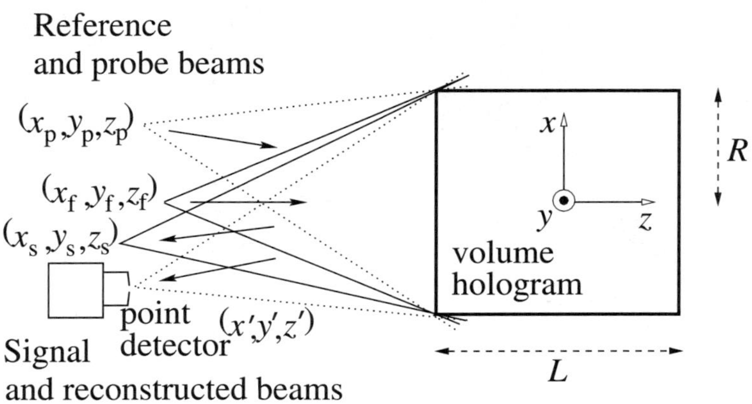

Figure 8.15: Reflection geometry with spherical wave reference and signal beams.

The diffracted field is again found starting from Eq. (8.4), and substituting Eq. (8.40) in the refractive index modulation $\Delta\varepsilon(\mathbf{r}) = E_{\mathrm{f}}^{*}(\mathbf{r})E_{\mathrm{s}}(\mathbf{r})$. Notice that the Fourier transform needed to focus the diffracted field in the case of a plane wave signal (Section 8.5.2) is not present now; hence, we obtain the diffracted field immediately as

$$E_{\mathrm{d}}(\mathbf{r}') = 2\pi R^2 \int_{-L/2}^{L/2} \exp\{i\pi C(z)\}\, \mathcal{L}(2\pi A(z)R^2, 2\pi B(z)R)\,\mathrm{d}z, \quad (8.41)$$

i.e., it is identical to the Fourier-transformed field $\tilde{E}_{\mathrm{d}}(\mathbf{r}')$ of Section 8.5.2, Eq. (8.33), but with

$$A(z) = \frac{1}{\lambda_{\mathrm{p}}(z-z_{\mathrm{p}})} + \frac{1}{\lambda_{\mathrm{p}}(z-z')} - \frac{1}{\lambda_{\mathrm{f}}(z-z_{\mathrm{f}})} - \frac{1}{\lambda_{\mathrm{f}}(z-z_{\mathrm{s}})}, \quad (8.42)$$

$$B_x(z) = -\frac{x_{\mathrm{p}}}{\lambda_{\mathrm{p}}(z-z_{\mathrm{p}})} - \frac{x'}{\lambda_{\mathrm{p}}(z-z')} + \frac{x_{\mathrm{f}}}{\lambda_{\mathrm{f}}(z-z_{\mathrm{f}})} + \frac{x_{\mathrm{s}}}{\lambda_{\mathrm{f}}(z-z_{\mathrm{s}})}, \quad (8.43)$$

$$B_y(z) = -\frac{y_{\mathrm{p}}}{\lambda_{\mathrm{p}}(z-z_{\mathrm{p}})} - \frac{y'}{\lambda_{\mathrm{p}}(z-z')} - \frac{y_{\mathrm{f}}}{\lambda_{\mathrm{f}}(z-z_{\mathrm{f}})} + \frac{y_{\mathrm{s}}}{\lambda_{\mathrm{f}}(z-z_{\mathrm{s}})}, \quad (8.44)$$

$$C(z) = \frac{x_{\mathrm{p}}^2+y_{\mathrm{p}}^2}{\lambda_{\mathrm{p}}(z-z_{\mathrm{p}})} + \frac{x'^2+y'^2}{\lambda_{\mathrm{p}}(z-z')} - \frac{x_{\mathrm{f}}^2+y_{\mathrm{f}}^2}{\lambda_{\mathrm{f}}(z-z_{\mathrm{f}})} - \frac{x_{\mathrm{s}}^2+y_{\mathrm{s}}^2}{\lambda_{\mathrm{f}}(z-z_{\mathrm{s}})} + 4\left(\frac{1}{\lambda_{\mathrm{p}}} - \frac{1}{\lambda_{\mathrm{f}}}\right) z\,. \quad (8.45)$$

The derivation of the degeneracy surface for this geometry is similar to that of the preceding section. We find that μ is the root, closest to 1, of

$$D\mu^2 + G\mu - H = 0, \quad \text{where}$$
$$D = \left(\frac{x_{\mathrm{f}}}{z_{\mathrm{f}}} + \frac{x_{\mathrm{s}}}{z_{\mathrm{s}}}\right)^2 + \left(\frac{y_{\mathrm{f}}}{z_{\mathrm{f}}} + \frac{y_{\mathrm{s}}}{z_{\mathrm{s}}}\right)^2, \qquad H = 4 - 2\frac{x'^2+y'^2}{F^2}, \quad (8.46)$$
$$G = 4 - 2\frac{x_{\mathrm{f}}x_{\mathrm{s}} + y_{\mathrm{f}}y_{\mathrm{s}}}{z_{\mathrm{f}}} + \frac{x'^2+y'^2}{z'^2} - \left(\frac{x_{\mathrm{f}}}{z_{\mathrm{f}}} + \frac{x_{\mathrm{s}}}{z_{\mathrm{s}}} + \frac{x'}{F}\right)^2 - \left(\frac{y_{\mathrm{f}}}{z_{\mathrm{f}}} + \frac{y_{\mathrm{s}}}{z_{\mathrm{s}}} + \frac{y'}{F}\right)^2.$$

The spatial coordinates of the degeneracy surface are obtained from

$$\frac{1}{z_{\mathrm{p}}} = \mu\left(\frac{1}{z_{\mathrm{f}}} + \frac{1}{z_{\mathrm{s}}}\right) - \frac{1}{z'},$$
$$\frac{x_{\mathrm{p}}}{z_{\mathrm{p}}} = \mu\left(\frac{x_{\mathrm{f}}}{z_{\mathrm{f}}} + \frac{x_{\mathrm{s}}}{z_{\mathrm{s}}}\right) - \frac{x'}{z'}, \qquad \frac{y_{\mathrm{p}}}{z_{\mathrm{p}}} = \mu\left(\frac{y_{\mathrm{f}}}{z_{\mathrm{f}}} + \frac{y_{\mathrm{s}}}{z_{\mathrm{s}}}\right) - \frac{y'}{z'}. \quad (8.47)$$

It is interesting to note that the degeneracy surface in this case is identical to the surface obtained in the case of a plane-wave signal (Section 8.5.2) with $x_s = -uz_s$, $y_s = 0$, and the output coordinates inverted ($x' \leftrightarrow -x'$, $y' \leftrightarrow -y'$). A numerical example, using $\mathbf{r}_f = (-100, 100, -2500)\lambda$, $\mathbf{r}_s = (0, 0, -3000)\lambda$, $-500\lambda \le x', y' \le 500\lambda$, $z' = -3000\lambda$, is given in Fig. 8.16. The darkest grid point corresponds to $\mu = 0.966$, and the brightest to $\mu = 1.0$.

8.5.4 90° geometry, plane-wave signal

The arrangement of the 90° geometry with a plane-wave signal beam is shown in Fig. 8.17. The computations in this case are complicated by the requirement that the paraxial approximation for the signal and diffracted beams be made along the $\hat{\mathbf{x}}$- rather than the $\hat{\mathbf{z}}$-axis. This leads to quantita

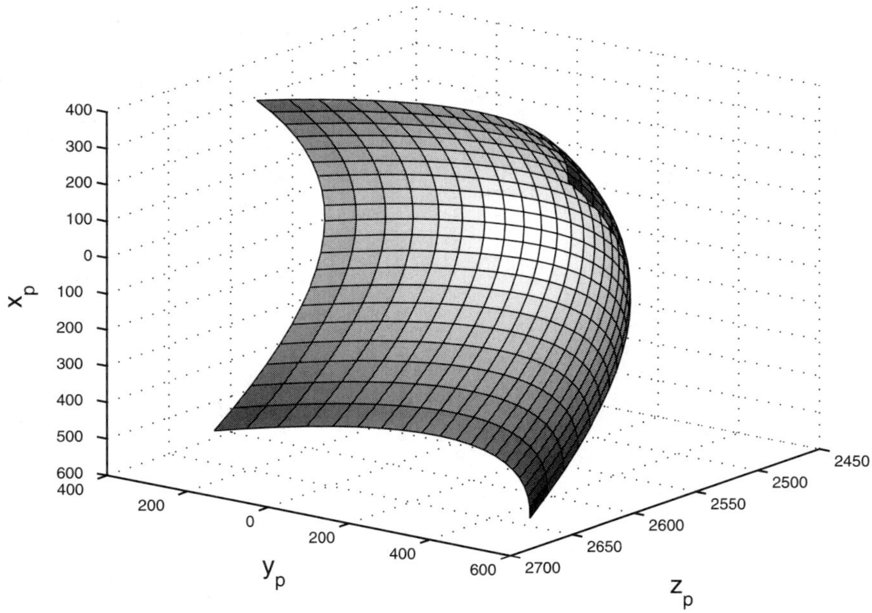

Figure 8.16: Computed degeneracy surface (space and color) of the reflection recording geometry with spherical wave reference and signal beam (see Fig. 8.15).

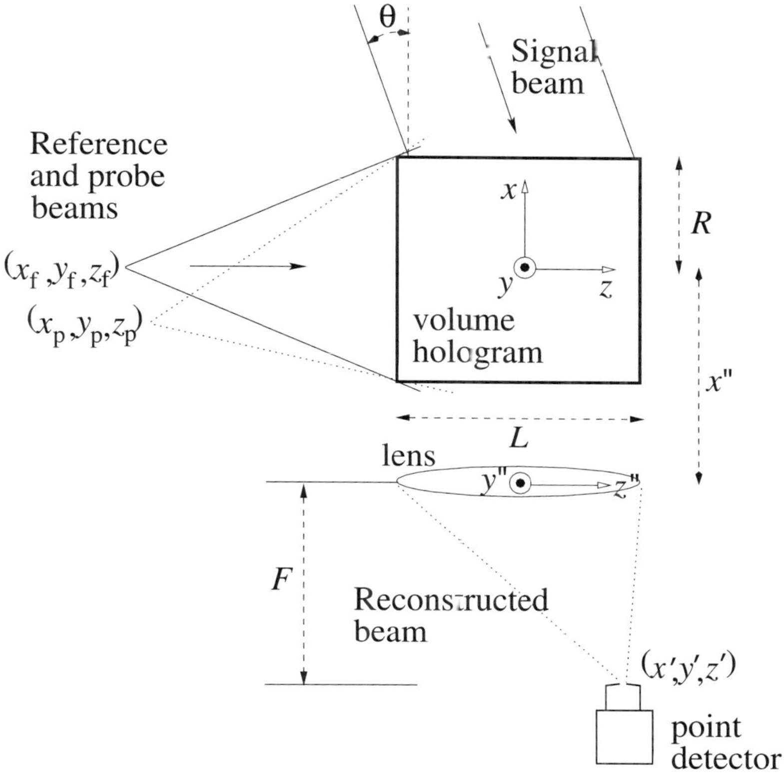

Figure 8.17: 90° geometry with spherical wave reference and plane-wave signal beams.

tively different expressions in the diffraction integrals. The reference and signal beams are

$$E_f(\mathbf{r}) = \exp\left\{ i2\pi \frac{z - z_f}{\lambda_f} + i\pi \frac{(x - x_f)^2 + (y - y_f)^2}{\lambda_f(z - z_f)} \right\}, \tag{8.48}$$

$$E_s(\mathbf{r}) = \exp\left\{ -i2\pi \left(1 - \frac{u^2}{2}\right) \frac{x}{\lambda} + i2\pi u \frac{z}{\lambda} \right\}. \tag{8.49}$$

For a probe field

$$E_p(\mathbf{r}) = \exp\left\{ i2\pi \frac{z - z_p}{\lambda_p} + i\pi \frac{(x - x_p)^2 + (y - y_p)^2}{\lambda_p(z - z_p)} \right\} \tag{8.50}$$

and an observation point $\mathbf{r}''$ near the $\hat{\mathbf{x}}$-axis, the paraxial diffraction integral is

$$
\begin{aligned}
E_{\mathrm{d}}(\mathbf{r}'') = \iiint E_{\mathrm{p}}(\mathbf{r})\Delta\varepsilon(\mathbf{r}) \operatorname{circ}\left(\frac{\sqrt{x^2+y^2}}{R}\right) \operatorname{rect}\left(\frac{z}{L}\right) \\
\times \exp\left\{ i2\pi \frac{x''-x}{\lambda_{\mathrm{p}}} + i\pi \frac{(y''-y)^2+(z''-z)^2}{\lambda_{\mathrm{p}}(x''-x)} \right\} \mathrm{d}^3\mathbf{r}.
\end{aligned} \tag{8.51}
$$

Performing the Fourier transform in y'', z'', and subsequently the integrals in cylindrical coordinates as in Section 8.5.2, we obtain yet again

$$
\tilde{E}_{\mathrm{d}}(\mathbf{r}') = 2\pi R^2 \int_{-L/2}^{L/2} \exp\{i\pi C(z)\}\, \mathcal{L}\,(2\pi A(z)R^2, 2\pi B(z)R)\,\mathrm{d}z \tag{8.52}
$$

with coefficients

$$
A(z) = \frac{1}{\lambda_{\mathrm{p}}(z-z_{\mathrm{p}})} - \frac{1}{\lambda_{\mathrm{p}}(z-z_{\mathrm{f}})}, \tag{8.53}
$$

$$
B_x(z) = -\frac{x_{\mathrm{p}}}{\lambda_{\mathrm{p}}(z-z_{\mathrm{p}})} + \frac{x_{\mathrm{f}}}{\lambda_{\mathrm{f}}(z-z_{\mathrm{f}})} + \frac{y'^2+z'^2}{2\lambda_{\mathrm{p}}F^2} - \frac{u^2}{2\lambda_{\mathrm{f}}} - \left(\frac{1}{\lambda_{\mathrm{p}}} - \frac{1}{\lambda_{\mathrm{f}}}\right), \tag{8.54}
$$

$$
B_y(z) = -\frac{y_{\mathrm{p}}}{\lambda_{\mathrm{p}}(z-z_{\mathrm{p}})} + \frac{y_{\mathrm{f}}}{\lambda_{\mathrm{f}}(z-z_{\mathrm{f}})} - \frac{y'}{\lambda_{\mathrm{p}}F}, \tag{8.55}
$$

$$
C(z) = \frac{x_{\mathrm{p}}^2+y_{\mathrm{p}}^2}{\lambda_{\mathrm{f}}(z-z_{\mathrm{p}})} - \frac{x_{\mathrm{f}}^2+y_{\mathrm{f}}^2}{\lambda_{\mathrm{f}}(z-z_{\mathrm{f}})} + 2\left(\frac{u}{\lambda_{\mathrm{f}}} - \frac{z'}{\lambda_{\mathrm{p}}F} + \frac{1}{\lambda_{\mathrm{p}}} - \frac{1}{\lambda_{\mathrm{f}}}\right)z. \tag{8.56}
$$

The degeneracy surface follows from the above expressions as in the reflection geometry cases. Similarly, it leads to a quadratic equation for μ of the form

$$
D\mu^2 + G\mu - H = 0, \quad \text{where}
$$

$$
D = 1 + \frac{x_{\mathrm{f}}^2+y_{\mathrm{f}}^2}{z_{\mathrm{f}}^2} - 2\frac{x_{\mathrm{f}}}{z_{\mathrm{f}}} - u^2, \qquad H = \left(1 - \frac{z'}{F}\right)^2, \tag{8.57}
$$

$$
G = \left(1 + \frac{x_{\mathrm{f}}}{z_{\mathrm{f}}}\right)^2 + \left(\frac{y_{\mathrm{f}}}{z_{\mathrm{f}}} + \frac{y'}{F}\right)^2 + \frac{z'^2}{F^2} + (u-1)^2 - 2\left(1 + \frac{x_{\mathrm{f}}^2+y_{\mathrm{f}}^2}{z_{\mathrm{f}}^2}\right).
$$

These expressions are consistent with the paraxial approximation, i.e., they contain up to quadratic terms in u, $\{x_f, y_f\}/z_f$, and $\{y', z'\}/F$, as with the entire analysis so far. After solving for μ as the root, closest to 1, of the quadratic, the spatial coordinates are obtained from

$$z_p = \frac{z_f}{\mu}, \qquad \frac{y_p}{z_p} = \mu\frac{y_f}{z_f} + \frac{y'}{F},$$
$$\frac{x_p}{z_p} = -\mu\left(1 - \frac{x_f}{z_f} - \frac{u^2}{2}\right) + \left(1 - \frac{y'^2 + z'^2}{2F^2}\right). \tag{8.58}$$

A numerical example, using $\mathbf{r}_f = (100, -100, -2500)\lambda$, $u = 0$, $-0.3 \leq x'/F, y'/F \leq 0.3$, is given in Fig. 8.18. The darkest grid point corresponds to $\mu = 0.619$, and the brightest to $\mu = 1.247$.

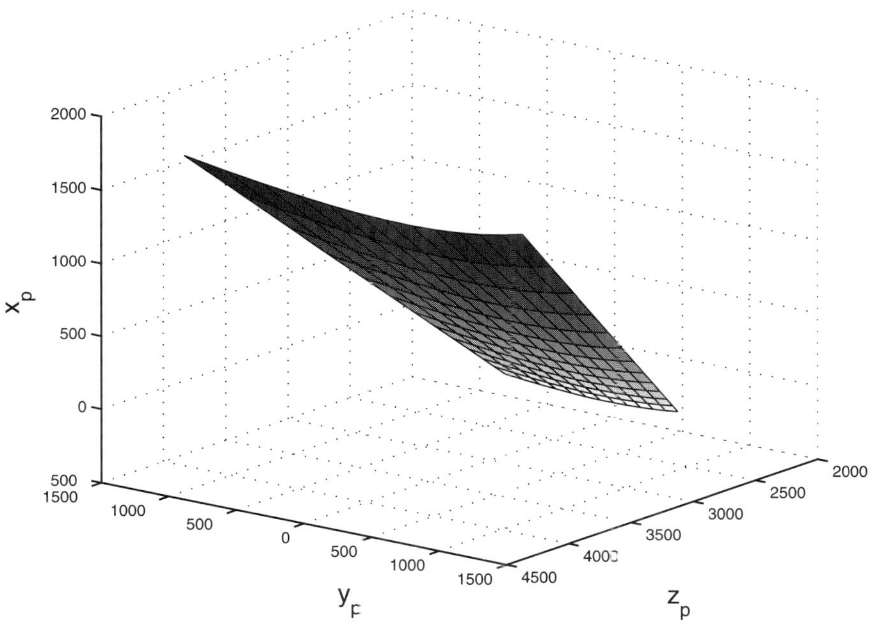

Figure 8.18: Computed degeneracy surface (space and color) of the 90° recording geometry with plane wave signal beam (see Fig. 8.17).

8.5.5 90° geometry, spherical wave signal

The geometry for this case is shown in Fig. 8.19. Because the paraxial approximations of the two perpendicular spherical waves produce mixed terms in all three coordinates (i.e., xy, yz, and zx), it is not possible to obtain a simplified 1-D diffraction integral as in the previous three cases. The diffracted field must rather be obtained from the full 3-D integral expression, which is not stable numerically. However, the degeneracy surface can still be obtained in simple form, without evaluating the actual diffracted field. We find that μ is the root, closest to 1, of the same quadratic expression

$$D\mu^2 + G\mu - H = 0, \qquad \text{where}$$

$$D = 1 + \frac{x_f^2 + y_f^2}{z_f^2} - 2\left(\frac{x_f}{z_f} + \frac{y_f\, y_s}{z_f\, x_s}\right) - \frac{z_s^2}{x_s^2}, \qquad H = 1 + \frac{z'^2}{x'^2}, \tag{8.59}$$

$$G = \left(1 + \frac{x_f}{z_f}\right)^2 + \left(\frac{y_f}{z_f} + \frac{y'}{x'}\right)^2 + \frac{z'^2}{x'^2} + \frac{y_s^2 + z_s^2}{x_s^2} - 1 - 2\frac{x_f^2 + y_f^2}{z_f^2} - 2\frac{y' y_s}{x' x_s}.$$

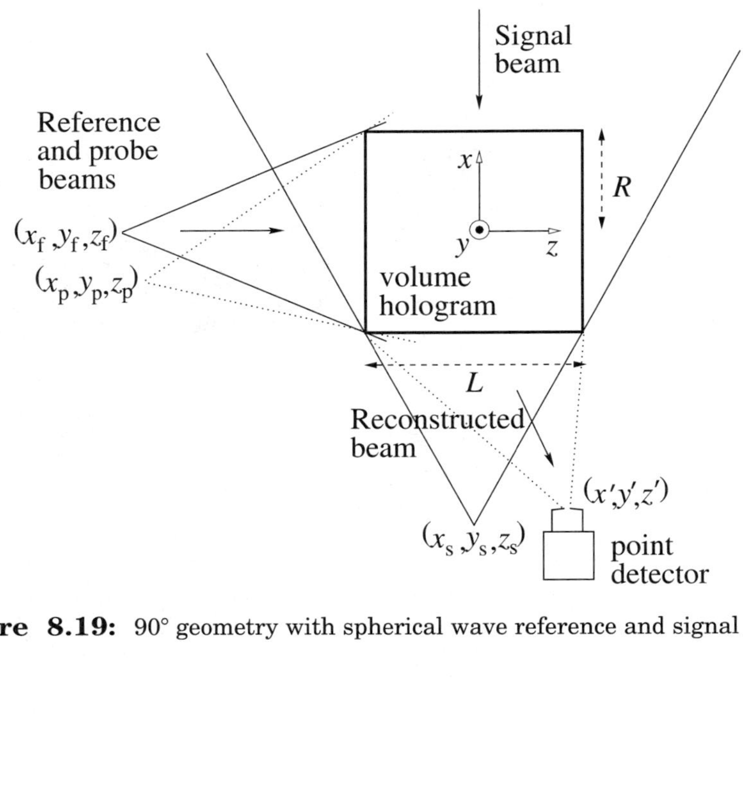

Figure 8.19: 90° geometry with spherical wave reference and signal beams.

Note that the correspondence with the case of a plane-wave signal is not exact as it was in reflection geometry. The spatial coordinates of the degeneracy surface are given by

$$z_p = \frac{z_f}{\mu}, \qquad \frac{y_p}{z_p} = \mu\left(\frac{y_f}{z_f} - \frac{y_s}{x_s}\right) + \frac{y'}{x'},$$
$$\frac{x_p}{z_p} = -\mu\left(1 - \frac{x_f}{z_f} - \frac{y_s^2 + z_s^2}{2x_s^2}\right) + \left(1 - \frac{y'^2 + z'^2}{2x'^2}\right). \tag{8.60}$$

A numerical example, using $\mathbf{r}_f = (-100, 100, -2500)\lambda$, $\mathbf{r}_s = (3000, 0, 0)\lambda$, $-500\lambda \leq y', z' \leq 500\lambda$, $x' = 3000\lambda$, is given in Fig. 8.20. The darkest grid point corresponds to $\mu = 0.979$, and the brightest to $\mu = 1.0$.

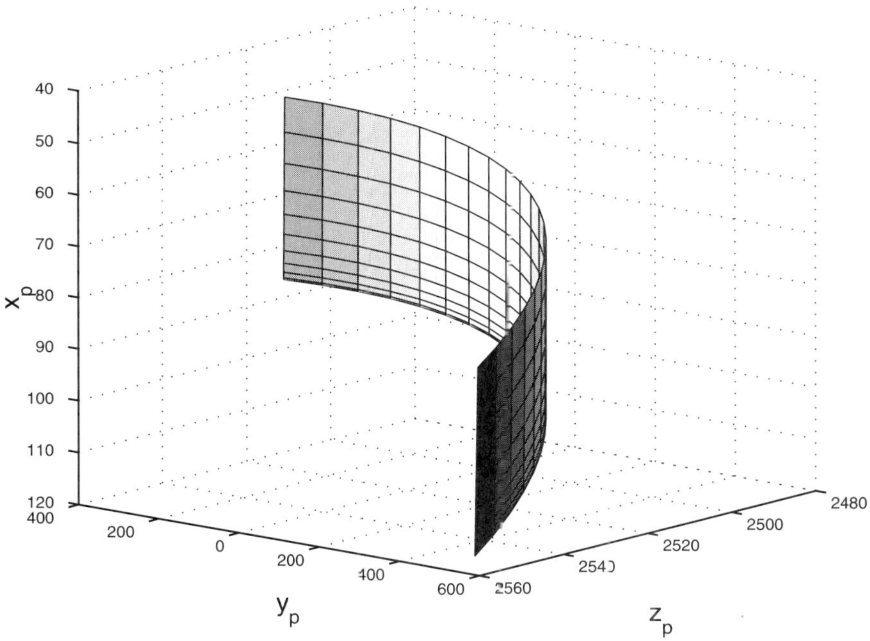

Figure 8.20: Computed degeneracy surface (space and color) of the 90° recording geometry with spherical wave signal beam (see Fig. 8.19).

8.6 Concluding remarks

We have studied in detail the volume diffraction properties of holograms recorded with spherical reference beams in the usual holographic recording arrangements: transmission, reflection, and 90° geometries. We showed that the reconstruction is selective to the relative position and wavelength of the probe and reference beams. If the relative translation suppresses the diffracted beam, a new page of information can be stored at the new position, leading to the shift-multiplexing method for holographic storage [18]. Alternatively, relative probe translation can be combined with appropriate color changes on the degeneracy surface to sustain high diffraction efficiency. If multiple probes and wavelengths are simultaneously present to reconstruct the hologram, a tomographic mapping results between the enseble of probes and the detector surface, isolating the probe sources that belong on the degeneracy surface [19]. It is striking that two applications in diverse areas such as data storage and 4-D imaging (i.e., imaging in the 3-D and color spaces) are based on the same physical principle. The primary advantage that makes shift-multiplexing attractive for holographic memories is the simplicity of the linear motion required to implement it. The simple version of tomographic imaging using volume holograms presented here has the advantage of reconstructing 4-D objects slicewise noninvasively on a flat detector surface, with no computational effort.

The spatiospectral behavior of volume holograms studied in this chapter is actually a small subset of the capabilities available to digital imaging from this class of optical elements. We are currently working on other architectures emphasizing superresolution through volume diffraction, as well as a generalized modal design approach that optimizes the volume hologram contents depending on the object statistics and inversion requirements. Experimental demonstrations of the proposed techniques are also under way.

References

1. P. J. van Heerden, "Theory of optical information storage in solids," *Appl. Opt.,* **2**(4), 393–400, 1963.
2. E. N. Leith, A. Kozma, J. Upatnieks, J. Marks, and N. Massey, "Holographic data storage in three-dimensional media," *Appl. Opt.,* **5**(8), 1303–1311, 1966.

3. H. Kogelnik, "Coupled wave theory for thick hologram gratings," *Bell Syst. Tech. J.,* **48**(9), 2909–2947, 1969.
4. A. Ashkin, G. D. Boyd, J. M. Dziedzic, R. G. Smith, A. A. Ballman, and K. Nassau, "Optically-induced refractive index inhomogeneities in $LiNbO_3$," *Appl. Phys. Lett.,* **9**, 72, 1966.
5. D. Psaltis, "Parallel optical memories," *Byte,* **17**(9), 179, 1992.
6. J. F. Heanue, M. C. Bashaw, and L. Hesselink, "Volume holographic storage and retrieval of digital data," *Science,* **265**(5173), 749–752, 1994.
7. D. Psaltis and F. Mok, "Holographic memories," *Sci. Am.,* **273**(5), 70–76, 1995.
8. D. Psaltis, D. Brady, X. G. Gu, and S. Lin, "Holography in artificial neural networks," *Nature,* **343**(6256), 325–330, 1990.
9. F. T. S. Yu and S. Jutamulia, *Optical Signal Processing, Computing, and Neural Networks*, Wiley-Interscience, 1992.
10. J. Hong, "Applications of photorefractive crystals for optical neural networks," *Opt. Quant. Electr.,* **25**(9), S551–S568, 1993.
11. D. J. Brady, A. G. S. Chen, and G. Rodriguez, "Volume holographic pulse shaping," *Opt. Lett.,* **17**(8), 610–612, 1992.
12. G. A. Rakuljic and V. Levya, "Volume holographic narrow-band optical filter," *Opt. Lett.,* **18**(6), 459–461, 1993.
13. F. H. Mok, M. C. Tackitt, and H. M. Stoll, "Storage of 500 high-resolution holograms in a $LiNbO_3$ crystal," *Opt. Lett.,* **16**(8), 605–607, 1991.
14. F. H. Mok, "Angle-multiplexed storage of 5000 holograms in lithium niobate," *Opt. Lett.,* **18**(11), 915–917, 1991.
15. J. H. Hong, I. McMichael, T. Y. Chang, W. Christian, and E. G. Pack, "Volume holographic memory systems — techniques and architectures," *Opt. Eng.,* **34**, 2193–2203, 1995.
16. M.-P. Bernal, H. Coufal, R. K. Grygglег, J. A. Hoffnagle, C. M. Jefferson, R. M. McFarlane, R. M. Shelby, G. T. Sincerbox, P. Wimmer, and G. Wittman, "A precision tester for studies of holographic optical storage materials and recording physics," *Appl. Opt.,* **35**(14), 2360–2374, 1996.
17. I. McMichael, W. Christian, D. Pletcher, T. Y. Chang, and J. Hong, "Compact holographic storage demonstrator with rapid access," *Appl. Opt.,* **35**(14), 2375–2379, 1996.
18. G. Barbastathis, M. Levene, and D. Psaltis, "Shift multiplexing with spherical reference waves," *Appl. Opt.,* **35**, 2403–2417, 1996.
19. G. Barbastathis and D. J. Brady, "Multidimensional tomographic imaging using volume holography", to appear in *Proc. IEEE.*

20. C. Denz, G. Pauliat, and G. Roosen, "Volume hologram multiplexing using a deterministic phase encoding method," *Opt. Commun.,* **85**, 171–176, 1991.
21. D. Psaltis, M. Levene, A. Pu, G. Barbastathis, and K. Curtis, "Holographic storage using shift multiplexing," *Opt. Lett.,* **20**(7), 782–784, 1995.
22. D. Gabor, "A new microscopic principle," *Nature,* **161**, 777, 1948.
23. D. Gabor, "Microscopy by reconstructed wavefronts," *Proc. Roy. Soc.*, A197, 454, 1949.
24. D. Brady and D. Psaltis, "Control of volume holograms," *J. Opt. Soc. Am.,* **A9**(7), 1167–1182, 1992.
25. D. Psaltis, D. Brady, and K. Wagner, "Adaptive optical networks using photorefractive crystals," *Appl. Opt.,* **27**(9), 1752–1759, 1988.
26. A. Pu, K. Curtis, and D. Psaltis, "Exposure schedule for multiplexing holograms in photopolymer films," *Opt. Eng.,* **35**(10), 2824–2829, 1996.
27. F. Mok, G. W. Burr, and D. Psaltis, "A system metric for holographic memory systems," *Opt. Lett.,* **21**(12), 896–898, 1996.
28. K. Meerholz, B. L. Volodin, B. Sandalphon Kippelen, and N. Peyghambarian, "A photorefractive polymer with high optical gain and diffraction efficiency near 100 percent," *Nature,* **371**(6497), 497–500, 1994.
29. G. J. Steckman, I. Solomatine, G. Zhou, and D. Psaltis, "Characterization of phenanthrenequinone-doped poly(methyl methacrylate) for holographic memory," *Opt. Lett.,* **23**(16), 1310–1312, 1998.
30. K. Büse, A. Adibi, and D. Psaltis, "Non-volatile holographic storage in doubly doped lithium niobate crystals," *Nature,* **393**(6686), 665–668, 1998.
31. D. L. Staebler, J. J. Amodei, and W. Philips, "Multiple storage of thick holograms in $LiNbO_3$," in *VII International Quantum Electronics Conference*, Montreal, 1972.
32. K. Curtis, A. Pu, and D. Psaltis, "Method for holographic storage using peristrophic multiplexing," *Opt. Lett.,* **19**(13), 993–994, 1994.
33. G. A. Rakuljic, V. Levya, and A. Yariv, "Optical data storage by using orthogonal wavelength-multiplexed volume holograms," *Opt. Lett.,* **17**(20), 1471–1473, 1992.
34. S. Yin, H. Zhou, F. Zhao, M. Wen, Y. Zang, J. Zhang, and F. T. S. Yu, "Wavelength-multiplexed holographic storage in a sensitive photorefractive crystal using a visible-light tunable diode-laser," *Opt. Commun.,* **101**(5–6), 317–321, 1993.
35. C. Gu, J. Hong, I. McMichael, R. Saxena, and F. Mok, "Cross-talk-limited storage capacity of volume holographic memory," *J. Opt. Soc. Am.,* **A9**(11), 1978–1983, 1992.

36. K. Curtis, C. Gu, and D. Psaltis, "Cross-talk in wavelength-multiplexed holographic memories," *Opt. Lett.*, **18**(12), 1001–1003, 1993.

37. A. Yariv, "Interpage and interpixel crosstalk in orthogonal (wavelength multiplexed) holograms," *Opt. Lett.*, **18**(8), 652–654, 1993.

38. K. Curtis and D. Psaltis, "Cross talk in phase-coded holographic memories," *J. Opt. Soc. Am.*, **A10**(12), 2547–2550, 1993.

39. K. Curtis and D. Psaltis, "Cross-talk for angle-multiplexed and wavelength-multiplexed image plane holograms," *Opt. Lett.*, **19**(21), 1774–1776, 1994.

40. M. C. Bashaw, J. F. Heanue, A. Aharoni, J. F. Walkup, and L. Hesselink, "Cross-talk considerations for angular and phase-encoded multiplexing in volume holography," *J. Opt. Soc. Am.*, **B11**(9), 1820–1836, 1994.

41. A. Pu and D. Psaltis, "High density recording in photopolymer-based holographic 3D disks," *Appl. Opt.*, **35**(14), 2389–2398, 1996.

42. S. Campbell, X. M. Yi, and P. Yeh, "Hybrid sparse-wavelength angle multiplexed optical data storage system," *Opt. Lett.*, **19**(24), 2161–2163, 1994.

43. G. Barbastathis and D. Psaltis, "Multiplexing methods," in H. Coufal, L. Hesselink, and D. Psaltis. (eds.), *Holographic Memories*, (in press).

44. D. A. Parthenopoulos and P. M. Rentzepis, "Two-photon volume information storage in doped polymer systems," *J. Appl. Phys.*, **68**(11), 5814–5818, 1990.

45. H.-Y. S. Li and D. Psaltis, "Three dimensional holographic disks," *Appl. Opt.*, **33**(17), 3764–3774, 1994.

46. A. Pu and D. Psaltis, "Holographic 3D disks using shift multiplexing," in *Summaries of Papers Presented at CLEO'96*, Baltimore, MD, 1996, p. 165.

47. J.-J. P. Drolet, E. Chuang, G. Barbastathis and D. Psaltis, "Compact, integrated dynamic holographic memory with refreshed holograms," *Opt. Lett.*, **22**(8), 552–554, 1997.

48. George Barbastathis, "Intelligent Holographic Databases," Ph.D. thesis, California Institute of Technology, 1998.

49. B. J. Goertzen and P. A. Mitkas, "Error-correcting code for volume holographic storage of a relational database," *Opt. Lett.*, **20**(15), 1655–1657, 1995.

50. M. A. Neifeld and J. D. Hayes, "Error-correction schemes for volume optical memories," *Appl. Opt.*, **34**(35), 8183–8191, 1995.

51. J. F. Heanue, K. Gurkan, and L. Hesselink, "Signal detection for page-access optical memories with intersymbol interference," *Appl. Opt.*, **35**(14), 2431–2438, 1996.

52. G. W. Burr, J. Ashley, H. Coufal, R. K. Grygier, J. A. Hoffnagle, C. M. Jefferson, and B. Marcus, "Modulation coding for pixel-matched holographic data-storage," *Opt. Lett.*, **20**(9), 639–641, 1997.

53. G. W. Burr, H. Coufal, R. K. Grygier, J. A. Hoffnagle, and C. M. Jefferson, "Noise reduction of page-oriented data storage by inverse filtering during recording," *Opt. Lett.,* **5**(15), 289–291, 1998.

54. R. M. Shelby, J. A. Hoffnagle, G. W. Burr, C. M. Jefferson, M. P. Bernal, H. Coufal, R. K. Grygier, H. Gunther, R. M. McFarlane, and G. T Sincerbox, "Pixel-matched holographic data storage with megabit pages," *Opt. Lett.,* **22**(19), 1509–1511, 1997.

55. C. Gu, J. Hong, and S. Campbell, "2-d shift-invariant volume holographic correlator," *Opt. Commun.,* **88**(4–6), 309–314, 1992.

56. F. T. S. Yu and S. Yin, "Bragg diffraction-limited photorefractive crystal-based correlators," *Opt. Eng.,* **34**(8), 2224–2231, 1995.

57. J. R. Goff, "Experimental realization of a multiproduct photorefractive correlation system for temporal signals," *Appl. Opt.,* **36**(26), 6627–6635, 1997.

58. F. T. S. Yu, "Optical neural networks: Architecture, design and models," *Progress in Optics,* **32**, 61–144, 1993.

59. H.-Y. S. Li, Y. Qiao, and D. Psaltis, "Optical network for real-time face recognition," *Appl. Opt.,* **32**(26), 5026–5035, 1993.

60. A. Pu, R. Denkewalter, and D. Psaltis, "Real-time vehicle navigation using a holographic memory," *Opt. Eng.,* **36**(10), 2737–2746, 1997.

61. G. T. Sincerbox, "Holographic storage — the quest for the ideal material continues," *Opt. Mat.,* **4**(2–3), 370–375, 1995.

62. M.-P. Bernal, G. W. Burr, H. Coufal, R. K. Grygier, J. A. Hoffnagle, C. M. Jefferson, R. M. McFarlane, R. M. Shelby, G. T. Sincerbox, and G. Wittmann, "Holographic-data-storage materials," *MRS Bulletin,* **21**(9), 51–60, 1996.

63. F. S. Chen, J. T. LaMacchia, and D. B. Fraser, "Holographic storage in lithium niobate," *Appl. Phys. Lett.,* **15**(7), 223–225, 1968.

64. N. V. Kukhtarev, V. B. Markov, S. G. Odulov, M. S. Soskin, and V. L. Vinetskii, "Holographic storage in electrooptic crystals, I. Steady state," *Ferroelectrics,* **22**, 949–960, 1979.

65. T. J. Hall, R. Jaura, L. M. Connors, and P. D. Foote, "The photorefractive effect — a review," *Progress in Quantum Electronics,* **10**(2), 77–145, 1985.

66. P. Yeh, *Introduction to Photorefractive Nonlinear Optics*, Wiley, 1993.

67. F. H. Mok, G. W. Burr, and D. Psaltis, "Angle and space multiplexed random access memory (HRAM)," *Optical Memory and Neural Networks,* **3**(2), 119–127, 1994.

68. G. Barbastathis, J. J. P. Drolet, E. Chuang, and D. Psaltis, "Compact terabit random-access memory implemented with photorefractive crystals," in F. T. S. Yu and S. Yin (eds.), *SPIE Proceedings: Photorefractive Fiber and*

Crystal Devices: Materials, Optical Properties and Applications III, San Diego, CA, 1997, pp. 107–122.

69. N. V. Kukhtarev, V. B. Markov, S. C. Odulov, M. S. Soskin, and V. L. Vinetskii, "Holographic storage in electrooptic crystals, II. Beam coupling — light amplification," *Ferroelectrics,* **22**, 961–964, 1979.

70. P. Yeh, "Two-wave mixing in nonlinear media," *IEEE J. Q. El.*, **25**, 484–519, 1989.

71. A. Yariv, "Phase-conjugate optics and real-time holography," *IEEE J. Q. El.*, **14**, 650–660, 1978.

72. J. Feinberg, "Self-pumped continuous-wave phase conjugator using internal reflection," *Opt. Lett.*, **7**, 486, 1982.

73. D. Z. Anderson and J. Feinberg, "Optical novelty filters," *IEEE J. Q. El.*, **25**(3), 635–647, 1989.

74. M. Segev, B. Crosignani, and A. Yariv, "Spatial solitons in photorefractive media," *Phys. Rev. Lett.*, **68**(7), 923–926, 1992.

75. M.-F. Shih, Z. Chen, M. Mitchell, M. Segev, H. Lee, R. S. Feigelson, and J. P. Wilde, "Waveguides induced by photorefractive screening solitons," *J. Opt. Soc. Am.*, **B14**(11), 3091–3101, 1997.

76. J. J. Amodei and D. L. Staebler, "Holographic pattern fixing in electrooptic crystals," *Appl. Phys. Lett.*, **18**(12), 540–542, 1971.

77. D. L. Staebler, W. J. Burke, W. Phillips, and J. J. Amodei, "Multiple storage and erasure of fixed holograms in Fe-doped $LiNbO_3$," *Appl. Phys. Lett.*, **26**(4), 182–184, 1975.

78. G. Montemezzani and P. Gunter, "Thermal hologram fixing in pure and doped $KNbO_3$ crystals," *J. Opt. Soc. Am.*, **B7**(12), 2323–2328, 1990.

79. D. Zhang, Y. Zhang, C. Li, Y. Chen, and Y. Zhu, "Thermal fixing of holographic gratings in $BaTiO_3$," *Appl. Opt.*, **34**(23), 5241–5246, 1995.

80. J. F. Heanue, M. C. Bashaw, A. J. Daiber, R. Snyder, and L. Hesselink, "Digital holographic storage system incorporating thermal fixing in lithium niobate," *Appl. Opt.*, **21**(19), 1615–1617, 1996.

81. A. Y. Liu, M. C. Bashaw, L. Hesselink, M. Lee, and R. S. Feigelson, "Observation and thermal fixing of holographic gratings in lead barium niobate crystal," *Opt. Lett.*, **22**(3), 187–189, 1997.

82. F. Micheron and G. Bismuth, "Electrical control of fixation and erasure of holographic patterns in ferroelectric materials," *Appl. Phys. Lett.*, **20**(2), 79–81, 1972.

83. Y. Qiao, S. Orlov, D. Psaltis, and R. R. Nourgaonkar, "Electrical fixing of photorefractive holograms in $(Sr_{0.75}Ba_{0.25})Nb_2O_6$," *Opt. Lett.*, **18**(12), 1004–1006, 1993.

84. M. Horowitz, A. Bekker, and B. Fischer, "Image and hologram fixing method with $(Sr_xBa_{1-x})Nb_2O_6$ crystals," *Opt. Lett.*, **18**(22), 1964–1966, 1993.

85. R. S. Cudney, J. Fousek, M. Zgonik, P. Günter, M. H. Garrett, and D. Rytz, "Photorefractive and domain gratings in barium titanate," *Appl. Phys. Lett.*, **63**(25), 3399–3401, 1993.

86. J. Ma, T. Chang, J. Hong, R. R. Neurgaonkar, G. Barbastathis, and D. Psaltis, "Electrical fixing of 1,000 angle-multiplexed holograms in SBN:75," *Opt. Lett.*, **22**(14), 1116–1118, 1997.

87. H. C. Külich, "A new approach to read volume holograms at different wavelengths," *Opt. Commun.*, **64**(5), 407–411, 1987.

88. H. C. Külich, "Reconstructing volume holograms without image field losses," *Appl. Opt.*, **30**(20), 2850–2857, 1991.

89. D. Psaltis, F. Mok, and H. Y. S. Li, "Nonvolatile storage in photorefractive crystals," *Opt. Lett.*, **19**(3), 210–212, 1994.

90. G. Barbastathis and D. Psaltis, "Shift-multiplexed holographic memory using the two-lambda method," *Opt. Lett.*, **21**(6), 429–431, 1996.

91. E. Chuang and D. Psaltis, "Storage of 1,000 holograms with use of a dual-wavelength method," *Appl. Opt.*, **36**(32), 8445–8454, 1997.

92. D. von der Linde, A. M. Glass, and K. F. Rodgers, "Multiphoton photorefractive processes for optical storage in $LiNbO_3$," *Appl. Phys. Lett.*, **25**(3), 155–157, 1974.

93. D. von der Linde, A. M. Glass, and K. F. Rodgers, "Optical storage using refractive index change induced by two-step excitation," *J. Appl. Phys.*, **47**(1), 217–220, 1976.

94. K. Buse, L. Holtmann, and E. Krätzig, "Activation of $BaTiO_3$ for infrared holographic recording," *Opt. Commun.*, **85**(2), 183–186, 1991.

95. K. Büse, F. Jermann, and E. Krätzig, "Infrared holographic recording in $LiNbO_3$:Cu," *Appl. Phys.*, **A58**(3), 191–195, 1994.

96. D. Psaltis, X. An, G. Barbastathis, A. Adibi, and E. Chuang, "Nonvolatile holographic storage in photorefractive materials," *SPIE Critical Reviews*, **CR65**, 181–213, 1997.

97. K. Curtis and D. Psaltis, "Recording of multiple holograms in photopolymer films," *Appl. Opt.*, **31**(35), 7425–7428, 1992.

98. U.-S. Rhee, H. L. Caulfield, and C. S. Vikram, and M. M. Mirsalehi, "Characteristics of the Du Pont photopolymer for angularly multiplexed page-oriented holographic memories," *Opt. Eng.,* **32**(8), 1839–1847, 1993.

99. K. Curtis and D. Psaltis, "Characterization of the Du-Pont photopolymer for 3-dimensional holographic storage," *Appl. Opt.,* **33**(23), 5396–5399, 1994.

100. A. Pu, K. Curtis, and D. Psaltis, "A new method for holographic data storage in photopolymer films," in *IEEE Nonlinear Optics: Materials, Fundamentals and Applications*, Waikoloa, Hawaii, 1994.

101. J. E. Ludman, J. Riccobono, J. Caulfield, J.-M. Fournier, I. Semenova, N. Rienhand, P. R. Hemmer, and S. M. Shahriar, "Porous-matrix holography for nonspatial filtering of lasers," *SPIE Proceedings*, **2406**, 76–85, 1995.

102. J. E. Ludman, J. R. Riccobono, N. O. Rienhand, I. V. Semenova, Y. L. Korzinin, S. M. Shahriar, H. J. Caulfield, J.-M. Fournier, and P. Hemmer, "Very thick holographic nonspatial filtering of laser beams," *Opt. Eng.,* **36**(6), 76–85, 1997.

103. C. Cohen-Tannoudji, B. Diu, and F. Laloë, *Quantum Mechanics*, Hermann & Wiley-Interscience, Paris, 1977.

104. M. Born and E. Wolf, *Principles of Optics*, 6th ed., Pergamon, 1980.

105. D. L. Marks and D. J. Brady, "Three-dimensional source reconstruction with a scanned pinhole camera," *Opt. Lett.,* **23**(11), 820–822, 1998.

106. D. Huang, E. A. Swanson, C. P. Lin, J. S. Schuman, W. G. Stinson, W. Chang, M. R. Hee, T. Flotte, K. Gregory, C. A. Puliafito, and J. G. Fujimoto, "Optical coherence tomography," *Science,* **254**(5035), 1178–1181, 1991.

107. W. H. Carter and E. Wolf, "Correlation theory of wavefields generated by fluctuating, three-dimensional, primary, scalar sources I. General theory," *Opt. Acta,* **28**, 227–244, 1981.

108. J. Rosen and A. Yariv, "General theorem of spatial coherence: Application to three-dimensional imaging," *J. Opt. Soc. Am.,* **A13**(10), 2091–2095, 1996.

109. D. Marks, R. Stack, and D. J. Brady, "Three-dimensional coherence imaging in Fresnel domain," *Appl. Opt.,* **38**(8), 1332–1342, 1999.

Chapter 9

Three-Dimensionally Photorefractive Bit-Oriented Digital Memory

Satoshi Kawata

Department of Applied Physics, Osaka University, Osaka, Japan

Abstract

To exceed the capacity limitation of the surface-recording method of current optical data storage, the third dimension is introduced with photorefractive materials. Photorefractive materials are suitable for three-dimensional (3-D) data storage in conjunction with a nonlinear optical system such as the two-photon absorption process of the material for recording and the confocal laser-scanning system for reading. In this chapter, the systems and the materials suitable for three-dimensional digital memory are discussed with a number of experimental results with use of photopolymers for read-only memory, lithium niobate crystals as erasable memory, and photochromic organic materials as rewritable photorefractive memory. The comparison between photorefractive digital 3-D memory with conventional holographic 3-D memory and near-field memory is also discussed in terms of dynamic range, noise, recording density, and accessibility.

PHOTOREFRACTIVE OPTICS
Materials, Properties and Applications

ISBN 0-12-774810-5/$30.00

9.1 Introduction: limitation and breakthrough of optical high-density data storage

Optical memory is advantageous with its removability, replicability, durability, lightness, and its inexpensive price, compared with magnetic data storage. However, the high-density or large-capacity capability of optical data storage may not be the major advantage in the near future unless some breakthrough technology is invented because the diffraction of laser light strictly limits the density of memory. To increase the density under the diffraction-limited optics, a short-wavelength laser with a high-numerical-aperture lens can be used. In the newly standarized digital video disk (DVD) system the capacity is increased by the factor of 7.3 from the compact disk (CD) standard by using a short-wavelength laser and a large-numerical-aperture pickup lens, resulting in a capacity large enough to contain a 2-h movie in its 12-cm-diameter disk. However, this factor is not high enough as a terabit data storage for the coming information era.

The shortest-wavelength of light source used for optical data storage is not determined by the laser technology but rather by the availability of an optical component without any photon absorption. Water in air even absorbs the light in the ultraviolet (UV) region. Near-field optics is another technology using a short-wavelength evanescent wave, and is currently attracting widespread attention, despite the fact that it entails losing the merits of optical data storage such as remote accessibility, high bit-rate, low cost, etc.

The use of the third dimension, or the use of the volume provided by the optical disk, is another way to realize ultra-high-capacity data storage. In the real world (not a computer world), a large amount of data is usually filed in a book with a number of pages rather than being printed on a single spread-out piece of paper. If a dictionary, for example, the famous Japanese *Kojien* dictionary which has 2858 pages, is unbound into pieces of paper and they are spread out on the ground, it covers the area of 100 square meters. Rather than finding a word from many pages spread out over 100 square meters, it is much easier to find a word just by turning the pages of a book on the desk. The book-format type of data storage thus saves much space in the recording format itself, and also cuts down greatly on the access time to locate a particular word.

The data in a CD or a DVD are stored in the volume with a depth of 50–100 nm from the surface of the 1.2-mm-thick disk. This corresponds to 0.004–0.008% of the total volume of the CD or DVD; 99.996–99.992% are used just for the substrate but not for data storage. If the third dimension is used for data recording in a CD / DVD disk, the data capacity increases by a factor of hundreds. Such a 3-D CD / DVD with the ordinal thickness (1.2 mm) can store 2.4 terabits in 400 layers with 3 μm between layers.

9.2 Materials and optics for three-dimensional digital optical memory

Figure 9.1 illustrates a basic configuration of 3-D digital photorefractive memory. The configuration is the same as that for the current surface-recording optical memory except that the laser beam is focused inside the volume. To record and read the digital data deeply in three dimensions from the surface, the material must be photorefractive — transparent and refractile for reading, and of course photosensitive for writing [1, 2].

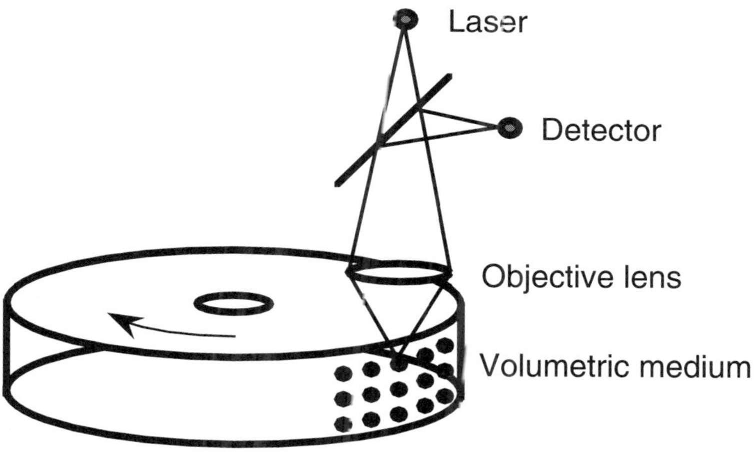

Figure 9.1: Basic configuration of three-dimensional photorefractive digital memory.

Table 9.1 lists the materials satisfying the requirement for 3-D optical memory [3, 4].

Besides photorefractive ferroelectric crystals, photopolymers known as holographic display materials can be regarded as photorefractive material because the polymerization changes the refractive index locally. Photopolymer has been investigated for the material of read-only 3-D memory [5, 6]. Glass is also a candidate for a photorefractive read-only memory; it changes the refractive index locally by the high-intensity laser beam at focus [7]. Photochromic molecules, recognized as absorptive memory material [8], can be also used as photorefractive memory materials [9]. With a near-infrared semiconductor laser, an isomer of photochromic molecule changes to the other isomer changing the refractive index, because the refractive index differs between the two isomers.

The requirements for the optics for 3-D photorefractive data storage are (1) nonlinearity in photosensitivity of material to select a specific data point from the data volume for both writing and reading and (2) capability of phase (or refractive index change) recording and reading. Table 9.2 shows a list of optical technologies for 3-D photorefractive digital memory [2]. The two-photon absorption process can be used for 3-D digital recording because of its nonlinear response to the light intensity. Only at

Silver halide film	ROM
Photopolymer	ROM
Photorefractive crystal	Erasable
PR + domain reversal	Erasable
Photochromic crystal	Rewritable
Phase-change	Rewritable

Table 9.1: Materials for Three-Dimensional Optical Data Storage

Replication and piling up	Write
Confocal system	Read
Two-photon absorption	Write/read
Phase-contrast	Read
Holographic	Write/read

Table 9.2: Optical Systems for Three-Dimensional Data Storage

the focus position of the near-infrared laser beam where photon density is high, two photons coherent with each other are simultaneously absorbed by photorefractive material, resulting in local photorefraction. Photopolymerizable monomers [6], ferroelectric photorefractive crystals [10], and photochromic molecules [3, 8] have been used for two-photon absorption 3-D memory.

In reading data, a confocal laser scanning microscope system can be used as a nonlinear optical system [1, 3]. The point spread function of confocal microscope optics is given by the multiplication of the point spread function of illumination optics and that of detection optics. As shown in Fig 9.2, the confocal optical system has a point detector (or a pinhole plus an area detector) and a point source (a laser), both focused at the same position in the medium. The light scattering except from the focus position is stopped at the screen with a pinhole. In transmission reading (Fig. 9.2a), the data are read as a phase-modulated signal. A differential interferometic-contrast object is used with a confocal pinhole [1] or without a

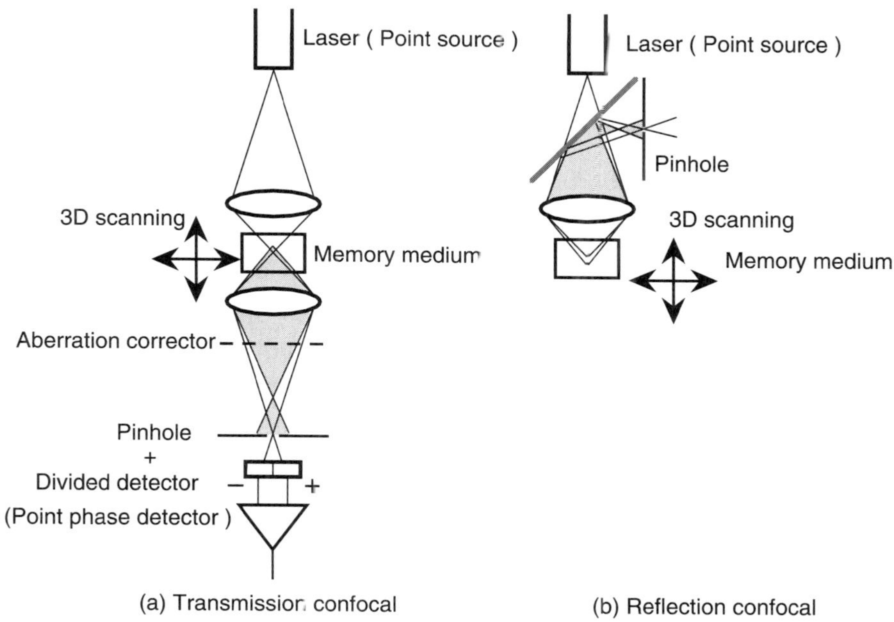

Figure 9.2: Confocal phase readout systems for 3-D memory. Shaded area shows the scattering from the layer currently not reading.

pinhole [6, 11]. A phase-contrast objective lens is also used with a phase-shifter in a pupil plane of illumination system and an annular pupil in detection system [12]. A divided detector can be also used to read a phase-modulated refractive-index information [9].

In reflection reading (Fig. 9.2b), a phase-modulator is not necessary because the photons scattered at the data point (local phase gradient) is detected by the detector through the pinhole as intensity [3]. In this case, we have to carefully choose appropriate parameters in the optical system such as the numerical aperture of the objective lens and the wavelengths of laser for writing and reading [3].

9.3 Three-dimensional photopolymer memory

Photopolymer, which has been used for volume hologram material for display [13, 14], can be used as the material for 3-D read-only digital memory. The advantages of using photopolymer for volume holography, in comparison with silver halides or dichromated gelatin, are (1) wet processing is not necessary for development, (2) there is small scattering of light (or high transparency) resulting in high-contrast output, and (3) high diffraction efficiency. Holographic data storage is not different from holographic display, including angular multiplexing, except that the original data to be recorded are binary.

Figure 9.3 shows bit data recorded in 30 different layers of photopolymer [15]. The bit data was written every 2 μm $\times$ 2 μm in a plane, and the longitudinal distance between the data planes was every 10 μm. The number of layers for the memory shown in Fig. 9.3 was 30 with a total thickness of 300 μm. The material used was a monomer mixture composed of methacryl compound and allyl compound with benzil as an initiator and Michler's ketone as a dye sensitizer [16]. The methacryl compound polymerizes faster than the allyl compound, while the refractive index of polymerized methacryl compound alone is higher ($n\sim1.60$) than that of polymerized methacryl compound ($n\sim1.50$) alone.

The photopolymerizable liquid is weakly photopolymerized (to solidify weakly) by preexposure of uniform illumination by an ordinary fluorescent lamp, and data are written with a focused beam of an argon ion laser at 488 nm with the power of 5 mW.

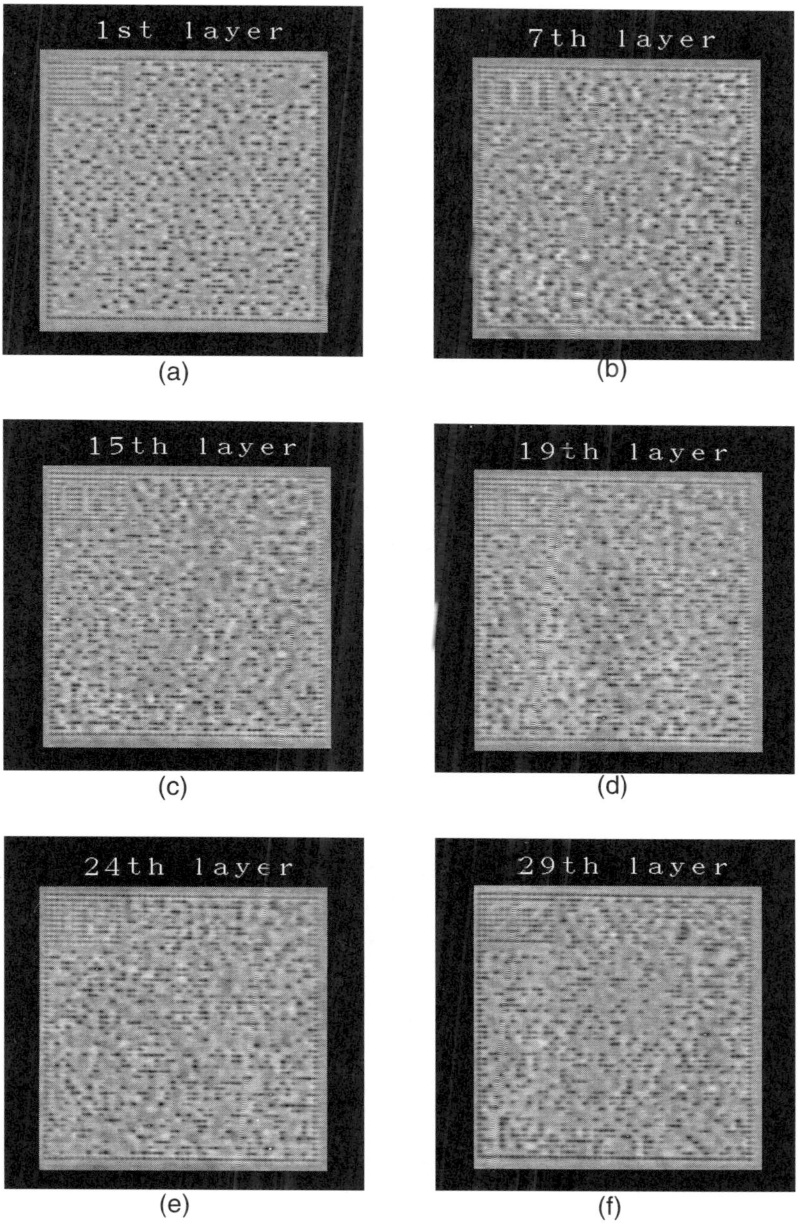

Figure 9.3: Bit data of different layers of 3-D photopolymer memory: (a) 1st, (b) 7th, (c), 15th, (d) 19th, (e) 20th, and (f) 29th layers.

Figure 9.4 shows a longitudinal cross section of another memory which was read by a laser beam scanning microscope with a phase-contrast objective (100×) [1]. This figure indicates that the separation between layers can be reduced by up to 3 μm at most. The reason for the good separation between layers is because the photopolymerization process exhibits high-pass spatial-filtering characteristics due to the limitation of the diffusion length of monomer in photopolymerization [17, 18]; a typical peak value in the spatial response of photopolymer is ~1 μm [19]. The spatially low frequency component or the defocused bit images of the out-of-focus layer cannot be written.

Figure 9.5a shows an example of readout data read by a homemade confocal system. A He-Ne laser (632.8 nm) was used for reading with a phase-contrast objective and annular pupil for phase-contrast (dark-field) imaging [1]. For comparison, the same segment of the data read by a conventional nonconfocal microscope with the same objective lens is shown in Fig. 9.5b. The result indicates that the confocal configuration exhibits much higher contrast and higher resolution. Figure 9.6 shows a photograph of a prototype 3-D digital memory in a credit card format [15].

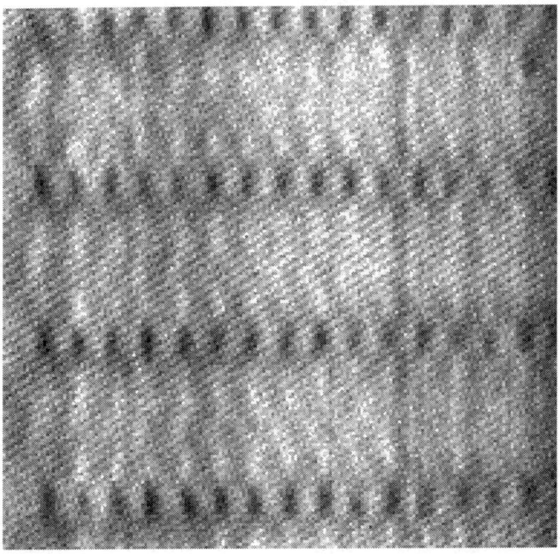

Figure 9.4: A longitudinal cross section of the recorded data which was read by laser-beam scanning microscope with a phase-contrast objective.

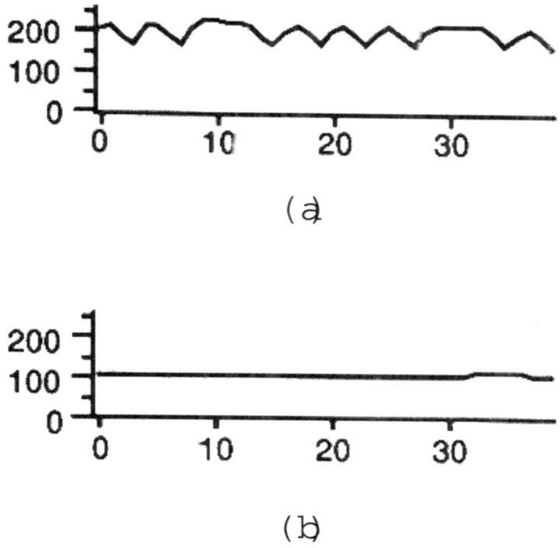

Figure 9.5: (a) Readout data by a confocal microscope, and (b) by a conventional microscope, both with a phase-contrast objective with an annular pupil for phase-contrast.

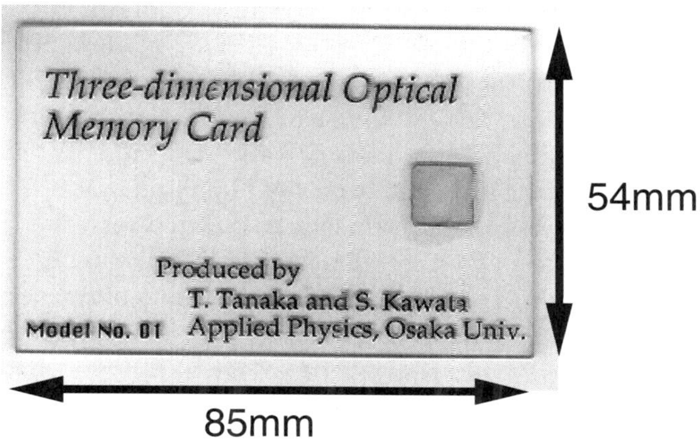

Figure 9.6: Credit card–size 3-D digital memory with 30 layers in a 300-μm-thick film.

9.4 Lithium niobate three-dimensional digital memory

Photopolymer can be successfully used for 3-D memory which is not rewritable but read-only. As a rewritable or erasable 3-D memory, photorefractive crystals, such as lithium niobate [10] and strontium barium niobate, are good candidates. Holographic data storage was reported by Chen in 1968 [20], while the digital recording of bit-oriented memory was first reported in 1995 by the authors [12, 21].

The mechanism of local photorefraction in ferroelectric crystal with a focused laser beam irradiation is explained as follows. When the focused light of a pulsed laser of near-infrared is incident on a particular point in a crystal, the electrons in the vicinity of the laser beam focus are excited by two-photon absorption from the donor level to the conduction band as shown in Fig. 9.7. The probability of excitation of electrons by two-photon absorption is proportional to the squared intensity of the light. The excited electrons are retrapped in vacant donor sites after movement in the conduction band due to photovoltaic effect. This movement of electrons yields a nonuniform distribution of charges, and the charge distribution produces an electric field to modulate the refractive index by the Pockels effect. As a result, 1 bit of data is recorded in crystal as a refractive index datum. By scanning the focused spot in the recording medium, data are three-dimensionally recorded. Figure 9.8a shows a calculated result (transaxial cut of the crystal) of refractive index distribution in y-cut $LiNbO_3$ crystal [21]. The calculation was done with use of Kukhtarev's material equation [22] and realistic values for the material and experimental parameters. The photon absorption is assumed to be single-photon process. The attainable refractive index change Δn, which is calculated with the value of the space-charge field Esc of iron-ion doped lithium niobate, is $1.1 \times 10^{-6.}$ The order of these values agrees with measurements of refractive indices induced by interference fringes [23].

Figure 9.8b shows a phase-contrast image of photorefractive datum calculated using the refractive index distribution shown in Fig. 9.8a. This image was obtained with calculation of the refractive datum in Fig. 9.8a, convolved in three dimensions with the point-spread function of the phase-contrast microscope with a practical experimental parameter set [21]. Figure 9.8c shows a real image of a bit datum written in $LiNbO_3$. This image was observed with a phase-contrast microscope. This experimental result well agrees with the calculated bit shown in Fig. 9.8b.

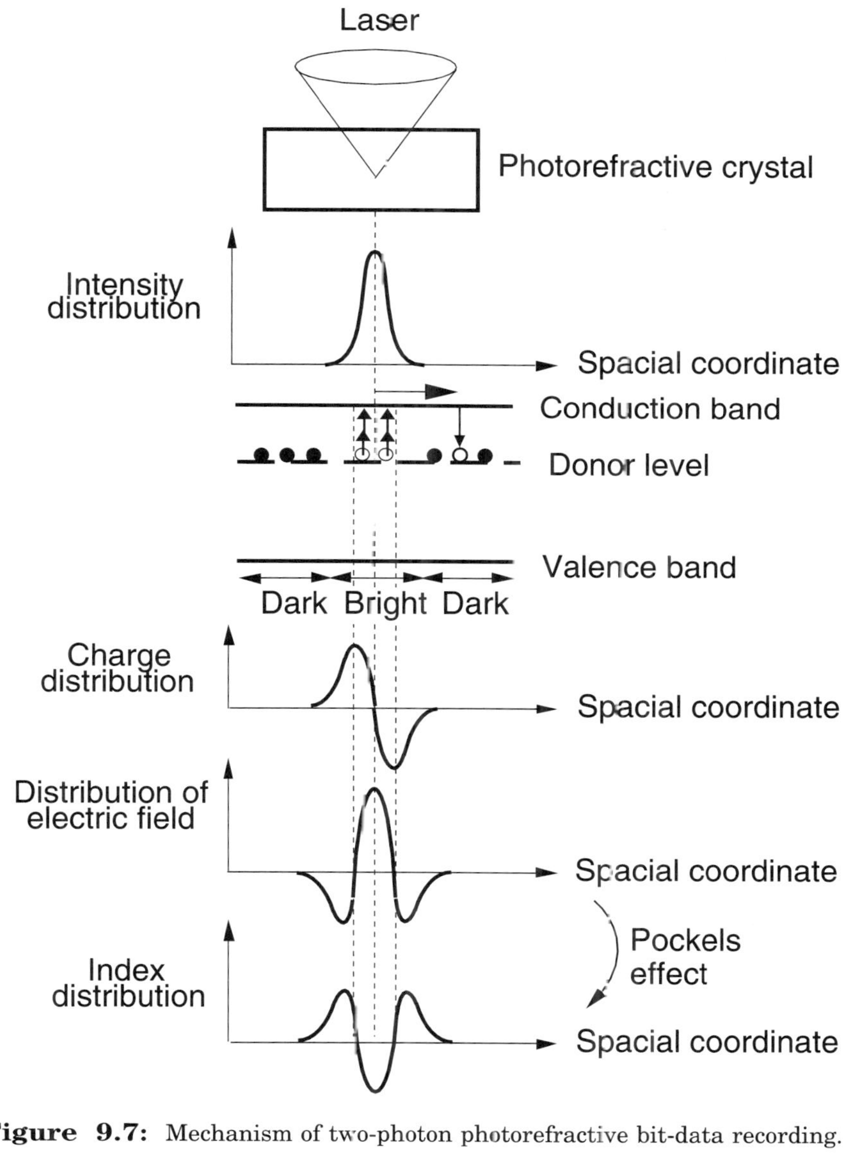

Figure 9.7: Mechanism of two-photon photorefractive bit-data recording.

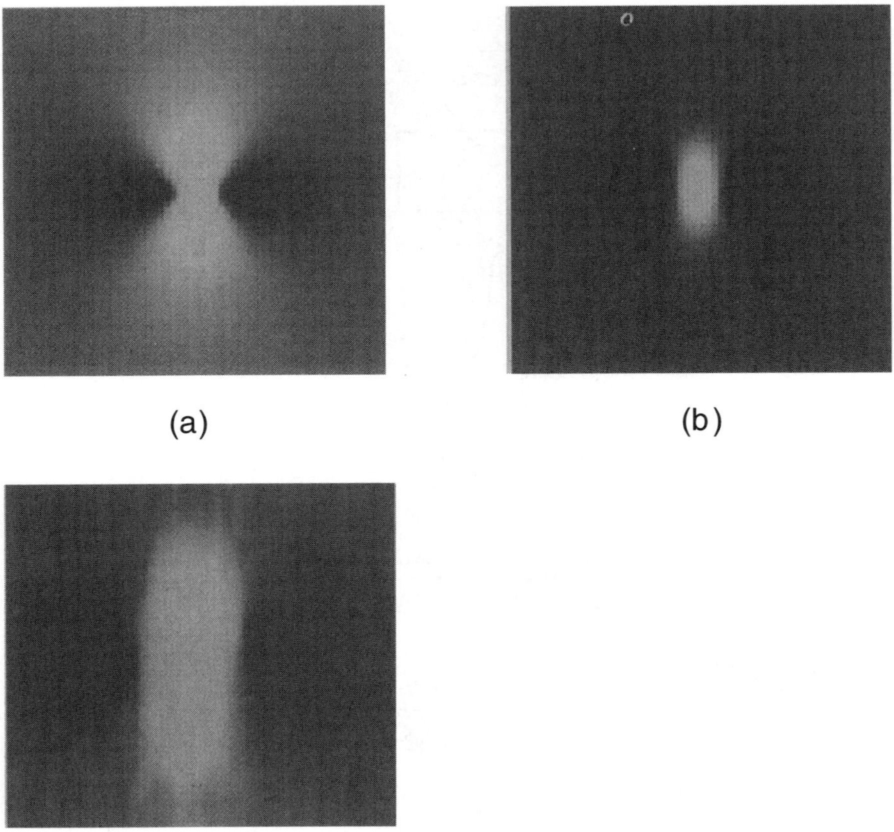

(c)

Figure 9.8: Calculated results (transaxial cut of the crystal) of (a) the refractive index distribution for a bit datum recorded in y-cut $LiNbO_3$ crystal, (b) its phase-contrast image, and (c) the real image of written data observed with phase-contrast microscope.

Figure 9.9 shows photographs of bit data recording in a $LiNbO_3$ crystal with different crystal orientations by single-photon process. The contrast values differ from each other because of different relative orientations between the optical axis and c-axis. A contrast value of 0.235 was obtained in Fig. 9.9c, which is 4.1 times larger than that obtained in Fig. 9.9a and 2.8 times larger than that obtained in Fig. 9.9b. The crystal

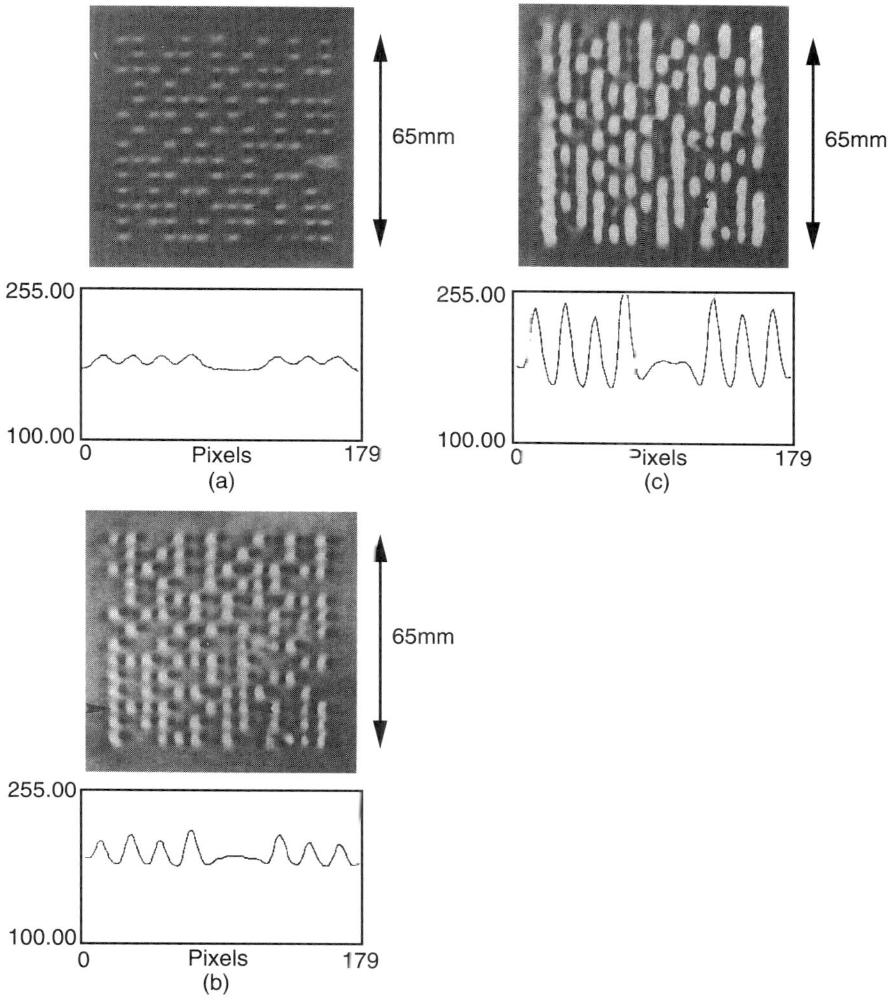

Figure 9.9: Observed images of bit data obtained for different relative orientations of the polarization vector: (a) the vector is perpendicular to the c-axis that is oriented parallel with the optical axis, (b) the vector is perpendicular to the c-axis that is oriented perpendicular to the optical axis, and (c) the vector is parallel to the c-axis. The curve below each image represents the intensity profile between two arrows.

orientation and the polarization of the laser beam for reading for Fig. 9.9a–c are shown in Fig. 9.10a, b, and c, respectively. The crystal in Fig. 9.10a is z-cut and that in Fig. 9.10b and c is y-cut. These experimental results are in good agreement with the predictions of our numerical analysis [21]. In conclusion, the use of a y-cut crystal with polarization of the reading beam in parallel to the c-axis is the best combination.

Irradiation of a uniform light beam erases the data, because the uniform beam homogenizes the charge density of the donor level or the refractive index of the region. As a result, this material is erasable and rewritable.

9.5 Two-photon recording in lithium niobate

Two-photon recording for 3-D digital memory was the first demonstrated by the authors in a photorefractive crystal [10]. In this experiment an undoped lithium niobate is used. It is completely transparent at the near-infrared region and absorptive at the UV-visible region. This is an essential requirement for two-photon absorption with a near-infrared laser. The absorption spectrum of the crystal used is shown in Fig. 9.11. If a Ti:sapphire laser with a wavelength of 762 nm is used for recording only at the beam focus position where photon density is extremely high, two photons coherent with each other are simultaneously absorbed by donor-level electrons to raise up to the conduction band with the energy gap (384 nm in wavelength), which is twice as large as the photon energy of laser light (768 nm). Except for the focus position, the crystal is transparent at 762 nm and the single-photon process is dominant.

Figure 9.12 shows a photograph of images of layers recorded and read in three dimensions in undoped lithium niobate [10]. Seven layers with 5 μm distance between bits in a layer and 20 μm distance between layers are recorded. The average contrast of the readout data is ~0.15. The contrast of data in the seventh layer is worse than that in the first layer. This degradation of the contrast in the deeper layer is due to spherical aberration as a result of mismatching in the refractive index between air and crystal ($n \sim 2.2$). The spherical aberration causes also the cross talk between layers in reading. The pulse width of the Ti:sapphire laser used for recording data is 130 fs with 4 mJ of power at a wavelength of 762 nm. An objective with a numerical aperture of 0.85 was used for writing data, and for reading a phase-contrast objective with a numerical

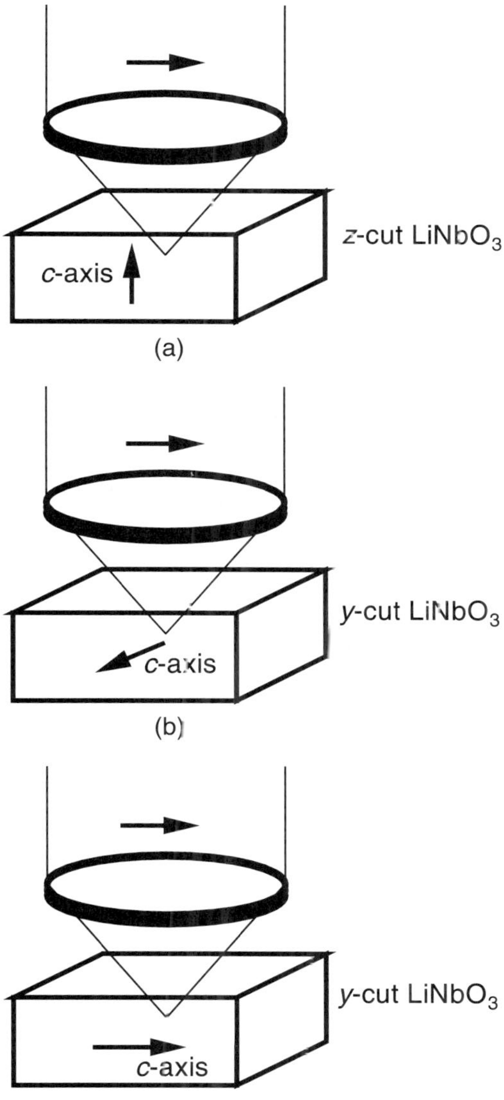

Figure 9.10: Possible orientations of the polarization vector relative to the c-axis of the crystal: (a) the vector is perpendicular to the c-axis that is oriented parallel with the optical axis, (b) the vector is perpendicular to the c-axis that is oriented perpendicular to the optical axis, and (c) the vector is parallel to the c-axis.

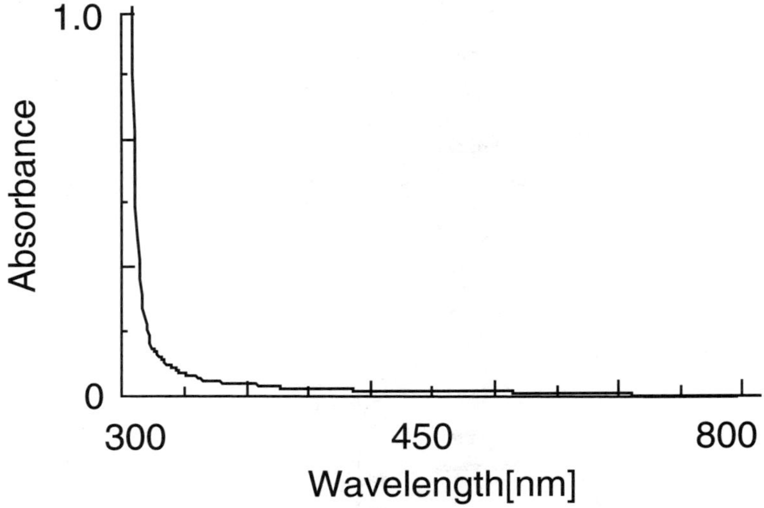

Figure 9.11: The absorption spectrum of undoped lithium niobate.

aperture of 0.75 was used. Figure 9.13 shows a readout of dots recorded at different peak powers of the laser [2]. Every single dot from the top left in clockwise rotation was recorded with a Ti:sapphire laser in mode-locked condition, while the others are mode-unlocked. It is found in the figures that when the laser is not mode-locked the refractive dot is not formed. The mean power of the laser beam in the mode-unlocked experiment is the same as that in the mode-locked experiment; while it is continuous in the former but is not compressed as a strong peak pulse. This result proves that the mechanism of recording is based on the two-photon process but not on the single-photon process.

9.6 Fixing the data

A main drawback with the use of photorefractive crystal for an optical data-storage application is the destruction (erasure) of prerecorded data in writing other layers and in reading the data. Even when a crystal is kept in a dark room without reading it, data are thermally erased in time. To protect the recorded data from erasure or from destruction, the recorded data should be fixed. There has been some work in holographic memory

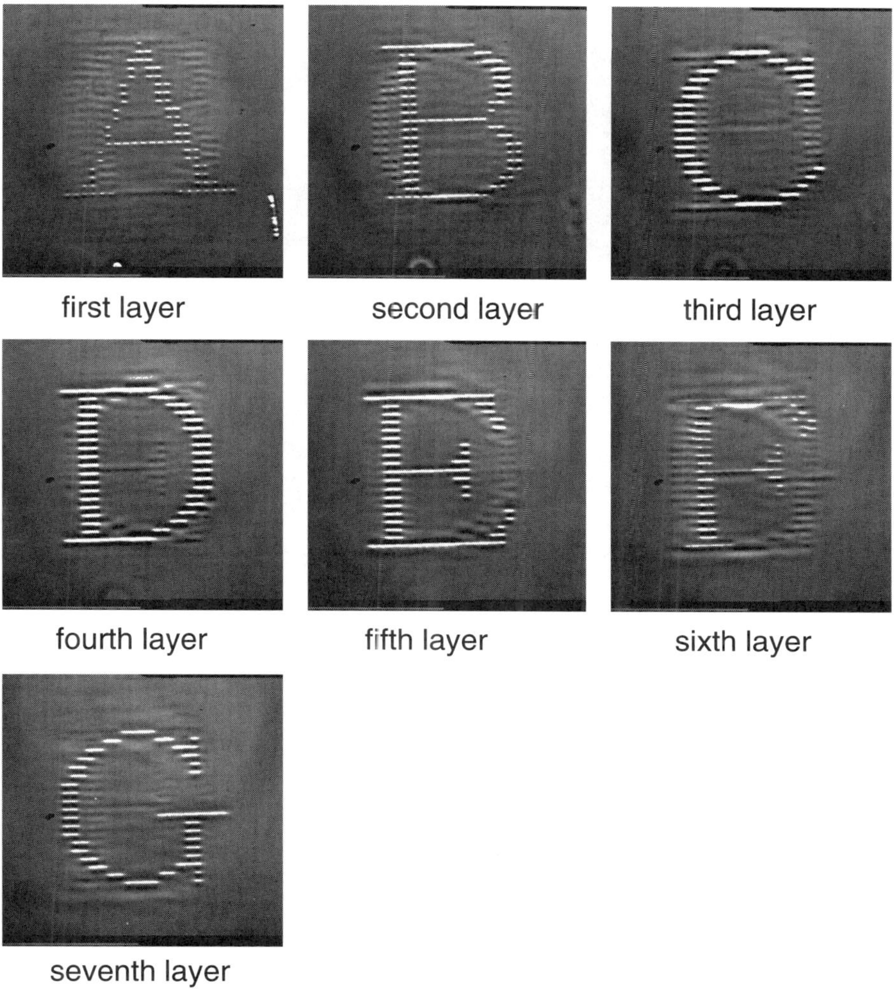

Figure 9.12: Seven layers recorded and read in an undoped $LiNbO_3$ crystal. The distance between dots in a layer is every 5 μm, with 20 μm between layers.

research on fixing data in photorefractive crystals [24–26]. It can be applied to digital 3-D memory.

An established method of fixing data in photorefractive crystal is domain reversal. Figure 9.14 shows the mechanism of domain reversal in a photorefractive crystal with a photoinduced electric field [27]. The

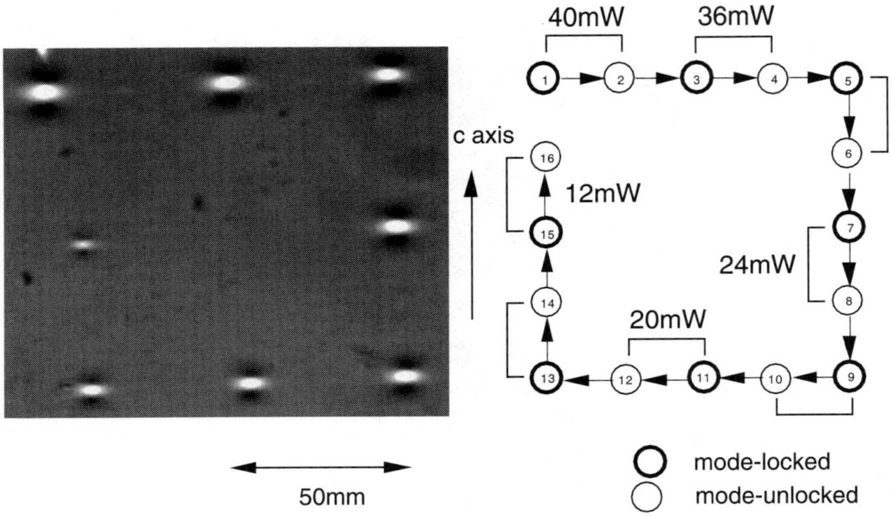

Figure 9.13: Readout of recorded dots in $LiNbO_3$ with different peak power of Ti:sapphire laser.

intensity distribution of the incident light produces the space-charge field in the crystal by photorefractive effect. By heating up the crystal temperature, coercive field *Ec* to reverse the *c*-axis decreases. At the beam spot where the photoinduced electric field exceeds the coercive field in a given temperature, the direction of the *c*-axis reverses to form a small local domain. As a result a bit is recorded as a single domain in a crystal. A spontaneous polarization of small domain produces an electric field and then modulates the refractive index by Pockels effect. The recorded data are again modulated as refractive-index change. Cooling down the crystal to room temperature makes the domain permanently stable.

Figure 9.15 shows the experimental result of domain-reversed bit data. Five dots are recorded in a dice-pip arrangement in a Ce-doped $Sr_{0.75}Ba_{0.25}NbO_6$ (SBN:75) of $5 \times 5 \times 5$ mm^3 by a focused beam of Ar ion laser at 488 nm at the power of 3.4 mW [27]. The crystal was heated at 38°C to reverse the domain, then cooled down to 15°C. From another experiment to measure the diffraction efficiency of grating recording and reading in a Ce-doped SBN crystal with fixing technology, it was found that the refractive-index change by photoinduced electrons is approximately 9.0×10^{-4} [27], while that of domain reversal is around $1.0 \times 10^{-4.}$ The experiment was carried out to estimate the lifetime of fixed

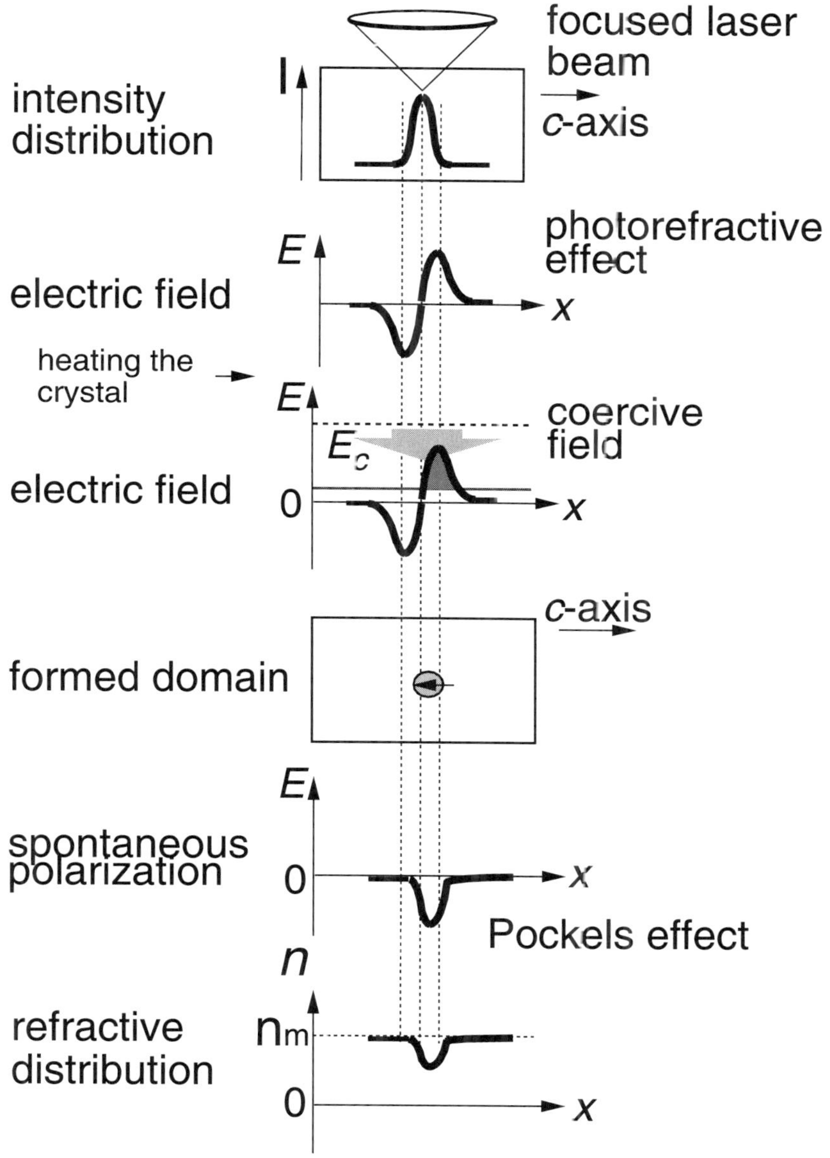

Figure 9.14: Mechanism of domain reversal in a photorefractive crystal by photoinduced electric field.

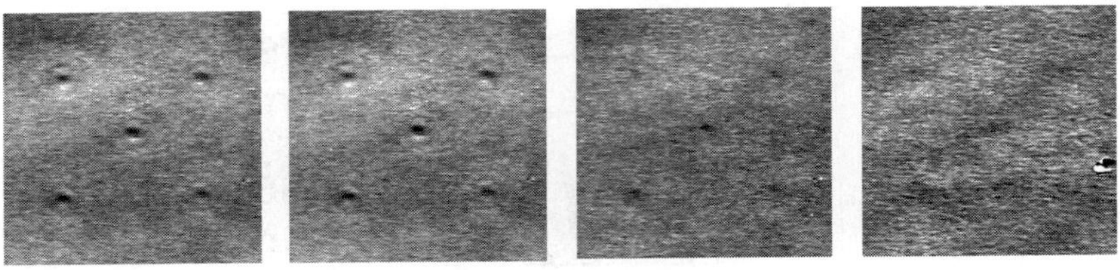

Figure 9.15: Bit image of domain-reversed memory after the high-beam exposure by Ar-ion laser at 3.4 mW of (a) 0 min, (b) 10 min, (c) 1 h, (d) 3 h.

reversed domain. To save time in this experiment, we purposely irradiated a beam of Ar ion laser at 3.4 mW onto the data-recorded crystal for 3 h. However, the recorded data were preserved for 3 h as shown in Fig. 9.15 even under this data-destructive experimental condition.

9.7 Photochromic recording in photorefractive crystals

A bismuth silicon oxide ($Bi_{12}SiO_{20}$) crystal, which is usually used as a photorefractive crystal with high electric-field application, can be also used as 3-D digital memory as rewritable memory. The mechanism is photochromism rather than photorefraction due to the Pockels effect. The merit of using photochromism for digital data storage is rewritability or overwrite capability of a single dot in volume.

Experimental results have been reported by the authors [28] in recording 3-D bit data in a BSO crystal as photochromic absorption change with a 488-nm Ar ion laser in room temperature without external voltage application. At the focus of the laser beam the impurities contained in crystal absorb the light, and the electrons trapped in the impurities are excited. The recorded bit datum can be erased with a He-Ne laser at 632.8 nm. Figure 9.16 shows experimental results of bit writing, erasing, and rewriting [28].

9.8 Photorefractive photochromic memory

The use of organic photochromic materials has been thoroughly investigated for 3-D digital memory [4, 8, 29, 30]. Even one single photochromic

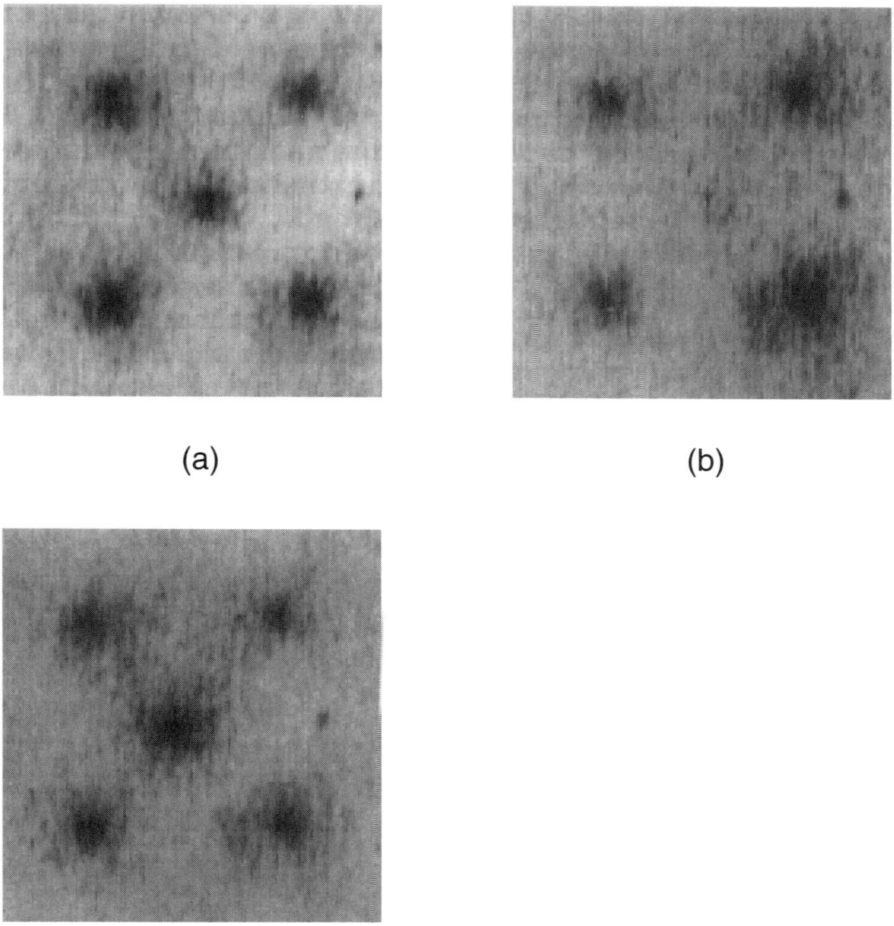

(a) (b) (c)

Figure 9.16: Experimental result of optical erasure of a bit datum. (a) Data are recorded by an Ar-ion laser at 488 nm. (b) A He-Ne laser beam erased only the center bit of five dots. (c) A bit datum is again recorded at the same position where datum was once erased.

molecule can store a binary datum by choosing its isomer structure. However, the problem with photochromic memory is that the recorded data are gradually erased in reading because the reading process is also associated with photon absorption, which is equivalent to very weak writing.

If the data can be recorded as refractive index–modulated data in photochromic material, they are not erased by reading because phase-reading does not involve absorption. For example, two isomers of a spiropyran derivative, 1,3,3-trimethylindolino-6′nitrobenzopyrylospiran (NSP), are different in refractive index by $\sim$ 0.02 at λ = 830 nm in near-infrared [9]. Here, the photochromic memory is regarded as photorefractive digital memory because bit data are recorded due to the photon absorption (precisely two-photon absorption) by a focused laser beam and they are recorded as a change in the volume of the refractive-index.

Figure 9.17a–c shows an example of multilayer data recording of digital bit data in photochromic material, while they are recorded as refractive data [3]. Photochromic molecules are *cis*-1,2-dicyano-1,2-*bis*(2,4,5-trimethyl-3-thienyl)ethene (B1536), which are doped in polymethylmetacrylate (PMMA) film. A red isomer of B1536 converts into a yellow isomer upon irradiation with UV light at 380 nm, and the yellow converts to a red isomer by irradiation at 532 nm (Fig. 9.18) [3]. For 3-D digital recording, a pulsed laser with a high peak with the wavelength twice larger than the absorption peak is used for two-photon recording. For the experiment shown in Fig. 9.17, an erbium-doped fiber laser at 775 nm in mode-locked pulse laser operation was used. In the areas where two-photon absorption does not take place, the material remains transparent at 775 nm.

The readout was done by using the reflection confocal system with a laser diode operating at 630 nm. The bit and layer interval were 5 and 20 μm, respectively. While both of the isomers of this photochromic molecule present no absorption at 630 nm, they have slightly different refractive indices at this wavelength. The objective lens used for reading the data has a numerical aperture equal to 1.4.

The written data can be selectively erased by irradiation at 532 nm as shown in the experiment in Fig. 9.19. This material does not exhibit any apparent fatigue even after 10^4 write / erase cycles, and the written data (e.g., the read isomer) is stable at 80°C for more than 3 months, and the thermal back-reaction (from the red isomer to the yellow isomer) does not occur at even 300°C [31].

9.9 Optical design for reflection confocal memory

For reading digital data written in 3-D space, different types of microscopes have been used including the phase-contrast microscope [12, 32],

Figure 9.17: Multilayer data recording of the bit data in photochromic material.

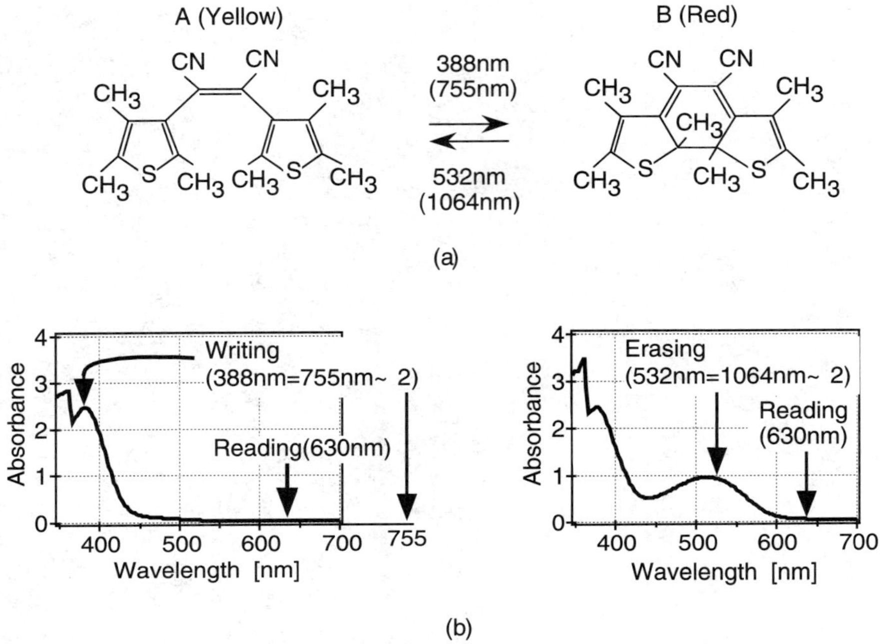

Figure 9.18: Photochromic molecules are *cis*-1,2-dicyano-1,2-*bis*(2,4,5-trimethyl-3-thienyl)ethene (B1536). A red isomer of B1536 converts into a yellow isomer upon irradiation with UV light of a 388 nm wavelength, and the yellow converts to a red isomer by irradiation at 532 nm.

the ordinary transmission microscope [7], the differential interference contrast microscope [6], and the differential phase-contrast scanning microscope [9, 11]. They are based on the transmission microscope to read phase-modulated signal. A confocal system with point detector (or a pinhole in front of the detector) and a laser is included in the transmission microscope to discriminate a layer from volumetric data storage. However, this microscope could deviate the beam spot from the pinhole due to the inhomogeneity of the refractive index and / or the thickness of the memory medium and the substrate (Fig. 9.20a). As a result, the signal detected has a varying background due to the local inhomogeneity of the medium and the substrate. Use of a reflection confocal microscope alleviates this problem because the beam reflected at the data bit must arrive at the pinhole even in the presence of inhomogeneity (Fig. 9.20b).

The 3-D spatial-frequency band of the reflection confocal microscope may not necessarily pick up the 3-D frequency component of a written

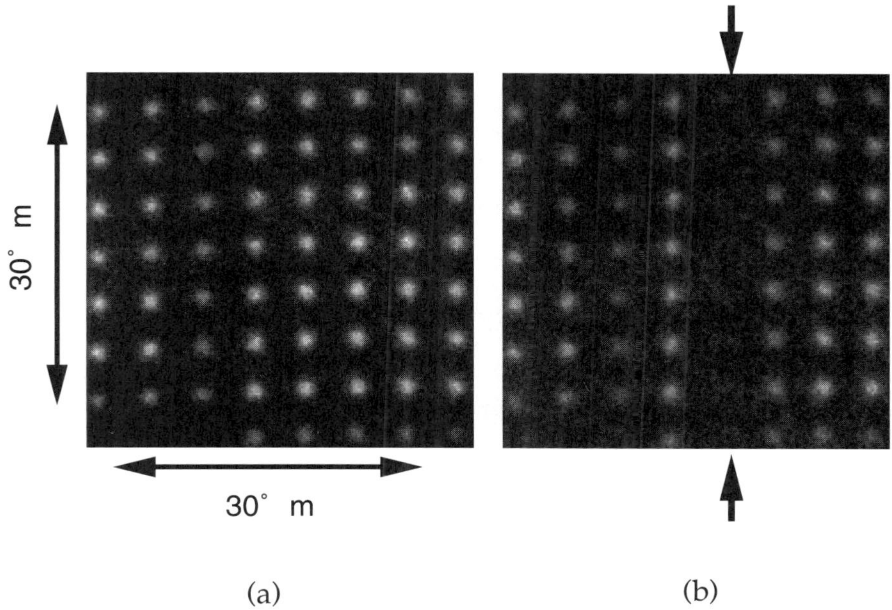

Figure 9.19: Experimental results for erasing written data: (a) before erasing and (b) after erasing.

bit datum formed in three dimensions, when a modulation of the phase is weak enough to approximate the system linear one [32]. In that case appropriate optical parameters should be chosen, including numerical apertures of the objective lens and wavelengths of writing and reading lights, so that there will be a common volume in 3-D spatial frequency space for both written data and reading optics. With such a parameter set, a bit datum written in three dimensions can be read out with a reflection confocal microscope. The determination of optics parameters for two-photon recording and reflection confocal readout was developed with 3-D Fourier analysis in k-vector space [3].

9.10 Concluding remarks: comparison with other advanced data storages

Compared with angularly multiplex 3-D holographic memory, in which data are recorded in analog format (Fourier-transformed or Fresnel-dif-

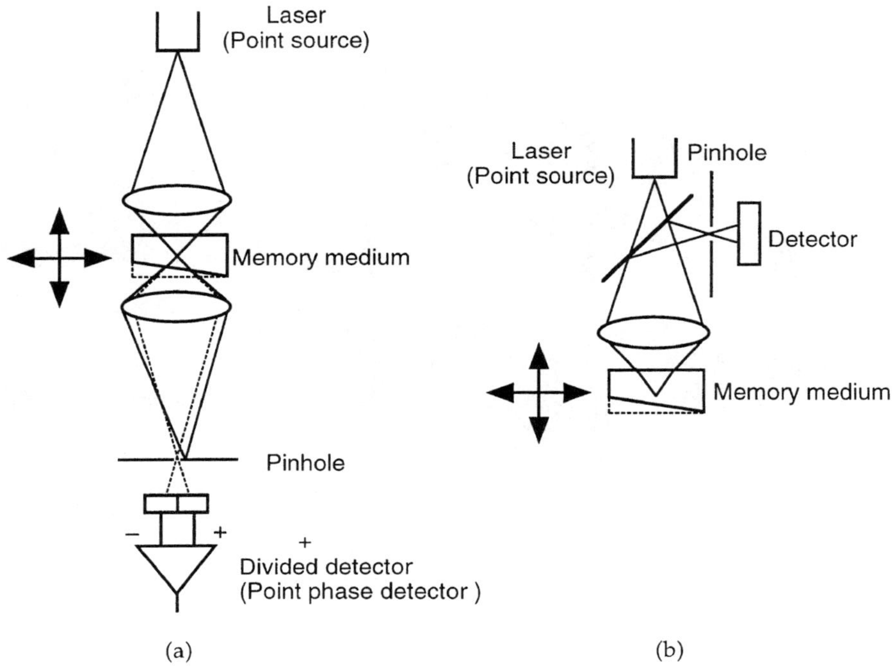

Figure 9.20: Schematics of (a) the transmission confocal microscope and (b) the reflection confocal microscope we utilized.

fracted signal of digital data) spread out in three dimensions, three-dimensional bit-oriented memory is essentially the technology of digital recording. Even a single photochromic molecule can record binary data, while a holographic memory requires a large dynamic range at every voxel to record all the information of mass data. The major difference between holographic and digital memory lies in the choice between analog or digital recording.

Three-dimensional digital memory, compared with holographic memory, is randomly data-accessible for writing and reading. It is erasable and rewritable at any bit at any timing, while holographic recording is essentially parallel recording. Parallel reading can be done even with current CD / DVD standards or 3-D digital memory with a Charge-Coupled Device (CCD) camera if necessary.

Concerning noise, laser scanning optics which has been established in CD / DVD data storage technology and in confocal laser-scanning microscopy, is strong against coherent noise because it is essentially a temporally and spatially incoherent imaging system [33].

Bit-oriented digital memory is suitable for enhancing the nonlinearity of material because the laser power focuses at a single position. Nonlinearity with holographic memory may generate spurious output due to the higher order of diffraction.

In addition, 3-D digital memory provides higher density with use of a high-numerical-aperture lens than a holographic system because in the former it is unnecessary to have a large field of view (simply to observe a single point) and to have space for a reference beam [34]. Density is limited even in holographic memory by the solid angle to read or write the signal (whether it be Fourier-transformed, Fresnel-diffracted, or image-holographic) from the memory medium to the lens or data-recorded plane.

Compared with near-field optical memory, which is another competing candidate for future high-density optical data storage, 3-D memory has the merits of (1) remote access and remote control of data, (2) ease of servo-control and tracking, (3) lack of necessity for dust cleaning because data are stored inside the medium, and (4) higher bit rate thanks to the extremely high optical throughput of the system. Using the photon process can be also coupled well with 3-D digital optical memory.

References

1. S. Kawata, T. Tanaka, Y. Hashimoto, and Y. Kawata, "Three-dimensional confocal optical memory using photorefractive materials," *Proc. SPIE, Photopolymers and Applications in Holography, Optical Data Storage, Optical Sensors and Interconnects,* **2042**, 1993, pp. 314–325.
2. S. Kawata, "Three-dimensional digital optical data storage with photorefractive crystal," *Proc. SPIE,* **3470**, *Photorefractive Fiber and Crystal Devices: Materials, Optical Properties, and Applications IV,* 56–63, 1998.
3. A. Toriumi, S. Kawata, and M. Gu, "A reflection confocal readout for three-dimensional photochromic optical data storage," *Opt. Lett.,* **23**, 1924–1926, 1998.
4. S. Kawata and A. Toriumi, "Three-dimensional optical memory using photopolymer, photorefractive crystals and photochromic materials," *Optical Data Storage '97, SPIE,* **3109**, 174–181, 1997.
5. Y. Hashimoto, Y. Kawata, and S. Kawata, "Three dimensional confocal optical memory with photorefractive materials," *Proceedings of Japan Optics '92,* 39–40, 1992.

6. J. H. Strickler and W. W. Webb, "Three-dimensional optical data storage in refractive media by two-photon point excitation," *Opt. Lett.,* **16**, 1780–1782, 1991.

7. E. N. Glezer, M. Milosavljevic, L. Huang, R. J. Finlay, T.-H. Her, J. P. Callan, and E. Mazur, "Three-dimensional optical storage inside transparent materials," *Opt. Lett.,* **21**, 2023–2025 1996.

8. D. A. Parthenopoulos and P. M. Rentzepis, "Three-dimensional optical storage memory," *Science,* **245**, 843–845, 1989.

9. A. Toriumi, J. M. Herrmann, and S. Kawata, "Nondestructive readout of a three-dimensional photochromic optical memory with a near-infrared differential phase contrast microscope," *Opt. Lett.,* **22**, 555–557, 1997.

10. Y. Kawata, H. Ishitobi, and S. Kawata, "Use of two-photon absorption in a photorefractive crystal for three-dimensional optical memory," *Opt. Lett.,* **23**, 756–758, 1998.

11. Y. Kawata, R. Juskaitis, T. Tanaka, T. Wilson, and S. Kawata, "Differential phase contrast microscope with a split detector for readout system of multilayered optical memory," *Appl. Opt.,* **35**, 2466–2470, 1996.

12. Y. Kawata, H. Ueki, Y. Hashimoto, and S. Kawata, "Three dimensional optical memory with a photorefractive crystal," *Appl. Opt.,* **34**, 4105–4110, 1995.

13. D. H. Close, A. D. Jacobson, J. D. Margerum, R. G. Brault, and F. J. McClung, "Holographic recording on photopolymer materials," *Appl. Phys. Lett.,* **14**, 159–160, 1969.

14. B. L. Booth, "Photopolymer material for holography," *Appl. Opt.,* **14**, 593–601, 1975.

15. T. Tanaka and S. Kawata, "Three-dimensional optical card storage with thirty recording-layers," *CLEO / Pacific Rim '95,* 1995, pp. 70–71.

16. H. Tanigawa, T. Ichihashi, A. Nagata, "Hologram recording on multicomponent monomer materials," *J. Opt. Soc. Jpn.,* **20**, 227–231, 1991.

17. S. Kawata, "Optics for near-field optical microscopy," *J. Opt. Soc. Jpn.,* **21**, 766–779, 1992.

18. B. L. Booth, "Photopolymer material for holography," *Appl. Opt.,* **14**, 593–601, 1975.

19. M. Kaneko, K. Aratani, and M. Ohta, "Multilayered magneto-optical disks for magnetically induced superresolution," *Jpn. J. Appl. Phys.,* **31**, 568–575, 1992.

20. F. S. Chen, J. T. LaMacchia, and D. B. Frase, "Holographic storage in lithium niobate," *Appl. Phys. Lett.,* **13**, 223, 1968.

21. H. Ueki, Y. Kawata, and S. Kawata, "Three-dimensional optical bit memory recording and reading with a photorefractive crystal: analysis and experiment," *Appl. Opt.,* **35**, 2457–2465, 1996.
22. N. V. Kukhtarev, V. B. Markov, S. G. Odulov, M. S. Soskin, and V. L. Vietskii, "Holographic storage in electrooptic crystals. I. Steady State," *Ferroelectrics,* **22**, 949–960, 1979.
23. P. Gunter and J.-P. Huignard (eds.), *Photorefractive Materials and Their Applications,* Vols. I and II, Springer-Verlag, Berlin, 1988.
24. A. Kewitsch, M. Segev, A. Yariv, R. Neurgaonkar, "Selective page-addressable fixing of volume holograms in $Sr_{0.75}Ba_{0.25}Nb_2O_6$ crystals," *Opt. Lett.,* **18**, 1262–1264, 1993.
25. Jian Ma, Tallis Chang, J. Hong, R. Heurgaonkar, G. Barbastathis, D. Psaltis, "Electrical fixing of 1000 angle-multiplexed holograms in SBN:75," *Opt. Lett.,* **22**, 1116–1118, 1997.
26. D. Lande, S.S. Orlov, A. Akella, L. Hesselink, R.R. Neurgaonkar, "Digital holographic storage system incorporating optical fixing," *Opt. Lett.,* **22**, 1722–1724, 1997.
27. M. Hisaka, Y. Kawata, and S. Kawata, "Three-dimensional optical recording by the ferroelectric domain reversal technique in a Ce-doped SBN:75 crystal," *Proceedings of 1997 Topical Meeting on Photorefractive Materials, Effects and Devices (PR'97),* 614–617, 1997.
28. Y. Kawata, T. Tanaka, and S. Kawata, "Randomly accessible multilayered optical memory using $Bi_{12}SiO_{20}$ crystal," *Opt. Lett.,* **35**, 5308–5311, 1996.
29. M. M. Wang, S. C. Esener, F. B. McCormick, I. Cokgör, A. S. Dvornikov, and P. M. Rentzepis, "Experimental characterization of a two photon memory," *Opt. Lett.,* **22**, 558–560, 1997.
30. A. Shih, S. J. Pan, W. S. Liou, M. S. Park, J. Bhawalkar, J. Swiatkiewicz, P. Prasad, and P. C. Cheng, "Three-dimensional image storage using two-photon induced photobleaching method," *Cell Vision,* **4**, 223–224, 1997.
31. M. Irie and M. Mohri, "Thermally irreversible photochromic systems. Reversible photocyclization of diarylethene derivatives," *J. Org. Chem.,* **53**, 803–808, 1988.
32. T. Wilson, Y. Kawata, and S. Kawata, "Readout of three-dimensional optical memories," *Opt. Lett.,* **21**, 1003–1005, 1996.
33. D. Kermisch, "Partially coherent image processing by laser scanning," *J. Opt. Soc. Am.,* **65**, 887–891, 1975.
34. T. Tanaka and S. Kawata, "Comparison of recording densities in three-dimensional optical storage systems: Multilayered bit-recording versus angularity multiplexed holographic recording," *J. Opt. Soc. Am. A,* **13**, 935–943, 1996.

Chapter 10

Conditions for Confocal Readout of Three-Dimensional Photorefractive Data Bits

Min Gu

Optoelectronic Imaging Group, School of Communications and Informatics, Victoria University of Technology, Australia

Abstract

Confocal scanning optical microscopy provides a powerful tool for imaging the structures within a thick object. In three-dimensional (3-D) optical bit data storage, data bits can be recorded as a result of the change in refractive index in thick photorefractive and photochromic materials under single-photon and two-photon excitation by a high-numerical-aperture objective. In principle, these photorefractive data bits recorded in a form of a 3-D array in a thick medium can be read out using reflection confocal microscopy because such a microscope exhibits an optical sectioning property for 3-D imaging. To achieve this aim, a reflection confocal microscope should satisfy a condition that the passband of the 3-D coherent transfer function for reflection confocal microscopy overlaps or partially overlaps the support region of spatial frequencies of 3-D data bits

PHOTOREFRACTIVE OPTICS
Materials, Properties and Applications

ISBN 0-12-774810-5/$30.00

recorded under single-photon and two-photon excitation. This condition is affected by the refractive-index mismatch between a thick recording medium and its immersion medium. To efficiently read out data bits in a reflection confocal microscope, one can also use two illumination beams in the recording process to increase the content of spatial frequencies of the recorded photorefractive data bits.

10.1 Introduction

Confocal scanning optical microscopy provides a powerful tool for imaging a thick object with depth structures [1–3]. In 3-D optical bit data storage, data bits can be recorded as a result of the change in refractive index in thick photorefractive and photochromic materials under single-photon [4–8] and two-photon excitation [9–12] by a high-numerical-aperture objective. In principle, these photorefractive data bits recorded in the form of a 3-D array can be read out using reflection confocal microscopy because such a microscope exhibits an optical sectioning property for 3-D imaging.

To use a confocal microscope system for reading 3-D data bits recorded by the photorefractive effect, the confocal microscope system should satisfy some conditions; otherwise the recorded 3-D data bits cannot be efficiently read out. In a reading process, the recorded 3-D data bits act as a 3-D object. In other words, the passband of the transfer function for reflection confocal microscopy should overlap, at least partially, the support region of spatial frequencies of 3-D data bits recorded by the photorefractive effect under single-photon and two-photon excitation. It has been experimentally demonstrated that the 3-D data bits recorded by the photorefractive effect under two-photon excitation can be read out using a reflection confocal microscope [9].

The aim of this chapter is to present a systematic description of the conditions under which the 3-D data bits recorded by the photorefractive effect under single-photon and two-photon excitation can be read out using a reflection-mode confocal microscope. This analysis is based on the consideration of the passband of the transfer function for a reflection confocal microscope readout system and the support region of spatial frequencies of 3-D data bits recorded by the photorefractive effect. In Section 10.2, 3-D bit data storage is briefly reviewed. The principle of confocal scanning microscopy is summarized in Section 10.3. The results regarding the passband of the 3-D transfer function for reflection confocal microscopy used in the reading process is described in Section 10.4. Sections 10.5 and 10.6 describe the support regions of spatial frequencies of the 3-D data bits recorded by the photorefractive effect under single-photon and two-photon excitation, respectively. Based on the support region and passband in the writing and reading processes, the conditions for reading the 3-D data bits recorded by the photorefractive effect are derived. The effect of the refractive-index mismatch between a thick recording material and its immersion material is discussed in Section 10.7. A conclusive discussion is presented in Section 10.8.

10.2 Three-dimensional bit data storage

Photorefractive material is one of the important recording media in optical memory including holographic memory and bit data storage. In optical bit data storage, a laser beam is usually focused by a high-numerical-aperture objective onto a recording material to produce a spot where physical or chemical properties of the material are changed. Information such as images, words, data, music, and so on is modulated with the laser beam, so that an array of spots (bits), determined by the logic state of the information, is formed in the recording material. Essentially, the size of the focal spot and the resolution of a recording medium determine the storage density. According to the diffraction effect of light in a far-field region, the shorter the illuminating wavelength the smaller the spot size. In conventional two-dimensional (2-D) optical bit data storage where data bits (spots) are recorded on the surface of a medium (Fig. 10.1a), a 2-D storage density has arrived at 0.1 Gbit/cm^2 for a given visible wavelength, which is close to the limit imposed by the diffraction by a circular objective lens. To break this limitation, one can perform optical bit data storage in the region where the effect of diffraction is not dominant. This idea has led to a new method for optical bit data storage involving a near-field probe [13].

An alternative way to achieve a high storage density in the far-field region is to record information bits in 3-D space of a thick medium (Fig. 10.1b). This method, called 3-D bit data storage hereafter, has an advantage of high signal-to-noise ratio and high recording density compared with the holographic storage method [14], and it has been expected that an achievable 3-D storage density can be as high as 3–4 Tbit/cm^3 for a circular objective of numerical aperture 1.4 [15].

A number of materials including photochromic materials [4, 5, 9, 16, 17], photorefractive crystals [6–8, 10, 11], photopolymerizable materials

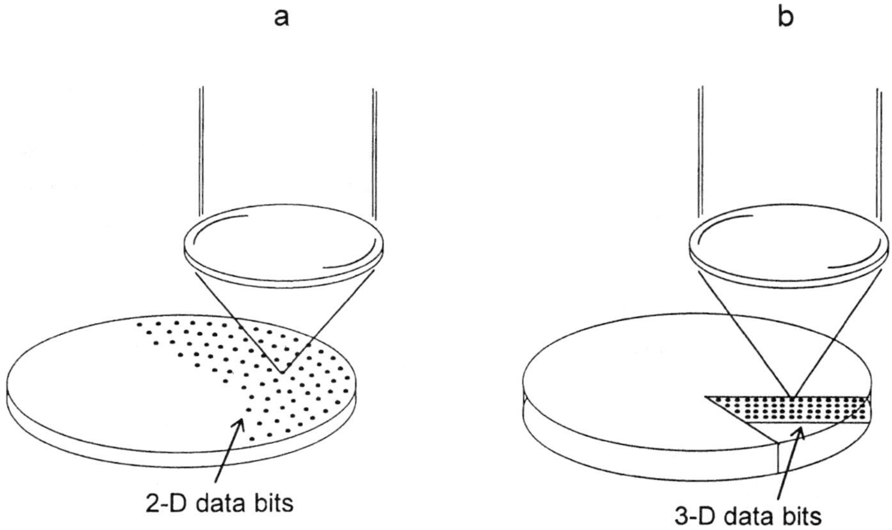

Figure 10.1: Schematic diagram for 2-D (a) and 3-D (b) optical data storage.

[7, 12], photobleaching materials [18, 19], and even fused silica glass [20] have been employed in 3-D bit data storage. Three-dimensional data bits can be recorded in those materials under single-photon excitation [4–8, 14, 20]. In this case a laser beam of a visible wavelength is focused deeply into a thick material and one incident photon is absorbed by one absorbing particle in the medium. Light scattering in the visible-wavelength region is so strong that the intensity in the focal spot reduces appreciably. Accordingly, information cannot be efficiently recorded at a deep position of a thick material. To solve this problem, one has used a two-photon process produced by an ultrashort pulsed laser beam of an infrared wavelength [9–12, 15–19]. *Two-photon excitation* means that two incident photons are absorbed simultaneously by an absorbing particle in a recording material. Using two-photon excitation can also help to reduce cross talk between two adjacent layers of recorded bits due to its nonlinear cooperation in the focal region of an objective.

For 3-D bit data storage, a localized physical and chemical reaction is needed at a certain depth of a volume recording material. One of the physical changes is an alteration of refractive index in the focal region of a writing objective. Such a change in refractive index can occur in photorefractive crystals [6–8, 10, 11], photochromic materials [4, 9], photopolymerizable materials [7, 12], and fused silica glass [20]. In this chapter, the term "photorefractive effect" is used to broadly refer to the change in refractive index caused by a focused laser beam. Except for fused silica glass, the change in refractive index in these materials mentioned above is small ($\Delta n = 10^{-6}$–10^{-3}). To read data bits recorded by the photorefractive effect, phase-contrast microscopy [4, 8, 21] and differential interference contrast microscopy [12] have been adopted. An alternative way to read 3-D data bits recorded by the photorefractive effect is to use a reflection-mode confocal microscope [9] which has a 3-D imaging ability due to its strong optical sectioning property [1–3]. The advantage of using a reflection confocal microscope readout system is that it gives a better depth sensitivity than phase-contrast microscopy and differential interference contrast microscopy.

10.3 Confocal scanning microscopy

The idea of confocal microscopy was first proposed by Minsky [22]. In this imaging system, a sample is illuminated by a small spot of light and the signal from the illuminated spot is collected by a small detector which is ideally a point detector. By moving the sample, a map of information from different parts in the sample can be recorded, which gives rise to the image of the sample (Fig. 10.2). There are three main advantages of confocal microscopy over conventional microscopy [1–3]. The first one is that transverse resolution in confocal microscopy is 1.4 times as large as that in a conventional microscope [23]. The improvement in resolution is obtained at the expense of the reduction of the field of view. However, this drawback can be easily compensated for by using the scanning mechanism. The second advantage is that confocal scanning microscopy has a strong depth discrimination property: the out-of-focus signal is detected much less strongly than the in-focus information whereas in a conventional system it is only blurred [24]. The third advantage is that the noise associated with flare and unwanted scattered light is significantly reduced because a small pinhole mask is employed in front of the detector [2].

Among these advantages, it is the depth discrimination property, which is also called the optical sectioning property [2], that allows one to record an image of the particular transverse section in a thick object and to form a 3-D image subsequently. In Fig. 10.2, the principle of optical

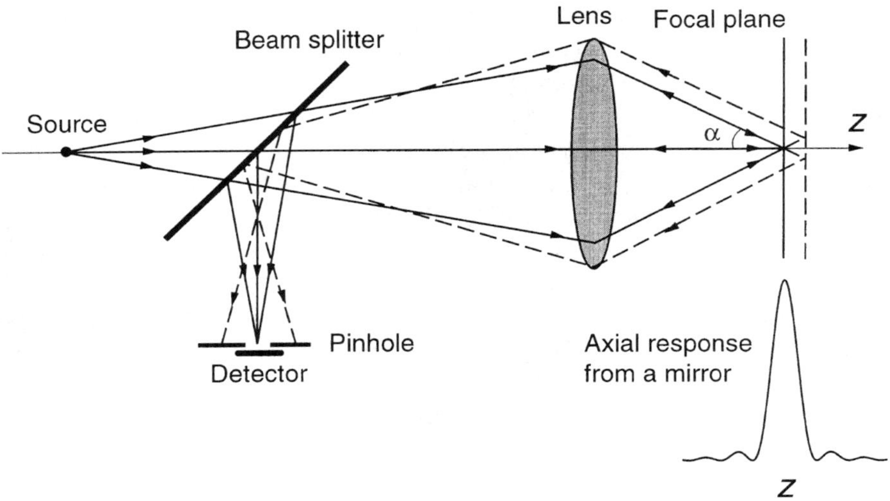

Figure 10.2: Schematic diagram of a reflection confocal scanning microscope. Such a microscope exhibits an optical sectioning property.

sectioning in a reflection-mode confocal scanning imaging system is demonstrated using a perfect reflector (e.g., a mirror) as an object. The reflector is scanned in the axial direction of an objective lens.

When the reflector is located in the focal plane of the objective, which is represented by the solid lines in Fig. 10.2, the reflected beam is focused exactly onto a point detector which collects most of the incident energy. However, if the reflector moves away from the focal plane (see the dotted lines in Fig. 10.2), the signal from the reflector is focused at a position either before or behind the point detector, depending on the scanning direction. Thus, the detector collects only a small amount of the incident energy. Therefore the signal strength detected when the reflector is at the defocus position is weaker than that at the in-focus position, which is shown in Fig. 10.2. It is clear that the larger the defocus distance the weaker the signal strength. This property implies that the location of the reflector can be determined according to the measured maximum signal. In other words, if there are details on the reflector, the image of the details is efficiently recorded only when the reflector is in the focal plane. For a real thick object, by recording a series of image sections at different depths of the object, a complete image of the thick object can be built up in a form of a stack of sections. Therefore, a 3-D image of the thick object can be recorded without necessarily slicing the object. It is obvious that the broadness of the axial response to the perfect reflector shown in Fig. 10.2 determines axial resolution of 3-D imaging because the narrower the axial response the weaker the cross talk between two adjacent image sections. It is this 3-D imaging ability that confocal scanning microscopy can apply to reading the 3-D data bits recorded in a thick photorefractive material.

10.4 Passband of the 3-D coherent transfer function for reflection confocal microscopy

Details on the analysis on confocal microscopy based on the concept of the 3-D transfer function can be found in standard references on confocal microscopy [1–3]. This section gives a brief summary of the results arising from this method.

For a high-numerical-aperture circular objective under the sine condition which a commercial objective satisfies [25, 26], its 3-D amplitude point spread function $h(r, z)$, which is the image amplitude of a single point object, is given by [25]

$$h(r, z) = A \int_0^{\alpha} \sqrt{\cos\theta}\, \exp(ikzn\cos\theta) J_0(krn\sin\theta) \sin\theta d\theta\,, \qquad (10.1)$$

where A is the constant of normalization, J_0 is a Bessel function of the first kind of order zero, θ is the angle of convergence of a ray, and α is the maximum semiangle of an objective. n is the refractive index of the immersion medium of an objective, and $k = 2\pi/\lambda$ is the wave number in vacuum. $r = \sqrt{x^2 + y^2}$ and z are radial and axial coordinates originating from the geometric focus of an objective (x and y are two orthogonal coordinates in the transverse plane).

The corresponding 3-D transfer function $c(l, s)$, called the 3-D coherent transfer function in a coherent imaging process [27], for an objective is given by the 3-D Fourier transform of $h(r, z)$ with respect to r and z and can expressed as [3]

$$c(l, s) = \frac{1}{[1 - (\lambda l)^2]^{1/4}}\, \delta\left(s - \sqrt{\left(\frac{n}{\lambda}\right)^2 - l^2}\right), \qquad (10.2)$$

where $l = \sqrt{m^2 + n^2} = n(\sin\theta)/\lambda$ is the radial spatial frequency. Here m, n, and s are the spatial frequencies in the x-, y-, and z-directions, respectively. The delta function in Eq. (10.2) implies that the 3-D coherent transfer function is nonzero only on a sphere given by

$$s^2 + l^2 = \left(\frac{n}{\lambda}\right)^2. \tag{10.3}$$

The significance of the transfer function for an objective is that it gives the efficiency with which spatial frequency components in an object are imaged. Due to the limit caused by the maximum semiangle α of an objective, the radial spatial frequency l and the axial spatial frequency s cut off at $l_c = n(\sin\alpha)/\lambda$ and $s_c = n(\cos\alpha)/\lambda$, respectively. As a result, the nonzero region of the 3-D coherent transfer function $c(l, s)$ for an objective is actually a cap of the sphere. In other words, a single objective has a poor ability of transferring the depth information of a 3-D object.

For a reflection confocal microscope consisting of a point source and a point detector (Fig. 10.2), imaging is purely coherent [1–3] as an image in a confocal system can be expressed as a superposition of amplitude, so that its amplitude point spread function is the square of Eq. (10.1) [1–3]. Thus the corresponding 3-D coherent transfer function $c_r(l, s)$ can be derived by the 3-D Fourier transform of $(h(r, z))^2$ [3]. An analytical expression for $c_r(l, s)$ can be found by a self-convolution of Eq. (2) and can be expressed as [3, 28]

$$c_r(l, s = \begin{cases} \dfrac{2|s|}{\pi\sqrt{l^2+s^2}}E(p) & , \quad \dfrac{2n(l\sin\alpha + |s|\cos\alpha}{\lambda} \le (l^2+s^2) \le \dfrac{4n^2}{\lambda^2}, |s| \ge \dfrac{2n\cos\alpha}{\lambda}, \\ \dfrac{2|s|}{\pi\sqrt{l^2+s^2}}E(\beta, p) & , \quad \dfrac{2n(l\sin\alpha + |s|\cos\alpha)}{\lambda} \ge (l^2+s^2), |s| \ge \dfrac{2n\cos\alpha}{\lambda}, \\ 0 & , \quad \text{otherwise,} \end{cases} \tag{10.4}$$

for an objective satisfying the sine condition. Here $E(x, y)$ is an incomplete elliptic integral of the second kind [33] and $E(x)$ is a complete elliptic integral [29]. β is defined as

$$\beta = \arcsin\left[\frac{1}{p}\left(1 - \frac{2n\cos\alpha}{\lambda|s|}\right)\right], \tag{10.5}$$

where

$$p = \frac{2nl}{|s|\lambda\sqrt{l^2 + s^2}}\left(1 - \frac{\lambda^2}{n^2}\frac{l^2 + s^2}{4}\right). \tag{10.6}$$

The 3-D view of Eq. (10.4) can be found elsewhere [3, 28]. It is, however, the passband of $c_r(l, s)$ in Eq. (10.4) that is important in reading recorded 3-D data bits and can be expressed as

$$s = \pm\begin{cases}\sqrt{\dfrac{4n_r^2}{\lambda_r^2} - l^2}, & 0 \le l < \dfrac{2n_r \sin\alpha_r}{\lambda_r},\\[2ex] \dfrac{2n_r \cos\alpha_r}{\lambda_r}, & l = \dfrac{2n_r \sin\alpha_r}{\lambda_r}.\end{cases} \tag{10.7}$$

Here n_r, λ_r, and α_r are the refractive index of the immersion material of an objective, the wavelength of a reading beam, and the maximum semiangle of an objective in the reading process, respectively. The passband in Eq. (10.7) cuts off at

$$l_c = \frac{2n_r \sin\alpha_r}{\lambda_r} \tag{10.8}$$

in the radial direction, and has low and high cutoff axial spatial frequencies of

$$s_c = \pm\frac{2n_r \cos\alpha_r}{\lambda_r} \tag{10.9}$$

and

$$s_c = \pm\frac{2n_r}{\lambda_r}. \tag{10.10}$$

It is clear from Eqs. (10.7)–(10.10) that the passband of $c_r(l, s)$ for a reflection confocal microscope is given by a cap of the solid spherical ball defined by Eq. (10.7). The projection of the cap in the l and s plane is shown in Figs. 10.3 and 10.6 (see dashed curves). The positive and negative signs in Eqs. (10.7), (10.9), and (10.10) imply that the cap of the 3-D coherent transfer function for a given reflection confocal microscope may occur on the either position or negative side of the axial spatial frequency axis depending on the direction of the beam incident upon the microscope.

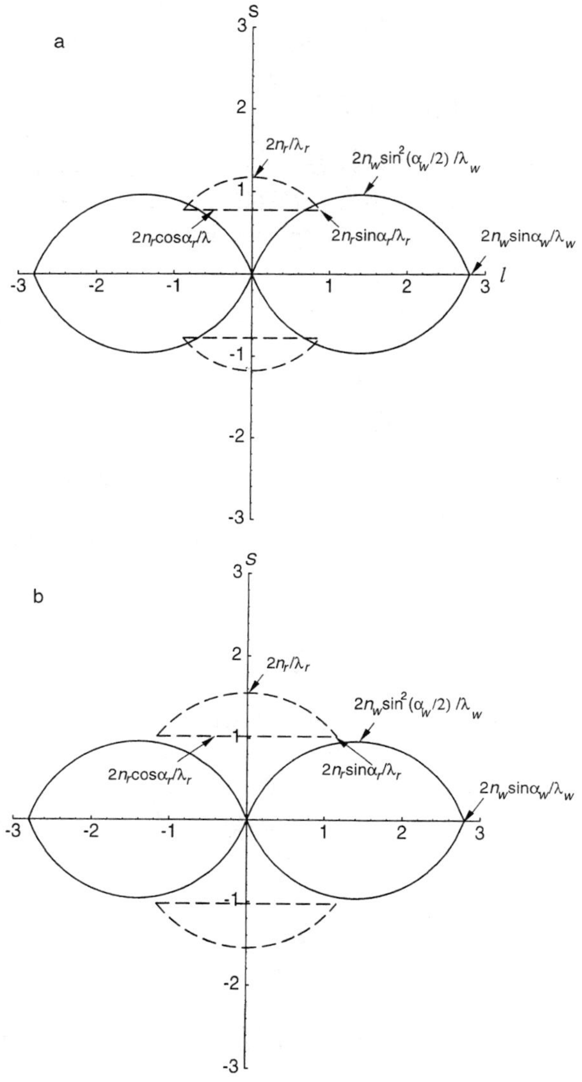

Figure 10.3: Support region (solid) of spatial frequencies of 3-D data bits recorded under single-photon excitation and the passband (dashed) of the 3-D coherent transfer function for reflection confocal microscopy: (a) $n_w = 1.5$, $n_r = 1$, $NA_w = 1.4$ (oil), $NA_r = 0.75$ (dry), $\lambda_w = 488$ nm, $\lambda_r = 830$ nm ($\beta = \lambda_r/\lambda_w = 1.7$); (b) $n_w = 1.5$, $n_r = 1$, $NA_w = 1.4$ (oil), $NA_r = 0.75$ (dry), $\lambda_w = 488$ nm, $\lambda_r = 630$ nm ($\beta = \lambda_r/\lambda_w = 1.29$).

10.5 Spatial frequency response of 3-D data bits recorded by the single-photon photorefractive effect

For the photorefractive effect discussed in this chapter, if the change in refractive index, Δn, is small, a reading beam may be weakly scattered due to the fact that multiple scattering is weak. Therefore, the reflectance of a recording medium consisting of recorded 3-D data bits can be assumed to be proportional to Δn [30]. In the case of single-photon excitation Δn is proportional to the incident intensity [30]. Let us consider the recording objective to be circular, so that we have from Eq. (10.1)

$$\Delta n \propto |h(r,z)|^2 . \tag{10.11}$$

The spatial frequency response of 3-D data bits is related to the material property and a recording optical system. However, the support region of spatial frequencies of the recorded 3-D data bits is given by the passband of the 3-D Fourier transform of $|h(r, z)|^2$ and can be expressed, according to Eq. (10.11), as

$$s = \pm \left[\sqrt{\frac{n_w^2}{\lambda_w^2} - \left(l - \frac{n_w \sin \alpha_w}{\lambda_w} \right)^2} - \frac{n_w \cos \alpha_w}{\lambda_w} \right] . \tag{10.12}$$

This expression has been derived by the convolution of Eq. (10.2) with its axially inverted function [3]. Here n_w, λ_w, and α_w are the refractive index of the immersion material of an objective, the wavelength of a writing beam, and the maximum semiangle of the objective in the writing process, respectively. It can be seen that the support region of spatial frequencies in Eq. (10.12) is cut off at

$$l_c = \frac{2n_w \sin \alpha_w}{\lambda_w} \tag{10.13}$$

and

$$s_c = \pm \frac{n_w (1 - \cos \alpha_w)}{\lambda_w} . \tag{10.14}$$

The support region of spatial frequencies of the 3-D data bits recorded by the single-photon photorefractive effect exhibits a doughnut shape in the l-s plane (see the solid curves in Fig. 10.3), which leads to a missing cone of spatial frequencies near the origin of the 3-D spatial-frequency space.

To be able to read 3-D data bits recorded by the single-photon photorefractive effect, the passband of the 3-D coherent transfer function in Eq. (10.7) should partially or completely overlap the support region of spatial frequencies of the recorded 3-D data bits given by Eq. (10.12). Figure 10.3 shows the passbands and the support regions in two practical cases, where the size of the passbands and the support regions has been normalized by the writing wavelength. It is noted that in the case of Fig. 10.3a in which the reading beam has an infrared wavelength, the passband and the support region have a small overlapping area. However, when a reading beam of a red wavelength is used, no overlapping occurs between the passband and the support region.

The first condition under which the passband (Eq. 10.7) and the support region (Eq. 10.12) start to overlap is that the low cutoff axial spatial frequency in Eq. (10.9) is equal to or smaller than the cutoff axial spatial frequency in Eq. (14) (assume that both the passband and the support region are along the positive axial spatial frequency axis), which leads to

$$NA_r \geq n_r \sin\left\{\arccos\left[\beta\frac{n_w}{2n_r}\left(1 - \cos\left(\arcsin\frac{NA_w}{n_w}\right)\right)\right]\right\}. \tag{10.15}$$

Here $\beta = \lambda_r / \lambda_w$. NA_r and NA_w are the numerical aperture for an objective in the reading and writing processes, respectively. With the equality sign, Eq. (10.15) is depicted in Fig. 10.4 for various values of β. Although the use of an oil-immersion objective may prove advantageous for reducing spherical aberration caused by the mismatch of refractive indices between a recording material and its immersion material (Section 10.7), it would not be a practically useful method if an oil-immersion objective is used in reading 3-D data bits. Therefore we do not consider oil-immersion objectives in the reading process.

Due to the existence of a missing cone of spatial frequencies in the 3-D data bits recorded by the single-photon photorefractive effect, the cutoff transverse spatial frequency of the 3-D coherent transfer function in Eq. (10.8) needs to be larger than the size of a missing cone of spatial frequencies. This condition can be expressed as

$$\cos(\alpha_r + \alpha_w) \leq \frac{n_r}{\beta n_w}. \tag{10.16}$$

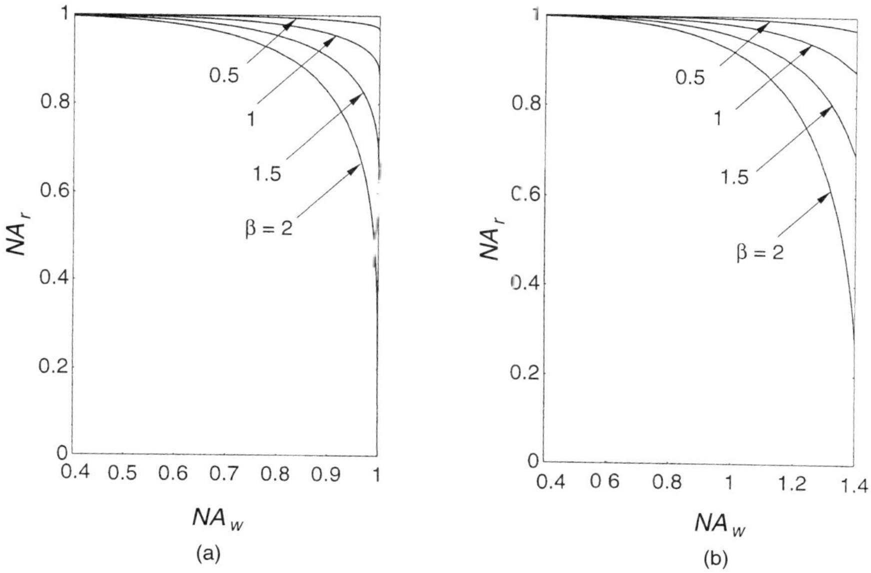

Figure 10.4: Critical condition in the axial direction for the reading numerical aperture NA_r as a function of the writing numerical aperture NA_w under single-photon excitation when $\beta = \lambda_r/\lambda_w = 0.5, 1, 1.5, 2$: (a) $n_w = 1.0$, $n_r = 1.0$; (b) $n_w = 1.5$, $n_r = 1.0$.

For an oil-immersion objective used in the writing process, the dependence of the factor $\cos(\alpha_r + \alpha_w)$ on the writing numerical aperture NA_w is shown in Fig. 10.5 when Eq. (10.15) is satisfied. It is clear that the left-hand side in Eq. (10.16) is always smaller than the right-hand side of Eq. (10.16). If a dry objective is used in the writing process, the value of the right-hand side in Eq. (10.16) is even larger. Therefore it is concluded that Eq. (10.16) always holds as long as Eq. (10.15) is satisfied.

It is seen from Figs. 10.4 and 10.5 that the conditions for the passband and the support region to overlap depend on the wavelength and the numerical aperture used in the writing and reading processes. It is shown from Fig. 10.4 that it is difficult to read 3-D data bits recorded by the single-photon photorefractive effect in a reflection confocal microscope even if the reading wavelength is longer than the writing wavelength because the overlapped area is small compared with the passband of the 3-D coherent transfer function for reflection confocal microscopy.

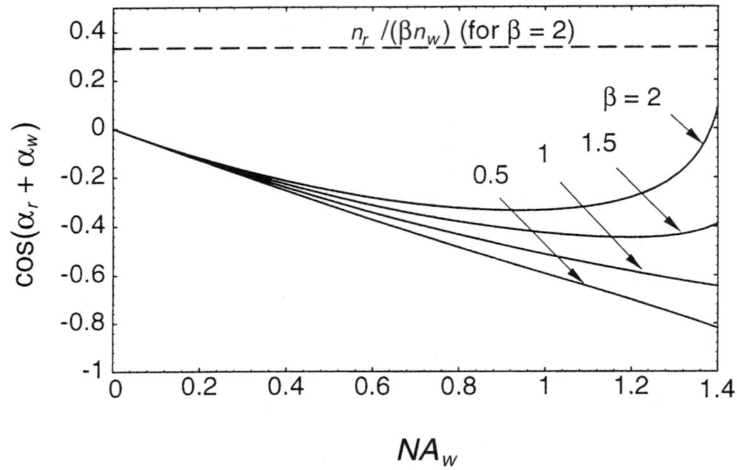

Figure 10.5: Critical condition in the transverse direction for the reading numerical aperture NA_r as a function of the writing numerical aperture NA_w under single-photon excitation when $\beta = \lambda_r/\lambda_w = 0.5, 1, 1.5, 2$; $n_w = 1.5$, and $n_r = 1.0$.

10.6 Spatial frequency response of 3-D data bits recorded by the two-photon photorefractive effect

In the case of the photorefractive effect under two-photon excitation, the change in refractive index, Δn, is proportional to the square of the incident intensity [31]. For a circular writing objective, we have

$$\Delta n \propto |h(r,z)|^4 . \tag{10.17}$$

The support region of spatial frequencies of 3-D data bits recorded under two-photon excitation is given by the passband of the 3-D Fourier transform of $|h(r, z)|^4$, which can be expressed as [3]

$$s = \pm \begin{cases} \dfrac{2n_w(1 - \cos\alpha_w)}{\lambda_w}, & 0 \le l < \dfrac{2n_w \sin\alpha_w}{\lambda_w}, \\ \sqrt{\dfrac{4n_w^2}{\lambda_w^2} - \left(l - \dfrac{2n_w \sin\alpha_w}{\lambda_w}\right)^2} - \dfrac{2n_w \cos\alpha_w}{\lambda_w}, & \dfrac{2n_w \sin\alpha_w}{\lambda_w} \le l \le \dfrac{4n_w \sin\alpha_w}{\lambda_w}, \end{cases} \tag{10.18}$$

where n_w, λ_w, and α_w are the refractive index of the immersion material of an objective, the wavelength of the writing beam, and the maximum semiangle of the objective in the writing process, respectively. Here l and s are cut off at

$$l_c = \frac{4n_w \sin \alpha_w}{\lambda_w} \tag{10.19}$$

and

$$s_c = \pm \frac{2n_w (1 - \cos \alpha_w)}{\lambda_w}. \tag{10.20}$$

Unlike the support region of spatial frequencies of 3-D data bits recorded by the single-photon photorefractive effect, the support region of spatial frequencies of 3-D data bits recorded by the two-photon photorefractive effect does not show any missing cone of spatial frequencies. Figure 10.6 shows the support regions and the passbands in two practical cases under two-photon excitation (the size of the passbands and the support regions has been normalized by the writing wavelength). In Fig. 10.6a a reading beam of a wavelength longer than the wavelength of the writing beam is used. In this case, for a given reading numerical aperture, resolution in a reflection confocal microscope can reach the diffraction-limited resolution because the passband of the 3-D coherent transfer function (dashed curve) completely overlaps with the support region of spatial frequencies of the recorded 3-D data bits (solid curve). If the reading wavelength becomes shorter than that of the writing beam (Fig. 10.6b), the passband and the support region become partially overlapping in the axial direction but the cutoff transverse spatial frequency of the 3-D coherent transfer function slightly increases compared with that in Fig. 10.6a. In other words, the axial resolution in this case becomes poor, while the corresponding transverse resolution is slightly improved.

Because there is no missing cone of spatial frequencies in the support region of spatial frequencies of the 3-D data bits recorded by the two-photon photorefractive effect, the only condition under which the passband of the 3-D coherent transfer function in Eq. (10.7) for reflection confocal microscopy and the support region of spatial frequencies in Eq. (10.18) start to overlap is that Eq. (10.9) is equal to or smaller than Eq. (10.20) (assume the positive axial spatial frequency axis). Consequently, we have

$$NA_r \geq n_r \sin\left\{\arccos\left[\beta\frac{n_w}{n_r}\left(1 - \cos\left(\arcsin\frac{NA_w}{n_w}\right)\right)\right]\right\}. \tag{10.21}$$

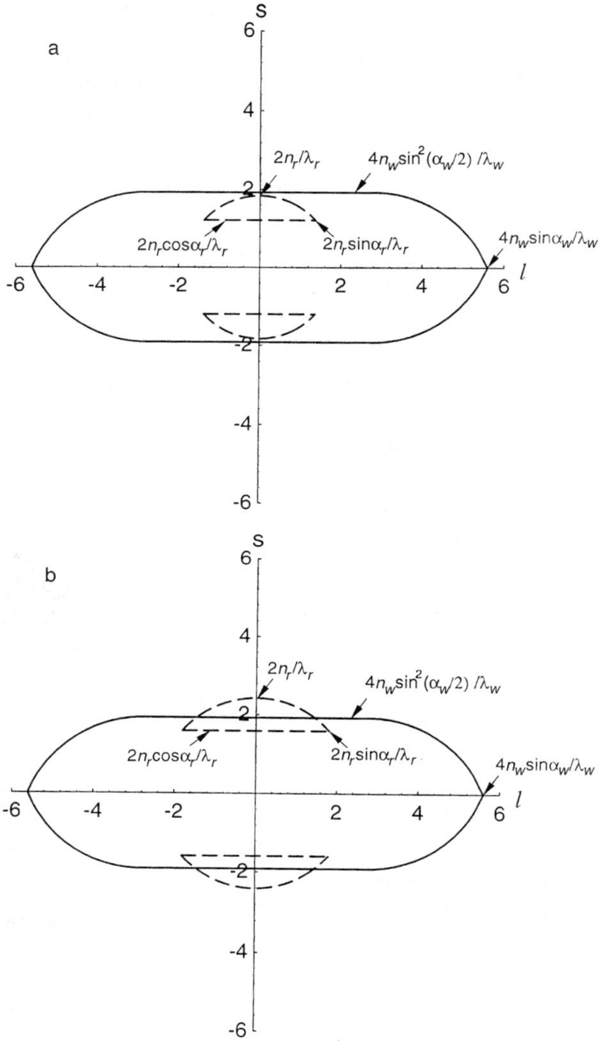

Figure 10.6: Support region (solid) of spatial frequencies of 3-D data bits recorded under two-photon excitation and the passband (dashed) for the 3-D coherent transfer function for a reflection confocal microscopy: (a) $n_w = 1.5$, $n_r = 1$, $NA_w = 1.4$ (oil), $NA_r = 0.75$ (dry), $\lambda_w = 760$ nm, $\lambda_r = 830$ nm ($\beta = \lambda_r/\lambda_w = 1.092$); (b) $n_w = 1.5$, $n_r = 1$, $NA_w = 1.4$ (oil), $NA_r = 0.75$ (dry), $\lambda_w = 760$ nm, $\lambda_r = 630$ nm ($\beta = \lambda_r/\lambda_w = 0.759$).

This expression with the equality sign is depicted in Fig. 10.7. It is seen from Fig. 10.7 that the condition for the passband and the support region to overlap depends on the wavelength and the numerical aperture used in writing and reading processes. It is shown that it is difficult to read 3-D data bits if the writing wavelength is longer than the reading wavelength. The figure also suggests that a reflection confocal microscope may read effectively 3-D data bits if a high-numerical-aperture oil-immersion objective and a high-numerical-aperture dry objective are used in the writing and reading processes, respectively.

The ratio of the overlapped axial passband of the 3-D coherent transfer function to the total axial passband of the 3-D coherent transfer function, γ, can be derived, under two-photon excitation, as

$$\gamma = \frac{\beta n_w \left[1 - \cos\left(\arcsin\dfrac{NA_w}{n_w}\right)\right] - n_r \cos\left(\arcsin\dfrac{NA_r}{n_r}\right)}{n_r\left[1 - \cos\left(\arcsin\dfrac{NA_r}{n_r}\right)\right]}, \quad \gamma \leq 1, \qquad (10.22)$$

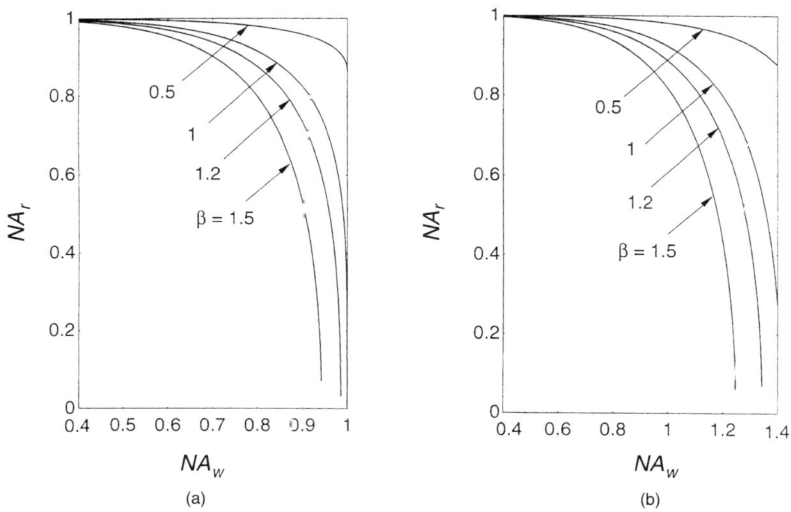

Figure 10.7: Critical condition in the axial direction for the reading numerical aperture NA_r as a function of the writing numerical aperture NA_w under two-photon excitation when $\beta = \lambda_r/\lambda_w = 0.5, 1, 1.5, 2$: (a) $n_w = 1.0$, $n_r = 1.0$; (b) $n_w = 1.5$, $n_r = 1.0$.

which indicates the achievable axial resolution in a reading process and is plotted in Fig. 10.8 for $\beta = 1$. When γ equals unity, the highest axial resolution can be achieved in a reading process using a reflection confocal microscope for a given reading objective.

Readout of 3-D data bits recorded by the photorefractive effect under two-photon excitation was successfully demonstrated in a reflection confocal microscope [9]. A reflection confocal microscope was achieved in a Zeiss LSM 410 scanning microscope with a He-Ne laser (3 mW) as a reading source. A dry objective of numerical aperture 0.75 was used. 3-D bit patterns were recorded in a photochromic medium. The recording optical setup was similar to that used before [4] except for the laser source. To produce two-photon excitation, an ultrashort pulsed laser (Coherent Scientific, Model 900), tuned at 760 nm and 130 fs, was adopted. An oil-immersion objective of numerical aperture 1.4 was used for writing data. It can be seen from Figs. 10.6–10.8 that this combination of the writing and reading objectives makes it possible to read the 3-D data bits in reflection confocal microscopy. Use of an oil-immersion objective can reduce the effect of spherical aberration (see Section 10.7). The laser power in the focus of the writing objective was approximately 10 mW and the writing time for each bit was 10 ms. Under these conditions, various patterns consisting of 24×24 bits and recorded at different depths, were successfully read out using the reflection confocal microscope. If, on the other hand, an oil-immersion objective of numerical aperture 0.85 was used, it was impossible to read out bit patterns recorded under two-photon excitation. These observations are consistent with the theoretical prediction given in the preceding discussion.

10.7 Effect of refractive index mismatch

The results and conditions presented in Sections 10.5 and 10.6 hold if the refractive-index mismatch between the volume recording medium and its immersion can be neglected. When the refractive-index mismatch becomes pronounced, they are applicable only in the region near the surface of a volume recording medium. As has been mentioned in the preceding discussion, the refractive-index mismatch may affect the performance in the writing and reading processes when a beam is focused into a deep position in a volume medium. When the refractive index of a recording material does not match that of its immersion material, the diffraction

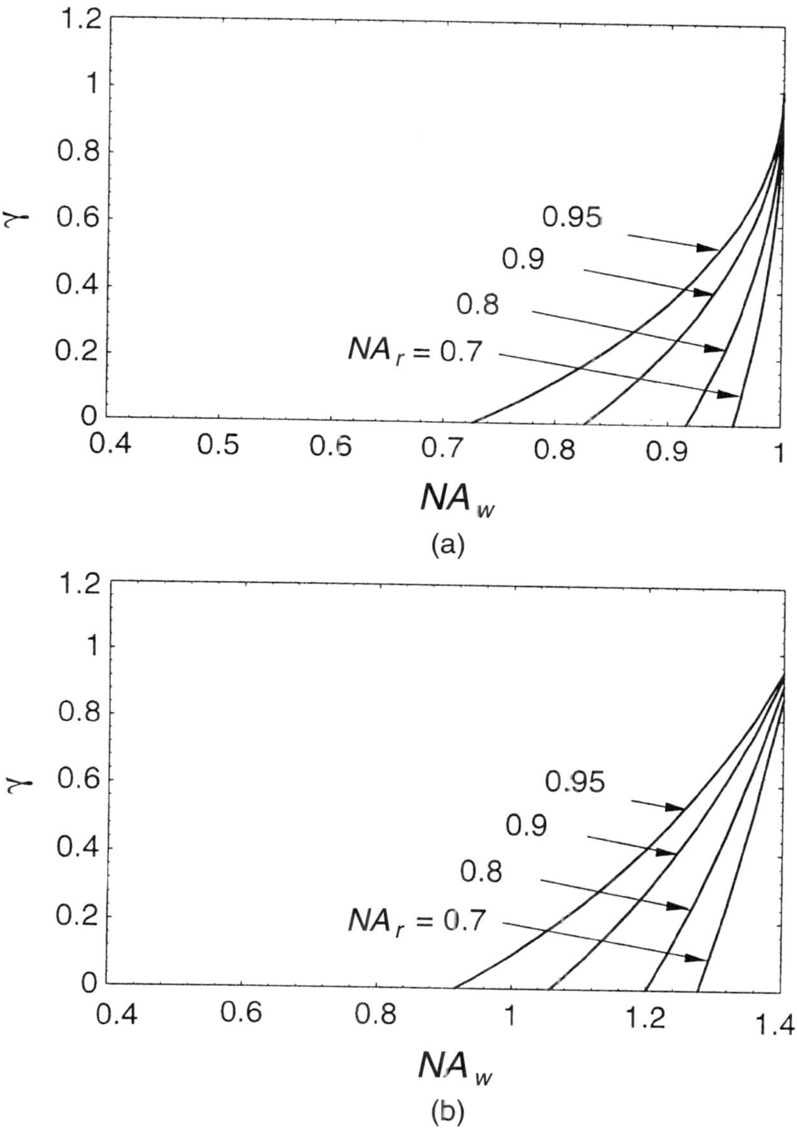

Figure 10.8: Axial overlapping ratio γ as a function of the writing numerical aperture NA_w under two-photon excitation when NA_r = 0.7, 0.8, 0.9, 0.95 from the bottom (β = 1): (a) n_w = 1.0, n_r = 1.0; (b) n_w = 1.5, n_r = 1.0.

pattern in the focal region of an objective is distorted compared with Eq. (10.1) which is the diffraction-limited pattern by an objective used in a uniform medium [32]. The distortion results from the fact that the refraction of a convergence ray depends on the angle of convergence. As a result, a high-numerical-aperture objective suffers from more distortion than a low numerical-aperture objective. According to the result by Török *et al.* [33], the 3-D amplitude point spread function for an objective satisfying the sine condition can be expressed, if an incident plane wave is focused from the first medium of refractive index n_1 into the second medium of refractive index n_2, as

$$h'(r,z) = A\int_0^{\alpha} \sqrt{\cos\theta_1}\,(\tau_s + \tau_p\cos\theta_2)J_0(krn_1\sin\theta_1) \exp(i\Phi + ikzn_2\cos\theta_2)\sin\theta_1 d\theta_1\,, \tag{10.23}$$

where θ_1 and θ_2 are the angles of a ray of convergence in the first and second media, respectively; r and z are radial and axial coordinates with an origin at the focus which would occur if there were no second medium; and τ_s and τ_p are the Fresnel transmission coefficients for s and p polarization states at the interface between the first and second media [26]. Both τ_s and τ_p are the function of angles θ_1 and θ_2.

In Eq. (10.23) Φ is given by

$$\Phi = -kd\,(n_1\cos\theta_1 - n_2\cos\theta_2)\,, \tag{10.24}$$

where d is the distance from the interface of the two media to the diffraction-limited focus [33]. It is clear that the function Φ acts as a spherical aberration source because of its dependence on the angle θ_1, which leads to a distortion of the diffraction pattern. If the refractive index of the immersion medium matches that of a recording medium, $\theta_1 = \theta_2 = \theta$, $n_1 = n_2 = n$ and $\Phi = 0$. Accordingly, Eq. (10.23) reduces to Eq. (10.1). In general, for a given value of d, the larger the difference of the refractive indices between the two media the stronger the effect of the spherical aberration.

It should be pointed out that the vectorial effect has been ignored in Eq. (10.23), which is equivalent to neglecting the depolarization effect of the objective. This assumption holds for the maximum convergence angle less than 45° [3]. Even for an objective of numerical aperture 1.4, the vectorial effect does not alter the shape of the 3-D amplitude point spread function appreciably [15].

Performing the 3-D Fourier transform of Eq. (10.23) with respect to r and z results in the 3-D coherent transfer function for an objective inside the second medium:

$$c(l, s) = F(l)\delta\left(s - \sqrt{\frac{n_2^2}{\lambda^2} - l^2}\right), \tag{10.25}$$

where $l = n_2\,(\sin\theta_2)/\lambda$. The function $F(l)$ includes the contribution from τ_s, τ_p and Φ. Clearly, the 3-D coherent transfer function is nonzero in a region which has the same form as Eq. (10.3) but is defined by n_2.

The corresponding 3-D coherent transfer function for reflection confocal microscopy is given by the 3-D Fourier transform of $(h'(r, z))^2$. As a result, the passband of the 3-D coherent transfer function for reflection confocal microscopy has the same form as Eq. (10.7) but n_r and α_r in Eq. (10.7) are the refractive index and the maximum semiangle of convergence in the recording material. Similarly, the support regions of spatial frequencies of the 3-D data bits recorded by the photorefractive effect under single-photon and two-photon excitation are the same as those in Eqs. (10.12) and (10.18), respectively, with a recognition of n_w and α_w being the refractive index and the maximum semiangle of convergence of an objective in the recording material.

These properties in the reading and writing processes indicate that as long as the numerical aperture of an objective is given, the spatial frequency response of the 3-D data bits recorded by oil-immersion and dry objectives is the same. However, the distribution of the spatial frequency responses for an oil-immersion objective and a dry objective is different because the effect of the refractive-index mismatch in the case of a dry objective may be stronger than that in the case of an oil-immersion objective. A similar conclusion can be reached for the passband of the 3-D coherent transfer function for reflection confocal microscopy. As a consequence, for a given numerical aperture of an objective in the reading process, both n_r and $\cos\alpha_r$ inside the volume medium are larger than those near the surface of the volume medium and thus the overlapping ratio γ inside the volume medium is smaller than that near the surface of the volume medium. Therefore, resolution in the reading process may be degraded when the reading beam is focused into a deep position of a volume material. In that sense, it may be difficult to use a reflection confocal microscope to read the 3-D data bits recorded in a photorefractive $LiNbO_3$ crystal under two-photon excitation because the refractive index of $LiNbO_3$ is so large that the passband and support region in the reading and writing processes do not overlap.

10.8 Conclusion

In the case of single-photon excitation in photorefractive materials, it is difficult to use reflection confocal microscopy to read out the recorded 3-D data bits because the passband of the 3-D coherent transfer function for reflection confocal microscopy does not effectively overlap the support region of spatial frequencies of the recorded 3-D data bits. To solve this problem, one can use two objectives which illuminate a volume medium from two opposite directions and produce a common focal spot [30, 34]. In this case, the support region of spatial frequencies of the recorded 3-D data bits consists of a region defined by Eq (10.12) and two other regions defined by

$$s = \pm \begin{cases} \sqrt{\dfrac{4n_w^2}{\lambda_w^2} - l^2}, & 0 \le l < \dfrac{2n_w \sin \alpha_w}{\lambda_w}, \\ \dfrac{2n_w \cos \alpha_w}{\lambda_w}, & l = \dfrac{2n_w \sin \alpha_w}{\lambda_w}, \end{cases} \tag{10.26}$$

These two regions have a similar form to the passband of the 3-D coherent transfer function for reflection confocal microscopy in Eq. (10.7) [3]. Therefore, the 3-D data bits recorded by the single-photon photorefractive effect can be efficiently read out using reflection confocal microscopy.

Another way to extend the content of spatial frequencies of the 3-D photorefractive data bits under single-photon excitation is to record the 3-D data bits in a series of thin layers of thickness less than 1 μm [35]. In this way, the spatial frequency content of the 3-D data bits is effectively enlarged in the axial direction due to the axial confinement of the data bits, and thus overlap the passband of the 3-D coherent transfer function for reflection confocal microscopy.

The 3-D data bits recorded by the two-photon photorefractive effect can be read out in a reflection confocal microscope as the conditions described in Section 10.6 may be satisfied. An oil-immersion objective of high numerical aperture should be employed for the writing process and therefore the recorded 3-D data bits can be read out using a high-numerical-aperture dry objective in reflection confocal microscopy. If the reading wavelength is slightly longer than the writing wavelength, axial resolution in the reading process becomes better. Of course, the two-beam illumination method [30] can significantly enlarge the support region of spatial frequencies of the 3-D data bits recorded by the two-photon photorefractive effect. In fact the support region in this case is composed by five areas [3]. This method may be particularly advantageous for reading 3-D data bits recorded by the two-photon photorefractive effect in $LiNbO_3$ crystals.

Acknowledgments

The author acknowledges the support from the Australia Research Council (ARC). Some of work in Section 10.5 was performed while the author visited the Department of Applied Physics, Osaka University, Japan. For this, he would like to thank the Japan Society for the Promotion of Science (JSPS) for supporting his trip to Japan, and Professor S. Kawata and Associate Professor O. Nakamura for their hospitality and valuable discussions during his visit. Figure 10.1 was prepared by Mr. D. Day at Victoria University of Technology.

References

1. T. Wilson and C. J. R. Sheppard, *Theory and Practice of Scanning Optical Microscopy,* Academic Press, London, 1984.
2. C. J. R. Sheppard, *Scanning Optical Microscopy,* in *Optical and Electronic Microscopy,* Academic Press, London, 1987.
3. M. Gu, *Principles of Three-Dimensional Imaging in Confocal Microscopes,* World Scientific, Singapore, 1996.
4. A. Toriumi, J. M. Herrmann, and S. Kawata, "Nondestructive readout of a three-dimensional photochromic optical memory with a near-infrared differential phase-contrast microscope," *Opt. Lett.,* **13**, 555–557, 1997.
5. Y. Kawata, T. Tanaka, and S. Kawata, "Randomly accessible, multilayered optical memory with $Bi_{12}SiO_{20}$ crystal," *Appl. Opt.,* **35**, 5308–5311, 1996.
6. H. Ueki, Y. Kawata, and S. Kawata, "Three-dimensional optical bit-memory recording and reading with photorefractive crystal: Analysis and experiment," *Appl. Opt.,* **35**, 2457–2465, 1996.
7. S. Kawata, T. Tanaka, Y. Hashimoto, and Y. Kawata, "Three-dimensional confocal optical memory using photorefractive material," *SPIE,* **2042**, 314–325, 1993.

8. Y. Kawata, H. Ueki, Y. Hashimoto, and S. Kawata, "Three-dimensional optical memory with a photorefractive crystal," *Appl. Opt.,* **34**, 4105–4110, 1995.
9. A. Toriumi, S. Kawata, and M. Gu, "Reflection confocal microscope readout system in three-dimensional photochromic optical data storage", *Opt. Lett.,* **23**, 1924–1926, 1998.
10. H. Ishitobi and S. Kawata, "Two-photon absorption for three-dimensional optical memory with a photorefractive crystal," *Technical Digest of International Conference on Optical MEMS and Their Applications,* Nara, Japan, November 18–21, 1997 pp. 33–35.
11. Y. Kawata, H. Ishitobi, and S. Kawata, "Use of two-photon absorption in a photorefractive crystal for three-dimensional optical memory," *Opt. Lett.,* **23**, 756–758, 1998.
12. H. Strickler and W. W. Webb, "Three-dimensional optical data storage in refractive media by two-photon point excitation," *Opt. Lett.,* **16**, 1780–1782, 1991.
13. E. Betzig, J. K. Trautman, R. Wolfe, E. M. Gyorgy, P. L. Finn, M. H. Kryder, and C. H. Chang, "Near-field magneto-optics and high density storage," *Appl. Phys. Lett.,* **61**, 142–144, 1992.
14. T. Tanaka and S. Kawata, "Comparison of recording densities in three-dimensional optical storage: Multilayered bit recording versus angularly multiplexed holographic recording," *J. Opt. Soc. Am. A,* **13**, 935–943, 1996.
15. D. Day and M. Gu, "Effects of refractive-index mismatch on three-dimensional optical data storage density in a two-photon bleaching polymer", *Appl. Opt.,* **37**, 6299–6304, 1998.
16. D. A. Parthenopoulos and P. M. Rentzepis, "Three-dimensional optical storage memory," *Science,* **245**, 843–845, 1989.
17. D. A. Parthenopoulos and P. M. Rentzepis, "Two-photon volume information storage in doped polymer systems," *J. Appl. Phys.,* **68**, 5814–5818, 1990.
18. P. Cheng, J. Bhawalkar, S. Pan, J. Wiatakiewicz, J. Samarabandu, W. Liou, G. He, G. Ruland, N. Kumar, and P. Prasad, "Two-photon generated three-dimensional photon bleached patterns in polymer matrix," *Scanning,* **18**, 129–131, 1996.
19. S. Pan, A. Shih, W. Lou, M. Park, J. Bhawalkar, J. Swiatkiewicz, J. Samarabandu, P. Prasad, and P. Cheng, Three-dimensional image recording by two-photon bleaching method, *Scanning,* **19**, 156–158, 1997.
20. E. Glezer, M. Milosavljevic, L. Huang, R. Finlay, T. Her, J. Callan, and E. Mazur, "Three-dimensional optical storage inside transparent materials," *Opt. Lett.,* **21**, 2023–2025, 1996.
21. Y. Kawata, R. Juškaitis, T. Tanaka, and S. Kawata, "Differential phase-contrast microscope with a split detector for the readout system of a multiplayered optical memory," *Appl. Opt,* **35**, 2466–2470, 1996.

22. M. Minsky, "Memoir on inventing the confocal scanning microscope," *Scanning,* **10**, 128–138, 1988.
23. C. J. R. Sheppard and A. Choudhury, "Image formation in the scanning microscope," *Optica Acta,* **24**, 1051–1073, 1977.
24. C. J. R. Sheppard and T. Wilson, "Depth of field in the scanning microscope," *Opt. Lett,* **3**, 115–117, 1978.
25. C. J. R. Sheppard and M. Gu, "Imaging by a high aperture optical system," *J. Modern Opt.,* **40**, 1631–1651, 1993.
26. M. Born and E. Wolf, *Principles of Optics,* Pergamon, Oxford, 1980.
27. J. W. Goodman, *Introduction to Fourier Optics,* 2nd ed., McGraw-Hill, New York, 1996.
28. C. J. R. Sheppard, M. Gu, Y. Kawata, and S. Kawata, "Three-dimensional transfer functions for a high aperture systems," *J. Opt. Soc. Am. A,* **11,** 593–598, 1994.
29. I. S. Gradstein and I. Ryshik, *Tables of Series, Products, and Integrals,* Herri Deutsch, Frankfurt, 1981.
30. T. Wilson, Y. Kawata, and S. Kawata, "Readout of three-dimensional optical memories," *Opt. Lett.,* **13**, 1003–1006, 1997.
31. W. Denk, J. Strickler, and W. W. Webb, "Two-photon laser scanning fluorescence microscopy," *Science,* **248**, 73–76, 1990.
32. B. Richard and E. Wolf, "Electromagnetic difraction in optical systems II: Structure of the image field in an aplanatic system", *Proc. Roy. Soc. A,* **253**, 358–379, 1959.
33. P. Török, P. Verga, Z. Laczik, and G. R. Booker, "Electromagnetic diffraction of light focused through a planar interface between materials of mismatched refractive indices: An integration representation," *J. Opt. Soc. Am. A,* **12**, 325–332, 1996.
34. S. Hell and E. Stelzer, "Properties of a 4Pi confocal fluorescence microscope," *J. Opt. Soc. Am. A,* **9**, 2159–2166, 1994.
35. M. Ishikawa, Y. Kawata, C. Egami, O. Sugihara, and N. Okamota, M. Tsuchimori and O. Watanabe, "Reflection-type confocal readout for multilayered optical memory", *Opt. Lett.,* **23**, 1781–1783, 1998.

Chapter 11

Three-Dimensional Photorefractive Memory Based on Phase-Code and Rotational Multiplexing

Xiangyang Yang

CompuSensor Technology Corporation, Gaithersburg, MD

11.1 Introduction

Mass memory systems serve computer needs in providing both archival and backup data storage as well as emerging applications involving network and multimedia services. There exist numerous applications in both commercial and military sectors that require data storage systems with huge capacity, high data rate, and fast (preferably parallel) access. To address such a need, three-dimensional (3-D) optical memories have been proposed [1, 2]. Since the data are stored in volume, 3-D optical memories have a much higher theoretical storage density than the present two-dimensional (2-D) memory devices. In addition, 3-D optical memories

PHOTOREFRACTIVE OPTICS
Materials, Properties and Applications

ISBN 0-12-774810-5/$30.00

have the potential for parallel access. The data are arranged in 2-D pages. An entire data page can be written or read in a single access operation.

In a holographic memory, the interference structure of a 2-D data page and a coherent reference beam is distributed in the whole holographic media. Due to the 3-D nature of the holographic medium, many pages of data may be multiplexed by changing the storage reference angle [3, 4], wavelength [5, 6], or phase-code [7, 8]. Among these multiplexing schemes, angle multiplexing is the most popular technique so far in the implementation of 3-D holographic memories. To change the angle of the reference beam, a mirror mounted on a rotating step motor can be utilized [3]. However, the slow mechanical movement involved substantially limits the access time in such systems. Acoustooptical deflectors (AODs) change the incident angle extremely fast, nevertheless the Doppler-shift in frequency decreases the coherence between the reference and object beams and must be compensated by some means, which increases the system complexity. For wavelength multiplexing, both the object and reference beams are fixed and only their wavelength is changed. The major disadvantage of wavelength multiplexing is the slow tuning of wavelength. For example, in a tunable laser diode with external cavity, changing the wavelength from one end of the wavelength range to the other (about 40 nm) takes more than 1 s.

To overcome these problems associated with angle and wavelength multiplexing, phase-code multiplexing has been proposed for the development of 3-D optical storage system. In the phase-code multiplexing scheme, the reference beam consists of multiple, e.g., N, plane wavefronts. The relative phases between all these N wavefronts are adjustable and represent the addresses of the stored data pages. Each data page is retrieved by illuminating the holographic medium by all the N plane waves with the exact same phase code as for the recording of the data page. It has been proven that, theoretically, phase-code multiplexing has the same storage capacity as angle multiplexing [7, 9]. Its merits include simplicity, compactness, fast access, high light efficiency, and moderate optical alignment requirements.

The storage density of phase-code-multiplexed holographic memory is primarily limited by the limited number of pixels of currently available spatial light modulators (SLMs). This limitation can be significantly alleviated by employing generalized Hadamard phase codes [10–12] and by combining rotation multiplexing method [13]. Using the devices currently available in our laboratory with relatively low performance level, hundreds of data pages have been successfully recorded in a small volume of photorefractive crystal [13, 14]. The stored images were later retrieved with high fidelity. With state-of-the-art and yet off-the-shelf optoelectronic devices (SLMs, CCD array detectors, optical array generators, etc.), thousands of data pages can be stored in the same volume of photorefractive crystal with the phase-code and rotation multiplexing method. This is close to the theoretical limitation of 3-D storage density imposed by the finite dynamic range of photorefractive crystal, the square law of diffraction efficiency, and the sensitivity of the array detector.

In the following sections, we first briefly review phase-code multiplexing and then discuss the construction of generalized Hadamard phase. Next, rotation multiplexing is introduced as a means to increase the storage density. Finally, a demonstration system made of all off-the-shelf devices is proposed that can be used to demonstrate the technical feasibility and evaluate the performance potential.

11.2 Phase-Code Multiplexing

In the phase-code multiplexing scheme, each data page is stored with the reference beam consisting of multiple plane wavefronts. The relative phases between all these wavefronts are adjustable and represent the addresses of the stored data pages. Assume M data pages are stored with N reference waves. Within the holographic storage medium (e.g., a photorefractive crystal or a photopolymer), the mth data page can be expressed as

$$D_m = D_m(x, y)\exp(j\vec{k} \times \vec{r}) \tag{11.1}$$

where we have omitted the constant terms for the sake of clarity and $D_m(x, y)$ is the distribution of the data-carrying wave at the entrance of the storage medium. The corresponding reference waves can be written as:

$$P = \sum_{n=1}^{N} P_n \exp(j\vec{k}_n \times \vec{r})\exp(j\phi_n^m) \tag{11.2}$$

The intensity inside the holographic medium during the recording of the mth data page is

$$\begin{aligned} H_m &= D_m(x,y)D_m^*(x,y) + \sum_{n=1}^{N} P_n P_n^* \\ &+ \sum_{n=1}^{N} \mathrm{D}_m(x,y)P_n^* \exp[j(\vec{k} - \vec{k}_n) \times \vec{r}] \exp(-j\phi_n^m) \\ &+ \sum_{n=1}^{N} P_n \sum_{l=1, l\neq n}^{N} P_l^* \exp[j(\vec{k}_n - \vec{k}_l) \times \vec{r}] \exp[j(\phi_n^m - \phi_l^m)], \end{aligned} \tag{11.3}$$

where the third term contributes to holographic storage of the data. After all M data pages are stored, each one of them can be reconstructed with the same reference waves by which it was recorded. When the hologram is illuminated by the reference waves corresponding to the pth data page, the reconstructed optical field is

$$\begin{aligned} R_p(x,y) = &\left\{\sum_{l=1}^{N} P_l \exp(j\vec{k}_l \times \vec{r}) \exp(j\phi_l^m)\right\} \cdot \\ &\left\{\sum_{m=1}^{M}\sum_{n=1}^{N} D_m(x,y) P_n^* \exp[j(\vec{k} - \vec{k}_n) \times \vec{r}] \exp(-j\phi_n^m)\right\} \end{aligned} \tag{11.4}$$

Due to the Bragg selectivity [3], the above equation can be simplified as

$$R_p(x,y) = \sum_{m=1}^{M}\sum_{n=1}^{N} D_m(x,y) P_n P_n^* \exp(j\vec{k} \times \vec{r}) \exp[j(\phi_n^p - \phi_n^m)] \tag{11.5}$$

To avoid the cross talk, the following conditions must be fulfilled:

$$\sum_{n=1}^{N} \exp[j(\phi_n^p - \phi_n^m)] = 0 \quad \text{for} \quad p \neq m, \tag{11.7}$$

$$\sum_{n=1}^{N} \exp[j(\phi_n^p - \phi_n^m)] = N \quad \text{for} \quad p = m. \tag{11.8}$$

Equations (11.7) and (11.8) mean that orthogonal phase codes have to be used to store and to retrieve data pages without cross talk.

In practice, binary phase coding devices are available and possess the performance level meeting the requirements of 3-D data storage system [7]. In this case, the phase code can be represented by a vector

$$V_m = [\exp(j\phi_1^m), \exp(j\phi_2^m), \ldots, \exp(j\phi_N^m)] = (\pm 1, \pm 1, \ldots, \pm 1) \qquad (11.9)$$

It has been proven that the maximum number of images that can be stored and reconstructed without cross talk is equal to N, the number of the wavefronts contained in the reference beam. The number of reference fronts is generally determined by the number of pixels of available spatial phase modulators. For example, if a ferroelectric liquid crystal spatial light modulator (FLC-SLM) with 128×128 pixels is used, more than ten thousand 2-D data pages can be multiplexed in the holographic medium. In practice, however, the number of holographically recorded data pages that can be retrieved without significant cross talk is limited by the phase representation accuracy of the SLM [15, 16].

11.3 Construction of Hadamard phase-codes for holographic memories

In phase-code multiplexing, each reference beam consists of a set of plane waves with a unique phase distribution across its component waves. To store N holograms, N orthogonal phase-codes must be used for data recording and retrieving. Due to its simplicity and other favorable properties, Hadamard matrix (H-matrix) has been solely used for the generation of orthogonal phase-codes [7, 8, 17]. Since the elements in an H-matrix has only two possible values ("+1" and "−1"), the Hadamard phase-codes can be readily implemented with SLMs. However, the well-documented generation method is only valid for H-matrices whose orders are a power of 2, i.e., $N = 2^n$, with n an integer [18]. The lack of general construction methods for H-matrices of other orders results in the inefficient utilization of currently available SLMs. For example, only 64 (2^6) pixels of a 100-pixel-SLM were used in a recent experimental demonstration [17]. The storage capacity could be increased by 56% (100/64 = 1.5625) if all the 100 pixels would be used which requires 100 orthogonal phase-codes. To accommodate to the available number of pixels of an SLM, H-matrices of order $N \neq 2^n$ often need to be generated.

In 1933, Paley described a variety of methods for constructing H-matrices of order $N = 4t$, with t an integer [19]. He listed examples of H-matrices for all $N = 4t \leq 200$ except six cases. The six exceptional values were 92, 116, 156, 172, 184, and 188. It has been proved that if the H-matrices of orders of M and N exist, then the H-matrix of order $M \times N$ exists and can be generated by tensor product [20]. The tensor product, denoted by $H_{m\times n} = D_m * G_n$, is defined as substituting each element in the second matrix by its multiplication with the first matrix. It should be noted that none of the three numbers (M, N, and $M \times N$) has to be the power of 2. As an example, the tensor product of two matrices of order 2 and 3 can be shown as

$$\begin{bmatrix} 3 & 4 \\ -5 & 6 \end{bmatrix} * \begin{bmatrix} 2 & -2 & 3 \\ 0 & 3 & 0 \\ 1 & 2 & 0 \end{bmatrix} = \begin{bmatrix} 6 & 8 & -6 & -8 & 9 & 12 \\ -10 & 12 & 10 & -12 & -15 & 18 \\ 0 & 0 & 9 & 12 & 0 & 0 \\ 0 & 0 & -15 & 18 & 0 & 0 \\ 3 & 4 & 6 & 8 & 0 & 0 \\ -5 & 6 & -10 & 12 & 0 & 0 \end{bmatrix}.$$

Using the tensor product extension method, a great number of H-matrices can be generated based on Paley's list. However, the six missing matrices in Paley's list result in the unavailability of many higher-order H-matrices. As the list extends beyond $N = 200$, more missing H-matrices are encountered. Thus an algorithm is required to find these missing members of the H-matrix set. It is noted that five of the six missing numbers (except 184) can be written as

$$M = 4p\,, \tag{11.10}$$

where p is an odd integer. Since the H-matrix of order p does not exist, M represents the order of a *basic* H-matrix and cannot be generated by tensor product of two smaller matrices.

Since Paley's work, the existence and generation of these H-matrices have attracted attentions of many scientists. In 1944, Williamson succeeded in constructing an H-matrix of order 172 [20]. With the help of a computer, Golomb *et al.* found an H-matrix of order 92 in 1963 [21]. Taking the tensor product of this matrix and the H-matrix of order 2, the H-matrix of order 184 can be generated. However, the author is unaware of any reported work on the other three missing H-matrices at the time of his work.

The rigorous proof of the existence of all H-matrices whose orders satisfy Eq. (11.10) may be difficult to achieve. However, in practice, the algorithm proposed by Williamson [20] can be modified and used for the generation of these H-matrices. To generate an H-matrix of order $M = 4p$, we assume that the H-matrix can be constructed with four $p \times p$ matrices as given by

$$H = \begin{bmatrix} A & B & C & D \\ -B & A & -D & C \\ -C & D & A & -B \\ -D & -C & B & A \end{bmatrix}. \tag{11.11}$$

The matrices A, B, C, and D are cyclically symmetrical, i.e., each row is just a cyclic permutation of its preceding row. Mathematically they can be expressed as

$$\begin{cases} \alpha_{ij} = \pm 1 \\ \alpha_{ij} = \alpha_{(i+1)(j+1)} \end{cases} \quad (\alpha = a, b, c \text{ or } d), \tag{11.12}$$

where the coefficients i, j, $i + 1$ and $j + 1$ are taken as module of p. To simplify the computation, we further assume that the first row in each cyclic matrix is symmetrical except for the first element

$$\begin{cases} \alpha_0 = 1 \\ \alpha_i = \alpha_{p-i} \end{cases} \quad \left(\alpha = a, b, c, \text{ or } d;\ i = 1, 2, 3, \ldots, \frac{p-1}{2}\right). \tag{11.13}$$

Thus there are only $2(p - 1)$ independent variables to be determined, which are known as the $2(p - 1)$ eigenelements of an H-matrix. Williamson has proved that the matrix H given by Eq. (11.11) is an H-matrix if and only if

$$A^2 + B^2 + C^2 + D^2 = 4pI, \tag{11.14}$$

where I denotes the unity matrix.

It can be proved that Eq. (11.14) will be true if and only if the following two conditions are satisfied:

$$\sum_{i=0}^{p-1} (a_i a_{i+r} + b_i b_{i+r} + c_i c_{i+r} + d_i d_{i+r}) = 0, \quad \left(r = 1, 2, \ldots, \frac{p-1}{2}\right), \tag{11.15}$$

and

$$\left(\sum_{i=0}^{p-1} a_i\right)^2 + \left(\sum_{i=0}^{p-1} b_i\right)^2 + \left(\sum_{i=0}^{p-1} c_i\right)^2 + \left(\sum_{i=0}^{p-1} d_i\right)^2 = 4p, \tag{11.16}$$

Equation (15) means that rows in the combined matrix $[\mathrm{ABCD}]_{p\times 4p}$ are orthogonal, with their inner product equal to zero. For any given combination of $\{a_0, a_1, a_2, \ldots, d_{p-1}, d_p\}$, this condition is used to check if an H-matrix can be generated. Since all elements in these row vectors are either "+1" or "−1", the inner product can be performed extremely fast in a computer by bitwise operations. Equation (16) provides the information that can significantly reduce the number of possible combinations of $\{a_0, a_1, a_2, \ldots, d_{(p-1)/2}\}$. It is well known that any positive integer can be decomposed as the sum of squares of four positive integers [22]. By decomposing $4p$ with four integers q_a, q_b, q_c, and q_d, and substituting Eq. (11.13) into Eq. (11.16), we have

$$\left(1 + 2\sum_{i=1}^{(p-1)/2} a_i\right)^2 + \left(1 + \sum_{i=1}^{(p-1)/2} b_i\right)^2 + \left(1 + \sum_{i=1}^{(p-1)/2} c_i\right)^2 + \left(1 + \sum_{i=1}^{(p-1)/2} d_i\right)^2 = q_a^2 + q_b^2 + q_c^2 + q_d^2 \,. \tag{11.17}$$

Let each term on the left side be equal to a corresponding term on the right, then

$$\alpha_1 + \alpha_2 + \cdots + \alpha_{(p-1)/2} = \frac{-1 \pm q_\alpha}{2}, \ (a = a, b, c, \text{ or } d)\,. \tag{11.18}$$

The number of positive and negative elements in the sequence $\{\alpha_0, \alpha_1, \alpha_2, \ldots, \alpha_{(p-1)/2}\}$ is determined by Eq. (11.18). For example, let $p = 29$, which corresponds to one of the missing H-matrix orders in Paley's list. Taking $4p = 116 = 9^2 + 5^2 + 3^2 + 3^2$, we have

$$a_1 + a_2 + \cdots + a_{14} = \frac{-1 + 9}{2} = 4\,. \tag{11.19}$$

Since the left side must be an even integer, the plus sign is taken in front of q_a, i.e., $q_a = 9$. Therefore among the 14 elements of $\{a_1, a_2, \ldots, a_{14}\}$, there must be 9 positive and 5 negative elements. The total number of possible combinations of $\{a_1, a_2, \ldots, a_{14}\}$, is then $\binom{14}{9} = 2002$, which represents a small vector set and can be searched by a computer with a reasonably small amount of computation. The searching spaces for $\{b_i\}$ and $\{c_i\}$ can be determined similarly. It has been proved that, if $a_0 = b_0 = c_0 = d_0$, exactly three of a_j, b_j, c_j, and d_j ($j \neq 0$) must have the same sign [12]. Thus after $\{a_i\}$, $\{b_i\}$ and $\{c_i\}$ are chosen, $\{d_i\}$ is automatically determined. These searching spaces can be further shrunk by taking into consideration of other obvious constraints.

A flowchart for computer search of H-matrix of order $M = 4p$ is illustrated in Fig. 11.1. Using this method, one of the H-matrix missing

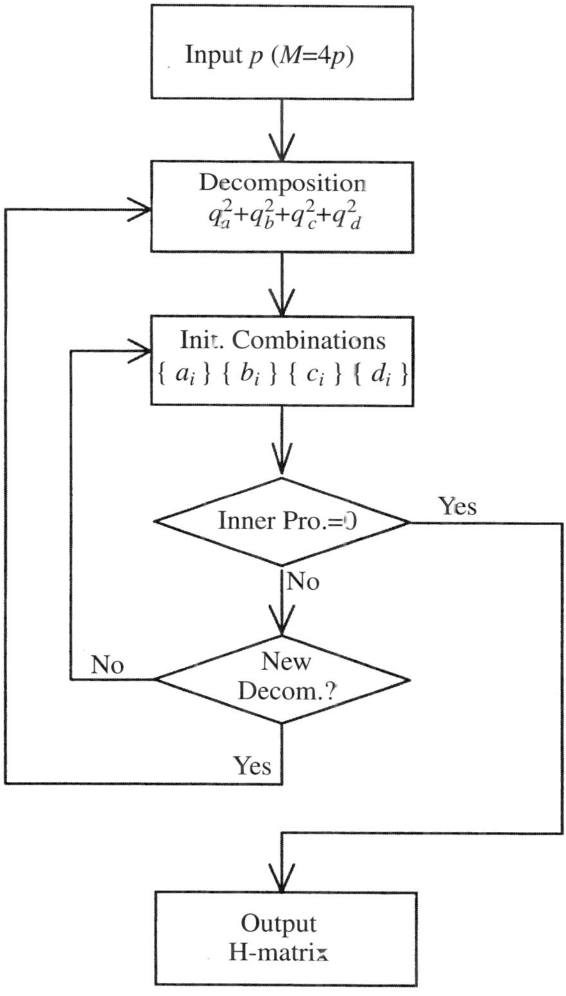

Figure 11.1: Flowchart for the computer search of H-matrices of order $M = 4p$.

in Paley's list ($M = 116$) was found within an hour with a Pentium PC. The sign for the eigenelements of an H-matrix of order 116 are listed in Table 11.1. Figure 11.2 illustrates the four matrices that construct the H-matrix, with black pixel denoting "+1" and white pixel denoting "−1". If more powerful computers are used, other missing H-matrices ($M = 156$

	a_1	a_2	a_3	a_4	a_5	a_6	a_7	a_8	a_9	a_{10}	a_{11}	a_{12}	a_{13}	a_{14}
A	−	−	−	+	+	+	−	+	+	+	+	+	+	−
B	+	+	+	+	−	−	+	+	−	−	−	+	−	+
C	−	+	−	−	−	−	+	+	+	+	−	−	+	−
D	−	+	−	+	−	−	+	−	+	+	−	+	+	−

Table 11.1: Signs of the Eigenelements of an *H*-Matrix of Order 116

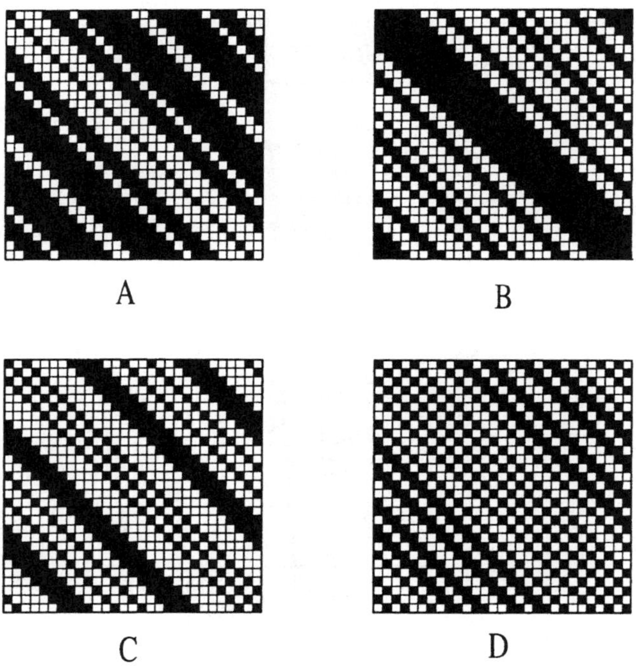

Figure 11.2: Four matrices *A B*, *C* and *D* that generate an H-matrix of order 116.

and $M = 188$) could be generated similarly that would complete the list of H-matrix for all $N = 4t \leq 200$.

We have noticed that the H-matrix for a given order is not unique. This is in contradiction with the statement made by Golomb *et al.* [21]. As an example, Table 11.2 lists two possible H-matrices of order $M = 4 \times 23 = 92$. They cannot be converted into an identical one by rearrangement of A, B, C, D and of row elements and column elements within A, B, C, D. The nonuniqueness, however, significantly shortens the searching time in practice.

The algorithm presented in this section is easy to implement. We have developed other algorithms for the generation of generalized Hadamard codes. However, the discussion of all these algorithms is beyond the scope of this chapter. Interested readers can find the complete list of Hadamard matrices for all orders up to 408, i.e., all orders of $n = 4m \leq 408$, in [12].

11.4 Utilization of Hadamard phase-codes of $m \neq 2^n$ in holographic memories

To accommodate to the available SLMs and to achieve the maximum possible storage capacity of a phase-code-multiplexed holographic memory, general Hadamard phase-code of order $m \neq 2^n$ should be used. In the previous section, algorithms were described for the construction of Hadamard phase-codes of order $m = 4p$. Using these methods, Hadamard matrices for all $m = 4p \leq 400$ can be constructed and most higher-order matrices can also be generated. This provides a great flexibility in the

	a_1	a_2	a_3	a_4	a_5	a_3	a_7	a_8	a_9	a_{10}	a_{11}
A_1	+	−	−	−	+	−	−	−	+	−	+
B_1	−	+	+	−	+	+	−	−	+	+	+
C_1	+	+	−	−	−	+	+	−	+	−	+
D_1	+	+	−	+	+	+	−	+	−	−	−
A_2	−	−	−	+	+	+	+	+	+	+	−
B_2	−	−	+	+	−	−	−	+	−	+	−
C_2	−	+	+	+	+	+	−	−	−	+	−
D_2	+	−	+	−	+	+	−	+	−	−	+

Table 11.2: Signs of the Eigenelements of Two H-Matrices of Order 92

system design of a holographic memory to achieve the maximum storage capacity with currently available SLMs. In a proof-of-concept experiment, phase-codes were implemented with a 10×10 pixel ferroelectric liquid crystal SLM (FLC-SLM) from DisplayTech, which is set at the bipolar pure phase modulation mode. Ideally a Hadamard matrix of $m = 100$ should be used to make the full use of the limited number of pixels.

The setup for the preliminary experimental studies is schematically illustrated in Fig. 11.3. A liquid crystal television (LCTV) panel (Kopin KVGA-AMLCD, 640×480 pixels) was converted into a SLM using the standard laboratory process [22], which displays the data pages to be stored. The laser beam from an Ar ion laser is divided by a beam splitter. The object beam is collimated and projected onto the LCTV panel. The Fourier spectrum of an input data page is generated at the entrance of the iron-doped $LiNbO_3$ crystal ($10 \times 10 \times 5\ mm^3$). The collimated reference illuminates a computer-generated hologram (CGH), which functions as a uniform array generator that converts the single reference beam into multiple beams with equal amplitude. These plane waves are focused by the lens L6 and the focused spots match the pixels of the FLC-SLM. After passing through the FLC-SLM, the phase modulated reference beams are projected onto the entrance of the photorefractive crystal by the lens L7. The angle between the object and reference beams is about 35°. A CCD

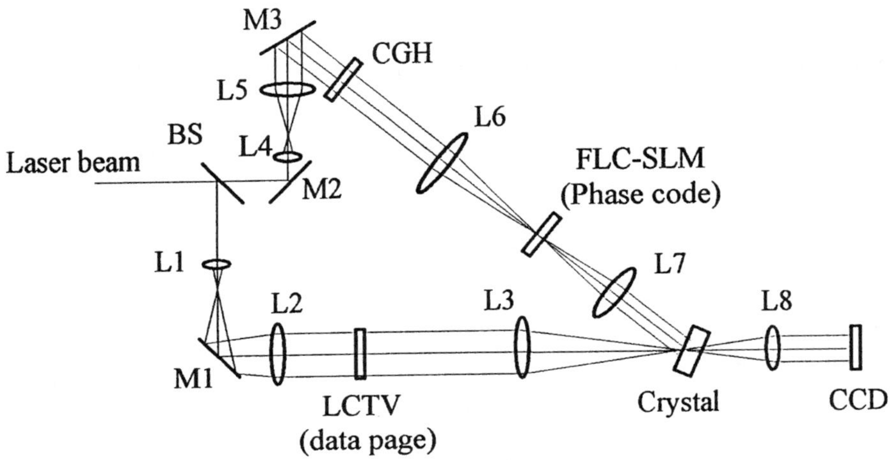

Figure 11.3: Preliminary experimental setup.

camera detects the reconstructed data pages and sends them to a computer via a frame grabber.

A 10×10 CGH array generator was not available when we conducted the preliminary experiment. A 2-D Dammann grating was instead used, which generates 9×9 reference beams. However, due to fabrication errors, the intensity of the zero-order (i.e., the central) beam is much higher than that of the other orders. To avoid the cross talk caused by the nonuniformity among reference beams [7], a rectangle aperture is placed in front of the FLC-SLM to block five rows of the focused spots generated by the Dammann grating. In other words, only four rows (36 beams) are used. The variation in the intensity among these 36 diffraction orders is less than 3%. The center-to-center distance between adjacent pixels of the FLC-SLM is 1 mm. To prevent degeneracy [24], the FLC-SLM as well as the Dammann grating are rotated around the optical axis by 7.2°, as illustrated in Fig. 11.4. In this way, the angular spacings between each pair of FLC-SLM pixels, projected to the plane formed by the object and reference axes, are about equal and are much larger than the Bragg

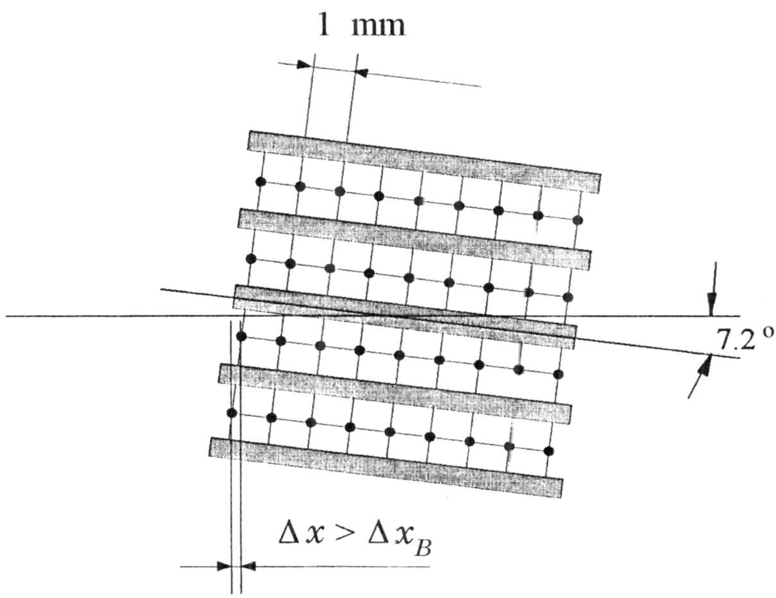

Figure 11.4: The Dammann grating and FLC-SLM are rotated by 7.2° to avoid degeneracy.

selectivity $\Delta\theta_B$. The angle spanned by the 36 reference beams is about 2.5°.

The signs of eigenelements of the Hadamard matrix ($m = 36$) used for the generation of 36 orthogonal Hadamard phase-codes are listed in Table 11.3. With these phase-codes, a total of 36 images were stored in the crystal with the scheduled exposure method [25]. The object beam intensity is about one-tenth of the total reference beam intensity. The diffraction efficiency of all the retrieved images is about equal. Two of the retrieved images are shown in Fig. 11.5. No cross talk was observed. This work was the first demonstration of the utilization of Hadamard phase-codes of $m \neq 2^n$ in holographic memories. In addition to the full use of the limited space-bandwidth-product of SLMs, the utilization of Hadamard phase-codes of $m \neq 2^n$ generally does not have a single "bad code" as discussed in [26]. As a consequence of the elimination of the single bad code, phase-code multiplexing may, at least theoretically, suffer less cross-talk than angle multiplexing scheme.

11.5 Increase storage density by rotation multiplexing

In the phase-code multiplexing scheme, the phase encoding can be implemented by dividing an incoming wavefront into N subwaves with an

	a_1	a_2	a_3	a_4
A	−	−	−	+
B	−	−	+	−
C	−	+	−	−

Table 11.3: Sign of the Eigenelements of the H-Matrices of Order 36

Figure 11.5: Two of the 36 retrieved data pages.

optical array generator and then modulating the phase of each subwave with a phase modulation SLM. To avoid the cross talk between different data pages, the N waves in a reference beam should have equal intensity. Lenslet arrays and CGHs (e.g., Dammann gratings), can be used as array generators. The number of beams that can be generated by these devices is limited when the stringent uniformity requirement is taken into consideration. The number of pixels of currently available high-quality pure phase modulation SLMs is also limited. As an example, a 10×10 pixel phase modulator (DisplayTech, Model FLC-10×10B) has been used in most recent demonstrations [10,17]. Even when the generalized Hadamard codes of order $m \neq 2^n$ are used, the maximum number of holograms that can be stored is only 100, which is much less than what has been demonstrated by the angular multiplexing approach [4].

Rotation multiplexing can be used to overcome the limitation imposed by the currently available SLMs and array generators. The geometry for phase-code and rotation multiplexing is similar to that shown in Fig. 11.6, in which only the principal rays of reference and object beams are shown. The N plane waves in the reference beam span an angle of $\Delta\theta$. After N holograms are recorded, the holographic medium is rotated within the incident plane by an angle of at least $\Delta\theta + \Delta\theta_B$, with $\Delta\theta_B$ being the angle corresponding to the Bragg selectivity. The rotation of the holographic medium causes the stored holograms to become non-Bragg-matched and

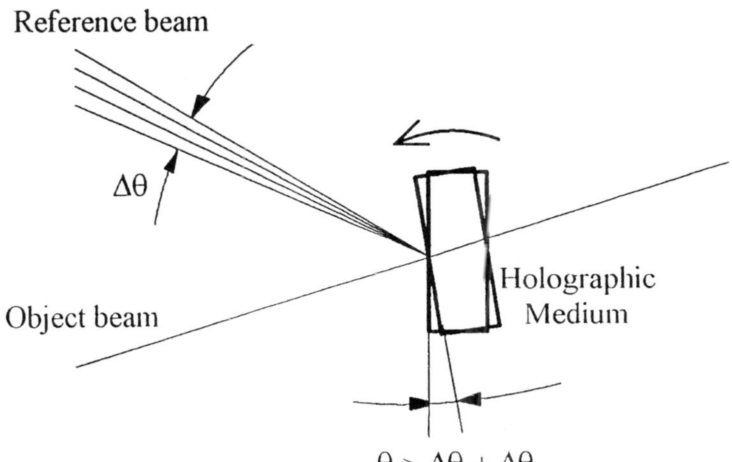

Figure 11.6: Geometry for rotation and phase-code multiplexing.

thus allows another set of N holograms to be recorded. The wavevector space description of holographic interaction is illustrated in Fig. 11.7. A set of gratings is written by N reference beams spanning $\Delta\theta$. As shown in Fig. 11.7b, these grating vectors are moved when the holographic medium is rotated. If the rotation angle is larger than $\Delta\theta + \Delta\theta_B$, these gratings are no longer phase-matched with any of the plane waves in the reference beam. Consequently, another N holograms can be stored without cross talk with the previously stored ones. The address of a data page is then represented by the rotation angle as well as the phase-code. All data pages can be retrieved with correct addresses without cross talk. Since $\Delta\theta + \Delta\theta_B \approx \Delta\theta$ is usually small (typically between 1° and 3°), at least 10 angular positions can be used for rotation multiplexing. This will boost the storage density by an order of magnitude.

The principle of rotation multiplexing is similar to that of angular multiplexing in terms of writing gratings within the holographic medium with wavevectors in different directions. However, the angle between the reference and object beams does not change in rotation multiplexing. In other words, the registration geometry is fixed that makes it possible to be combined with the phase-code-multiplexing scheme. Unlike peristrophic multiplexing [9], the rotation of holographic medium is within the incident plane and all the holograms are recorded in approximately the same volume of the medium. If the total rotation is less than 40° (i.e., ±20°), the

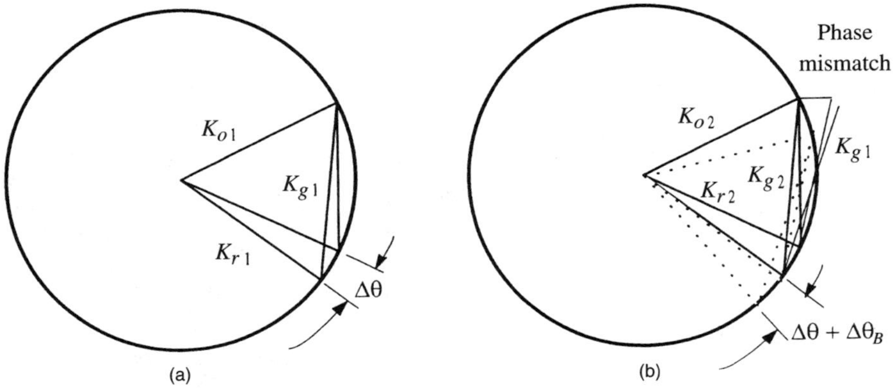

Figure 11.7: Wavevector space description of holographic interaction. (a) Wavevectors before rotation. (b) Wavevectors after rotation.

illumination area of the reference beam on the surface of the holographic medium has little change and space multiplexing [27] can still be combined to further increase the storage capacity.

Experiments have been conducted to demonstrate the feasibility of holographic storage using rotation and phase-code multiplexing. The experimental setup is similar to that schematically illustrated in Fig. 11.3, except that the photorefractive crystal is installed on a rotating table. The laser beam from an argon ion laser is divided by a beam splitter. The object beam is collimated and projected onto the page composer, which is an LCTV panel with 640 × 480 pixels. The Fourier spectrum of the input data page is generated at the entrance of the iron-doped $LiNbO_3$ crystal ($10 \times 10 \times 5\ mm^3$, z-cut) by the Fourier transform lens L3 ($f = 300$ mm). The collimated reference beam illuminates a 9 × 9 2-D Dammann grating, which converts the single reference beam into multiple beams with approximately equal amplitude. These plane waves are focused by the lens L6 and the focused spots match the pixels FLC-SLM which has 10 × 10 pixels with a center-to-center distance of 1 mm. The FLC-SLM operates at the pure phase modulation mode. After passing through the FLC-SLM, the phase-modulated reference beams are projected onto the entrance of the crystal by the lens L7 ($f = 200$ mm). The angle between the object and reference beams is about 35°. A high-sensitivity CCD camera detects the reconstructed data pages and sends them to a computer via a frame grabber.

Ideally a 10 × 10 Dammann grating should be used to match the number of pixels of the FLC-SLM. A 9 × 9 Dammann grating was instead used because of its availability. As described in the Section 11.4, the intensity of the zero-order (i.e., central) beam generated by the Dammann grating is much higher than that of other orders due to fabrication errors. To avoid the cross talk caused by the nonuniformity among reference waves, an aperture is placed in front of the FLC-SLM to block five rows of the focused spots generated by the Dammann grating, as shown in Fig. 11.4. In other words, only 36 beams (four rows) are used. The variation in the intensity among these 36 diffraction orders is less than 3%. To prevent the degeneracy, the FLC-SLM as well as the Dammann grating are rotated around the optical axis by 7.2°. In this way, the angular spacings between each pair of FLC-SLM pixels, projected to the plane formed by the object and reference axes, are about equal and are much larger than the Bragg selectivity $\Delta\theta_B$. The angle spanned by the 36 reference beams is about 2.5°.

A generalized Hadamard matrix of $m = 36$ was used for the generation of 36 orthogonal Hadamard phase-codes. With these phase-codes and three rotation positions (separated by 3°), three sets of images (36 pages of English letters, 36 pages of binary data and 36 pages of cartoon patterns) were stored in the $LiNbO_3$ photorefractive crystal by the incremental exposure method [28]. The object beam intensity is about one-fifth of the total reference beam intensity. The diffraction efficiency of all the retrieved images is about equal. Global contrast enhancement and thresholding operations have been performed to enhance the quality of the reconstructed images detected by a CCD camera. Two of the retrieved images from each of the rotation positions are shown in Fig. 11.8. This is the first experimental demonstration that more than 100 holograms are stored in a phase-code-multiplexed holographic memory. If a 10×10 array generator had been available, as many as 300 phase-coded holograms could have been recorded and retrieved.

11.6 Demonstration with off-the-shelf devices

To demonstrate the feasibility as well as performance potentials of phase-code-multiplexed photorefractive 3-D memory, a demonstrator has been designed with all off-the-shelf devices. The design objective was to achieve a storage capacity of 64 Gbits, readout data transfer rate of 15 Mbit/s, and a page access time of about 10 ms. Error correction code was planed to be used to improve the bit error rate. It combines phase-code, rotation, and space multiplexing schemes to achieve the design goal. The details of the system design are discussed in the following sections.

11.6.1 Demonstration system design

Although bulky optical and optoelectronic components are used, the optical system of the designed demonstration system can be packaged within a $1 \times 1 \times 0.67$ ft^3 ($30.5 \times 30.5 \times 20$ cm^3) housing. The layout of the optical components is shown in Fig. 11.9. Figure 11.10 illustrates the schematic diagram of the optical system, with most of the mirrors removed for clarity.

A diode-pumped double-frequency solid state $Nd{:}YOV_4$ laser is used as the light source. A pair of lenses (L1 and L2) forms a telecentric system

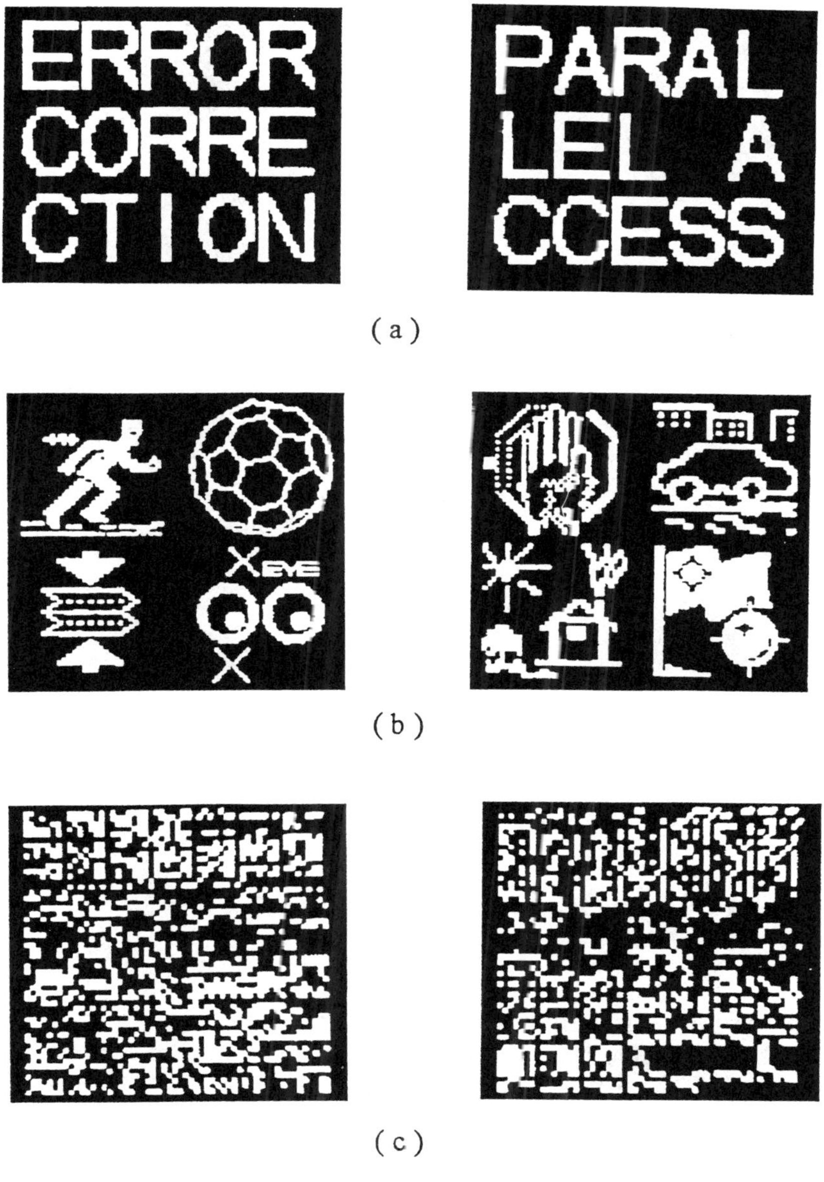

Figure 11.8: Experimental results: (a) two images retrieved from the 36 images stored in the first rotation position; (b) two images retrieved from the 36 images stored in the second rotation; (c) two images retrieved from the 36 images stored in the third rotation position.

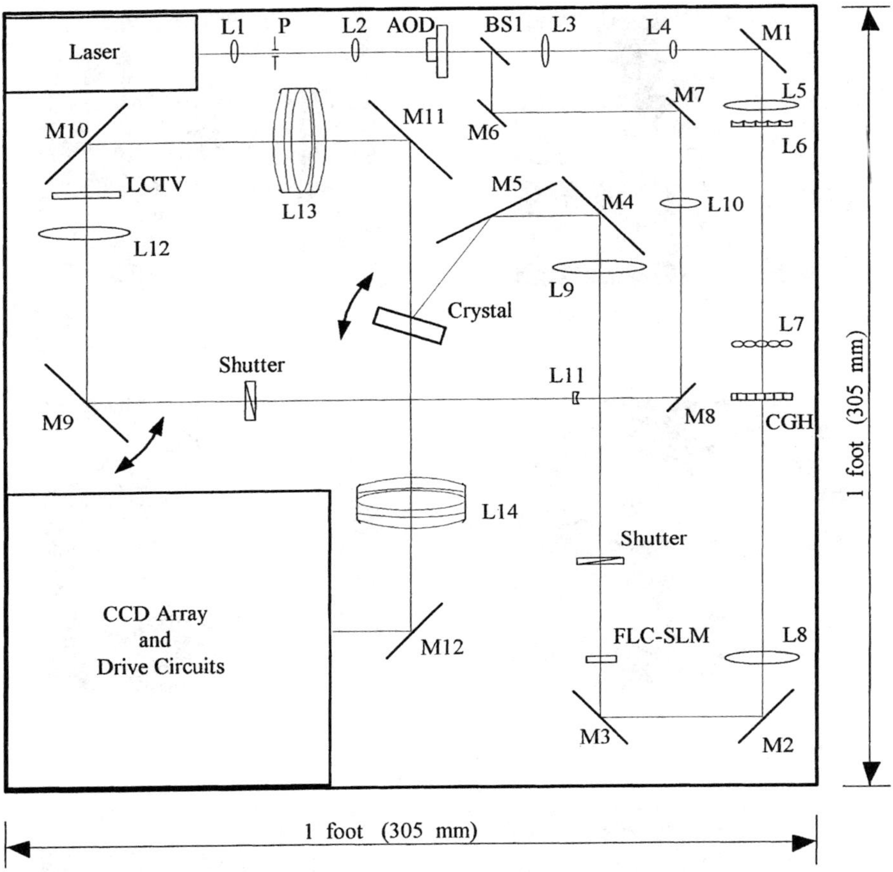

Figure 11.9: Optical layout of the 3-D photorefractive memory demonstrator.

and generates a 1-mm-diameter laser beam at the entrance pupil of the 2-D AOD. A pinhole placed at the common focal plane of the two lenses serves as a spatial filter to improve the uniformity of the laser beam. The 2-D AOD deflects the laser beam in both horizontal and vertical directions, implementing the space multiplexing for the reference beam. The light transmitted by the beam-splitter BS1 enters the reference arm. As shown in Fig. 11.10a, a pair of lenses (L3 and L4) magnifies the angular range of the AOD by a factor of 6. At the same time they image the exit pupil of the AOD at the front focal plane of the lens L5. A pair of lenslet arrays

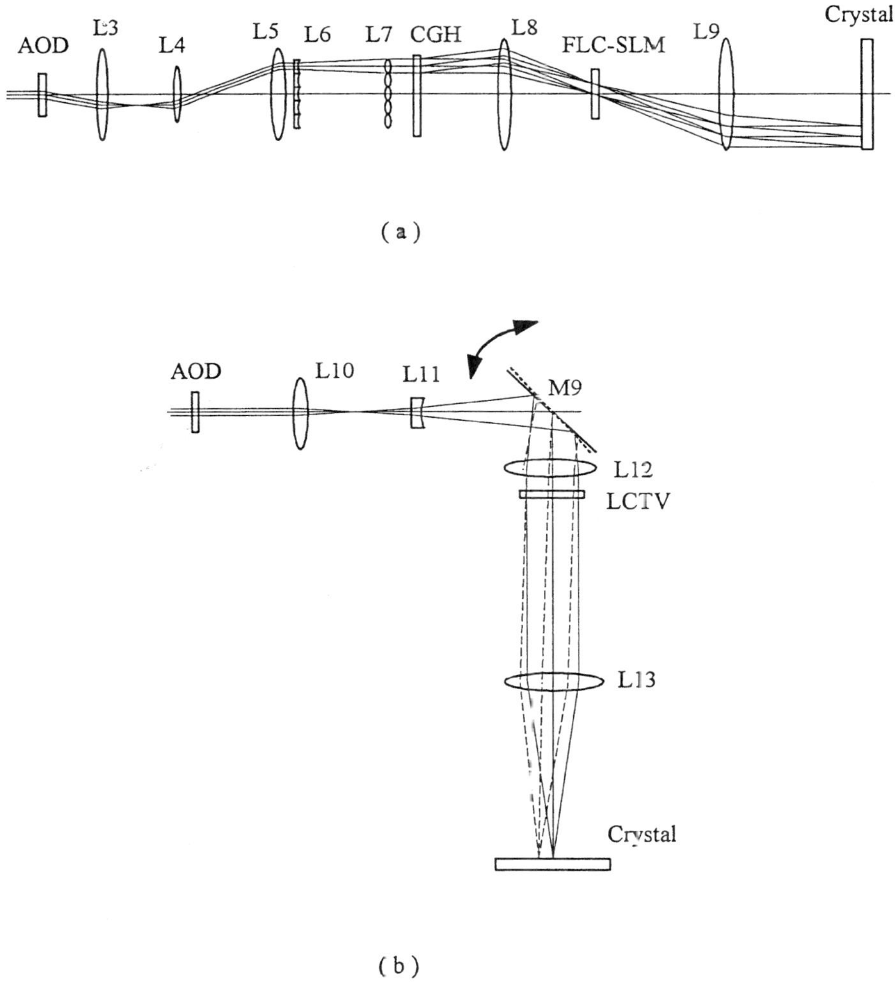

Figure 11.10: Optical system of the demonstrator: (a) reference arm, (b) object arm.

behind the lens L5 collimates the deflected reference beam and projects it onto a specific portion of the CGH corresponding to the space multiplexing scheme. The CGH functions as a uniform array generator that converts the single reference beam into multiple beams with equal amplitude. An example of such CGH is the Dammann grating [29], although other types of CGH may be used. These diffracted beams are focused by the lens L8 on to the surface of the FLC-SLM operating at the pure phase modulation mode. The focused spots match exactly the pixels of the FLC-SLM and are modulated according to the Hadamard phase-codes. After passing through the FLC-SLM, the phase-modulated reference beams are projected onto a specific location of the photorefractive crystal. By changing the driving frequency of the AOD, the reference beams can be delivered into any of the 5×5 locations of the photorefractive crystal for space multiplexing.

Light from the AOD that is reflected by the beam splitter enters the object arm. As shown in Fig. 11.10b, a compact beam expander (L11 and L12) expands and collimates the laser light and illuminates the high-resolution LCTV panel which is used as page composer. A field lens L10 images the exit pupil of the AOD to the entrance pupil of the beam expander so that it will not walk off the field of view when the beam out of the AOD is steered. A custom-designed high-quality Fourier transform lens (L13) projects the Fourier spectrum of the input data page into the photorefractive crystal where it interferes with the reference beams and is recorded in the form of a volume hologram. The deflection range of the AOD is too small to implement space multiplexing of the object arm. Therefore a rotating mirror (M9) is inserted between the lenses L11 and L12 to steer the object beam to different locations in the crystal. Due to the relatively long time required to write a thousand holograms in each location, the time for the mechanical motion required by the mirror M9 does not cause a severe limitation on write times. After all the data pages are recorded, the object arm is not needed and the 2-D AOD provides rapid random access to any space multiplexed locations. To increase storage density, rotation multiplexing is employed. The cost for the increased density is the mechanical movement (rotation) of the crystal. Due to the light weight of the crystal (less than 20 grams) and small angle required, a low inertia galvanometric driver can be used and a random access time of less than 10 ms can be achieved [30].

Another customer-designed Fourier transform lens (L14) reimages the reconstructed data pages onto the CCD array. Our analysis showed that the combination of the customer-designed Fourier transform lens pair is capable of resolving the LCTV pixels smaller than $10 \times 10\ \mu m^2$. It has also been confirmed that if 4 (2×2) CCD pixels are used to detect one LCTV pixel, the distortion of such an imaging system does not cause registration errors.

The parameters of all optical components are listed in Table 11.4. Table 11.5 lists the specifications of other key components. The selection of these devices was primarily based on their availability. As the technology advances rapidly, devices with a higher performance level will become available soon. The use of new devices will certainly boost the overall performance of the system

11.6.2 Performance potential

The performance of the demonstration system is heavily dependent upon the performance of the key components such as the laser, the LCTV, the AOD, the FLC-SLM, and the CCD array. The three critical parameters are storage capacity, readout data transfer rate, and data page access time.

The storage capacity is determined by the number of space-multiplexed holographic locations, the number of holograms in each location, and the number of bits stored in each hologram. In the proposed demonstrator, the aperture of the reference beams is 4 mm in diameter and the center-to-center distance between adjacent locations is 4.2 mm. Holograms are spatially multiplexed in 5 × 5 locations in a Fe-doped $LiNbO_3$

Item	Description	Parameters	
L1	Beam shaping lens	f = 15 mm	D = 8 mm
L2	Beam shaping lens	f = 30 mm	D = 8 mm
L3	Deflection range amplifier	f = 42 mm	D = 8 mm
L4	Deflection range amplifier	f = 7 mm	D = 3 mm
L5	Field lens	f = 50 mm	D = 30 mm
L6	5 × 5 lenslet array	f = −3 mm	D = 2 mm
L7	5 × 5 lenslet array	f = 87 mm	D = 3.2 mm
L8	Telescope objective	f = 100 mm	D = 35 mm
L9	Telescope objective	f = 130 mm	D = 50 mm
L10	Field lens	f = 75 mm	D = 15 mm
L11	Beam expander lens	f = −3 mm	D = 2 mm
L12	Collimating lens	f = 200 mm	D = 60 mm
M9	Rotating mirror	W = 50 mm	H = 35 mm
L13	Customer-designed FT lens	f = 100 mm	D = 65 mm
L14	Customer-designed FT lens	f = 75 mm	D = 50 mm

Table 11.4: Parameters of Optical Elements

Item	Specifications
Laser	$Nd:YVO_4$ diode-pumped solid state laser Diameter: 30 mm; Length: 72 mm Power: 150–250 mW
or	Double-frequency Nd:YAG Diameter: 40 mm; Length: 100 mm Power: 200 mW
Page composer	Super VGA LCTV No. of pixels: 1280 × 1024 Pixel size: 24 × 24 μm^2 Contrast ratio: 200:1 Speed: 30 frames/s
Phase modulator	FLC-SLM No. of pixels: 16 × 16 Pixel size: 1 × 1 mm^2 Speed: 1000 frames/s
Crystal rotation driver	Osc. frequency: > 1000 Hz Rotation Range: 40° Wobble: $< 5 \times 10^{-6}$ rad
Acousto-optical deflector	2-D AO deflector Deflection range: 2.5°–3.5° Drive frequency: KHz–MHz
CCD array	No. of pixels: 2048 × 2048 Pixel size: 14 × 14 μm^2 Speed: 15 frames/s
or	No. of pixels: 2048 × 2048 Pixel size: 9 × 9 μm^2 Speed: 2 frame/s

Table 11.5: Specifications of Key Components

photorefractive crystal measuring $25 \times 25 \times 6$ mm^3. The number of holograms that can be recorded at each location without cross talk is determined by the number of pixels of the phase modulator and the number of angular positions for rotation multiplexing. A 16 × 16 FLC-SLM is proposed. Each set of phase-code-multiplexed holograms is recorded in one of the 10 angular positions (with a total span of ±19.6°). A super Video Graphics Adaptor (VGA) LCTV is used as page composer which has 1280 × 1024 pixels. However, since the CCD has only 2048 × 2048 pixels and since the oversampling scheme will be used to match each LCTV pixel with 2 × 2 CCD pixels, only 1024 × 1024 pixels of the LCTV can be utilized. Based on these design parameters, the demonstration system should have a nominal storage capacity of 67.1 Gbits ($5 \times 5 \times 10 \times 16 \times 16 \times 1024 \times 1024 = 6.71 \times 10^{10}$ bits).

In the readout process, the space multiplexing is implemented by a 2-D AOD. The switch time of an AOD is typically between 50 to 100 μs. The frame speed of the 16 × 16 pixel FLC-SLM is 1 KHz, which corresponds to a switch time of 1 ms. The rotation multiplexing is implemented by a low-inertia galvanometric driver that has a switch time of less than 10 ms. Therefore, the data page access time during the readout process will be about 10 msec. This is less than the time required by the CCD to transfer a whole frame of data and thus does not impose any limitations on the operation speed of the demonstrator. The data transfer rate is primarily determined by the frame speed of the CCD array. The CCD array chosen for the demonstrator has a frame speed of 15 frame/s, which corresponds to a readout data transfer rate of 15.7 Mbit/s ($1024 \times 1024 \times 15 = 1.57 \times 10^7$ bit/s).

The photorefractive data storage system described in this section is a fast, compact, rugged, low-cost demonstration system based on off-the-shelf devices. If developed, it would provide a platform for feasibility and performance study. Unfortunately, the development was not completed because of lack of funding.

11.7 Conclusions

Optical and electronic computers will soon require memories with capacities beyond those of magnetic or electronic systems. Three-dimensional storage and parallel access are needed for densities greater than 1 Tbit/cm^3, and input/output (I/O) bandwidth in excess of 1 GBit/s. Large 3-D optical memories promise to meet such requirements.

To overcome certain problems associated with angle and wavelength multiplexing, we have chosen phase-code multiplexing in the proposed 3-D optical storage system. Although it has been proven that, theoretically, phase-code multiplexing has the same storage capacity as angle multiplexing, the storage density of a phase-code-multiplexed holographic memory is primarily limited by the limited number of pixels of currently available SLMs. Our study has shown that this limitation can be significantly alleviated by employing generalized Hadamard phase codes and by combining rotation multiplexing. With state-of-the-art and yet off-the-shelf optoelectronic devices (SLMs, CCD array detectors, optical array generators, etc.), thousands of data pages can be stored in the same volume of photorefractive crystal with the phase-code and rotation multiplexing methods. This approaches the theoretical limitation of 3-D storage density.

To achieve terabits storage capacity, it becomes necessary to employ space multiplexing. In the proposed demonstrator, the space multiplexing is implemented by a 2-D AOD during the data readout process to ensure high data transfer rate and fast page access. The AODs modulate both the object and reference beams therefore, the Doppler-shift does not cause any problems and need not be compensated.

A fast, compact, rugged, low-cost, and high-density 3-D photorefractive memory, if successfully developed, will surely have a significant impact on today's data storage community and will have wide applications in the areas of supercomputer systems, information superhighway, virtual reality, video-on-demand services, artificial intelligence, fingerprint analysis, medical diagnostics, large-scale databases, etc.

Acknowledgments

This work was supported in part by Air Force Rome Laboratories, the University of New Orleans, and Quantex Corporation. Most of the experiments were conducted while the author was affiliated with the University of New Orleans. The author thanks Z. Wen, Y. Xu, N. Li, and Z. Gu for their helpful discussion and laboratory assistance.

References

1. A A Jamberdino and X. Yang (eds.), "Special issue on three dimensional optical memories: Materials, components, and system architectures," *Opt. Mem. Neu. Net.,* **3**t(2), 1994.
2. W. J. Miceli, J. A Neff and S. T. Kowel, (eds.), "Photonics for computers, neural networks, and memories," *Proc. SPIE,* **1773**, 1992.

3. F. H. Mok, M. C. Tackitt and H. M. Stoll, "Storage of 500 high resolution holograms in a LiNbO3 crystal," *Opt. Lett.,* **16**,605, 1991.
4. F. H. Mok, "Angle-multiplexed storage of 5000 holograms in lithium niobate," *Opt. Lett.,* **18**, 915, 1993.
5. G. A. Rakuljic, V. Leyva and A. Yariv, "Optical data storage by using orthogonal wavelength-multiplexed volume holograms," *Opt. Lett.,* **17**, 1471, 1992.
6. F. T. S. Yu, S. Wu, A. W. Mayers and S. Rajan, "Wavelength-multiplexed reflection matched spatial filters using $LiNbO_3$," *Opt. Commun.* **81**, 343, 1991.
7. C. Denz, G. Roosen and T. Tschudi, "Potentialities and limitations of hologram multiplexing by using the phase-encoding technique," *Appl. Opt.,* **31**, 5700, 1992.
8. X. Yang and Z.-H. Gu, "Three dimensional optical data storage and retrieval system based on phase-code and space multiplexing," *Opt. Eng.,* **35**, 452, 1996.
9. K. Curtis, A. Pu and D. Psaltis, "Method for holographic storage using peristrophic multiplexing," *Opt. Lett.,* **19**, 993, 1994.
10. X. Yang, Z. Wen, Y. Xu and N. Li, "Use of Hadamard codes of $m \neq 2^n$ in phase-code multiplexed holographic memories," *Opt. Eng.,* **35**, 3639, 1996.
11. X. Yang, Y. Xu and Z. Wen, "Generation of Hadamard matrices for phase-code multiplexed holographic memories," *Opt. Lett.,* **21**, 1067, 1996.
12. X. Yang, Z. Wen and Y. Xu, "Construction of Hadamard phase-codes for holographic memories," in F. T. S. Yu and S. Yin, (eds.), "Photorefractive fiber and crystal devices: Materials, optical properties, and applications II," *Proc. SPIE,* **2829**, 217–228, 1996.
13. X. Yang, N. Li, Z. Wen and Y. Xu, "Holographic storage using phase-code and rotation multiplexing," in F. T. S. Yu and S. Yin (eds.), "Photorefractive fiber and crystal devices: Materials, optical properties, and applications II," *Proc. SPIE,* **2829**, 125–132, 1996.
14. X. Yang and Z. Gu, "Compact 3-D optical data storage and retrieval system — Phase I final report," U.S. Air Force Rome Laboratory, 1996.
15. Z. Wen, Y. Tao and X. Yang, "Crosstalk in phase-coded holographic memories using different orthogonal codes," in F. T. S. Yu and S. Yin, (eds.), "Photorefractive fiber and crystal devices: Materials, optical properties, and applications III," *Proc. SPIE,* **3137**, 123–133, 1997.
16. Z. Wen and Y. Tao, "Orthogonal codes and cross-talk in phase-code multiplexed volume holographic data storage," *Opt. Commun.,* **148**, 11, 1998.
17. C. Alves, G. Pauliat and G. Roosen, "Dynamic phase-encoding storage of 64 images in a BaTiO3 photorefractive crystal," *Opt. Lett.,* **19**, 1894, 1994.
18. R. C. Gonzalez and R. E. Woods, *Digital Image Processing*, Addison-Wesley, Reading, Massachusetts, 1993, pp. 136–143.
19. R. E. A. C. Paley, "On orthogonal matrices," *J. Mathematics & Physics,* **12**, 311, 1933.

20. J. Williamson, "Hadamard's determinant theorem and the sum of four squares," *Duke Math. J.* **11**, 65, 1944.
21. S. W. Golomb and L. D. Baumert, "The search for Hadamard matrices," *Am. Math. Monthly,* **70**, 12, 1963.
22. C. V. Eynden, *Elementary Number Theory,* Random House, New York, 1987, pp. 238–240.
23. F. T. S. Yu and X. Yang, *Introduction to Optical Engineering,* Cambridge University Press, Cambridge, UK, 1997 pp. 122–124.
24. H. Lee, X. Gu and D. Psaltis, "Volume holographic interconnections with maximal capacity and minimal cross talk," *J. Appl. Phys.* **65**, 2191, 1989.
25. A. C. Starrier, E. S. Maniloff, K. M. Johnson and S. D. D. Goggin, "Procedure for recording multiple-exposure holograms with equal diffraction efficiency in photorefractive media," *Opt. Lett.*, **14**, 6, 1989.
26. K. Curtis and D. Psaltis, "Cross talk in phase-coded holographic memories," *J. Opt. Soc. Am. A,* **10**, 2547, 1993.
27. F. H. Mok, G. W. Burr, D. Psaltis, "Angle and space multiplexed holographic random access memory," *Opt. Mem. Neu. Net.,* **3**, 119, 1994.
28. Y. Taketomi, J. E. Ford, H. Sasaki, J. Ma, Y. Fainman and S. H. Lee, "Incremental recording for photorefractive hologram multiplexing," *Opt. Lett.,* **16**, 1774, 1991.
29. N. Streibl, "Beam shaping with array generators," *J. Mod. Opt.,* **30**, 1559, 1989.
30. J. I. Montagu, "Galvanometric and resonant low-inertia scanners," in G. E. Marshall (ed.), *Optical Scanning*, Marcel Dekker, New York, 1991.

Chapter 12

Compact Holographic Memory Module

Ernest Chuang

Sony Corporation, Tokyo, Japan

Jean-Jacques P. Drolet

MicroDisplay Corporation, San Pablo, California

Wenhai Liu and Demetri Psaltis

Electrical Engineering Department, California Institute of Technology, Pasadena, California

Abstract

We examine the present state of holographic compact memory module systems and address the primary challenges that face this technology, specifically size, speed, and cost. We show that a fast holographic random access memory (HRAM) system can be implemented with a compact architecture by incorporating conjugate readout, a smart-pixel array, and a linear array of laser diodes. Preliminary experimental results support the

PHOTOREFRACTIVE OPTICS
Materials, Properties and Applications

ISBN 0-12-774810-5/$30.00

feasibility of this architecture. Our analysis shows that for the HRAM to become competitive, the principal tasks will be to reduce SLM pixel sizes to 1 μm, increase the output power of compact visible-wavelength lasers to several hundred milliwatts, and develop ways to raise the sensitivity of holographic media to the order of 1 cm/J.

12.1 Introduction

Holographic memory has received attention in recent years as a technology that can provide large storage density and high speed [1–5]. Figure 12.1 shows a typical angle-multiplexed holographic memory in the 90° geometry. Information is recorded in the holographic medium through the interference of two coherent beams of light. We refer to the information-

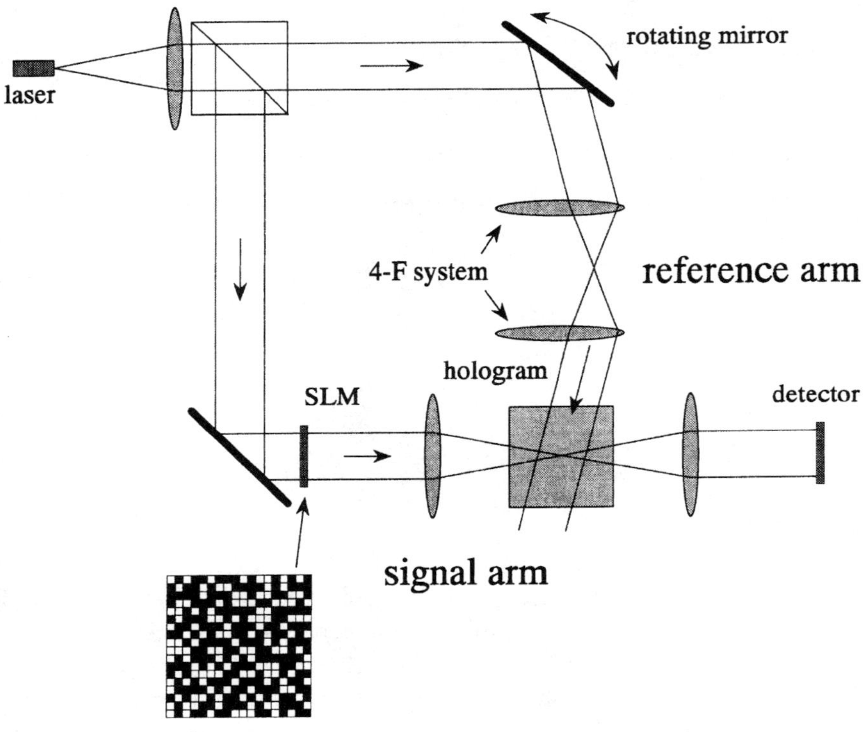

Figure 12.1: Typical angle-multiplexed holographic memory.

carrying beam as the signal beam, and the interfering beam as the reference beam. The resulting interference pattern causes an index grating (the hologram) to be written in the material. When the hologram is subsequently illuminated with one of the original writing beams, light is diffracted from the grating in such a way that the signal beam is reproduced.

We can multiplex many holograms within the same volume of the material by slightly changing the angle of the reference beam with each new data page. Thousands of holograms can be multiplexed this way in a small volume of crystal, offering the potential of high storage densities. Furthermore, holography has the inherent advantage of massive parallelism. Unlike conventional storage media such as magnetic hard disks and CD-ROMs which access only 1 bit at a time, each access of a holographic memory yields an entire data page — more than a megabit at a time.

In this paper we describe a holographic random access memory (HRAM) design that leads to the implementation of compact and inexpensive memory modules that can be used to construct large read-write memories. The potential place for such devices in the memory hierarchy is between magnetic disks and semiconductor memories [dynamic random access memory (DRAM)]. Specifically, we believe that HRAM can become a competitive technology if optoelectronics technology can achieve the following three milestones in the next few years.

1. SLM and detector pixel sizes on the order of 1 μm
2. Recording sensitivity of the holographic material at most 1 J/cm^2 to reach saturation
3. Inexpensive laser diodes with at least 500 mW of output power in the near-infrared or red wavelength

We will review the HRAM design and present experimental results from this architecture. We will estimate the performance that can be achieved with currently available technologies, and we justify the need to meet the milestones listed above in order for HRAM to become competitive. Specifically, if these developments take place, then HRAM can succeed as a memory that is less expensive than DRAM and faster than magnetic disks.

12.2 Conjugate readout method

Despite the high theoretical limit on the storage density of volume holographic storage (1 bit per cubic wavelength of material [6]), the practical implementation of holographic systems is often bulky due to the large space occupied by the various components that are necessary to provide the recording and readout mechanisms for the crystal. The system of Fig. 12.1 is fairly simple with a relatively small number of components, however, the spacing requirements of the imaging lenses imposes constraints on how closely these components can be placed. For example, assuming SLM and detector array dimensions of 1 cm and high-quality lenses with $F/\# = 1$, the focal distance between the arrays, lenses, and crystal must also be at least 1 cm. The system of Fig. 12.1 would then occupy a volume of approximately 6 cm $\times$ 5 cm $\times$ 1 cm which is 30 times larger than the volume of the recording material.

The reason we normally need to place lenses within the signal path is to undo the effects of diffraction. When we record a hologram of the signal beam diverging from the input SLM and reconstruct it with the original reference beam, we produce a virtual image of the input data page and thus require a lens to refocus it onto the detector array. We can eliminate the lens system between the SLM and detector array if we reconstruct a real image instead of a virtual one. One way to do this is to use phase conjugate readout [7–9] as illustrated in Fig. 12.2. Using this method, a hologram is recorded in the usual manner between the signal and reference beams, but the hologram is read out with the phase conjugate of the reference beam, propagating in the opposite direction as the one used for recording. This causes the signal reconstruction from the hologram to propagate back along the direction from which it originally came, reversing the original signal diffraction, and refocusing exactly at the plane of the SLM array. To generate the conjugate reference we may use a phase-conjugate mirror [8], or in the case of a planar reference beam, we may simply use a counterpropagating plane wave at each angle.

We compared experimentally the reconstructed image fidelity that can be obtained with conventional reconstruction using high-quality, custom-designed lenses to the image fidelity we get with the conjugate readout method. An SLM and detector array, each with pixel spacing of 24 μm, were used for these tests, allowing one-to-one matching of the SLM and detector pixels. Both methods yielded signal-to-noise ratio (SNR) values ranging from about 3.8–4.5, verifying that the conjugate readout method produces results that can only be achieved with quality lenses, while using a much more compact and inexpensive optical system.

Conjugate readout eliminates the lenses and associated path lengths that are normally required in the signal path, but it requires that both

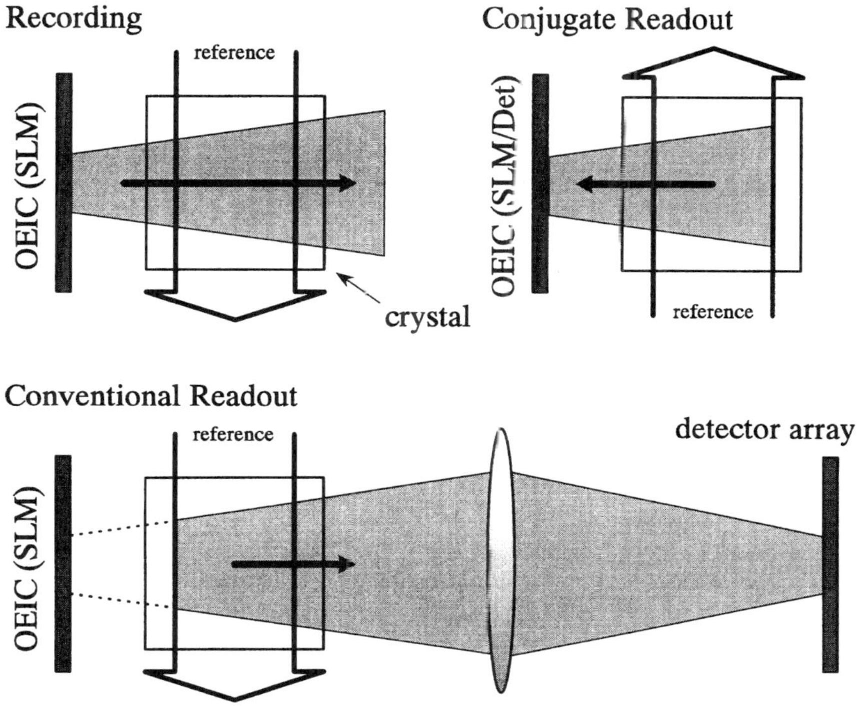

Figure 12.2: Comparison of phase conjugate readout method with conventional readout using imaging lenses.

the input and output devices be located on the same side of the crystal, as shown in Fig. 12.2. One approach is to employ a smart-pixel array that combines at each pixel the functions of a light modulator and detector in a single optoelectronic integrated circuit (OEIC).

12.3 Dynamic hologram refresher chip

In the experimental demonstrations that we describe below we implemented smart-pixel arrays by merging liquid crystal and silicon technologies [10, 11]. Figure 12.3a shows the cross-section of our device [12]. It is composed of silicon circuitry overlaid with hybrid-aligned nematic liquid crystal [13, 14] and a glass cover plate with a transparent electrode. The circuitry contains a photodetector, a static memory element to hold the data that is read out or to be written, and liquid crystal driving circuitry. Each pixel contains an exposed metal pad on top of the silicon that is used to modulate field across the liquid crystal layer, which in turn modulates the polarization of the light reflected from the metal pad. As a result, with a properly oriented polarizer at the output, the reflectance from each SLM pixel can be turned on or off.

Figure 12.3b shows a picture of the prototype array. It contains an array of 20 × 24 pixels which was designed to appear as a static random-access memory to a controlling microprocessor. The photodetector in each pixel is a PNP phototransistor with active-well-substrate structure. The pixel size in this prototype is 132 μm × 211 μm. The dimensions of the reflective SLM pads within each pixel is 49 μm square, with 18 μm square photodetector pads adjacently located. For reasons that will become apparent in the next section we refer to this device as the dynamic hologram refresher (DHR) chip.

12.4 Periodic copying

If we wish to use a holographic system as a rewritable memory, we must preserve the dynamic nature of the recorded gratings. For this purpose, photorefractive crystals (e.g., lithium niobate or barium titanate) are currently the most promising type of holographic media. Unfortunately, when a photorefractive crystal is used as the recording material, the recorded gratings decay when illuminated by the readout beam. One way to compensate for this is to use copying techniques [15–18] to periodically refresh the recorded holograms. Using this approach, stored holograms must be intermittently read out and rewritten into the memory to strengthen the gratings.

We conducted an experiment to test the periodic copying technique by building a holographic memory using the conjugate readout method and the DHR chip. A schematic and photograph of the experimental setup is shown in Fig. 12.4. The photorefractive medium was a cube of $BaTiO_3$, cut 30° with respect to its *c*-axis as shown in the figure. We used the DHR chip to serve as both input SLM and output detector. We used light from an argon laser at a wavelength of 488 nm for these experiments. The crystal was mounted on a rotation stage for angular multiplexing. We recorded 25 holograms (the letters "CIT" displayed on the DHR). Each

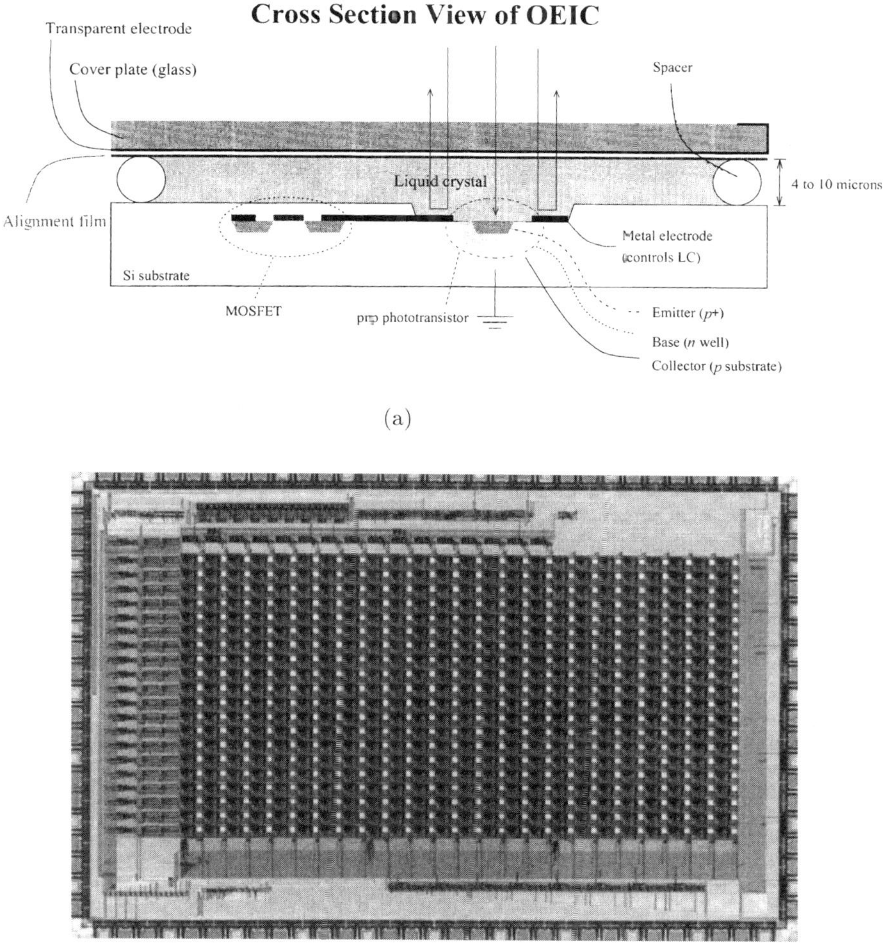

Figure 12.3: (a) Cross-section diagram of dynamic hologram refresher chip. (b) Picture of actual device array.

hologram was initially recorded for 4 s with the input data page. Upon completion of the recording cycle, we immediately returned to the first hologram to begin the refresh cycle. The purpose of the refreshing is to restore the strength of the holograms to their original recorded strength

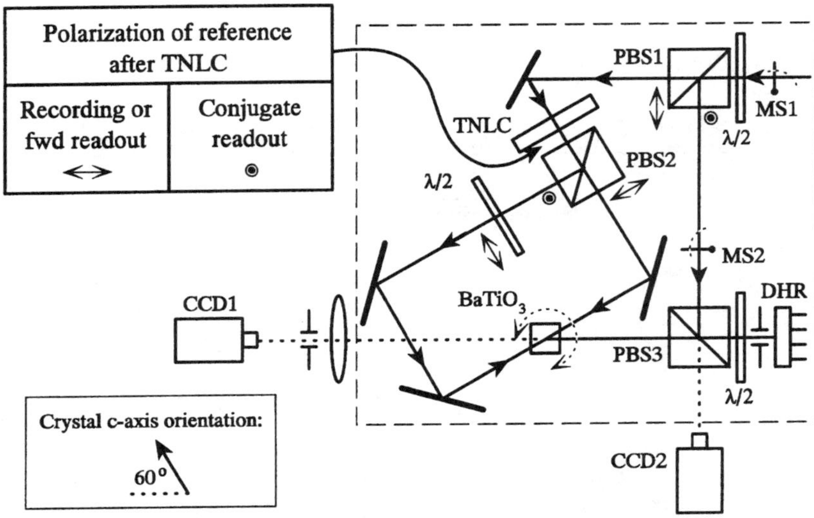

(a)

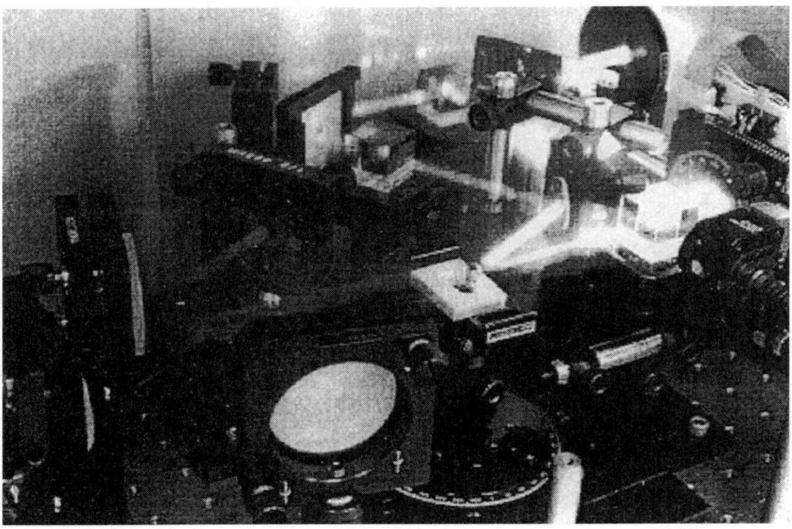

(b)

Figure 12.4: (a) Schematic of experimental setup for conjugate readout with periodic copying in the transmission geometry. (b) Photograph of the setup.

after they decay when they are read out. To refresh each hologram, the stored hologram was read by the conjugate beam, and this retrieved data was used to rewrite the hologram for the same recording time as the initial recording, 4 s. This refresh was repeated for each of the 25 holograms, and the entire refresh cycle was repeated for 100 cycles.

Figure 12.5 shows the diffraction efficiencies measured from the experiment for all 25 holograms. The curves for all of the holograms are superimposed in this graph to demonstrate the consistency among the hologram grating strengths. A few sample reconstructions from this experiment are shown in Fig. 12.6. Figure 12.6a shows a sample conjugate reconstruction after the initial recording; Figures 6b–d are conjugate reconstructions of holograms #1, #13, and #25 at the end of the 100 refreshings. The conjugate diffraction efficiency was measured to be 77% of the forward diffraction efficiency in this experiment.

Results of analyzing these images for SNR and probability of error are summarized in Table 12.1. Both visually and analytically we observed no appreciable loss in image quality from the refresh operations. The higher SNR and lower probability of error at the end of the experiment is consistent with the fact that after 100 refreshes, all of the holograms

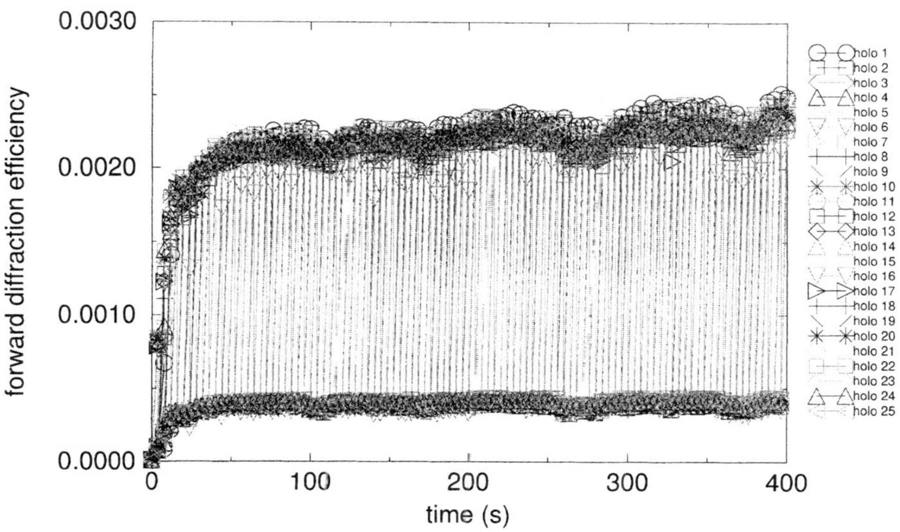

Figure 12.5: Evolution of diffraction efficiencies of 25 angle-multiplexed holograms over 100 cycles in the transmission geometry.

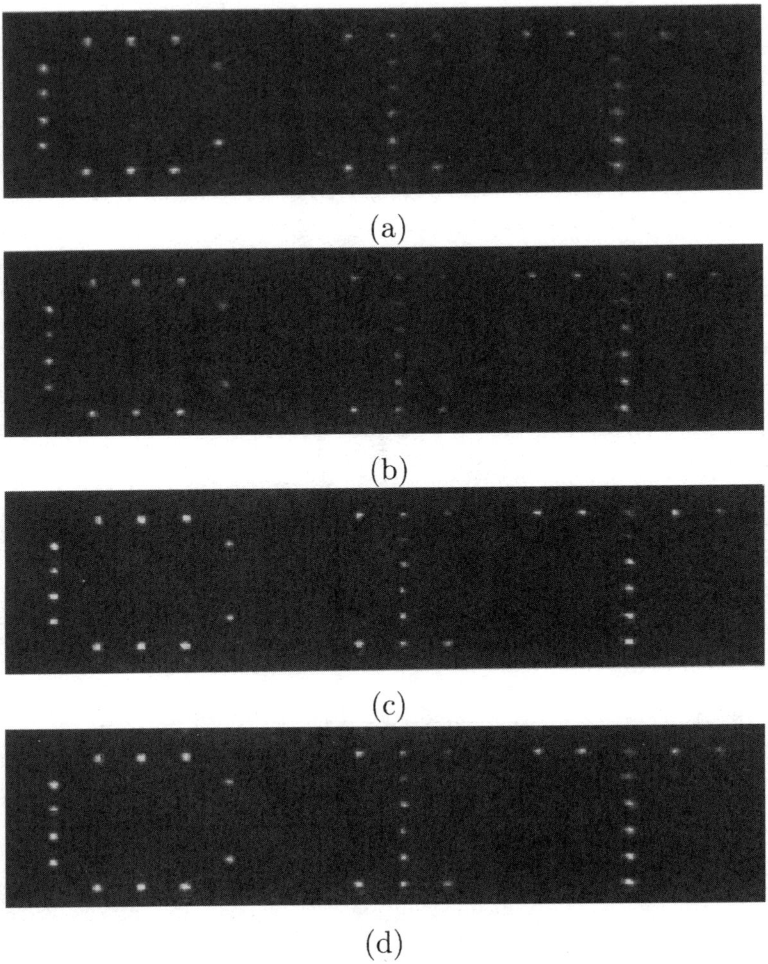

Figure 12.6: Sample reconstructions from 25-hologram experiment: (a) after initial recording, (b) hologram #1, (c) hologram #13, and (d) hologram #25 after 100 refreshes.

are well into the steady-state region where the diffraction efficiencies are significantly higher than after the initial recording. Also, among the three holograms examined at the end of the experiment, the holograms toward the end of the cycle yield better values because their diffraction efficiencies are higher, having been the most recently refreshed and thus the strong-

Image	SNR	PE
Conjugate reconstruction, 1 cycle	3.94:1	8.2×10^{-4}
Hologram #1, 100 cycles	4.28:1	1.0×10^{-4}
Hologram #13, 100 cycles	4.69:1	5.3×10^{-5}
Hologram #25, 100 cycles	5.03:1	2.9×10^{-5}

Table 12.1: Signal-to-Noise Ratio and Probability of Error Corresponding to the Images Shown in Fig. 12.6

est. No errors were detected in any pixel in any hologram during the course of the experiment.

12.5 Compact fast-access architecture

While conjugate readout eliminates the lenses in the signal path of the memory system, we still require a compact design to rapidly deflect the reference beam for multiplexing purposes. The 4-F system shown in Fig. 12.1, while reliable, is bulky and slow due to the limited mechanical speed of the rotating mirror.

With the recent development of compact laser emitters, such as laser diodes and vertical-cavity surface-emitting laser (VCSEL) devices [19, 20], it has become feasible to consider the possibility of incorporating arrays of hundreds of microscopic laser sources in a holographic memory. We can then design a system in which each angle multiplexed hologram is addressed by a dedicated laser source. This architecture is shown in Fig. 12.7. A Fourier transforming lens is used to convert the spatial shifts between the laser elements into angularly offset plane waves incident on the crystal. In this implementation, the time it takes to produce the proper readout reference beam is determined by the switching time of the laser sources, which is in the nanosecond regime. Using a 1-cm-thick crystal and a wavelength of 630 nm, the first null of the angular selectivity function occurs at an angular spacing of 0.0036°. Using a lens with a focal length of 2 cm would require the laser elements to be placed only 1.3 μm apart to produce this angular separation. In practice, the separation is 10 μm or more to reduce interpage crosstalk while also making the array easier to fabricate.

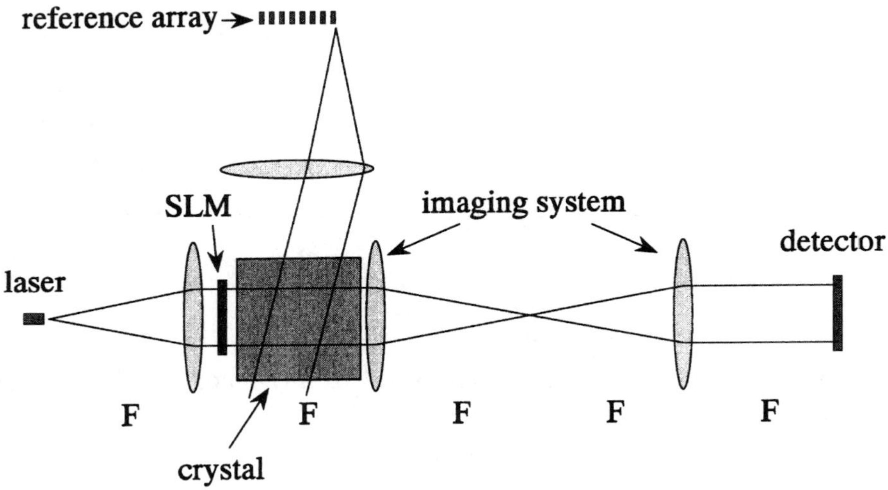

Figure 12.7: Use of a laser array in the reference arm of an angle-multiplexed memory for fast page access.

This approach is also compatible with the conjugate readout method as shown in Fig. 12.8a. With a properly aligned laser array and a mirror placed on the opposite face of the crystal such that it lies at the focus of the Fourier transforming lens, the proper conjugate beam can be generated with the symetrically opposite laser source. A beamsplitter must

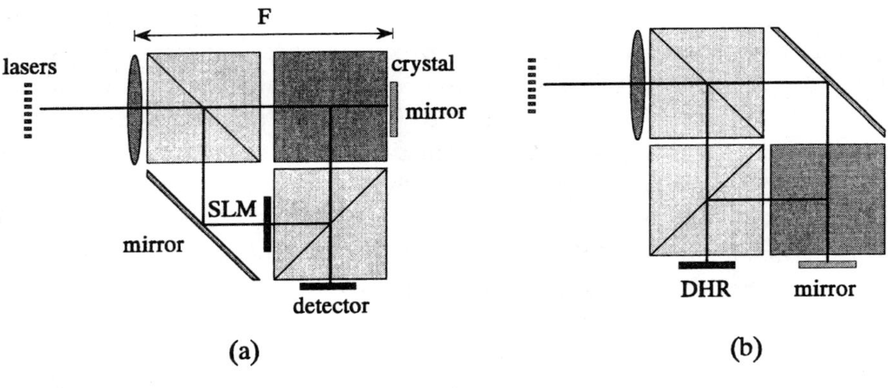

Figure 12.8: Variations of compact memory module incorporating (a) separate SLM and detector devices or (b) using a smart pixel array (DHR) combining SLM and detector functions.

also be introduced to accommodate both the SLM and detector devices. The combination of conjugate readout in the signal beam path and laser diode arrays in the reference beam path results in a compact holographic memory module with fast access. It is not completely lensless, since one lens still remains in the system, but such a lens would be required to collimate the laser source in any optical system that uses plane waves. This module design can be easily modified to accommodate a smart pixel array, such as the DHR chip, as shown schematically in Fig. 12.8b and visually by the model in Fig. 12.9.

12.5.1 Readout

Since the laser diode array discussed in the previous section allows us to switch between multiplexed data pages with negligible delay (on the order

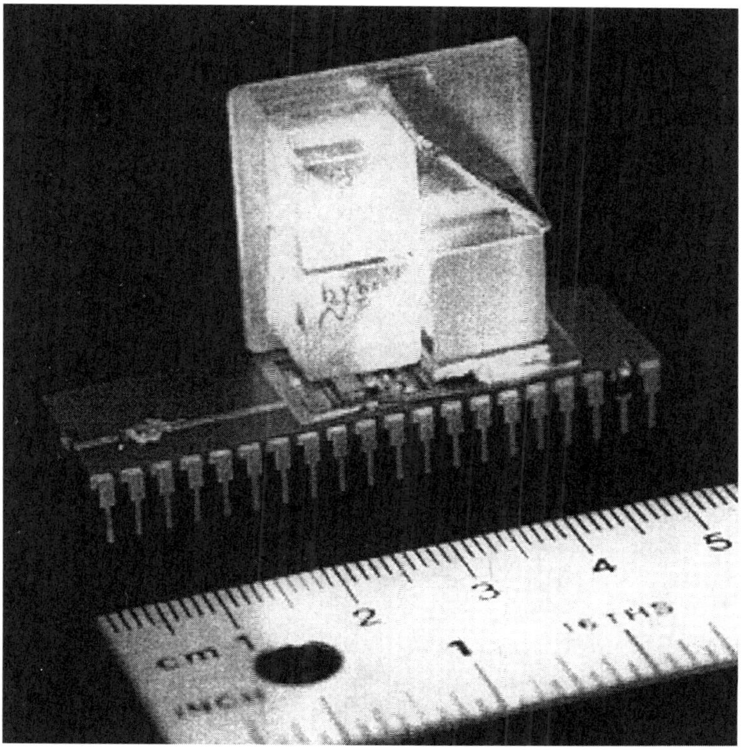

Figure 12.9: Model of the HRAM module with the DHR chip.

of nanoseconds), the random access time and the readout rate become limited by the required integration time of the detector. We can write the integration time as

$$\text{Detector integration time} = \frac{N_e h\nu N^2}{\left(\frac{M/\#}{M}\right)^2 P_i}, \tag{12.1}$$

where N_e is the number of electrons per pixel that we need to integrate for the given detector sensitivity and level of background noise, h is Planck's constant (6.63×10^{-34} J·s), ν is the light frequency, N^2 is the total number of pixels in the detector array, *M*/# is the system metric [21] of the holographic medium, M is the number of multiplexed holograms, and P_i is the incident readout power. For example, if we use a crystal of $M/\# = 10$ to record 500 holograms of a 1000×1000 pixel array, and we read out with 100 mW of laser power, requiring 300 electrons per pixel, the integration time, and hence the random access time, would be 2.4 μs. This corresponds to a sustained readout transfer rate, from the hologram to the silicon detectors, of 53 GB/s (b = bit, B = byte).

12.5.2 System volume density

An analysis of the system storage density (including the recording medium and all the optical components) of the holographic memory module of Fig. 12.8a shows that the module storage density peaks at about 40 Mbit/cm^3 for an optimum pixel size of 5 μm [22]. There is an optimum pixel size because as the pixel size decreases, the light in the signal path spreads more due to diffraction, causing us to use larger apertures for the crystal and beamsplitters.

A more aggressive concept for minimizing the volume is shown in Fig. 12.10. This design relies on total internal reflection to contain the beam diffraction within the boundaries of the module, so that the optical elements can be made the same size as the SLM array. Preliminary experiments indicate that accurate recordings are obtained using the internally reflected light. In this case, the system density can be raised to the order of 2 Gbit/cm^3, if SLM pixel sizes fall to 1 μm. At this density, a gigabyte of data could be stored in a single module with a volume of $1 \times 2 \times 2$ cm^3. The challenges in achieving such high densities are several: development of SLM and detectors with 1-μm pixels, designing the optical system

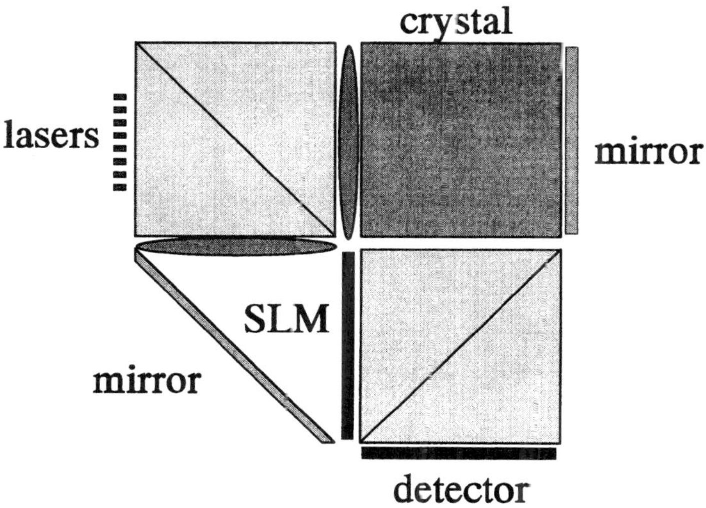

Figure 12.10: Variation of compact memory module for minimum volume.

so that we have uniform illumination throughout, and further characterization of the performance of the module when the light is allowed to undergo total internal reflection.

12.5.3 Recording rate

We can write the recording rate of the memory module as

$$\text{Recording rate} = \frac{N^2 ISLp}{(M/\#)/M} \tag{12.2}$$

where N^2 is the total number of pixels per data page, I is the incident recording intensity, S is the sensitivity per unit length of the recording medium, L is the crystal thickness, and p is the light efficiency of the SLM. Again assuming a crystal of $M/\# = 10$ to record 500 holograms of a 1000×1000 pixel array, with $I = 100$ mW/cm^2, $S = 0.1$ cm/J, $L = 1$ cm, and $p = 50\%$, we obtain a recording rate of 31 kB/s. This is typical for experiments currently performed. Increasing the recording rate to make it comparable to the readout rate is highly desirable for a practical system. We will discuss possible methods for achieving this goal in the following.

12.5.4 Cost

The cost is perhaps the most important metric for accessing the commercialization prospects of HRAM. We will compare the costs of HRAM and DRAM with reference to Fig. 12.11. We can think of HRAM as a holographic module which sits on top of a page of DRAM. The ability of the HRAM to multiplex holograms essentially allows us to store M DRAM data pages, hence saving us the cost of fabricating M-1 additional DRAM pages in silicon. However, it is not quite that simple. First, the silicon device in the HRAM is not really a DRAM page, but rather the DHR chip described previously or an SLM/detector pair. Because of the necessity of fabricating SLM and detector pixels (either in the same OEIC or in two separate devices), the page density of the DHR will be less than that of a true DRAM. We call this ratio of the page densities $R > 1$. Therefore, we have a storage of M/R pages DRAM data in a holographic module with the detector interface of the same area as one page of DRAM. If we have a larger number of multiplexed data pages M and use a smaller

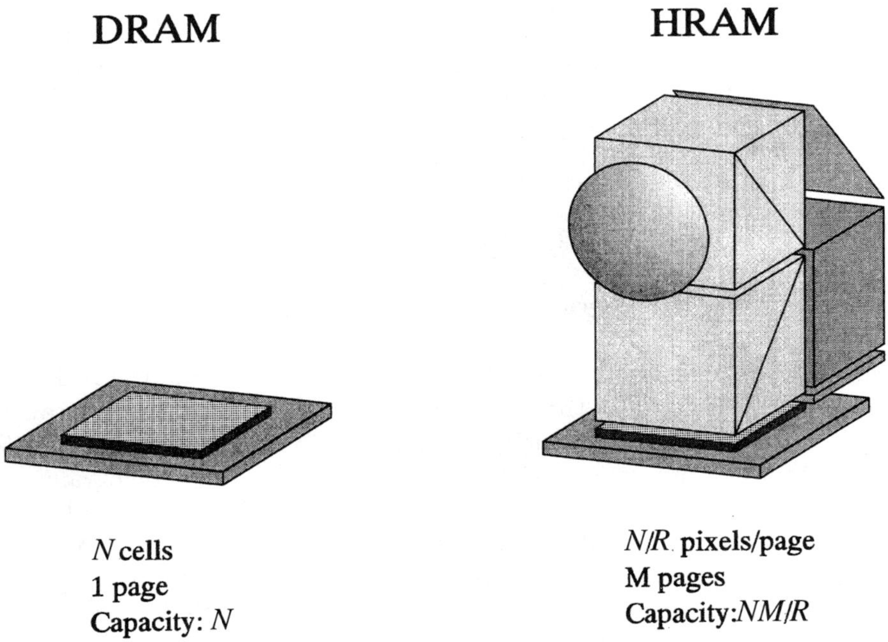

Figure 12.11: Model for cost comparison between HRAM and DRAM.

SLM/detector pixel area R, the holographic module will have a larger storage capacity with similar cost, which increases the final cost efficiency.

Moreover, the cost of the holographic module also includes the cost of the optical elements (C_{Opt}) and laser diode array (C_{LD}), in addition to that of the silicon (C_{Si}). The projected costs of the optical elements (assuming production in large quantities) are summarized in Table 12.2. We assume the silicon cost to be purely based on area, and therefore will be identical to that for an equal-sized DRAM. The cost of the laser array is not well known at this time, since large arrays have not yet been produced for visible wavelengths; however, we estimate the cost to be in the range of \$25–\$100 per array.

We can then write the cost-per-bit ratio of the HRAM to DRAM as

$$\text{Cost ratio} = \frac{C_{\text{Si}} + C_{\text{Opt}} + C_{\text{LD}}}{C_{\text{Si}}} \cdot \frac{R}{M} \tag{12.3}$$

For current commercial SLMs and detector arrays, the smallest available pixel pitch is on the order of 4 μm, with the spacing of DRAM cells at 1 μm, leading to a value of $R = 16$. Hence, for HRAM to have a cost advantage over DRAM by a factor of 10 or more, we need to record at least 200–300 holograms in each HRAM module. Since this can be readily achieved, cost emerges as the major competitive advantage of HRAM.

12.6 Pixel size limit for holograms

As we discussed in the last section, there are two factors to decrease the cost-per-function for the holographic module, a larger page number M

Component	**Estimated cost**
$LiNbO_3$ (1 cm^3)	\$10
Liquid crystal	\$5
Beamsplitters and lens	\$6
Silicon (1 cm^2)	\$115
Laser diode array (500)	\$25–\$100
Total	\$161–\$236

Table 12.2: Estimated Cost of Components in the Holographic Memory Module, Assuming Production in Large Quantities

and a smaller SLM/detector pixel size R. The number of multiplexed data pages M is limited by the dynamic range of the material $M/\#$, sensitivity S, and the readout accessing time. Excessively large M will increase the detector integration time and complicate multiplexing technology. On the other hand, decreasing R by decreasing the pixel size of SLM and the detector array, we have a larger number of pixels per page N^2 without increasing system volume. Thus, the system volume density, the readout and recording rate are increased. To take the advantages of a holographic parallism, it is essential to decrease the pixel size of SLM and the detector down to the available limit of the holographic module technology.

Besides the physical fabrication limits of SLM and the detector array, the fundamental limit for the pixels of a hologram in a holographic module is the angle-dependent diffraction efficiency in the thick recording medium [23]. It has been shown theoretically that the $LiNbO_3$ in 90° geometry has the holographic recording bandwidth for holograms with pixel size down to 1 μm [24]. Experimentally, we have used a mask fabricated with e-beam lithography to record and reconstruct data pages with 1-μm pixels holographically with good image fidelity. Figure 12.12 shows a phase

Figure 12.12: Phase conjugate hologram reconstruction for binary random pixels of 1×1 μm^2.

conjugate reconstructed sample of 1×1-μm^2 pixels, which yields values of SNR = 4.8 and the bit error rate 7×10^{-5}. Figure 12.13 shows the results from an experiment in which we measured the SNR from the conjugate reconstruction of various pixel sizes from 8 down to 1 μm.

12.7 Roadmap for a competitive HRAM technology

From the preceding discussion, we can summarize the current parameters for the HRAM system as shown in Table 12.3. For comparison, we also show the specifications projected for DRAM by the year 2006. DRAM access times should fall to 10–40 ns; the DRAM transfer rates can reach 10 GB/s, assuming, e.g., 800 pins, each with a bandwidth of 100 MHz. The cost is projected to be $0.40/MB [25]. Although the holographic read-

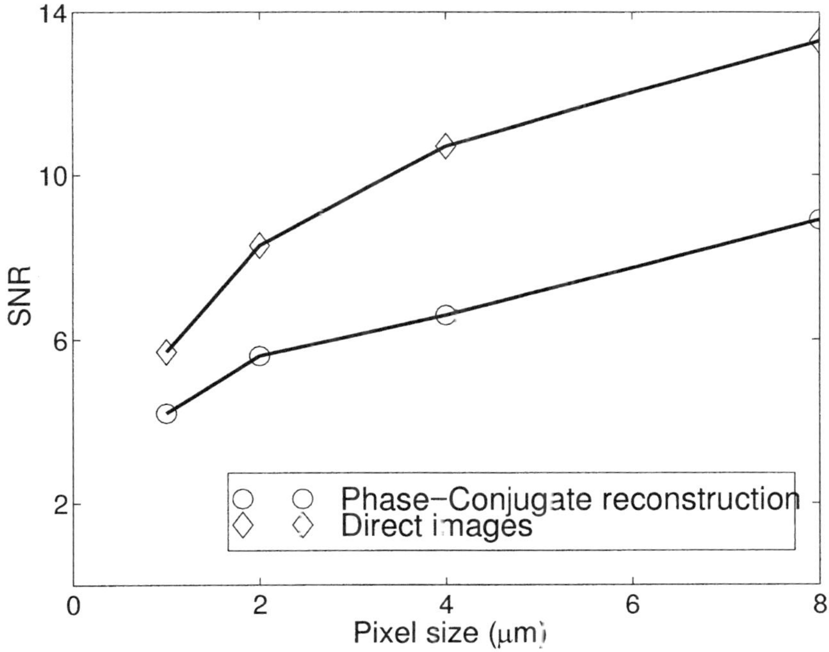

Figure 12.13: SNR versus pixel size measured for both direct imaging and for conjugate hologram reconstructions.

	HRAM (current)	DRAM (2006)	HRAM (target)
Access time	2.4 μs	10–40 ns	530 μs
Recording rate	31 kbit/s	10 Gbit/s (pin limited)	1.0 Gbit/s
Readout rate	53 Gbit/s (optical limit)		24 Gbit/s (optical limit)
	10 Gbit/s (pin limited)	10 Gbit/s (pin limited)	10 Gbit/s (pin limited)
Cost	NA	\$0.40/MB	\$0.04/MB

Abbreviations: NA, not applicable. B = byte

Table 12.3 Comparison Between DRAM and HRAM

out rate of the HRAM system is nominally 53 GB/s, the fact that its readout interface is through silicon (the DHR) limits the transfer rate to that of DRAM, 10 GB/s.

Presently, the greatest challenge for the HRAM is to raise its recording rate by several orders of magnitude. To achieve this, we must rely in part on improvements in SLM technology to bring the pixel sizes down to 1 μm. This will allow us to increase the size of each data page to 10,000 × 10,000 pixels while still holding the array size to about 1 cm^2. By increasing the page size in this way, we immediately gain two orders of magnitude in the sustained recording rate due to the increased parallelism.

Reducing the pixel sizes to 1 μm is not only necessary for raising the recording rate, but also for maintaining the cost advantage of HRAM over DRAM. By 2006, the DRAM cell pitch is expected to fall to 0.2 μm [25]. By bringing the SLM pixel pitch down to 1 μm, we can hold the factor R in Eq. (12.3) at 25, and still beat the cost of DRAM by an order of magnitude by recording only 350–500 holograms.

Because the HRAM readout rate is limited by the electronic transfer rate out of the detector chip, we can afford to give up some readout speed in favor of increasing the recording speed. We do this by intentionally reducing the strength of the holograms so that we can record with shorter exposures, at the cost of increasing the detector integration time. In Eqs. (12.1) and (12.2), this is equivalent to recording in a medium with lower M/#, but without sacrificing sensitivity. Unfortunately, as we increase the required integration time, we increase at the same time the random access time of the memory. To maintain an advantage of at least an order of magnitude over magnetic disks in random access time, we can only afford to increase the integration time to several hundreds of microseconds.

Other opportunities for increasing the recording rate can arise from improvements in laser output powers or from improving the sensitivity of the recording materials. Compact laser arrays with outputs of 500 mW per emitter may be possible by 2006, or if not, we may consider sharing a larger, more powerful tunable laser among multiple HRAM modules. Increasing material sensitivity presents more of a challenge. The sensitivity of $LiNbO_3$:Fe, by far the most commonly used recording material today, is typically around 0.02 cm/J in the 90° geometry. To get recording rates on the order of 1 GB/s, we must find ways to boost the material sensitivity to about 1 cm/J by improving lithium niobate's properties. For instance, switching to transmission geometry and increasing the doping level results in substantial increases in M/#, which can be traded for better sensitivity as we discussed previously. Alternatively, we can switch to materials such as barium titanate which we measured to have $S = 0.55$ cm/J in the 90° geometry. Even higher sensitivities are possible in the transmission geometry. However, this is a relatively untested material compared to lithium niobate and much more expensive at present.

12.8 Conclusion

In order to develop a competing HRAM technology, three main challenges must be met: reducing pixel size to 1 μm, producing arrays of high-power laser diodes, and increasing the sensitivity of holographic recording media. Each of these tasks is difficult, but if they can be achieved in the next few years, then the projected HRAM performance levels shown in Table 12.3 become feasible. These values assume an array size of 10,000 × 10,000 pixels, 500 holograms recorded to diffraction efficiencies in a 1-cm-thick material with effective M/# = 3 and 2 cm/J sensitivity, laser power of 500 mW, and 300 electrons required for detection. Attaining these goals will position the HRAM as a viable alternative memory technology to magnetic storage, offering performance that is at least one order of magnitude better in terms of random access and transfer rate than magnetic hard disks, and no more than one tenth the cost compared to fabricating an equivalent memory in DRAM.

Acknowledgments

The authors would like to thank George Barbastathis, Xu Wang, and Ali Adibi for helpful discussions. Support for this effort was funded in part by Jet Propulsion Laboratory-Hybrid Technology Multithreaded Architecture (JPL-HTMT) (#49-220-85603-0-3950) and Rome Laboratories (#F30602-97-C-0049).

References

1. D. Psaltis and F. Mok, "Holographic memories," *Scientific American,* **273**, 70–76, Nov. 1995.
2. J. H. Hong, I. McMichael, T. Y. Chang, W. Christian, and E. G. Pack, "Volume holographic memory-systems: Techniques and architectures," *Optical Engineering,* **34**, 2193–2203, 1995.
3. J. F. Heanue, M. C. Bashaw, and L. Hesselink, "Recall of linear combinations of stored data pages based on phase-code multiplexing in volume holography," *Optics Lett.,* **19**(14), 1079–1081, 1994.
4. A. Pu and D. Psaltis, "High-density recording in photopolymer-based holographic 3-dimensional disks," *Applied Optics,* **35**, 2389–2398, 1996.
5. D. Psaltis and G. W. Burr, "Holographic data-storage," *Computer,* **31**, 52–60, 1998.
6. P. J. van Heerden, "Theory of optical information storage in solids," *Applied Optics,* **2**(4), 393–400, 1963.
7. J.-J. P. Drolet, G. Barbastathis, J. S. Patel, and D. Psaltis, "Liquid crystal devices for volume holographic memories," *OSA Annual Meeting*, Portland, OR, Sept. 1995.
8. J.-J. P. Drolet, G. Barbastathis, and D. Psaltis, "Optoelectronic interconnects and packaging," *SPIE Critical Reviews,* **CR62**, 106–131, 1996.
9. Z. O. Feng and K. Sayano, "Compact read-only memory with lensless phase-conjugate holograms," *Optics Letters,* **21**, 1295–1297, 1996.
10. L. K. Cotter, T. J. Drabik, R. J. Dillon, and M. A. Handschy, "Ferroelectric-liquid-crystal/silicon-integrated-circuit spatial light modulator," *Optics Letters,* **15**, 291–293, 1990.
11. K. M. Johnson, D. J. McKnight, and I. Underwood, "Smart spatial light modulators using liquid crystal on silicon," *IEEE Journal of Quantum Electronics,* **29**, 699–714, 1993.

12. J. Drolet, G. Barbastathis, J. Patel, and D. Psaltis, "Liquid crystal devices for volume holographic memories," *OSA Annual Meeting*, Sept. 1995.

13. J. P. Drolet, J. Patel, K. G. Haritos, W. Xu, A. Scherer, and D. Psaltis, "Hybrid-aligned nematic liquid-crystal modulators fabricated on VLSI circuits," *Optics Letters,* **20**, 2222–2224, 1995.

14. J.-J. P. Drolet, E. Chuang, G. Barbastathis, and D. Psaltis, "Compact, integrated dynamic holographic memory with refreshed holograms," *Optics Letters,* **22**, 552–554, 1997.

15. D. Brady, K. Hsu, and D. Psaltis, "Periodically refreshed multiply exposed photorefractive holograms," *Optics Letters,* **15**, 817–819, 1990.

16. H. Sasaki, Y. Fainman, J. Ford, Y. Taketomi, and S. Lee, "Dynamic photorefractive optical memory," *Optics Letters,* **16**, 1874–1876, 1991.

17. S. Boj, G. Pauliat, and G. Roosen, "Dynamic holographic memory showing readout, refreshing, and updating capabilities," *Optics Letters,* **17**, 438–440, March 1992.

18. Y. Qiao and D. Psaltis, "Sampled dynamic holographic memory," *Optics Letters,* **17**, 1376–1378, 1992.

19. J. L. Jewell, K. F. Huang, K. Tai, Y. H Lee, R. J. Fischer, S. L. Mccall, and A. Y. Cho, "Vertical cavity single quantum well laser," *Applied Physics Letters*, **55**(5), 424–426, 1989.

20. W. W. Chow, K. D. Choquette, M. H. Crawford, K. L. Lear, and G. R. Hadley, "Design, fabrication, and performance of infrared and visible vertical-cavity surface-emitting lasers," *IEEE Journal of Quantum Electronics,* **33**, 1810–1824, 1997.

21. D. Psaltis, D. Brady, and K. Wagner, "Adaptive optical networks using photorefractive crystals," *Applied Optics,* **27**(9), 1752–1759, 1988.

22. E. Chuang, Methods and Architecture for Rewritable Holographic Memories, Ph.D. thesis, California Institute of Technology, 1998.

23. H. Zhou, F. Zhao, and F. Yu, "Angle-dependent diffraction efficiency in a thick photorefractive hologram," *Applied Optics,* **34**, 1303–1309, 1995.

24. W. Liu and D. Psaltis, "The storage density limit of a holographic memory system," in *OSA Annual Meeting*, Baltimore, MD, Sept. 1998.

25. "The national technology roadmap for semiconductors," Technical Report, Semiconductor Industry Association, 1997.

Chapter 13

Dynamic Interconnections Using Photorefractive Crystals

Osamu Matoba

Institute of Industrial Science, University of Tokyo, Tokyo, Japan

Kazuyoshi Itoh

Department of Applied Physics, Osaka University, Suita, Osaka, Japan

Kazuo Kuroda

Institute of Industrial Science, University of Tokyo, Komaba, Meguro-ku, Tokyo, Japan

13.1 Introduction

Optical interconnection is one of the key techniques for the next-generation computers because of its inherent parallelism and extremely wide bandwidth for two-dimensional data processing. If we can modify dynamically the optical interconnection patterns, we will benefit much more than the static optical interconnection. The optical dynamic interconnections

PHOTOREFRACTIVE OPTICS
Materials, Properties and Applications

ISBN 0-12-774810-5/$30.00

offer unique applications such as real-time image-processors [1–3], novelty filter [4], image amplification [5], pattern recognition [6], and optical neural networks with real-time learning ability [7–15]. Implementation of the optical dynamic interconnections needs active devices for manipulating optically or electrically the amplitude, phase, or polarization of an optical wave. It is well known that the photorefractive effect [16–19] is useful for the simple implementation of the optical dynamic interconnections. The photorefractive effect is slow, but shows a large refractive index change even if we use an optical beam with an optical power level of a milliwatt. There are two optical interconnection approaches based on the photorefractive effect. One is based on free-space optics and the other is based on guided optics.

Optical dynamic free-space interconnections have been mainly investigated until recently. The photorefractive effect offers many distinctive devices such as real-time volume holograms [9, 20, 21], spatial light modulators [22–24], and phase conjugators [25–30]. Photorefractive holographic interconnection is the most well-known optical dynamic or programmable free-space interconnection technique because the photorefractive material can store the large number of interconnection patterns per unit volume. This research is closely related to that in the holographic memory system [31–33]. However, these are still many problems to overcome for the immediate practical use.

Optical interconnections based on the guided optics are suited for present optoelectrical integrated technology. However, the number of interconnections is restricted because conventional fabrication methods of optical waveguides can create structures in the two-dimensional plane such as the surface of the bulk material or the thin films. Several methods for fabricating waveguide structures by optical irradiation in the volume of various materials such as ultraviolet (UV) epoxy [34, 35], glass [36, 37], and photorefractive materials [38–40] have been proposed. These techniques enable the implementation of a large number of waveguide interconnections in the volume of the material. They can be used for massively parallel interconnections between fibers or between fibers and planar waveguides. At present, glass or UV epoxy is not rewritable by optical illumination, thus cannot be used for the dynamic interconnection. Furthermore, the glass needs an extremely high peak intensity such as a femtosecond pulse for the fabrication of the waveguide structure [36, 37]. The photorefractive effect requires small laser power to create the refractive index change for the waveguides. The refractive index change can be erased optically, which opens the way for dynamic interconnections. Note that it is also possible to fix the waveguide structures by thermal [41, 42] and electrical [43, 44] treatment. There are two novel fabrication techniques of optical waveguides based on the photorefractive effect. Photorefractive spatial solitons [45–49] is one of the techniques to create the waveguide structure [39, 50–53]. In this waveguide, a soliton with a small optical power can control a strong guided beam with an insensitive wavelength to the photorefractive effect. Shih *et al.* observed that multimode beams were successfully guided through the soliton-induced waveguide [39].

We focus here on another technique of the optical dynamic interconnections using the photorefractive waveguide [38]. The photorefractive waveguide is simply fabricated by scanning a focused laser beam in the photorefractive material. The fabricated waveguide can be modified by its guided beam, which changes the refractive index distribution of the waveguide [54]. This modification of the waveguide structure leads to change in the output power of a signal beam. This dynamic modification of the output power of the signal beam can be used as adaptive connections in optical neural network systems with learning ability. Weights in the interconnections are stored as the optical power transmittances of the photorefractive waveguides.

In this chapter, we will provide an overview of the optical dynamic interconnection based on the photorefractive waveguides. This will include the fabrication of a single straight and a curved waveguide, and a waveguide with nonuniform structures. We will discuss also the optical modification of the waveguide structure and the experimental fabrication of an array of photorefractive waveguides.

13.2 Photorefractive waveguides

A photorefractive waveguide is simply fabricated by scanning a focused laser beam in a photorefractive material as shown in Fig. 13.1. Figure 13.2 shows an illustration of a large number of integrated interconnections using the photorefractive waveguides. The waveguide structure is determined by the illumination pattern of the focused laser beam and the photorefractive property of the material. In the experiments, an undoped lithium niobate (LN) crystal was used as the photorefractive material. The LN crystal is one of the most popular photorefractive materials, and

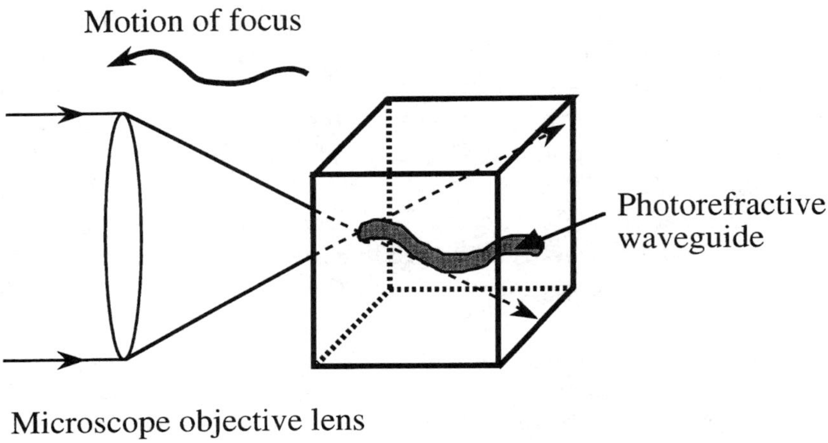

Figure 13.1: Illustration of a fabrication method of a photorefractive waveguide.

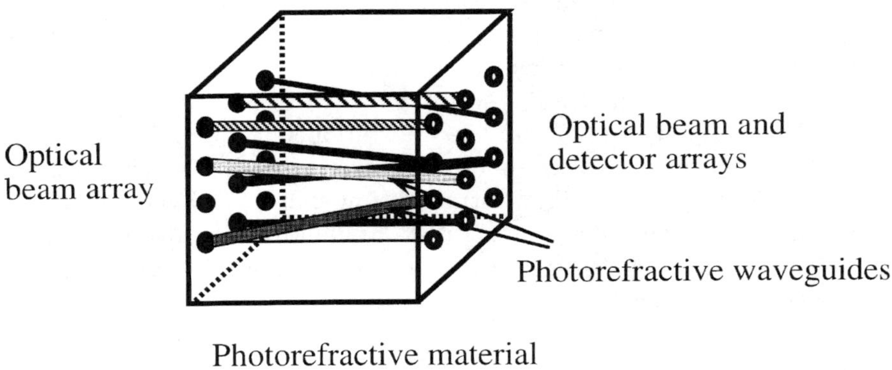

Figure 13.2: High-density optical interconnection using the photorefractive waveguides.

shows a large refractive-index variation along the *c*-axis of the order of 10^{-3} and a long storage time (over a few months). This refractive-index variation is enough to create the waveguide structure. Figure 13.3 illustrates the process of the refractive-index change when an LN crystal is illuminated by the focused beam [38,53]. The mechanism of the refractive index change due to the photorefractive effect in the LN crystal is consid-

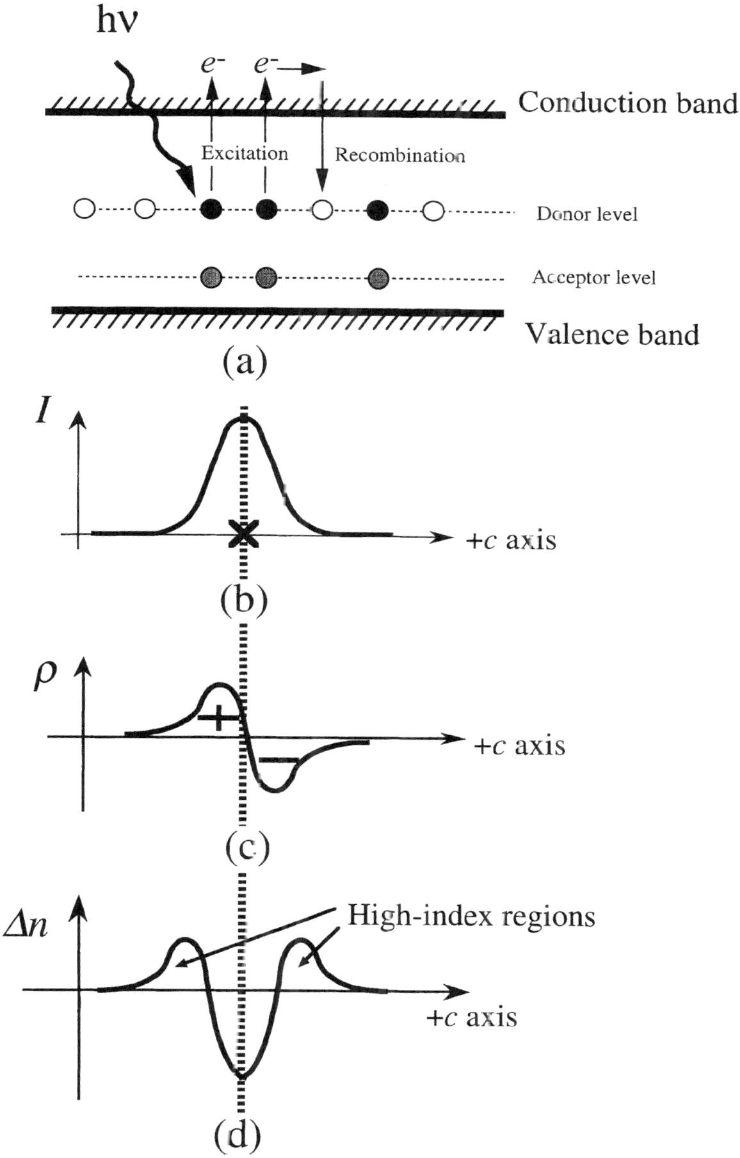

Figure 13.3: Illustration of the profile of the refractive-index change caused by a focused beam in a LN crystal: (a) model of the photorefractive effect, (b) the intensity distribution, (c) the space-charge density, and (d) the distribution of the refractive index change.

ered as follows [56–58]. Photoexcited electrons generated by the focused laser illumination move unidirectionally to the positive side of the c-axis due to the bulk photovoltaic effect [56]. The electrons drift out of the laser spot and are finally trapped in the dark region near the beam periphery. As a result, the space-charge field is created around these trapped electrons and positive ionized donors, mainly along the c-axis. This field induces the refractive index change via the Pockels effect. In the LN crystal, the largest electrooptical coefficient is r_{33}. Thus the largest variation of the refractive-index is obtained when one uses an incident light with extraordinary polarization.

13.2.1 Fabrication

Figure 13.4 shows the experimental setup for fabrication, guiding tests, and measurement of the refractive index distribution of the photorefrac-

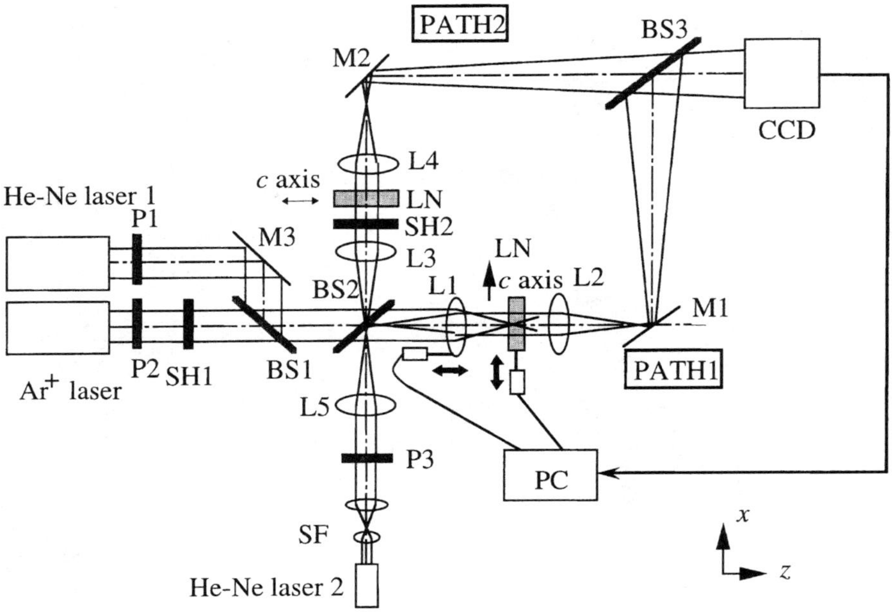

Figure 13.4: Experimental setup for fabricating and testing the photorefractive waveguides; Ps, polarizers; SHs, shutters; Ls, lenses; SF, spatial filter; Ms, mirrors; BSs, beamsplitters.

tive waveguides. An undoped LN crystal (LN1) was placed between two objective lenses so that the *c*-axis was perpendicular to the optical axis of an Ar^+ laser beam. The thickness of the crystal was 2.0 mm. The Ar^+ laser beam (λ = 514.5 nm) with ordinary polarization was focused by a microscope objective lens, L1, to fabricate the waveguide structure in the LN crystal. The effective numerical aperture of the focused fabrication beam was approximately 0.1. To scan the focus into the crystal, the longitudinal position of the focus was controlled within an accuracy of submicrometers by a translator coupled with a personal computer (PC). The lateral position of the crystal was also controlled in the same manner.

In the guiding test, a He-Ne laser beam (He-Ne laser1) with extraordinary polarization was mainly used. A guided beam was excited at the input end of the waveguide when the focus of L1 was located on the front face of the crystal. The optical power of the beam was made sufficiently low to avoid further refractive index change. In some guiding tests, the Ar^+ laser beam was used as a guided beam after a half-wave plate was inserted to select the extraordinary ray. The optical power of the guided beam was reduced to approximately 1/1000 of the fabrication beam to keep the waveguide unchanged. In the practical use of the waveguides for the optical computing and the optical communication, other light beams having wavelengths insensitive to the photorefractive medium may be used. A microscope objective lens, L2, was placed behind the LN crystal for observation of near-field patterns on the rear face of the crystal. The near-field patterns were observed by a charge-coupled device (CCD) through the optical path (PATH1).

The refractive-index distribution of the straight waveguide can be measured *in situ* by an interferometric technique by another collimated He-Ne laser beam (He-Ne laser2) with extraordinary polarization. If the refractive index change is sufficiently small, the phase front deformation of the collimated laser beam is a replication of the modified refractive index integrated along the optical axis. An identical LN crystal (LN2) was placed in the second optical path (PATH2) for compensation of the optical effects, except those caused by exposure to the fabrication beam.

A straight waveguide was fabricated for observing its refractive-index distribution. Since the present optical configuration allows us to measure small refractive index changes, the optical power of the fabrication beam was reduced to 0.22 mW. After numerical analysis based on the Fourier transform method [59], the integrated refractive index profile of the fabricated waveguide as shown in Fig. 13.5 was obtained. Figure 13.5a shows

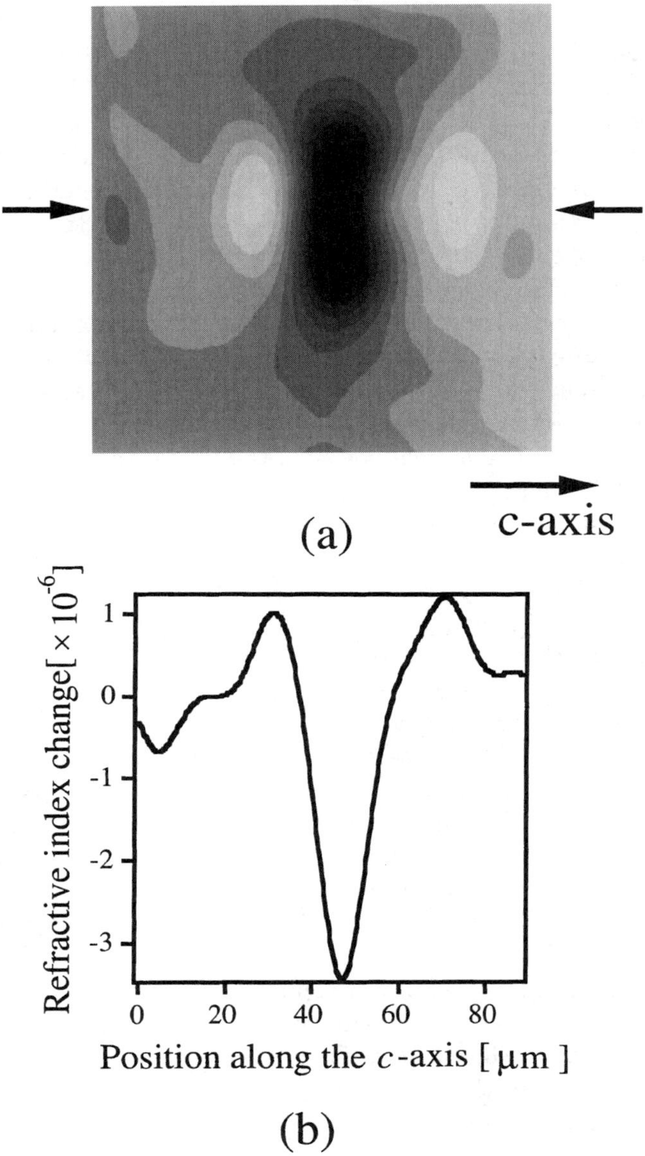

Figure 13.5: Refractive-index distribution of a straight photorefractivewave-guide: (a) the cross section of the refractive index distribution of the waveguide and (b) its profile along the *c*-axis.

roughly the distribution of the refractive index perpendicular to the optical axis of Ar^+ laser beam. The field of view of this figure covers a rectangular area of 130 μm × 130 μm. Bright areas indicate a high refractive index. The horizontal axis is parallel to the *c*-axis. Figure 13.5b shows the detailed index distribution along a line parallel to the *c*-axis. It is already known that the refractive index profile is not proportional to the intensity distribution in the LN crystal [38, 55]. This result also shows that areas of higher refractive index are present at two distinct locations adjacent to the center of the illumination. Thus two parallel waveguides are simultaneously created by scanning a focused spot through the LN crystal.

A straight waveguide was fabricated for the guiding experiment. The beam power was 70 mW, and it took 20 s to scan the focused beam from the rear to the front face of the crystal. The focus of the fabrication beam was not moved continuously, to create a favorable refractive index profile [60]. Since the focus is followed by the defocused part, the defocused part illuminates again the area that has already been illuminated, and destroys the previously formed distribution of space-charge density. The distribution of the space-charge density is destroyed by the reexcitement of trapped electrons and the recombination of positive ionized donors. A discontinuous movement of the focus may reduce the overlap between the successive exposures. In this experiment, the interval was 112 μm and the exposure time at each illumination was 1/125 s. The depth of focus was 60 μm. This fabrication technique may fabricate a nonuniform structure along the propagation direction of the waveguide. The refractive index variation along the propagation direction is small because the interval between successive exposures is comparable with the depth of focus of the fabrication beam.

After fabrication, one of the waveguides located at the negative side along the *c*-axis was used in the guiding test. The focused Ar^+ laser beam with extraordinary polarization was launched into the waveguide. The numerical aperture of the excitation beam was approximately 0.1. Before fabrication, a broad and weak intensity pattern was seen due to the defocused beam. Figure 13.6 shows the output pattern after fabrication. This figure covers an area of 65 μm × 65 μm. The lateral axis is parallel to the *c*-axis of the LN crystal. A sharp and intense profile of a localized intensity distribution was observed. Forty percent of the input power was transmitted. The fabricated waveguide was single-mode because no variations of the near-field pattern were observed by the variations of input positions of the excitation beam.

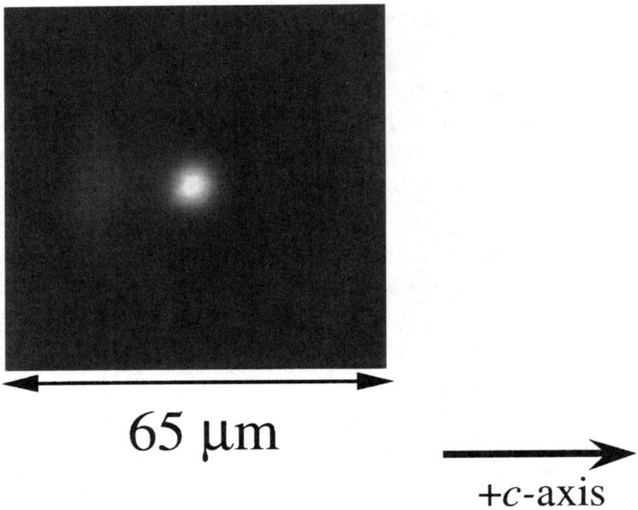

Figure 13.6: Near-field pattern in the straight photorefractive waveguide.

Curved waveguides were fabricated to demonstrate the feasibility of optical fabrication of three-dimensional waveguides. The curved waveguides were bent in the horizontal plane of the experimental setup as shown in Fig. 13.7a. The gray line in this figure indicates the distribution of exposure by the fabrication beam. A curved waveguide located at the negative side along the *c*-axis was used for the guiding test. The output pattern of the guided beam and its profile along the *c*-axis are presented in Figs, 13.7b and c, respectively. A sharp and intense pattern was observed at appropriate position.

13.2.2 Model of Photorefractive waveguides

The accurate knowledge of the three-dimensional refractive index distribution of the photorefractive waveguide is necessary to evaluate the optical characteristics of the waveguide and to design high-density interconnections. However, the distribution of the refractive index change caused by the focused beam in the LN crystal is not understood yet either analytically or experimentally. We describe briefly a parametric model of the refractive index distribution of the photorefractive waveguide that was made from the index distribution of the straight waveguide as shown

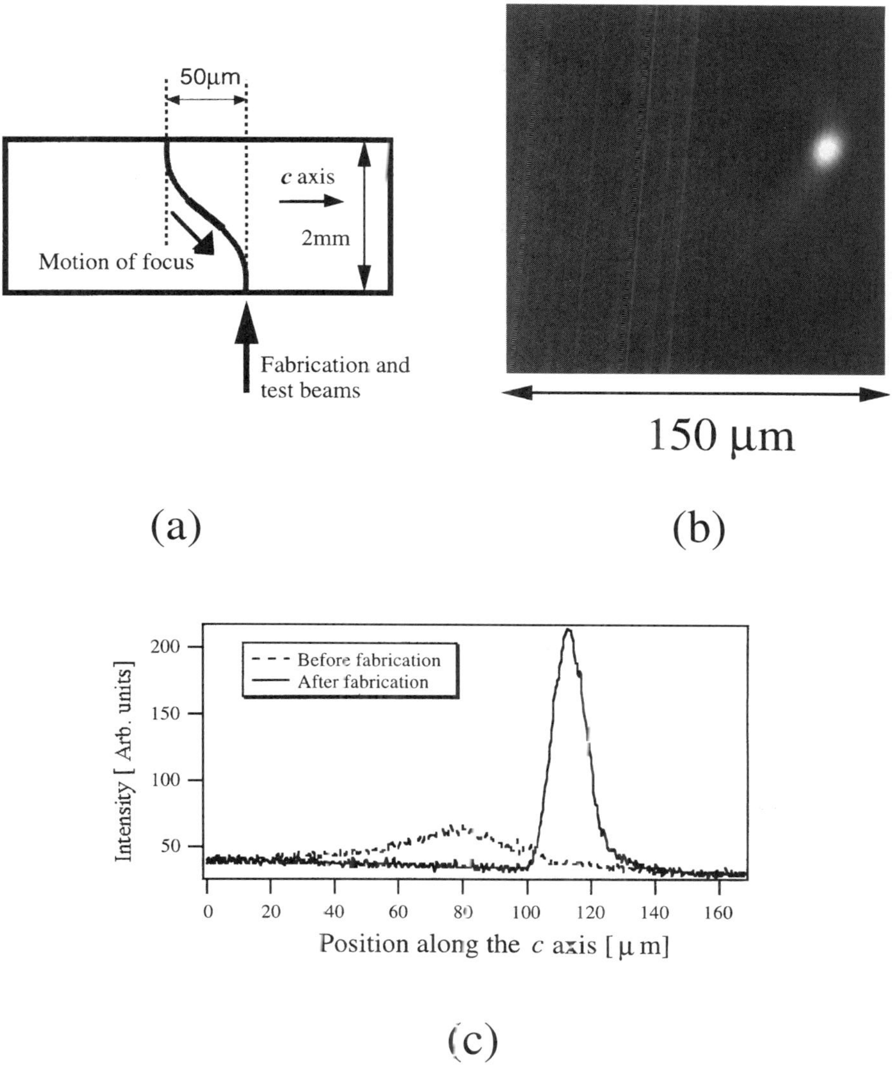

Figure 13.7: Fabrication of the curved waveguide: (a) shape of the curved waveguide, (b) near-field pattern after fabrication of the curved waveguide and (c) its cross section along the c-axis.

in Fig. 13.5a. In the model, it is assumed that the photorefractive waveguides fabricated by scanning the focused beam have same profile in their cross sections except the value. The value of the positive refractive-index change is an unknown parameter and depends on the optical power of the fabrication beam.

For simplicity, we will describe the two-dimensional model of the photorefractive waveguide [61,62]. Let the x-axis be parallel to the c-axis in the LN crystal and the z-axis be parallel to the optical axis of Ar^+ laser beam in Fig. 13.4. The fabrication beam has a Gaussian profile. The model of the refractive index distribution of the photorefractive waveguide that is composed of three different Gaussian functions is expressed as

$$\Delta n(x,z) = \alpha \sum_{i=1}^{3} a_i \exp \frac{-(x + c_i + f(z))^2}{b_i^2}, \qquad (13.1)$$

where α is an unknown parameter and depends on the power of the fabrication beam; a_i, b_i, and c_i are determined by fitting $\Delta n(x, z)$ to the experimental data shown in Fig. 13.5 by using the least-squares method and are summarized in Table 13.1 in the case of $\alpha = 1$; and $f(z)$ denotes the variation of the focused spot along the x-axis. If $f(z)$ is constant with respect to z, the model of refractive index distribution is the same as that of the straight waveguide. Note that the model is based on two assumptions: (1) that the maximum index change, Δn_{max}, is the only unknown parameter and is determined by adapting α to the experimental results; and (2) that the profile along the x-axis is unchanging except that the position of the center of the profile changes in accordance with the locus of the focused spot.

The maximum value of the refractive-index change can be estimated by the comparison between numerical and experimental results of the near-field patterns in the curved waveguides. In the curved waveguide, a part of guided light may be radiated around the curved portion by

i	$ai[\times 10^{-6}]$	$bi[\mu m]$	$ci[\mu m]$
1	0.96	7.1	−17.1
2	−3.63	7.4	0.0
3	1.23	7.8	23.2

Table 13.1: Constant Values in Eq. (1).

directional change of the waveguide. Thus, an accurate index distribution of the waveguide can be obtained by this comparison. Three curved waveguides were fabricated, whose transverse displacements between the input positions and output positions of the waveguides along the x-axis were 30, 50, and 70 μm, respectively. The cross sections along the x-axis of experimental near-field patterns of these curved waveguides are shown in Fig. 13.8a. The radiation loss is almost negligible. In numerical simulation, various profiles of near-field patterns were calculated using the beam propagation method (BPM) [63–65]. The calculated field covered a rectangular area of 204.8 μm $\times$ 2000 μm in the x-z-plane and was represented by a 512 $\times$ 4096 computational grid. The step sizes of the grid along the x- and z-beam focused by lens L1 was assumed to have a Gaussian profile with $1/e^2$ intensity radius of 1.7 μm. Figure 13.8b shows the profiles of near-field patterns in three curved waveguides in the case of $\Delta n_{\max} = 1.24 \times 10^{-3}$ ($\alpha = 1.3 \times 10^3$). Judging from the profiles of the near-field patterns, these results were in good agreement with the experimental results. The intensity distribution of propagating light in the curved waveguide whose transverse displacement along the x-axis was 50 μm is shown in Fig. 13.9a. In cases with a lower value of $\Delta n_{\max}$, most of the guided light is radiated at the second curved portion of the waveguide. Figure 13.9b shows the intensity distribution of propagating light in the case of $\Delta n_{\max} = 0.5 \times 10^{-3}$. To make a more accurate index distribution model, the effect of the saturation of the refractive-index change and the time-evolution of the photorefractive medium exposed to the moving fabrication beam should be included in the model. Recently, Ueki *et al.* [66] proposed a technique to solve numerically the refractive index distribution from Kukhtarev's equations when the focused beam is incident in the photorefractive material.

13.2.3 Modification of waveguide structure for dynamic interconnections

For the implementation of the dynamic interconnection, the modification of a waveguide structure was demonstrated [54, 67]. Optically fabricated waveguide can be modified by a guided beam itself. The guided beam with a strong optical power can modify the refractive-index of the photorefractive waveguide due to the photorefractive effect. This modification of the waveguide structure leads to the change of the output power of the guided signal beam. The experimental setup was the same as shown in

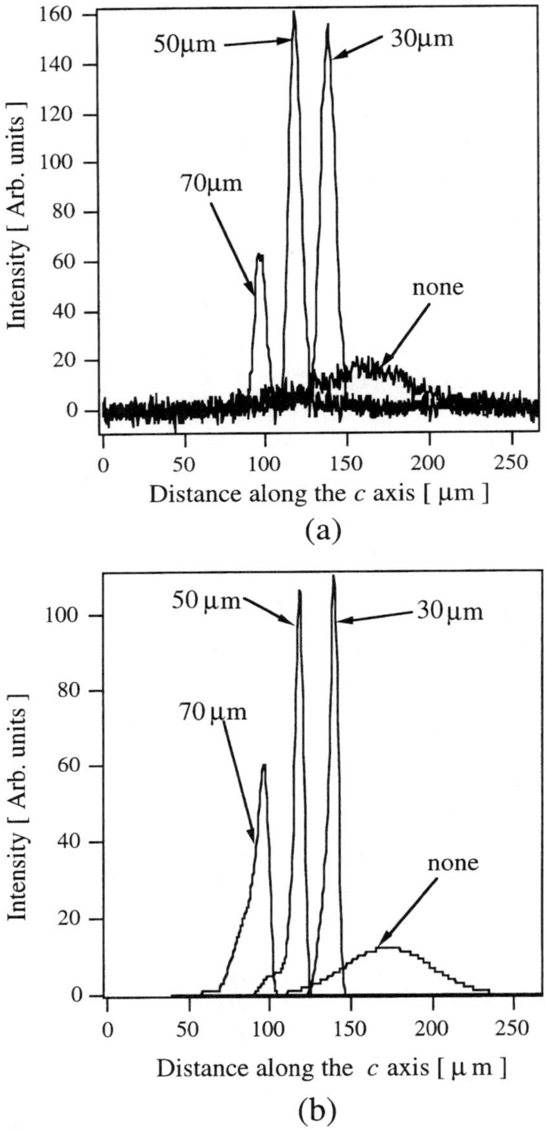

Figure 13.8: Cross sections of near-field patterns in the cases with no waveguide(none), and three curved waveguides: (a) experimental and (b) numerical results.

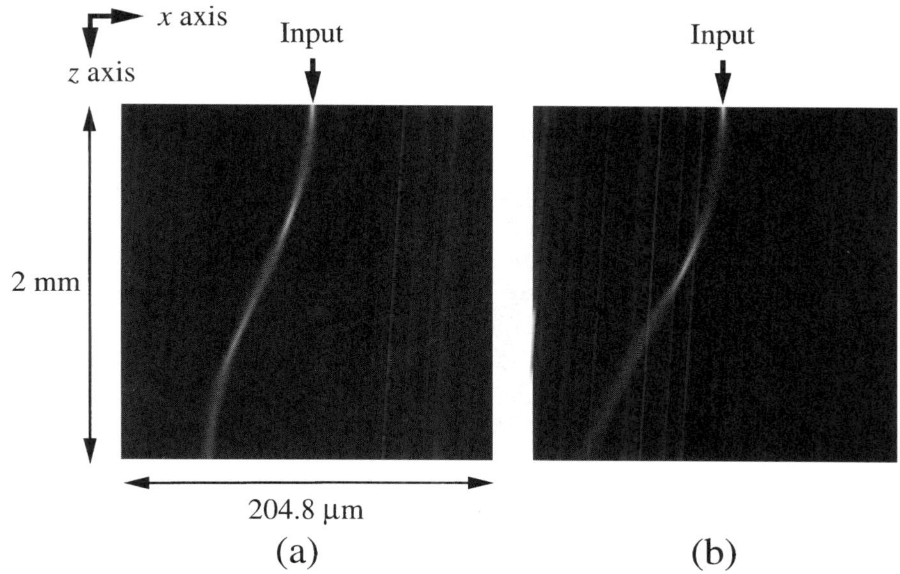

Figure 13.9: Numerical results of the intensity distribution of the propagating light in the curved waveguide where (a)$\Delta n = 1 \times 10^{-3}$ and (b) $\Delta n = 0.5 \times 10^{-3}$.

Fig. 13.4. We describe briefly a modification process of the photorefractive waveguide. The Ar^+ laser beam with extraordinary polarization was used to excite a strong guided beam in the photorefractive waveguide. This strong guided beam is called as a modification beam. The modification beam was excited at the input end of the photorefractive waveguide. A half-wave plate was inserted between the polarizer, P2, and the shutter, SH1, to select the extraordinary ray. The shutter SH1 was used to control the exposure energy of the modification beam. To investigate the modified waveguide, near-field patterns of a probe beam that propagates in the photorefractive waveguide were measured. A He-Ne laser beam (He-Ne laser 1) with extraordinary polarization was used to excite the probe beam. The probe beam was excited at the input end of the photorefractive waveguide in the same manner as the modification beam. The output power of the He-Ne laser 1 was adjusted to a sufficiently low level to prevent further modifications. The near-field pattern of the probe beam was captured by the CCD.

The change of the probe beam caused by the modification beam in a straight waveguide was investigated. The fabrication parameters were as follows: the optical power of the fabrication beam was 40 mW, the spatial interval between two successive positions of illumination was 45 μm, and the exposure time at each position of illumination was 1/125 s. After fabrication, an initial near-field pattern of the probe beam was obtained. Then, the modifications of the waveguide by illumination of the modification beam and observations of the nearfield patterns of the probe beam were repeated alternately. Figure 13.10 shows the changes of the near-field patterns of the probe beams in the waveguide located at the negative side of the *c*-axis. Figure 13.10a shows the initial near-field pattern. Figures 13.10b and c show the near-field patterns obtained when the total exposure energies of the modification beam were approximately 0.48 and 0.96 mJ, respectively. The output power of the modification beam was 10 mW and the exposure time at a single shot of illumination of modification beam was 1/125 s. As a result, the optical energy of 0.08 mJ was launched at the single shot of illumination of the modification beam. It is clearly shown that the intensity of the probe beam decreased with illumination of the modification beam.

Figure 13.11 shows the normalized peak intensities of the probe beam versus the total energy of the modification beam. Three different intensities of the modification beam of 4, 6, and 10 mW were tested. All the peak intensities were normalized by each of initial peak intensity. The normalized peak intensity decreased monotonously as a function of the total energy of the modification beam and showed a total decrease of approximately 73% against the initial value. The total decrease is an important factor for the optical dynamic interconnections. The solid line in this figure indicates a curve expressed by

$$I = \alpha \exp(-\beta E) + \gamma, \tag{13.2}$$

where I is the normalized peak intensity of the probe beam, E is the total energy of the modification beam, $\alpha = 0.59$, $\beta = 1.31$, and $\gamma = 0.27$. It is shown that the peak intensity of the probe beam can be controlled by the total energy of the modification beam independent of the intensity of the modification beam.

The mechanism of the decay of the probed power was numerically investigated by using the model of the refractive index distribution of the photorefractive waveguide. For simplicity, two-dimensional distribution of the refractive index in the straight waveguide was investigated. An

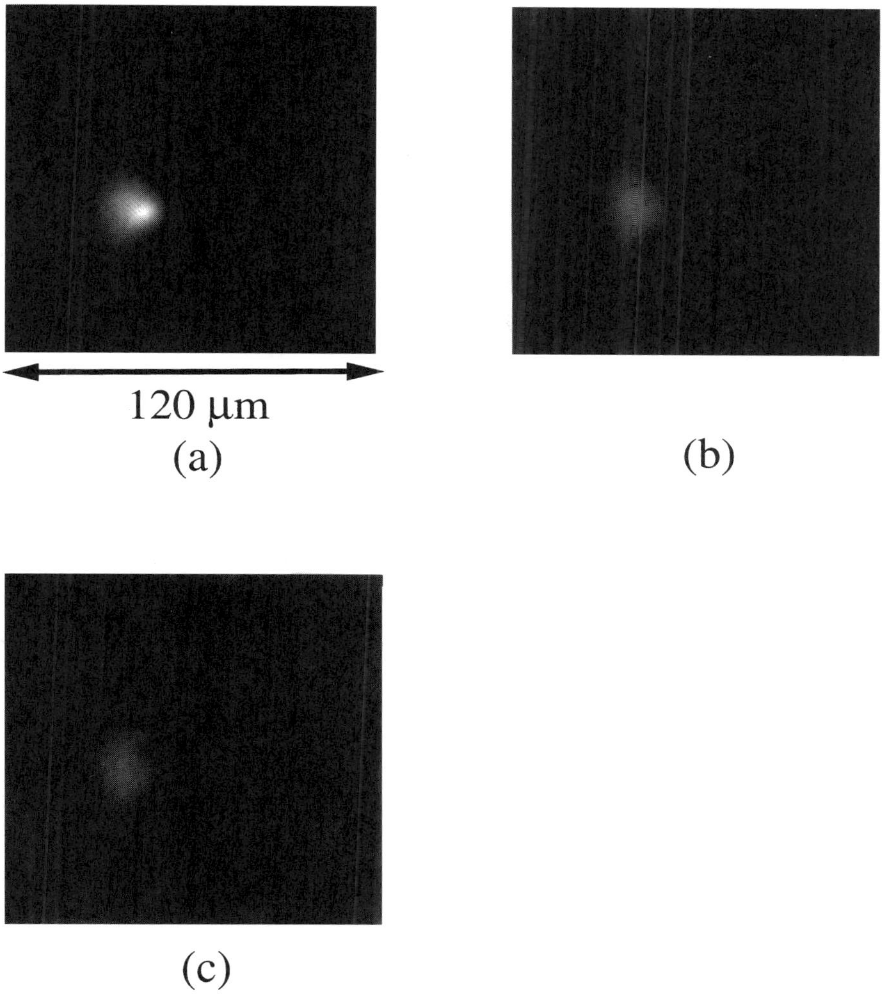

Figure 13.10: Near-field patterns of the probe beam when total energy of the modification beam are (a) 0 mJ, (b) 0.48 mJ, and (c) 0.96 mJ, respectively.

initial distribution of the refractive index of the straight waveguide was calculated in the case that the maximum refractive index change was assumed to be 1.0×10^{-3}. The intensity distribution of the modification beam, including the guided and radiated modes, was calculated by using the BPM. The profile of the modification beam at the input end of the

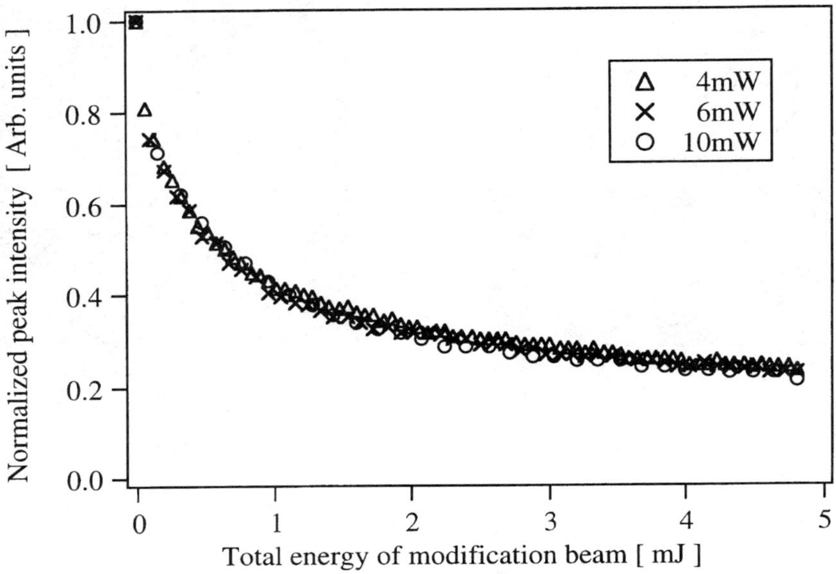

Figure 13.11: Peak intensity of the probe beam as a function of total energy of the modification beam.

waveguide was assumed to have a Gaussian profile with a $1/e^2$ intensity radius of 1.7 μm. For the modification of the refractive index, a simple model suggested in [68] was applied. The refractive-index change, $\Delta n(x,z)$, is, then, expressed by

$$\Delta n(x,z) = -\kappa T \times I(x,z), \tag{13.3}$$

where κ is a positive constant, T is the exposure time, and $I(x,z)$ is the intensity distribution of the modification beam. Equation (13.3) indicates that the refractive index decreases in proportion to the intensity of the modification beam in the LN crystal. The modified distribution of the refractive index of the waveguide is expressed as

$$n_{t+1}(x,z) = n_t(x,z) + \Delta n(x,z), \tag{13.4}$$

where $n_t(x,z)$ denotes the distribution of the refractive index when the number of computational iteration is t.

The intensity distribution of the probe beam that propagates the resultant model waveguide was, then, calculated by using the BPM. The profile of the incident probe beam at the input end of the waveguide was assumed to have a Gaussian shape with a $1/e^2$ intensity radius of 2.0 μm. The near-field profile was obtained on the rear face of the crystal. The peak intensities were calculated from all the near-field profiles. The numerical results were compared with the experimental results.

Figure 13.12 shows the normalized peak intensity as a function of the total energy of the modification beam. In this case, κ was chosen to be 5.0 J^{-1}. Numerical data were given the bias term estimated in the experiment and were normalized by the peak intensity of the initial near-field pattern. The numerical results showed good agreement with the experimental results and demonstrated that the normalized peak intensity decreased exponentially as a function of the total energy of the modification beam.

The numerical calculations showed the refractive-index change in the straight waveguide caused by the modification beam. Figure 13.13 shows the changes in the refractive index along the central axis of the

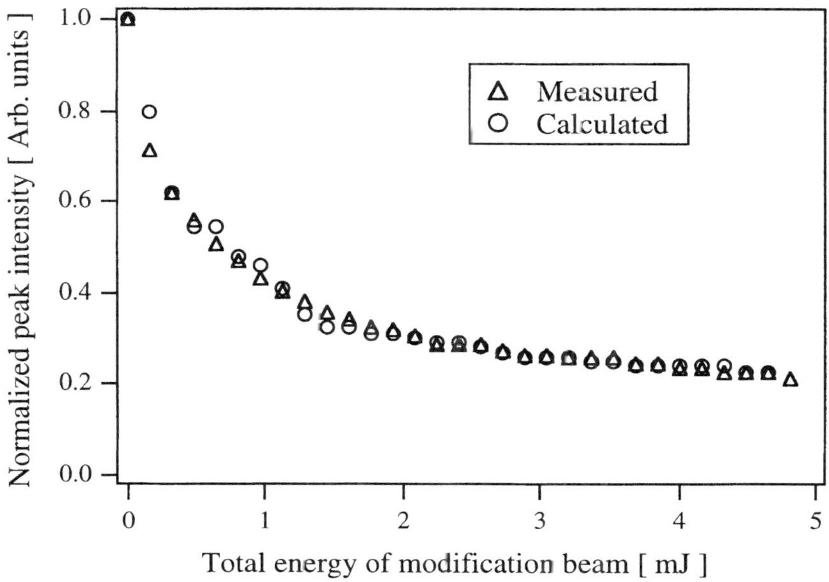

Figure 13.12: Comparison of the numerical results with the experimental results. Calculated and measured data are plotted simultaneously as a function of the total energy of the modification beam.

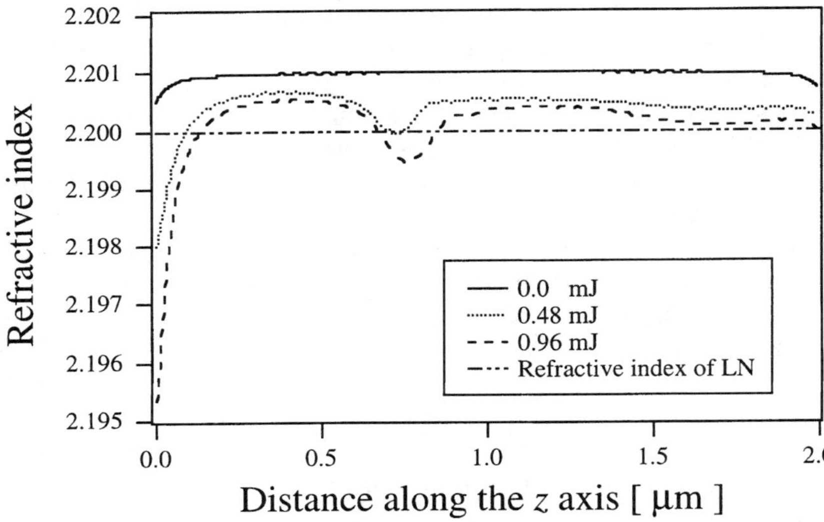

Figure 13.13: Refractive-index profiles along the central axis of the waveguide.

waveguide. In Fig. 13.13, we can see that the change of the refractive index is a decrease and is mainly induced around the input end of the waveguide, which leads to lower coupling efficiency of the incident probe beam to the guided mode. Large decrease of the refractive index is caused by the localization of light flux near the input end of the waveguide. At the initial stage of the experiments, the fractional power of the guided beam was approximately 40% of the input power. The rest of the input power was radiated mainly at the input end. This is why the decrease of the refractive index at the input plane was larger than that at the other part in the waveguide. The decrease of the refractive index can be seen in the middle way of the z-axis. This is caused by the radiated modes that occasionally form high-intensity portions during the propagation. These results agree with the fact that the refractive index decreases at the center of the focused beam as shown in Fig. 13.5a.

13.2.4 Application

In [69], the learning ability of a neural network using the photorefractive waveguides was numerically evaluated. The network consists of three layers. A weight of an interconnection was encoded as a power transmittance of the photorefractive waveguide. In Fig. 13.11, it was shown that the optical power transmittance of the signal beam in the photorefractive waveguide decreased monotonously as a function of total exposure of the modification beam. Thus two channels of waveguides and a differential detection scheme [70, 71] were used in one connection to realize the positive and negative signs of weights. In the simulated network, the positive and negative optical signals of a neuron propagated through respective waveguides and were detected at the end face of the crystal. The final output signal was the difference between the positive and negative signal obtained electrically. To decrease the weight, the transmittance of positive weight was decreased by the modification beam. Oppositely, to increase the weight, the transmittance of negative weight was decreased. The noise caused by the radiated flux gives no serious effects because the radiated flux can hardly couple again with the propagation modes and the intensity of the radiated beam becomes small near the plane of detectors. To implement a parallel modification of weights in the learning phase, a modified anti-Hebbian local algorithm [72] that uses local signals was employed. Numerical results showed that the network using the photorefractive waveguides can solve the exclusive OR problem better than the conventional network based on the backpropagation algorithm [73].

13.3 Segmented photorefractive waveguide

Another type of waveguide structure, called segmented waveguide [74, 75], has been used for the quasi-phase matching in the second harmonic generation or the wavelength filter. The segmented waveguide consists of periodically aligned high refractive-index regions, so its axial structure is not uniform. This waveguide works as a kind of lens array waveguide. Optical interconnections based on the segmented waveguides in the photorefractive material may be more flexible than continuous waveguides. Various interconnection paths can be taken by tilting an input angle or shifting an input position when there are many high-refractive-index regions as shown in Fig. 13.14a. Each high-refractive-index region can be shared by many waveguides. Furthermore, the structure of the segmented waveguide can be modified by adding or erasing the high-refractive-index

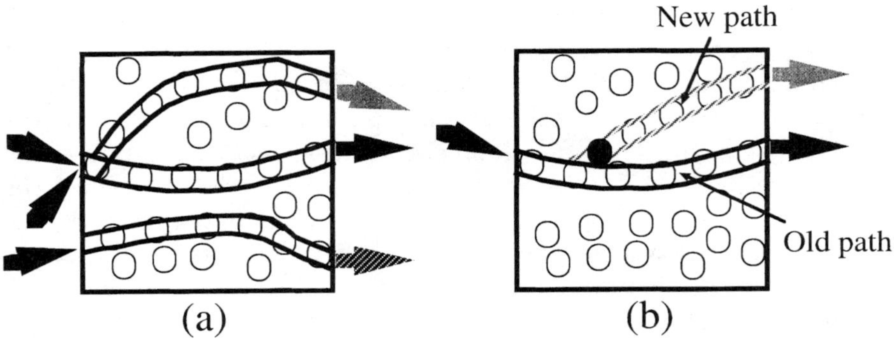

Figure 13.14: Illustrations of optical interconnections by use of SPWs. (a) One can select various interconnection paths by shifting an input position or tilting an incident angle. (b) Description of the modification of waveguide structures. If a new high-refractive-index region represented by a solid circle is created by optical illumination, a new interconnection path can be created. When the high-refractive-index region is erased, the path can be eliminated.

regions, thus resulting in the dynamic changes of waveguide functions (Fig. 13.14b). The Fresnel loss caused at the boundary between the high and low refractive index region is small because the refractive index change due to the photorefractive effect is less than 10^{-2}. In this section the fabrication of segmented waveguides in the LN crystal and a dynamic modification of the waveguide structure for an adaptive interconnection are described [76]. The fabricated waveguide is called as a segmented photorefractive waveguide (SPW).

The SPW consists of many localized high-refractive-index regions. The high-refractive-index region is called a cell. The cell is created by a single exposure of a focused laser beam in a photorefractive material. The SPW is fabricated by exposing the photorefractive material point by point by shifting the focused beam. Periodically aligned cells behave like a uniform waveguide whose refractive index is equal to an average index over a period of segmented waveguide [75].

13.3.1 Fabrication

Figure 13.15 shows the experimental setup for the fabrication and the guiding experiments of the SPWs. A LN crystal doped with iron ion (0.03 mol%) was placed between two microscope objective lenses, OL2 and OL3.

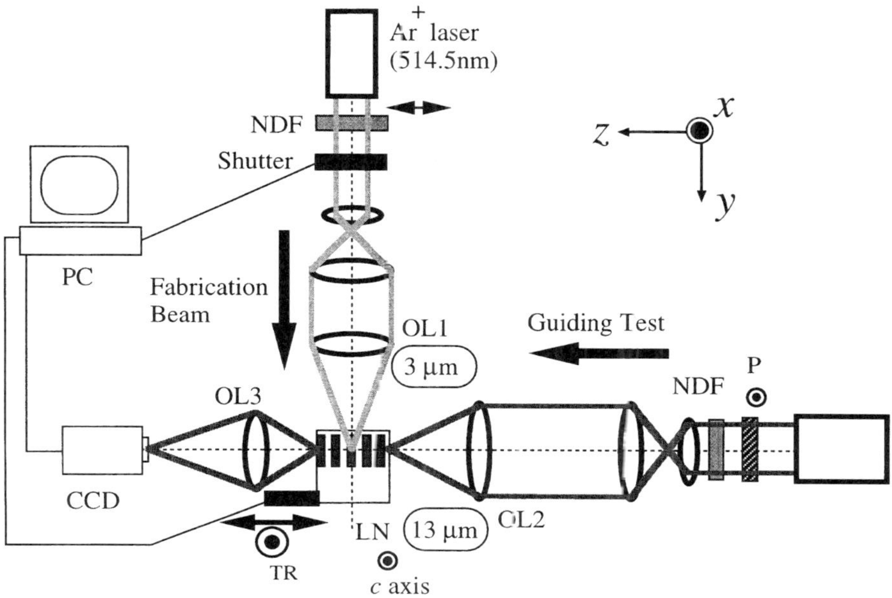

Figure 13.15: Experimental setup: OLs, microscope objective lenses; NDF, neutral-density filter; TR, translator; P, polarizer; PC, personal computer.

The iron-doped LN crystal is more sensitive than undoped crystal at the wavelength of the Ar^+ laser. The c-axis was perpendicular to both of the optical axes of Ar^+ and He-Ne laser beams. The dimensions of the LN crystal along both the optical axes were 5.0 mm. The y- and z-axes are taken to be parallel to the optical axes of Ar^+ and He-Ne laser beams, respectively. The x-axis is parallel to the c-axis of the LN crystal. The Ar^+ laser beam (λ = 514.5 nm) with ordinary polarization was used as a fabrication beam. The fabrication beam was focused through a microscope objective lens OL1 with a numerical aperture of 0.42. The crystal was illuminated by the focused beam step by step to create the segmented structure. The crystal was moved along x- and z-axes on two translation stages controlled by a personal computer. A shutter controlled the exposure at each illumination. The distribution of the refractive index change caused by the focused beam in the LN crystal was measured by another Mach-Zehnder interferometer. The cross section of the distribution of the refractive index change was the same as shown in Fig. 13.5a. Thus two cells are simultaneously created by a single exposure. Each cell has an asymmetrical structure along the c-axis. A symmetric and large index change cell is created by illuminating the crystal at two positions separated along the c-axis by the appropriate distance, δ. This illumination method is called a sandwich illumination. A central high-index region is used as a segment in the waveguide.

In the guiding experiments, an optical beam with extraordinary polarization emitted from the He-Ne laser at 632.8 nm was launched into the waveguide from the front face of the crystal and near-field patterns on the rear face were observed by a CCD. The diameter of the input beam was 13.2 μm. If a waveguide is not fabricated, a broad and weak output intensity pattern by a defocused beam is observed. After fabrication, the output pattern changes to a sharp and intense pattern by the guided beam.

A straight SPW with a periodic structure was fabricated. Fabrication parameters were as follows: The optical power of the fabrication beam was 2.3 mW, the exposure time at each illumination was 1/31 s, the period along the z-axis, Λ, was 50 μm, and δ in the sandwich illumination was 25 μm. Figure 13.16a shows the near-field pattern after fabrication of the straight SPW. This figure covers a rectangular area of 300 μm × 280 μm. The vertical and horizontal axes are parallel to the c-axis of the crystal and the y-axis, respectively. Figures 13.16b and c show the intensity profiles taken along the central lines parallel to the vertical and horizontal axes in Fig. 13.16a. Dashed lines indicate the intensity profiles before fabrication of the SPW. This result showed that the waveguide structure was successfully fabricated. A fraction of 2.5% of the input power was guided and widths of the guided beam at $1/e^2$ of the peak intensity were 5.5 and 81.5 μm along the vertical and horizontal axes, respectively. This low output power is due to the low coupling of the input beam to guided modes at the input plane. The input beam from the He-Ne laser was a circular spot with the radius of 6.6 μm. On the other hand, Figure 13.16a shows that the guided mode has a narrow profile along the vertical axis (c-axis), but has a wide profile along the horizontal axis (y-axis) because of the large depth of focus along the y-axis of the fabrication beam. The width along the y-axis of the used fabrication beam was 40 times as large as that along the x- or z-axes.

The dark decay time of the straight SPW at room temperature was measured. The straight SPW was fabricated by the optical power of the fabrication beam of 3.9 mW, the exposure time at each illumination of

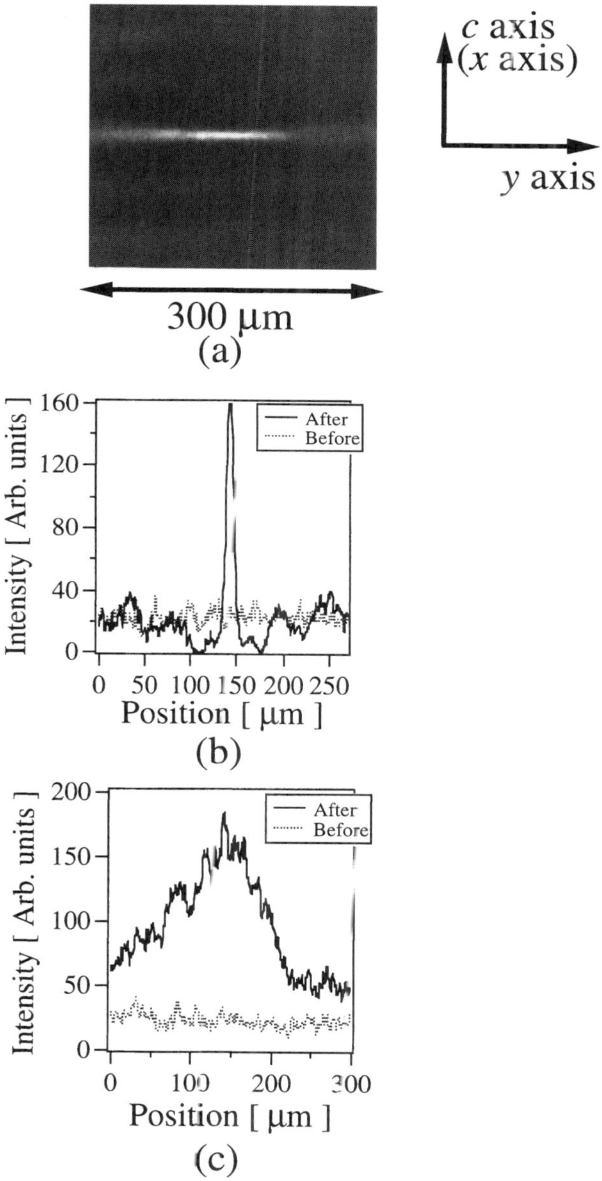

Figure 13.16: Experimental results in the straight SPW: (a) near-field pattern after fabrication, (b) and (c) their cross-sectional profiles of intensity distributions along the vertical and horizontal axes, respectively.

1/8 s, $\Lambda = 100\ \mu m$, and $\delta = 30\ \mu m$. Figure 13.17 shows an average power of the guided beam, $D(t)$, as a function of time, t. The average power was calculated as an average intensity over an effective area where the intensity exceeded half of the peak intensity. The dark decay time, t_d, was estimated to be 81 days by fitting the equation given by $D(t) = \exp(-2t/t_d)$ to the experimental data of Fig. 13.17.

An output power of a guided beam can be changed by controlling a period of the cell. Output powers of the straight SPWs with various periods were measured. All the cells have the same structure. The optical power of the fabrication beam was 3.2 mW, the exposure time at each illumination was 1/31 s, and $\delta = 25\ \mu m$. Figure 13.18 shows the output power of the guided beam as a function of the period of the cell. The output powers were calculated from the near-field patterns. The output power was defined as a total intensity over an effective area where intensity exceeded half of the peak intensity in the waveguide with the period of 50 μm. This result shows that short periods result in high power of the guided beam. This is because the loss of the optical power between successive cells decreases.

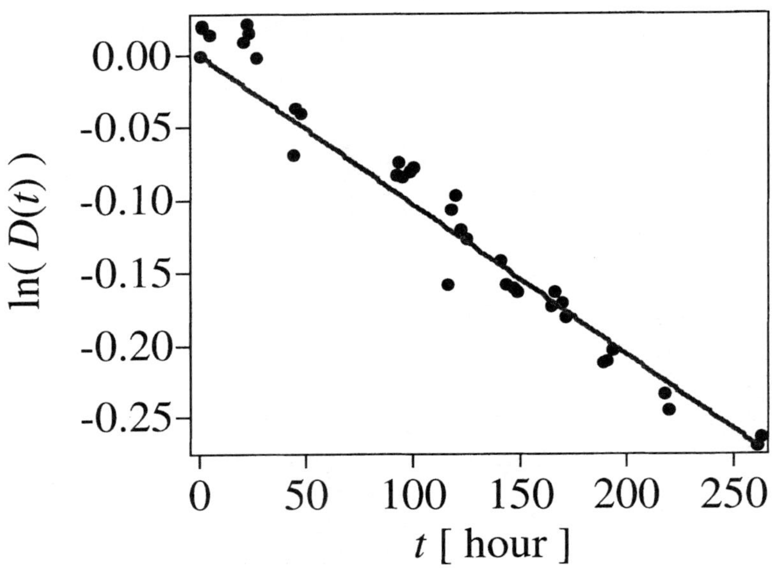

Figure 13.17: Output power density of the guided beam in the straight SPW as a function of time, t, at room temperature in the dark.

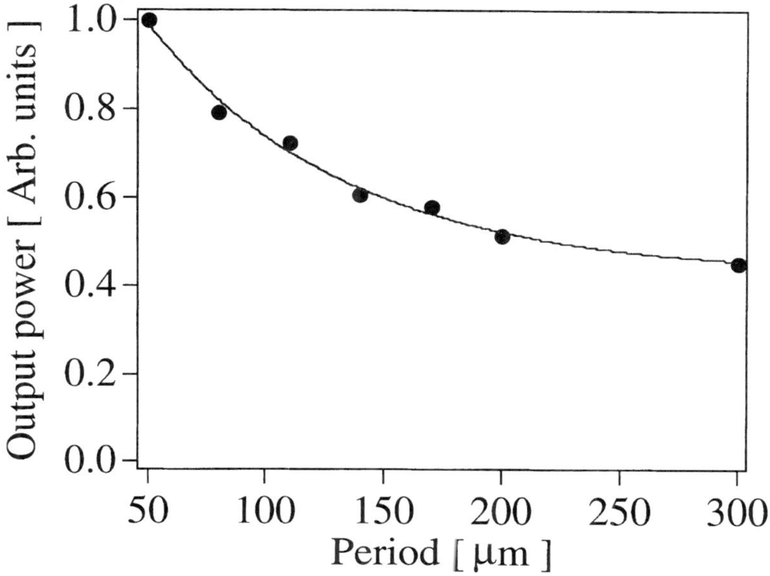

Figure 13.18: Output power of the guided beam as a function of period in straight SPWs.

13.3.2 Tolerance for fabrication errors

Tolerance for fabrication errors in the straight SPWs was investigated both experimentally and numerically. All the cells were not created in the right position of the periodic array and each cell was irregularly shifted along the z- or x-axes. The position error of each cell was assumed to obey the normal distribution with the mean of zero and the variance, ρ^2. Differences between intensity profiles of the guided beams in the SPWs with and without the random position errors were evaluated by using an error, E, that is defined as

$$E = \frac{\sum_x \sum_y \{In(x,y) - Iw(x,y)\}^2}{\sum_x \sum_y \{In(x,y) + Iw(x,y)\}^2}, \tag{13.5}$$

where $Iw(x,y)$ and $In(x,y)$ denote the output intensity distributions with and without the position errors, respectively.

At first the SPWs with the position errors along the z-axis were experimentally investigated. Fabrication parameters were as follows: The period was 50 μm, the exposure at each illumination was 0.08 mJ, and $\delta = 25$ μm. The random position errors were introduced intentionally by a personal computer. Figure 13.19a shows the transmitted power of the guided beam as a function of ρ. $E = 0.007$, 0.003, and 0.003 when $\rho = 10$, 20, and 40 μm, respectively. We can see that the output power of the guided beam in the SPWs with the longitudinal position errors is not changed.

Figure 13.19b shows the output power of the guided beams in the SPWs with the random position errors along the x-axis. The period was 50 μm, the exposure at each illumination was 0.09 mJ, and $\delta = 25$ μm. When $\rho = 10$, 16 μm, the intense output patterns by the guided beam cannot be observed. $E = 0.06$, 0.42, and 0.39 when $\rho = 4$, 10, and 16 μm, respectively. This result shows that large position errors along the x-axis cannot create the waveguide structure. These experimental results show that in the fabrication of the straight SPW the strict control of the positions of the cells along the propagation direction is not important. However, the positions along the lateral direction should be precisely controlled.

A numerical analysis was made by using the simple two-dimensional model of the SPW and the BPM. The structure of the cell was a rectangle with a uniform refractive index. The numerical result shows that the high tolerance for the longitudinal position errors is due to the small beam divergence of the guided beam. The guided mode has a broad profile with a radius about 10 μm along the x-axis. The light with the small beam divergence is relatively insensitive to the longitudinal position errors of the cells.

13.3.3 Transformation of waveguide structure for dynamic interconnections

The structure of the SPW can be readily changed by adding or erasing cells. In Sec. 13.3.1, it was shown that the guided power can be changed by controlling the period of the cell. Here the modification of the waveguide structure in a SPW was described. A curved structure was fabricated and was subsequently transformed into a Y-branch structure as shown in Fig. 13.20. In Fig.13.20a, hatched squares indicate centers of the illumination and closed ones indicate used cells. The sandwich illumination technique was not used here. After a single scan of a focused beam, two curved

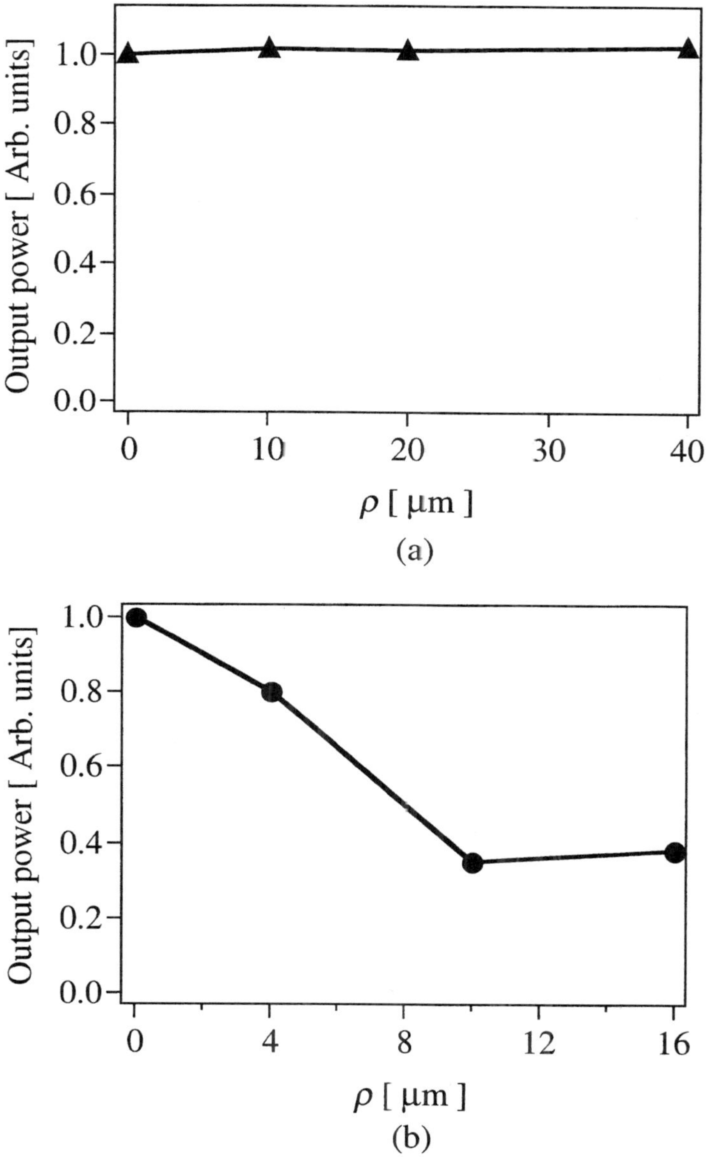

Figure 13.19: Output power of the guided beam as a function of ρ in the SPWs with the position errors along (a) the z- and (b) the x-axes.

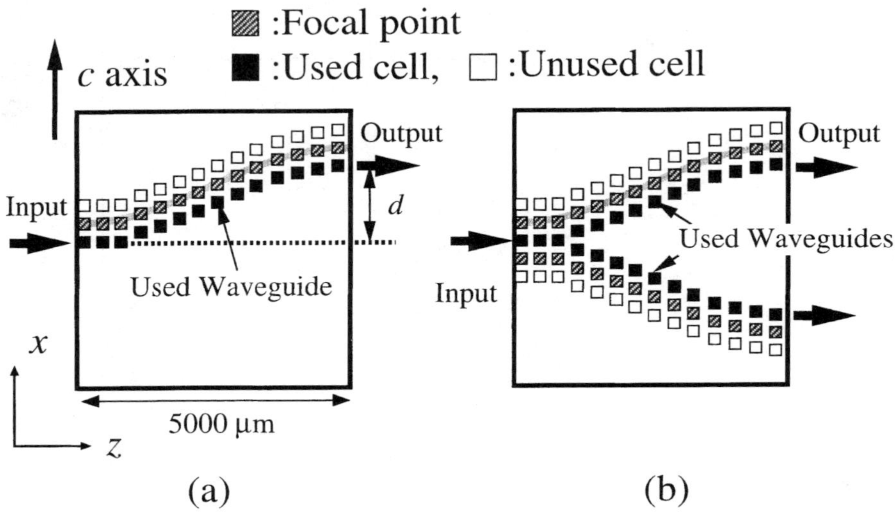

Figure 13.20: Illustration of (a) a curved and (b) a Y-branch structure.

waveguides bent in the x-z plane were fabricated as shown in Fig. 13.20a. Each cell has the asymmetric structure along the x-axis (c-axis) because it was created by a single exposure. The Y-branch structure consists of two curved waveguides that are symmetrical with respect to the z-axis as shown in Fig.13.20b. Figure 13.21 shows near-field patterns in the curved and the Y-branch structures. The optical power of the fabrication beam was 2.5 mW, the exposure time at each illumination was 1/8 s, and the period, $\Lambda = 100$ μm. The position shift, d, along the x-axis between the input and output planes was 57 μm. Figure 13.21a indicates that the curved structure was successfully fabricated. In the experiments, the curved structure with 250 μm shift along the x-axis was successfully fabricated. In Fig. 13.21b we can see two intense output beams that are observed at appropriate positions. This result shows that after creation of the curved structure, its structure was successfully changed into the Y-branch structure by adding the array of cells. We can also see no remarkable decrease in the output power at the curved waveguide between Fig. 13.21a and b. This is because in the Y-branch waveguide the coupling power between the input beam and guided modes increases due to the large and symmetrical refractive index variation along the c-axis at the input. Note that the cell structure at the input in the Y-branch waveguide is equivalent to the structure created by the sandwich illumination.

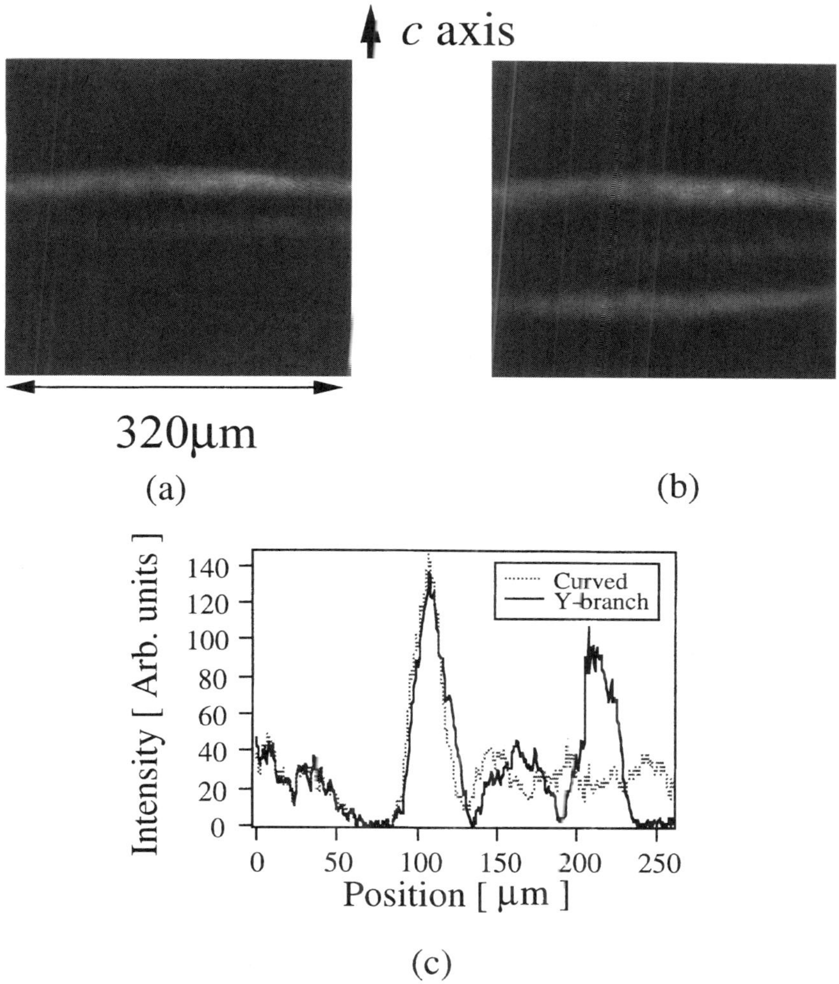

Figure 13.21: Near-field patterns in (a) the curved waveguide, (b) the Y-branch waveguide, and (c) its cross section along the c-axis.

13.4 Array of photorefractive waveguides

For massively parallel interconnections, fabrication methods of an array of the photorefractive waveguides were proposed [77–79]. A two-dimensional array of photorefractive waveguides can be optically fabricated in the photorefractive material by illumination of a checkered pattern composed of an array of nondiffracting beams.

13.4.1 Fabrication technique

Figure 13.22 shows an example of optical systems to fabricate the array of three-dimensional photorefractive waveguides. Light emitted from four coherent point sources, namely, $\delta(x' - a, y' - b)$, $\delta(x' + a, y' - b)$, $\delta(x' + a, y' + b)$, and $\delta(x' - a, y' + b)$, passes through a Fourier transform lens, L. When one of the sources has a phase delay of π rad, a checkered intensity pattern is formed in the x-y plane and is uniform along the z-axis in the interference region of the four plane waves. The checkered intensity pattern is described by

$$I(x, y) = 1 - \sin(2a\phi x)\sin(2b\phi y), \tag{13.6}$$

$$\phi = \frac{2\pi}{\lambda\sqrt{a^2 + b^2 + f^2}}, \tag{13.7}$$

where λ is the wavelength and f is the focal length of the Fourier-transform lens. It is important to note that $I(x, y)$ is uniform along the optical axis

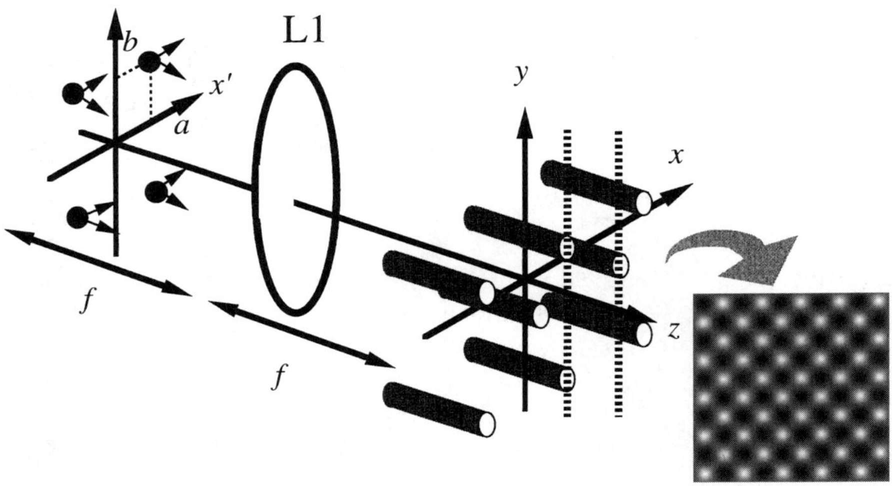

Figure 13.22: Illustration of a fabrication method of an array of photorefractive waveguides.

because the point sources are coplanar and on a single circle. Thus the field is composed of an array of nondiffracting beams [80]. Let a LN crystal be placed so that the c-axis is parallel to the x-axis. In the LN crystal, the light-induced space-charge field is formed mainly along the c-axis of the crystal and creates the refractive-index change via the Pockels effect. In the simplified model described in [68], the distribution of the refractive index change along the x-axis for the light with the extraordinary polarization is given by

$$\Delta n(x,y) = -\gamma / I(x,y)\,, \tag{13.8}$$

where γ is a constant that depends on the properties of the crystal such as electrooptic coefficients, the extraordinary refractive index, and the characteristic constant of the impurity. Equation (13.8) shows that the spatial variation of the refractive index is proportional to the intensity distribution of light. Since the refractive-index distribution along the z-axis is uniform, the array of straight photorefractive waveguides can readily be fabricated. Unlike a fiber bundle, the photorefractive waveguide has the advantage that the output power of a guided signal beam can be dynamically controlled by another guided beam with a strong optical power.

13.4.2 Experiments

Figure 13.23a shows the experimental setup for fabricating the array and observing the refractive index distribution of the array. Figure 13.23b shows the setup for guiding tests of fabricated waveguides. An Ar^+ laser ($\lambda = 514.5$ nm) with ordinary polarization was used as a light source in the fabrication. A beam emitted from the Ar^+ laser was divided into four beams by two Mach-Zehnder configurations. The two interferometers made vertical and horizontal interference fringes, respectively. A $20 \times 20 \times 2$ mm^3 LN crystal doped with iron ions (0.03 mol%) was placed in the interference region of four beams and was located at the focus of lens L1. The thickness of the crystal along the optical axis was 2 mm. The transmission of the crystal at the Ar^+ line used was 77.5%. The four beams were slightly converged by lens L1 ($f = 1000$ mm) to increase the power densities. The phase fronts of these beams, however, can be regarded substantially as plane waves in the LN crystal. The full width of the focused beam at $1/e^2$ of maximum was 0.69 mm. The phase shift of π rad of one beam was achieved by the path difference. The refractive index

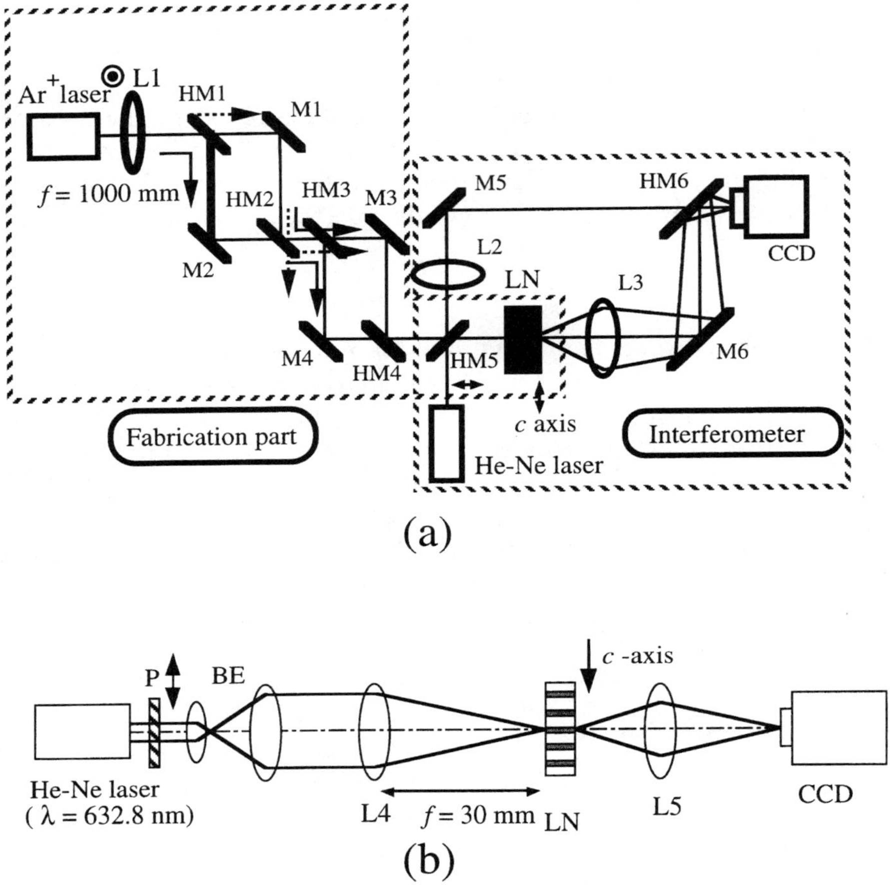

Figure 13.23: Experimental setup for (a) the fabrication of an array of two-dimensional photorefractive waveguides and (b) the guiding tests: Ls, lenses; HMs, half-mirrors; Ms, mirrors; P, polarizer; BE, beam expander.

distribution of the array can be measured by a Mach-Zehnder interferometer.

An array of the photorefractive waveguides was fabricated for observing the refractive index distribution. To create small variation of the refractive index distribution, the optical power of the Ar^+ laser was 24 mW and the exposure time was 25 s. The intensity pattern was stable during the exposure. After numerical analysis of the fringe pattern, the refractive index distribution of the cross section of the array as shown in Fig. 13.24a was obtained. The horizontal direction is parallel to the c-axis of the crystal. Periods along the vertical and horizontal axes are both 30 μm. Figure 13.24b shows the profile of the refractive index along the vertical axis. Maximum variation of the refractive index was 0.6×10^{-5}. This experimental result shows that the technique can create the two-dimensional array of photorefractive waveguide structures.

Two hundred segments of photorefractive waveguides with a 30 μm × 30 μm period were fabricated for the guiding tests. The optical power of the Ar^+ laser was 45 mW and the exposure time was 40 s. After fabrication of the array, the LN crystal was set in the position that was located at the focus of lens L4. A test beam with extraordinary polarization emitted from another He-Ne laser was focused by lens L4 and near-field patterns were observed by a CCD. The spot size of the test beam was estimated to be 2.6 μm and the optical power was made sufficiently low (submicrowatt) to avoid further refractive index change during the guiding tests. The results of guiding tests are shown in Fig. 13.25. The horizontal axis of the figure is parallel to the c-axis of the crystal. Figure 13.25a shows the near-field pattern when the test beam impinged on the center of a waveguide in the array. This figure shows that the guided light is well confined within the waveguide and the optical coupling between the adjacent waveguides is not seen. However, the small leakage of optical power can be seen along two diagonal directions. This is because the refractive index variations along the diagonal directions are slightly small as shown in Fig. 13.24a. Figures 13.25b and c show the near-field patterns when the test beam was shifted along the c-axis by 15 and 30 μm, respectively. In Fig. 13.25b, the test beam was located in between the successive waveguides. The test beam did not excite the guided modes in the two waveguides. The same results were observed at all the locations on the array.

13.4.3 Maximum density of photorefractive waveguides

The maximum density of the waveguide array was numerically estimated when the transferred power between adjacent waveguides was negligible in the array. If the density of the array is too high, the coupling of light wave between the adjacent waveguides may occur [81]. If each waveguide is to be used as an isolated signal channel, the transfer of the optical

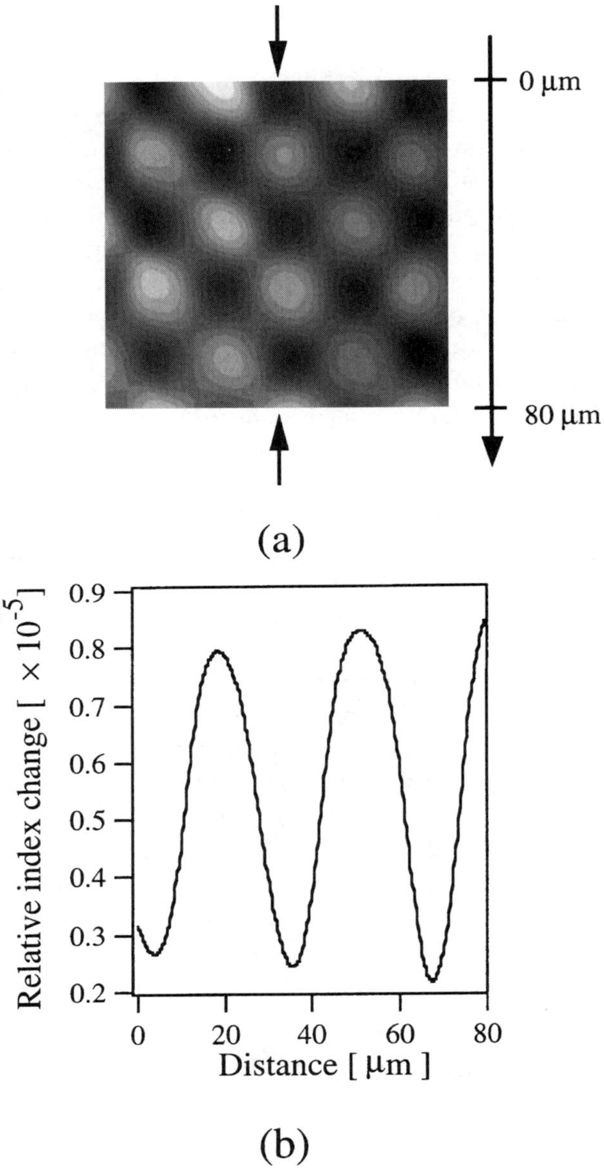

Figure 13.24: Cross section of the refractive-index distribution of the array created by the checkered intensity pattern and its profile along the vertical axis.

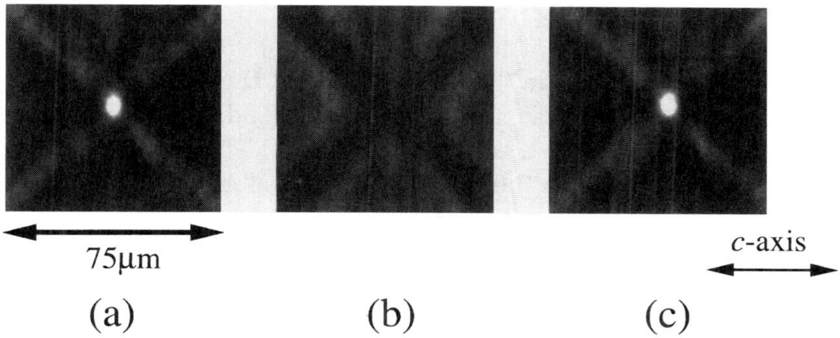

Figure 13.25: Near-field patterns when the center of the test beam at the input plane is shifted by (a) 0 μm, (b) 15 μm, and (c) 30 μm along the c-axis.

power of the guided beam due to this coupling between the adjacent waveguides is a serious problem. The coupling can be reduced by an increase in the period of the array. In this analysis, it is assumed that the coupling is negligible when the transferred power is smaller than 0.5% of the incident power. One can derive this critical waveguide distance between the centers of a pair of waveguides from the coupled mode theory [82] if the waveguides are slab waveguides. The minimum waveguide distance is estimated to be 12.4 μm when the width of the slab waveguide is 5 μm, the propagation distance is 2 mm, the wavelength of light is 632.8 nm, and the refractive indices in the core and cladding layers are 2.201 and 2.2, respectively. The propagation distance 2 mm is the thickness of the LN crystal used in the experiment. In the case of the sinusoidal index profile as in Eq. (13.8), it is difficult to solve analytically the period of the array with negligible coupling. By using the finite-difference BPM, the appropriate period of the array that has little coupling between the guided modes in the specified propagation distance was numerically calculated. Two-dimensional waveguides were used in this analysis. The period of the array was varied from 1 to 20 μm, in 1 μm steps. The calculated field covered a rectangular area of 81.9 μm × 2000 μm in the x-z plane defined in Fig. 13.22, and was represented by a 8192 × 16,384 computational grid. To estimate the transferred power, the intensity distribution of light that passes through a waveguide in the array was calculated. In this calculation, a fundamental mode of the slab waveguide whose width was half the period of the array was used as the incident light. The wavelength of light was 632.8 nm. Figure 13.26 shows the intensity distributions of propagating light for $\Delta n_{max} = 1 \times 10^{-3}$ when the periods of the array were 6 and 15 μm. It can be seen that the short period causes a large transfer of the power to the other waveguides, whereas the long period causes a negligible transfer of the power. The normalized transferred power (NTP) due to the coupling is defined as

$$\text{NTP} = 1 - \frac{P_a}{P_i}, \tag{13.9}$$

where P_a and P_i are optical powers passing through a waveguide with and without adjacent waveguides, respectively. The optical power was calculated as the total intensity within one period of the sinusoidal index profile. Figure 13.27 shows NTP versus the period of array. If NTP must be smaller than 5×10^{-3}, the period must be longer than 15 or 19 μm depending on the maximum refractive index change that is $\Delta n_{max} = 1.0 \times 10^{-3}$ or $\Delta n_{max} = 0.5 \times 10^{-3}$. 4.4×10^5 interconnections may be

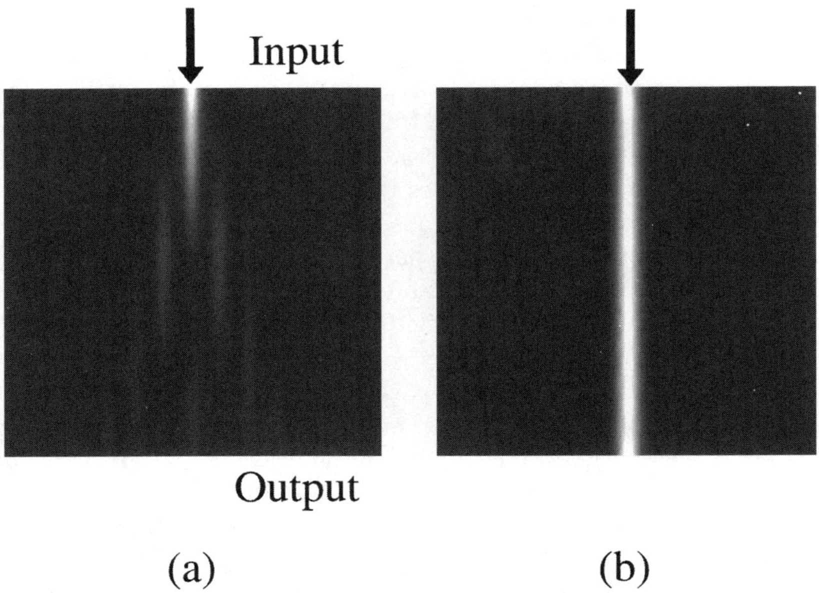

Figure 13.26: Intensity distributions in arrays of waveguides with periods of (a) 6 μm and (b) 15 μm.

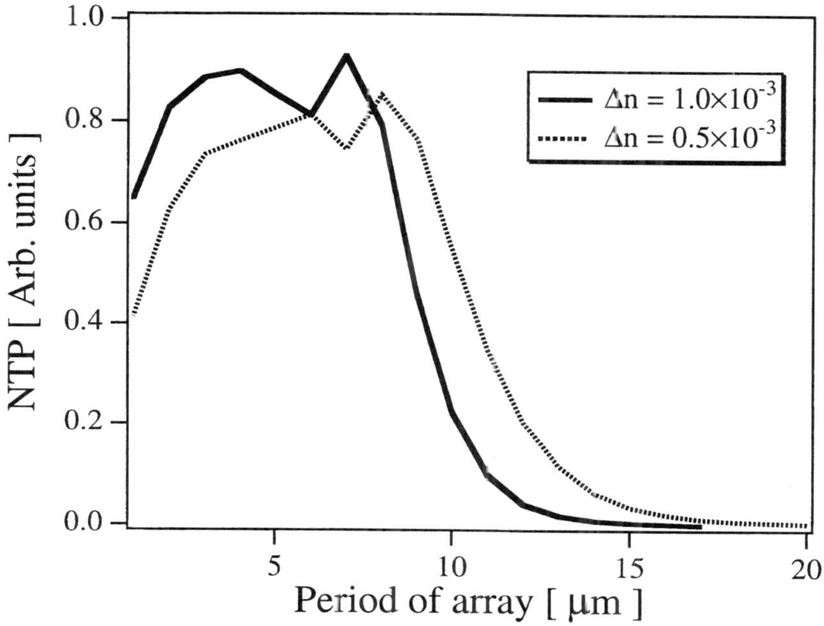

Figure 13.27: Normalized transferred power as a function of period of the array.

achieved in the LN crystal with a square centimeter for $\Delta n_{\max} = 1.0 \times 10^{-3}$.

13.5 Summary

In this chapter, we have described the optical interconnections based on the photorefractive waveguides. The photorefractive waveguides can be simply fabricated by scanning a focused beam in the photorefractive material. The structures fabricated in the LN crystal described in this chapter were straight, curved, and Y-branched. These structures remained during a few months at room temperature in the dark. We can fix the structure permanently by thermal or electrical treatment. Another technique such as two-photon absorption [83] or two-step recording [84–86] may be useful in the fabrication to avoid the erasure of the waveguide structure in the guiding test. One of the most distinct features of the photorefractive waveguide is the possibility of modifiying the waveguide structure by its strong guided beam or other external beam. The interconnections using the photorefractive waveguides are thus adaptive and we expect that this technique may realize a self-adaptive or a self-organizing network such as optical learnable neural networks. The interconnection technique based on the photorefractive waveguides can be used in the optical communication networks by developing photorefractive multiple quantum well devices [87–89] that have sufficient sensitivity at a wavelength of 1.3 or 1.55 μm.

References

1. J. W. Goodman, *Introduction to Fourier Optics*, 2nd ed., McGraw-Hill, New York, 1996.
2. J. Feinberg, "Real-time edge enhancement using the photorefractive effect," *Opt. Lett.,* **5**, 330–332, 1980.
3. A. E. Chiou and P. Yeh, "Parallel image subtraction using a phase-conjugate Michelson interferometer," *Opt. Lett.,* **11**, 306–308, 1986.
4. D. Z. Anderson and J. Feinberg, "Optical novelty filters," *IEEE J. Quantum Electron.,* **25**, 635–647, 1989.
5. T. Tschudi *et al.*, "Image amplification by two- and four-wave mixing in $BaTiO_3$ photorefractive crystal," *IEEE J. Quantum Electron.,* **QE-22**, 1493–1502, 1986.
6. F. T. S Yu and D. A. Gregory, "Optical pattern recognition: Architectures and techniques," *Proc. IEEE,* **84**, 733–752, 1989.
7. D. Psaltis and N. Farhat, "Optical information processing based on an associative-memory model of neural nets with thresholding and feedback," *Opt. Lett.,* **10**, 98–100, 1985.
8. K. Wagner and D. Psaltis, "Multilayer optical learning networks," *Appl. Opt.,* **26**, 5061–5076, 1987.
9. D. Psaltis *et al.,* "Adaptive optical networks using photorefractive crystals," *Appl. Opt.,* **27**, 1752–1759, 1988.
10. H. J. Caulfield, J. Kinser, and S.K. Rogers, "Optical neural networks," *Proc. IEEE,* **77**, 1573–1583, 1989.
11. M. Ishikawa, N. Mukohzaka, H. Toyoda and Y. Suzuki, "Optical association: A simple model for optical associative memory," *Appl. Opt.,* **28**, 291–301, 1989.
12. D. Psaltis, D. Brady, X.-G. Gu, and S. Lin, "Holography in artificial neural networks," *Nature,* **343**, 325–330, 1990.
13. J. H. Hong, S. Campbekk, and P. Yeh, "Optical pattern classifier with Perceptron learning," *Appl. Opt.,* **29**, 3019–3025, 1990.

14. H.-Y. S. Li, Y. Qiao, and D. Psaltis, "Optical network for real-time face recognition," *Appl. Opt.,* **32**, 5026–5035, 1993.
15. Y. Owechko, "Nonlinear holographic associative memories," *IEEE J. Quantum Electron.,* **QE-25**, 619–634, 1989.
16. A. Ashkin, G. D. Boyd, J. M. Dziedzic, R. G. Smith, A. A. Ballman, J. J. Levinstein, and K. Nassau, "Optically-induced refractive index inhomogeneities in $LiNbO_3$ and $LiTaO_3$," *Appl. Phys. Lett.,* **9**, 72–74, 1966.
17. P. Günter and J.-P. Huignard (eds.), *Photorefractive Materials and their Applications* Vols I and II, Springer-Verlag, Berlin, 1988.
18. P. Yeh, *Introduction to Photorefractive Nonlinear Optics,* Wiley-Interscience, New York, 1993.
19. L. Solymar, D. J. Webb, and A. Grunnet-jepsen, *The Physics and Applications of Photorefractive Materials,* Clarendon, Oxford, 1996.
20. F. S. Chen, J. T. Macchia and D. B. Fraser, "Holographic storage in lithium niobate," *Appl. Phys. Lett.,* **13**, 223–225, 1968.
21. H. Kurz, "Lithium niobate as a material for holographic information storage," *Philips Tech. Rev.,* **37**, 109–120, 1977.
22. Y. Shi, D. Psaltis, A. Marrakchi, and A. R. Tanguray, Jr., "Photorefractive incoherent-to-coherent optical convertor," *Appl. Opt.,* **22**, 3665–3667, 1983.
23. E. Voit and P. Günter, "Photorefractive spatial light modulation by anisotropic self-diffraction in $KNbO_3$," *Opt. Lett.,* **12**, 769–771 1987.
24. J. H. Hong, F. Vachss, S. Campbell, and P. Yeh, "Photovoltaic spatial light modulator," *J. Appl. Phys.,* **69**, 2835–2840, 1991.
25. S. Sternklar, S. Weiss, M. Segev, and B. Fischer, "Beam coupling and locking of lasers using photorefractive four-wave mixing," *Opt. Lett.,* **11**, 528–530, 1986.
26. S. Weiss, M. Segev, S. Sternklar, and B. Fischer, "Photorefractive dynamic optical interconnects," *Appl. Opt.,* **27**, 3422–3428, 1988.
27. M. Cronin-Golomb, "Dynamically programmable self-aligning optical interconnect with fan-out and fan-in using self-pumped phase conjugation," *Appl. Phys. Lett.,* **54**, 2189–2191, 1989.
28. N. Wolffer, V. Vieux, V. Royer, G. Picoli, and P. Gravey, "Self-aligned holographic interconnects between single-mode fibers with mutually pumped phase conjugate mirrors," *Int. J. Opt. Com.,* **2**, 245–257, 1991.
29. B. Fischer and S. Sternklar, "Self Bragg matched beam steering using the double color pumped photorefractive oscillator," *Appl. Phys. Lett.,* **51**, 74–75 1987.
30. S. Ashihara, O. Matoba, T. Shimura, K. Kuroda, J. Kato, and I. Yamaguchi, "Mutually pumped phase conjugators with picosecond pulses," *J. Opt. Soc. Am.,* **B15**, 1971–1976, 1998.
31. J.F. Heanue, M. C. Bashaw, and L. Hesselink, "Volume holographic storage and retrieval of digital data," *Science,* **265**, 749–752, 1994.

32. F. H. Mok, " Angle-multiplexed storage of 5000 holograms in lithium niobate," *Opt. Lett.*, **18**, 915–917, 1993.
33. J.-J. P. Drolet, E. Chuang, G. Barbasahis, and D. Psaltis, "Compact, integrated dynamic holographic memory with refreshed holograms," *Opt. Lett.*, **22**, 552–554 1997.
34. S. J. Frisken, "Light-induced optical waveguide uptapers," *Opt. Lett.*, **18**, 1035–1037, 1993.
35. A. S. Kewitsch and A. Yariv, "Self-focusing and self-trapping of optical beams upon photopolymerization," *Opt. Lett.*, **21**, 24–26, 1996.
36. K. M. Davis, K. Miura, N. Sugimoto, and K. Hirao, "Writing waveguides in glass with a femtosecond laser," *Opt. Lett.*, **21**, 1729–1731, 1996.
37. K. Miura, Q. Jianrong, H. Inouye, T. Mitsuyu and K. Hirao," Photowritten optical waveguides in various glasses with ultrashort pulse laser," *Appl. Phys. Lett.* **71**, 3329–3331, 1997.
38. K. Itoh, O. Matoba, and Y. Ichioka, "Fabrication experiment of photorefractive three-dimensional waveguides in lithium niobate," *Opt. Lett.*, **19**, 652–654, 1994.
39. M. Shih, M. Segev, and G. Salamo, "Circular waveguides induced by two-dimensional bright steady-state photorefractive spatial screening solitons," *Opt. Lett.*, **21**, 931–933, 1996.
40. A. Bekker, A. Peda'el, N. K. Berger, M. Horowitz, and B. Fischer, "Optically induced domain waveguides in $Sr_xBa_{1-x}Nb_2O_6$ crystals," *Appl. Phys. Lett.*, **72**, 3121–3123, 1998.
41. J. J. Amodei and D. L. Staebler, "Holographic pattern fixing in electro-optic crystals," *Appl. Phys. Lett.*, **18**, 540–542, 1971.
42. D. L. Staebler, W. J. Burke, W. Phillips, and J. J. Amodei, "Multiple storage and erasure of fixed holograms in Fe-doped $LiNbO_3$," *Appl. Phys. Lett.*, **26**, 182–184, 1975.
43. F. Micheron and G. Bismuth, "Electrical control of fixation and erasure of holographic patterns in ferroelectric materials," *Appl. Phys. Lett.*, **20**, 79–81 1972.
44. A. S. Kewitsch, M. Segev, A. Yariv, and G. J. Salamo, "Ferroelectric domain gratings in strontium barium niobate induced by photorefractive space charge fields," *Phy. Rev. Lett*, **73**, 1174–1177, 1994.
45. M. Segev, B. Crosignani, and A. Yariv, "Spatial solitons in photorefractive media," *Phy. Rev. Lett.*, **68**, 923–926, 1992.
46. G. C. Valley, M. Segev, B. Crosignani, A. Yariv, M. M. Fejer, and M. C. Bashaw, "Dark and bright photovoltaic spatial solitons," *Phy. Rev.*, **A50**, R4457–R4460, 1994.
47. M. Segev, G. C. Valley, B. Crosignani, P. DiPorto, and A. Yariv, "Steady-state spatial screening solitons in photorefractive materials with external applied field," *Phy. Rev. Lett.*, **73**, 3211–3214, 1994.

48. M. Segev, M. Shih, and G. C. Valley, "Photorefractive screening solitons of high and low intensity," *J. Opt. Soc. Am.*, **B13**, 706–718, 1996.
49. G. Duree, M. Morin, G. Salamo, M. Segev, B. Crosignani, P. D. Porto, E. Sharp, and A. Yariv, "Dark photorefractive spatial solitons and photorefractive vortex solitons," *Phy. Rev. Lett.*, **74**, 1978–1981, 1995.
50. M. Morin, G. Duree, G. Salamo, and M. Segev, "Waveguides formed by quasi-steady-state photorefractive spatial solitons," *Opt. Lett.*, **20**, 2066–2069, 1995.
51. M. Taya, M. C. Bashaw, M. M. Fejer, M. Segev, and G. C. Valley, "Observation of dark photovoltaic spatial solitons," *Phy. Rev.*, **A52**, 3095-3100, 1994.
52. S. Lan, M. Shih, and M. Segev, "Self-trapping of one-dimensional and two-dimensional optical beams and induced waveguides in photorefractive $KNbO_3$," *Opt. Lett.*, **22**, 1467–1469, 1997.
53. M. Shih, Z. Chen, M. Mitchell, M. Segev, H. Lee, R. S. Feigelson, and J. P. Wilde, "Waveguides induced by photorefractive screening solitons," *J. Opt. Soc. Am.*, **B14**, 3091–3100, 1997.
54. O. Matoba, *et al.*, "Modification of photorefractive waveguides in lithium niobate by guided beam for optical dynamic interconnection," *Opt. Rev.*, **2**, 438–443, 1995.
55. F. S. Chen, "Optically induced change of refractive indices in $LiNbO_3$ and $LiTaO_3$," *J. Appl. Phys.*, **40**, 3389–3396, 1969.
56. A. M. Glass, D. V. D. Linde, and T. J. Negran, "High-voltage bulk photovoltaic effect and the photorefractive process in $LiNbO_3$," *Appl. Phys. Lett.*, **25**, 233–235, 1974.
57. A. M. Glass, "The photorefractive effect," *Opt. Eng.*, **17**, 470–479, 1978.
58. D. V. D. Linde and A. M. Glass, "Photorefractive effects for reversible holographic storage of information," *Appl. Phys.*, **8**, 85–100, 1975.
59. M. Takeda, H. Ina, and S. Kobayashi, "Fourier-transform method of fringe-pattern analysis for computer-based topography and interferometry," *J. Opt. Soc. Am.*, **72**, 156–160, 1982.
60. O. Matoba, K. Itoh, and Y. Ichioka, "Nonuniform and off-axis strctures for photorefractive waveguides in lithium niobate," *Opt. Eng.*, **35**, 2175–2181, 1996.
61. O. Matoba, K. Itoh, and Y. Ichioka, "Analysis of photo-induced waveguide in lithium niobate crystal," *Opt. Rev.*, **1**, 73–75, 1994.
62. K. Itoh, O. Matoba, and Y. Ichioka, "Optical characteristics of photorefractive three-dimensional waveguides in lithium niobate," *Proc. Int. Conf. on Optical Computing, Heriot-Watt University,* 641–644, 1995.
63. M. D. Feit and J. A. Fleck, Jr., "Light propagation in graded-index optical fibers," *Appl. Opt.*, **17**, 3990–3998, 1978.
64. R. Scarmozino and R. M. Osgood, Jr., "Comparison of finite-difference and Fourier-transform solutions of the parabolic wave equation with emphasis on integrated-optics applications," *J. Opt. Soc. Am.*, **A8**, 724–731, 1991.

65. Y. Chung and N. Dagli, "An assessment of finite difference beam propagation method," *IEEE J. Quantum Electron,* **QE-26**, 1335–1339, 1990.
66. H. Ueki, Y. Kawata, and S. Kawata, "Three-dimensional optical bit-memory recording and reading with a photorefractive crystal: Analysis and experiment," *Appl. Opt.,* **35**, 2457–2465, 1996.
67. K. Itoh, O. Matoba, and Y. Ichioka, "ODINN in LiN; Optical dynamic interconnections for neural networks in lithium niobate," *Proc. SPIE,* **2529**, 71–81, 1995.
68. Q.W. Song, C.-P. Zhang, and P.J. Talbot, "Self-defocusing, self-focusing, and speckle in $LiNbO_3$ and $LiNbO_3$:Fe crystals," *Appl. Opt.,* **32**, 7266–7271, 1993.
69. O. Matoba, K. Itoh, and Y. Ichioka, "Optical learning neural network using photorefractive waveguides," *Opt. Rev.,* **4**, 465–470, 1997.
70. N.H. Fahat, D. Psaltis, A. Prata, and E. Paek, "Optical implementation of the Hopfield model," *Appl. Opt.,* **24**, 1469–1475, 1989.
71. P.J. de Groot and R.J. Noll, "Adaptive neural network in a hybrid optical/ electronic architecture using lateral inhibition," *Appl. Opt.,* **28**, 3852–3859, 1989.
72. Y. Qiao and D. Psaltis, "Local learning algorithm for optical neural networks," *Appl. Opt.,* **31**, 3285–3288, 1992.
73. D.E. Rumelhart. G.E. Hinton and R.J. Williams, "Learning internal representaions by error propagation," in D.E. Rumelhart and J.L. McClelland (eds.) *Parallel Distributed Processing*, MIT Press, Cambridge, Massachusetts, 1986, pp. 318–364.
74. L. Li and J.J. Burke, "Linear propagation characteristics of periodically segmented waveguides," *Opt. Lett.,* **17**, 1195–1197, 1992.
75. Z. Weissman and A. Hardy, "Modes of periodically segmented waveguides," *IEEE J. Lightwave Technol.,* **LT-11**, 1831–1838, 1993.
76. Matoba, T. Inujima, T. Shimura, and K. Kuroda, "Segmented photorefractive waveguides in $LiNbO_3$:Fe," *J. Opt. Soc. Am.,* **B15**, 2006–2012, 1998.
77. O. Matoba, K. Itoh, and Y. Ichioka, "Array of photorefractive waveguides for massively parallel optical interconnections in lithium niobate," *Opt. Lett.,* **21**, 122–124, 1996.
78. O. Matoba, T. Itano, K. Itoh, and Y. Ichioka, "Fabrication of two-dimensional array of photorefractive waveguides," *Proc. SPIE,* **2778**, 527–528, 1996.
79. O. Matoba, K. Kuroda, and K. Itoh, "Fabrication of a two-dimensional array of photorefractive waveguides in $LiNbO_3$:Fe using non-diffracting checkered pattern," *Opt. Commun.,* **145**, 150–154, 1998.
80. J. Durnin, J.J. Miceli, Jr., and J.H. Eberly, "Diffraction-free beam," *Phy. Rev. Lett.,* **58**, 1499–1501, 1987.
81. S. Somekh, E. Garmire, A. Yariv, H.L. Gravin and R.G. Hunsoerger, "Channel optical waveguide directional couplers," *Appl. Phys. Lett.,* **22**, 46–47, 1973.

82. D. Marcuse, *Light Transmission Optics*, 2nd ed., Van Nostrand Reinhold, New York, 1982, Chapter 10.3.
83. M. Horowitz, B. Fischer, Y. Barad, and Y. Silberberg, "Photorefractive effect in a $BaTiO_3$ crystal at the 1.5-μm wavelength regime by two-photon absorption," *Opt. Lett.*, **21**, 1120–1122, 1996.
84. Y.S. Bai, R.R. Neurganonkar, and R. Kachru, "High-efficiency nonvolatile holographic storage with two-step recording in praseodymium-doped lithium niobate by use of continuous-wave lasers," *Opt. Lett.* **22**, 334–336, 1997.
85. M.A. Noginov, N. Noginova, M. Curley, N. Kukhtarev, H.J. Caulfield, and P. Venkateswarlu, and G.B. Loutts, "Optical characterization of Mn:$YAlO_3$: material for holographic recording and data storage," *J. Opt. Soc. Am.*, **B15**, 1463–1468, 1998.
86. L. Paraschis, M.C. Bashaw, A. Liu, and L. Hesselink, "Resonant two-photon processes for nonvolatile holography in photorefractive crystals under continuous-wave illumination," *J. Opt. Soc. Am.*, **B14**, 2670–2680, 1997.
87. D.D. Nolte, D.H. Olson, G.E. Doran, W.H. Knox, and A.M. Glass, "Resonant photodiffractive effect in semi-insulating multiple quantum wells," *J. Opt. Soc. Am.*, **B7**, 2217–2225, 1990.
88. C. De Matos, A. Le Corre, H. L'Haridon, B. Lambert, S. Salaün, J. Pleumeekers, and S. Gosselin, "Photorefractive *p-i-n* diode quantum well operating at 1.55 μm," *Appl. Phys. Lett.*, **68**, 3576–3578, 1996.
89. C. De Matos, A. Le Coree, H. L'Haridon, S. Gosselin, and B. Lambert, "Fe-doped InGaAs/InGaAsP photorefractive multiple quantum well devices operating at 1.55 μm," *Appl. Phys. Lett.*, **70**, 3591–3593, 1997.

Chapter 14

Self-Pumped Phase Conjugation in $BaTiO_3$:Rh for Dynamic Wavefront Correction of Nd:YAG Lasers

Nicolas Huot, Jean-Michel Jonathan, and Gérald Roosen

Laboratoire Charles Fabry de l'Institut d'Optique, Unité Mixte du Centre National de la Recherche Scientifique, Orsay Cedex, France

High-power Nd:YAG lasers delivering a diffraction-limited TEM_{00} (transverse electro-magnetic) mode are of interest for many applications such as pumping of optical parametric oscillator (OPO) or laser manufacturing. However, the laser beam quality is often affected by thermal aberrations induced in amplifier rods. Photorefractive self-pumped phase conjugation is one of the techniques that has been proposed to compensate for these aberrations. Since the early 1980s, it has been widely studied [1, 2] and has led to several applications such as the optical gyroscope [3], linewidth narrowing of lasers [4] and laser diodes [5], or dynamic wavefront correction of lasers [6]. However, the development of applications at near-infra-

PHOTOREFRACTIVE OPTICS
Materials, Properties and Applications

ISBN 0-12-774810-5/$30.00

red wavelengths was limited by the spectral sensitivity of the photorefractive materials used in the phase conjugate mirrors. Barium titanate ($BaTiO_3$), one of the most interesting photorefractive crystals for phase conjugation, has large electrooptical coefficients and provides high reflectivity phase conjugation but its spectral sensitivity was mainly limited to the visible range up to the 1990s. In 1994, intentional rhodium doping extended significantly its sensitivity to near-infrared wavelengths. Indeed, self-pumped phase conjugation was demonstrated in rhodium-doped barium titanate ($BaTiO_3$:Rh) in a total internal reflection geometry up to 0.99 μm in 1994 [7]. Most of the demonstrations were then made at laser diode wavelengths using $BaTiO_3$:Rh: optical feedback from a phase conjugate mirror was used to narrow the linewidth of a laser diode emitting at 813 nm [8], double-phase conjugation was performed at 800 nm [9], beam clean-up was achieved with powerful laser diodes emitting at 860 nm [10]. The photorefractive response of $BaTiO_3$:Rh has also been investigated at 1.06 μm [7, 11]. The encouraging results obtained at this wavelength stimulated further research on self-pumped phase conjugation and made possible the dynamic wavefront correction of pulsed Nd:YAG laser sources.

This chapter is devoted to the results obtained with $BaTiO_3$:Rh at 1.06 μm. In the first section, photorefractive characterizations of $BaTiO_3$:Rh are presented, under both cw and nanosecond conditions of illumination. In the second part, self-pumped phase conjugation is described. The two geometries which proved to be successful at 1.06 μm are detailed and compared. Finally, the introduction of photorefractive self-pumped phase conjugate mirrors using $BaTiO_3$:Rh in master-oscillator power-amplifier (MOPA) laser sources is presented in the third section. In conclusion, the performances of the photorefractive self-pumped phase conjugate mirrors using $BaTiO_3$:Rh are compared to other possible techniques able to perform dynamic wavefront correction at 1.06 μm, such as photorefractive beam cleanup, stimulated Brillouin scattering (SBS), or adaptative optics.

14.1 Characterization of the materials

In 1993, near-infrared sensitivity of blue $BaTiO_3$ was reported in the literature [12]. Self-pumped phase conjugate reflectivities of 76% were measured between 860 and 1004 nm, but the reasons for the blue color of the crystal and its near-infrared sensitivity were unknown. In 1994, rhodium was identified as the photorefractive trap responsible for this near-infrared significant response and intentionally doped $BaTiO_3$:Rh started to be investigated under cw illumination [7].

The following results have been obtained with $BaTiO_3$:Rh crystals from FEE (Forschunginstitut für mineralishe und metallishe Werkstoffe Edelsteine/Edelmetalle GmbH) and Deltronics. Throughout the chapter, the samples are respectively designated by FEE "crystal number" or Deltronic's "crystal number."

14.1.1 Characterization with continuous-wave illumination

Two kinds of experiments are conducted to investigate the photorefractive properties of $BaTiO_3$:Rh. Spectroscopic characterizations enable us to determine the nature of the photorefractive traps. Photorefractive two-wave mixing experiments provide a measurement of the photorefractive gain. Additional experiments such as light-induced absorption measurements, may discriminate between various photorefractive band transport models (one-center [13], two-center [14, 15], or three-charge state model [16, 17]).

Spectroscopic characterizations (electron spin resonance, absorption spectra, and pump/probe experiments) succeeded in identifying the active impurities, lying in the band gap of as-grown $BaTiO_3$:Rh. Two principal species have been identified: iron under two valence states (Fe^{3+} and Fe^{4+}) and rhodium under three valence states (Rh^{3+}, Rh^{4+}, and Rh^{5+}) [18]. The thermal levels of these impurities have been found to be of 1 eV for $Rh^{3+/4+}$, 0.9 eV for $Fe^{3+/4+}$, and 0.7 eV for $Rh^{4-/5+}$ above the valence band. Optical transitions (corresponding to the absorption of a photon and creation of a free hole in the valence band) and thermal levels (corresponding to recombination of holes) are different because of the Franck-Condon shift which takes into account lattice relaxation. Indeed, following [18], the optical transitions are lower than 1.9 eV for $Rh^{3+/4+}$, lower than 2.8 eV for $Fe^{3+/4+}$, and lower than 1.6 eV for $Rh^{4+/5+}$. As a consequence, at 1.06 μm (energy of 1.2 eV), Fe^{4+} is unlikely to be photoionized. The relevant band diagram for $BaTiO_3$:Rh illuminated at 1.06 μm therefore contains only one center under three charge states (Rh^{3+}, Rh^{4+}, Rh^{5+}). The corresponding bookkeeping model is given in Fig. 14.1

The photorefractive band transport model used here is the three-charge state model, which was initially published in 1995 [16]. It accounts

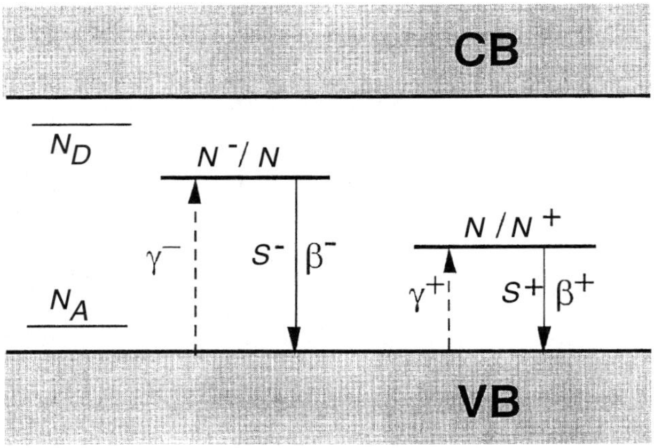

Figure 14.1: "Bookkeeping" diagram for the three-charge state model. N^-, N, and N^+ are the densities of Rh^{3+}, Rh^{4+}, Rh^{5+} respectively. $S^{\pm}$ are the photoexcitation cross sections, $\gamma^{\pm}$ are the recombination rates, and $\beta^{\pm}$ are the thermal excitation coefficients. N_A and N_D are optical inactive traps which ensure the electric neutrality of the crystal. Optical and thermal excitation as well as recombination of holes occur to and from the valence band.

for light-induced absorption, which is observed in such crystals at 1.06 μm [19]. Initially, light-induced absorption was explained by a model with two different centers [14, 19] which appears later to be inappropriate for BaTiO$_3$:Rh at 1.06 μm [17].

In BaTiO$_3$:Rh, the dopant is substituted to Ti^{4+}. In the following equations, Rh^{4+}, which has the same charge state as Ti^{4+}, is considered as the neutral level, of volume density N. Rh^{3+} appears as a hole acceptor of volume density N^-, Rh^{5+} as a hole donor of volume density N^+. Only hole conductivity is considered [7]. The electrical neutrality of the crystal is achieved by shallow donor and acceptor densities N_A and N_D. The main results of this model are presented below [16, 17].

Considering that the crystal is illuminated by a sinusoidal interference pattern of the form:

$$I = I_0\left[1 + \frac{m}{2}\left(e^{ik_g x} + \text{c.c.}\right)\right] \tag{14.1}$$

where I_0 is the average illumination, m is the modulation of the system of fringes, and k_g its wavenumber, one may assume that m is small enough to write any quantity involved in the equations as:

$$Q = Q_0 + \frac{1}{2}\left(Q_1 e^{ik_g x} + \text{c.c.}\right) \tag{14.2}$$

The resolution of the usual rate equations, continuity, and Poisson equations at zero order leads to the determination of the light-induced absorption $\Delta\alpha$ which can be written in terms of the steady-state average density of Rh^{4+}, $N_{0\text{Stat}}$:

$$N_{0\text{Stat}} = \frac{1}{1-\kappa}\left\{\left[\kappa N_T^2 + \kappa(\kappa - 1)\left(N_A - N_D\right)^2\right]^{1/2} - \kappa N_T\right\} \tag{14.3}$$

with

$$\kappa = \frac{S^+ I_0 + \beta^+}{S^- I_0 + \beta^-}\frac{\gamma^-}{4\gamma^+} \tag{14.4}$$

where S^+, S^-, γ^+, γ^-, β^+, and β^- are defined in Fig. 14.1. N_T is the total density of dopants: $N_T = N^- + N + N^+$.

Assuming that the absorption of a photon by Rh^{4+} or Rh^{5+} leads to the excitation of a hole with a quantum yield of 1, the steady-state value $\Delta\alpha_{\text{Stat}}\ (I_0)$ of the induced absorption is a function of the illumination:

$$\Delta\alpha_{\text{Stat}}(I_0) = \alpha_{\text{Stat}}(I_0) - \alpha_{\text{Stat}}(0) = \frac{hc}{\lambda}(S^+ - 2S^-)\frac{N_{0\text{Stat}}(0) - N_{0\text{Stat}}(I_0)}{2} \tag{14.5}$$

where hc/λ is the energy of an incident photon. As soon as $\frac{S^+}{S^-} << \frac{\beta^+}{\beta^-}$, the slope at the origin of the kinetics of the light-induced absorption (response to a step of illumination) can be expressed as follows:

$$\left(\frac{\partial\alpha}{\partial t}\right)_{t=0} \approx \frac{hc}{\lambda}(S^+ - 2S^-)S^- I_0 \frac{\gamma^+(N_T + N_A - N_D)^2}{\gamma^+(N_T + N_A - N_D) - \gamma^-(N_A - N_D)}. \tag{14.6}$$

This will be a relevant number in the characterization of the crystals.

Solving the rate, continuity, and Poisson equations at first order yields the photorefractive spatially varying space-charge field:

$$E_1 = imk_g \frac{k_B T}{e}\frac{1}{1 + k_g^2/k_0^2(I_0)}\eta(I_0) \tag{14.7}$$

with

$$k_0^2(I_0) = k_0^{-2} + k_0^{+2} \tag{14.8}$$

and

$$\eta(I_0) = 1 - \left(\frac{\beta^-}{S^- I_0 + \beta^-}\frac{k_0^{-2}}{k_0^2} + \frac{\beta^+}{S^+ I_0 + \beta^+}\frac{k_0^{+2}}{k_0^2}\right) \tag{14.9}$$

where k_0^- and k_0^+ are the Debye wavenumbers:

$$k_0^{\pm 2}(I_0) = \frac{e^2}{k_B T \varepsilon_S} N_{\text{eff}}^{\pm}(I_0) \tag{14.10}$$

with

$$N_{\text{eff}}^{+}(I_0) = N_0^{+}(I_0)\frac{(N_T + N_D - N_A)}{N_T} \quad \text{and}$$
$$N_{\text{eff}}^{-}(I_0) = N_0^{-}(I_0)\frac{(N_T - N_D + N_A)}{N_T} \tag{14.11}$$

where $\eta(I_0)$ is a saturation factor, which increases with I_0 to a value of 1 for high enough illuminations. Similarly, the effective Debye wave number k_0 varies with I_0 and saturates at high intensities. Equation (14.7) is similar to the one obtained in the case of a two-site photorefractive material [14]. However, the behaviors for $k_0(I_0)$ and $\eta(I_0)$ are different and the three-charge state model cannot be described as a subcase of the two-site model. In the three-charge state model, k_0^- and k_0^+ are not independent. There is a strong coupling between the populations of the three-charge states, and k_0 may be defined in terms of the population N_0 of the intermediate state alone:

$$k_0^2(I_0) = \frac{e^2}{k_B T \varepsilon_S}\left[N_T - N_0(I_0) - \frac{(N_D - N_A)^2}{N_T}\right] = \frac{e^2}{k_B T \varepsilon_S} N_{\text{eff}} \tag{14.12}$$

where N_{eff} is the intensity-dependent effective density of traps.

A 0°-cut BaTiO$_3$:Rh crystal referenced to as FEE-X16 was characterized by measuring the steady-state light-induced absorption (Fig. 14.2) together with its slope at $t = 0$ (Fig. 14.3). Additional two-wave mixing experiments were performed in the counterpropagating geometry [20] with ordinary polarized beams to determine N_{effsat}, the effective density of traps at saturation in intensity ($N_{\text{effsat}} = 5 \cdot 10^{15}$ cm $^{-3}$ for sample FEE-X16). All these data are fitted simultaneously and values of the internal parameters of the crystal with their error bars are deduced (Table

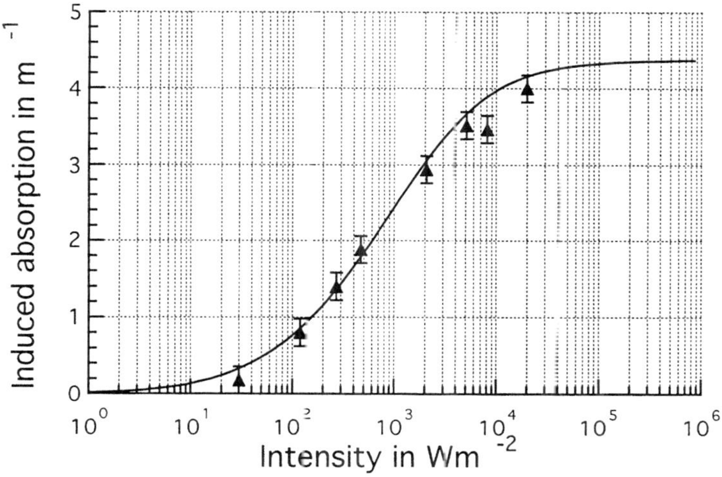

Figure 14.2: Steady-state light-induced absorption versus intensity for crystal FEE-X16. The full curve is a theoretical fit using Eq. 14.5 and parameters of Table 14.1.

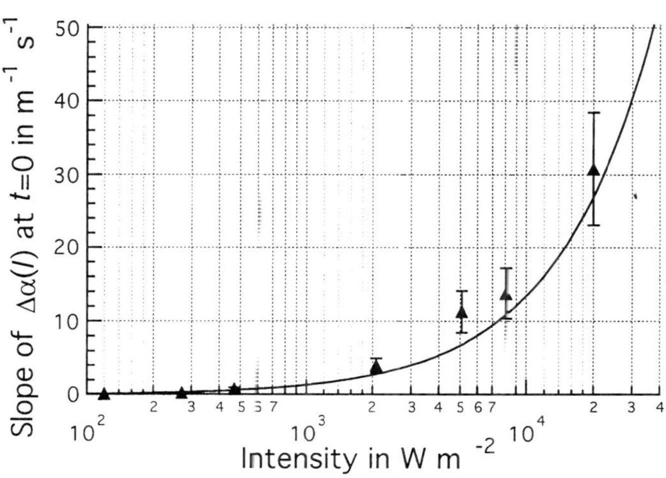

Figure 14.3: Slope at $t = 0$ of the light-induced absorption versus intensity for crystal FEE-X16. The entire curve is a theoretical fit using Eq. 14.6 and parameters of Table 14.1.

14.1). One can remark that the photoionization cross section from $Rh^{4+/5+}$ is 75 times higher than the one from $Rh^{3+/4+}$. This explains the light-induced absorption: holes are photoionized from $Rh^{3+/4+}$ (Rh^{4+} is the main charge state present in the crystal). Some of these holes, retrapped in $Rh^{4+/5+}$, create Rh^{5+}. Then Rh^{5+} is more likely to be photoionized than Rh^{4+}. Thus, absorption increases with illumination. As we will show in the following discussion, the determination of the internal parameters of the crystal is a key point for further optimization of nonlinear functions like optical phase conjugation.

Various two-beam coupling gain values have been reported in the literature using $BaTiO_3$:Rh at 1.06 μm with continous-wave (cw) illumination. Gains of 9.3 cm^{-1} and 11 cm^{-1} have been measured in 0°-cut samples [11, 21]. A gain of 23 cm^{-1} is obtained in a 45°-cut crystal referenced to as FEE-Y32-B. In this latter experiment, the sensitivity of the material is deduced using the rise time τ_{ph} of the photorefractive effect for a given incident intensity I and the photoinduced index modulation δn. A sensitivity of $S = \dfrac{\Delta n}{I\tau_{ph}} = 1.710^{-7} cm^2 J^{-1}$ is reported [20], which is 1000 times smaller than these determined at 532 nm and 4 times smaller than these obtained at 670 nm with $BaTiO_3$:Fe or $BaTiO_3$:Co [22]. The relatively poor sensitivity of $BaTiO_3$:Rh at 1.06 μm is mainly due to the long photorefractive response time at 1.06 μm, which is as expected given the low value of the absorption coefficient (0.1 cm^{-1}).

For applications, the photorefractive characteristics of $BaTiO_3$:Rh crystals at 1.06 μm must be reproducible. Even when they are cut from the

$\dfrac{\gamma^-}{\gamma^+}$	1.5 ± 0.3
S^-	$(1.2 \pm 0.5)\ 10^{-5}\ m^2J^{-1}$
S^+	$(9.0 \pm 0.1)\ 10^{-4}\ m^2J^{-1}$
β^-	$\beta^- << \beta^+$
β^+	$2.3 \pm 0.4\ s^{-1}$
N_T	$(3.3 \pm 0.6)\ 10^{23}\ m^{-3}$
$\lvert N_A - N_D \rvert$	$< 4.4\ 10^{21}\ m^{-3}$

Table 14.1: Parameters of crystal FEE-X16 at 1.06 μm deduced from the simultaneous fit of N_{effsat}, the steady-state light-induced absorption, and its slope on the origin, using the three-charge-state-model[a]

[a] See caption of Fig. 14.1 for meaning of parameters.

same boule, photorefractive crystals may present significant variations in their performances due to inhomogeneities in the boule. The reproducibility is checked by investigating two $BaTiO_3$:Rh crystals referred to as FEE-X16-5/2 (0°-cut) and FEE-X16-45 (45°-cut) from the same boule grown with 1000 ppm of rhodium in the melt. As derived from the three-charge state model, the steady-state light-induced absorption, its slope at $t = 0$, and the effective density of traps at saturation in intensity are relevant parameters. As the photorefractive gain Γ at saturation in intensity is related to N_{effsat}, comparing the different values of Γ measured in several samples in the same photorefractive configuration allows us to follow the eventual variations of N_{effsat}. Three parameters are given in addition: the photorefractive time constant τ_{ph} in the counterpropagating geometry for a given intensity, the erasure time in the dark τ_{dark} divided by τ_{ph}, and the absorption coefficient at saturation of the light-induced absorption. The experimental data given in Table 14.2 (ordinary polarizations and counterpropagating beams) show that the two samples have reproducible characteristics and that the homogeneity of the boule is correct. For sample FEE-X16-45, Γ and τ_{ph} are given for an angle of $\beta = 45°$ between the grating wavevector and the c-axis. Consequently, these data cannot be directly compared to these obtained in 0°-cut crystals. Then, a second boule (FEE-X17), also with 1000 ppm of rhodium in the melt, has been grown on purpose in the same conditions as the previous one, to reproduce the properties of FEE-X16-5/2 and FEE-X16-45 crystals. Two 0°-cut crystals from this boule FEE-X17-6/3 and FEE-X17-6/4 are investigated. The data reported in Table 14.2 indicate that these two crystals have similar properties and that the two boules are nearly equivalent. The only difference between the four crystals concerns the ratio τ_{dark}/τ_{ph}. Nevertheless, for the application to dynamic wavefront correction, the high value of this ratio for all the samples ensures that the photorefractive gain is well saturated.

14.1.2 Performances of oxidized crystals

As detailed in the previous section, one major feature of $BaTiO_3$:Rh at 1.06 μm is a photoexcitation cross section 75 times higher for the $Rh^{4+/5+}$ than for the $Rh^{3+/4+}$ level. This indicates that the photorefractive rise time might be reduced by lowering the Fermi level. Indeed, it has been shown that the Fermi level lies around the $Rh^{3+/4+}$ level in as-grown $BaTiO_3$:Rh [18]. When subjected to oxidation, the $Rh^{3+/4+}$ level becomes

	X16	X16-45	X17 6/3	X17 6/4
α	$(0.07 \pm 0.02)\ cm^{-1}$	$(0.12 \pm 0.02)\ cm^{-1}$	$(0.06 \pm 0.02\ cm^{-1}$	$(0.05 \pm 0.02)\ cm^{-1}$
$\Delta\alpha(1.1\ W/cm^2)$	$(3.9 \pm 0.2) m^{-1}$	$(5.7 \pm 0.2)\ m^{-1}$	$(2.8 \pm 0.2)\ m^{-1}$	$(3.6 \pm 0.2)\ m^{-1}$
$\left(\frac{\partial \Delta\alpha}{\partial t}\right)_{t=0, I=1.1\ W/cm^2}$	$(14 \pm 4)\ m^{-1}\ s^{-1}$	$(11 \pm 3)\ m^{-1}s^{-1}$	$(11 \pm 3)\ m^{-1}s^{-1}$	$(13 \pm 4)\ m^{-1}s^{-1}$
τ_{ph} at $6.7\ W/cm^2$	(8 ± 1) s	(2 ± 0.5) s $\beta = 45°$	(11 ± 1) s	(8 ± 1) s
τ_{dark}/τ_{ph}	240 ± 50	200 ± 40	90 ± 15	390 ± 70
Γ	$2.1\ cm^{-1}$	$0.6\ cm^{-1}$ $\beta = 45°$	$2.6\ cm^{-1}$	$2.5\ cm^{-1}$

Table 14.2: Relevant parameters for $BaTiO_3$:Rh crystals characterized at 1.06 μm with continuous-wave illumination (ordinary polarized beams and counterpropagating geometry)

full of holes. Consequently, Rh^{3+} vanishes and the density of Rh^{4+} rises. Moreover, Rh^{5+} is populated. This goes along with a decrease of the Fermi level, which drops from 1 eV (around the $Rh^{3+/4+}$ level) down to 0.7 eV ($Rh^{4+/5+}$ level). In oxidized crystals, it has also been proved in [18] that an additional level is present ($Fe^{4+/5+}$), which comes from oxidation of the $Fe^{3+/4+}$ level. But, as the optical transition of $Fe^{4+/5+}$ is larger than 2.5 eV, it is unlikely to be photoionized at 1.06 μm [18]. So, at 1.06 μm in oxidized crystals, only one level remains with a relatively high photoexcitation cross section: $Rh^{4+/5+}$.

The influence of the oxidation of $BaTiO_3$:Rh on its photorefractive properties is predicted by the three-charge state model [23]. Indeed, during the process of oxidation, the total density of dopants remains constant:

$$[Rh^{3+}] + [Rh^{4+}] + [Rh^{5+}] = N_T. \tag{14.13}$$

Moreover, the relative densities of Rh^{3+}, Rh^{4+}, and Rh^{5+} are controlled by the neutrality equation:

$$[Rh^{5+}] - [Rh^{3+}] \approx N_{AD} \tag{14.14}$$

where N_{AD} stands for $N_A - N_D$. Rh^{4+} does not appear in this equation as it is considered as the neutral level. Consequently, oxidizing the sample results in increasing N_{AD}. Equations (14.11)–(14.12) are still valid, but this time, N_{AD} is the relevant parameter to be varied. The main results are the following: the absolute absorption increases with oxidation and the photorefractive rise time in a given photorefractive configuration decreases, which is suggested by the high photoexcitation cross section from $Rh^{4+/5+}$; light-induced absorption vanishes, which is consistent with a one-level, two-charge state model; the dark decay time drops sharply with oxidation, which is consistent with a high thermal ionization rate from $Rh^{4+/5+}$ compared to $Rh^{3+/4+}$; and the two-wave mixing gain in a given photorefractive configuration remains nearly unchanged. The corresponding theoretical curves showing these evolutions are presented in the following discussion, along with experimental results (Fig. 14.4). No free parameter is used in these simulations and the parameters of Table 14.1 are employed. It should be noted that an excessive oxidation would suppress any photorefractive effect, as the only charge state present in the crystal would be Rh^{5+}. No modulated space-charge field could be created. This is why the two-wave mixing gain drops for high values of N_{AD}.

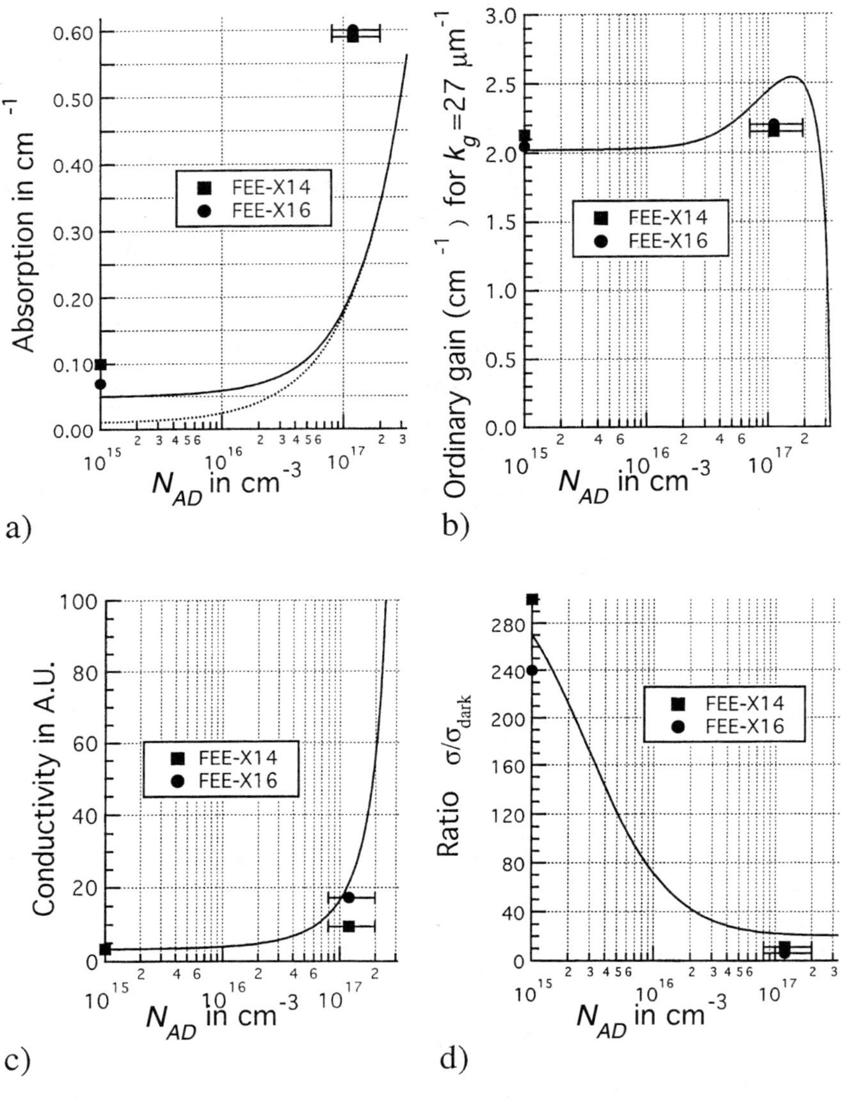

Figure 14.4: Theoretical predictions and experimental results of an oxidation of two $BaTiO_3$:Rh crystals (FEE-X14 and FEE-X16). (a) Absorption versus N_{AD}. Solid line: with saturated light-induced absorption. Dotted line: without light-induced absorption. (b) Ordinary gain versus N_{AD} for counterpropagating beams ($k_g = 27\ \mu m^{-1}$). (c) Photoconductivity at $I = 6{,}7$ W/cm^2 versus N_{AD} normalized to its value for $N_{AD} = 10^{15}$ cm^{-3}. (d) Ratio σ/σ_{dark} versus N_{AD}.

Experimental verification of these statements have been conducted through a collaboration work between FEE (Germany), the University of Osnabrück (Germany), and the Laboratoire Charles Fabry de l'Institut d'Optique (France) [23]. Two 0°-cut crystals of $BaTiO_3$:Rh referenced to as FEE-X14 and FEE-X16 and doped with the same quantity of rhodium (1000 ppm) have been characterized before and after oxidation. Absolute absorption, light-induced absorption, the photorefractive time constant τ_{ph} for an intensity of 6.7 W cm^2, the ratio of the dark decay time τ_{dark} divided by τ_{ph}, and the photorefractive gain Γ in the counterpropagating geometry for ordinary polarizations are measured. The results are reported in Table 14.3. In Fig. 14.4, experimental data are plotted together with theoretical curves with no free parameter. Note that $\tau_{dark}/\tau = \sigma/\sigma_{dark}$. The new photorefractive characterizations fulfill the predictions of the three-charge state model: a 2.8–5.2 time decrease of the photorefractive rise time, along with a reduction by a factor of 30 of the ratio τ_{dark}/τ_{ph} and a sixfold increase of the permanent absorption. The light-induced absorption vanishes for sample FEE-X14 and is turned into light-induced transparency in FEE-X16, which may result from some shallow weakly absorbent level. Moreover, the photorefractive two-wave mixing gain remains constant. The estimated corresponding value of N_{AD} that is reached by oxidation is $8 \cdot 10^{16}$ $cm^{-3} < N_{AD} < 2\ 10^{17}$ cm^{-3}.

These results show that the photorefractive rise time can be significantly reduced by oxidation at the expense of absorption. The strong increase of absorption will limit the use of oxidized $BaTiO_3$:Rh crystals for powerful nanosecond neodymium:yttrium-aluminum-garnet (Nd:YAG) wavefront correction because the damage thermal threshold will be reduced at least by a factor of 6 compared to as-grown crystals. Nevertheless, oxidized samples show the same behavior at 850 nm and at 1.06 μm. Therefore, oxidized $BaTiO_3$:Rh would be of great interest for applications at low-power laser diode wavelengths, where damage threshold is not a problem.

14.1.3 Characterization with nanosecond illumination

In $BaTiO_3$, for pulse durations larger than 10 ns, the recombination time of free holes is shorter than the pulse duration. Consequently, the photorefractive effect builds up during the pulses. Nothing happens between the pulses, as long as thermal excitation can be neglected during the pulse interval.

	X14		X16	
	As grown	**Oxidized**	**As grown**	**Oxidized**
α	(0.1 ± 0.02) cm^{-1}	(0.6 ± 0.02) cm^{-1}	(0.07 ± 0.02) cm^{-1}	(0.06 ± 0.02) cm^{-1}
$\Delta\alpha$ (1.2 W/cm^2)	(4.5 ± 0.2) m^{-1}	No light-induced absorption	(3.9 ± 0.2) m^{-1}	(-11 ± 2) m^{-1}
τ_{ph} at 6.7 W/cm^2	6.3 s	2.2 s	8.3 s	1.6 s
τ_{dark}/τ_{ph}	300	11	240	6
Γ	2.1 cm^{-1}	2.1 cm^{-1}	2.0 cm^{-1}	2.2 cm^{-1}

Table 14.3: Relevant parameters for $BaTiO_3$:Rh crystals before and after oxidation (ordinary polarized beams and counterpropagating geometry)

Let us summarize the observed effects when $BaTiO_3$ is illuminated with nanosecond visible light. Under pulsed illumination, the saturation of the free carrier density was theoretically brought to evidence [24]. This effect causes a nonconstant intensity × photorefractive rise time product, contrary to the cw illumination at high enough illumination ($SI >> \beta$). It was also shown that the density of free carriers could not be neglected compared to that of the traps [24]. As a consequence, the optimum grating spacing (highest photorefractive gain) is shifted. Moreover, an intensity-dependent electron-hole competition might occur for pulsed illumination [25, 26]. As a result, the photorefractive gain is widely intensity-dependent. It may decrease, vanish at the so-called compensation intensity, and even change its sign for high enough illuminations. This intensity-dependent electron-hole competition under pulsed illumination has been observed in the visible range by two-wave mixing gain measurements [26, 27] and through photorefractive time constant measurements [28] for various pulse intensities. Thus, in the visible range, the photorefractive characteristics (gain and time constant) depend widely on the used pulsed intensity, which is unfavorable for applications. What happens with nanosecond illumination at 1.06 μm?

To answer this question, two-wave mixing experiments have been performed using a 10-ns Nd:YAG laser at a repetition rate of 10 Hz and the setup shown in Fig. 14.5 [29]. Counterpropagating extraordinary polarized beams and a 45°-cut $BaTiO_3$:Rh crystal referred to as FEE-Y32-B have been used. For intensities (average value on the pulse duration) up to 20 MW cm^2, the photorefractive gain keeps nearly constant (Fig. 14.6), whereas such experiments with visible illumination clearly show an inversion of the sign of the gain for this range of intensities [26, 27]. Therefore this crystal does not show electron-hole competition at this wavelength for this range of intensities. This can be simply explained by the fact that a photon does not carry enough energy to excite an electron to the conduction band. The same conclusions are also reached with another $BaTiO_3$:Rh crystal referred to as FEE-X14. Free carrier density saturation has been looked for by measuring the photorefractive time constant in the same photorefractive configuration for various pulsed intensities. For pulsed average intensities up to 20 MW cm^2, the inverse of the time constant is a linear function of the intensity. As the time constant is inversely proportional to the free carrier density, Fig. 14.7 also indicates that no free carrier density saturation appears in this range of intensities. This conclusion is in agreement with the model of [26]. Indeed, the authors

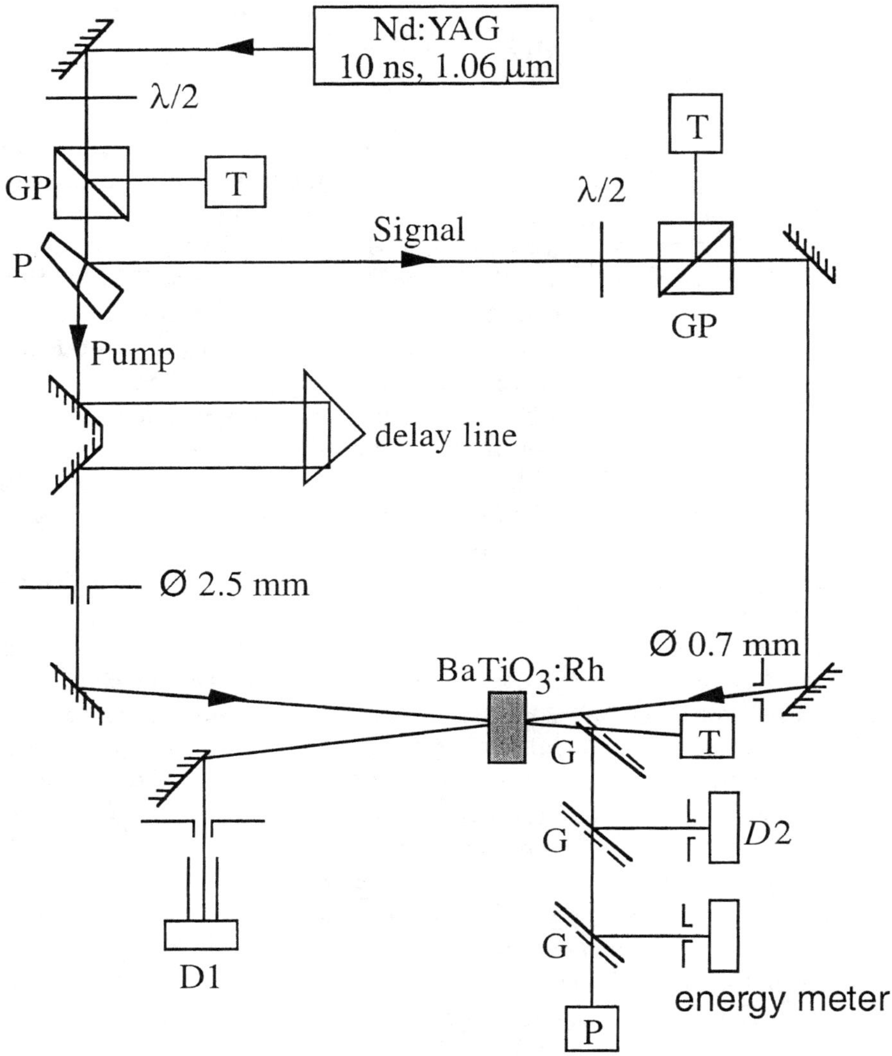

Figure 14.5: Experimental setup for two-wave mixing experiments in the counterpropagating geometry. D1, D2, detectors; G = glass plates; P = prismatic glass plate; Ts = light traps.)

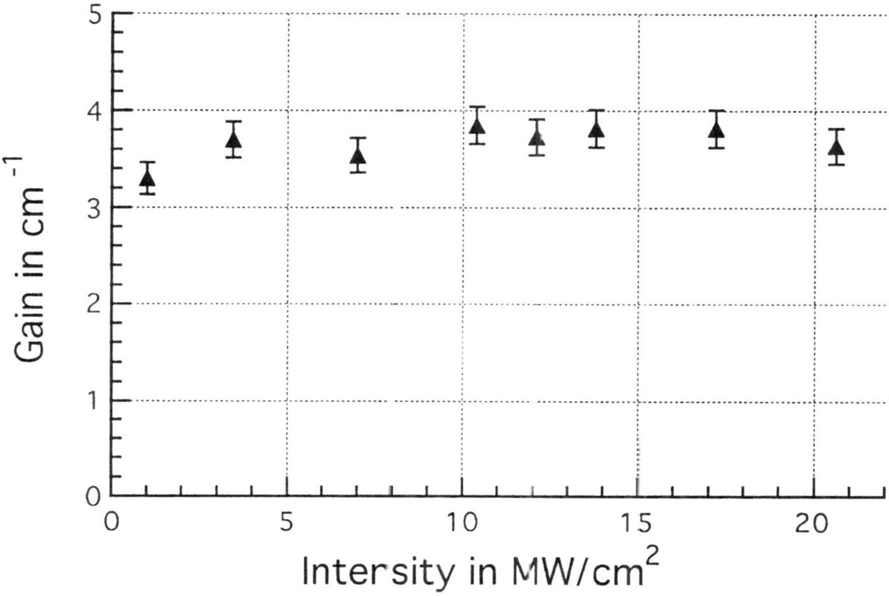

Figure 14.6: Photorefractive two-wave mixing gain measured versus intensity in the counterpropagating geometry with extraordinary polarized beams in sample FEE-Y32-B.

introduce a saturation parameter $f_h = S_h I \tau_h$ which defines the relative number of photoionized photorefractive traps. S_h is the photoexcitation cross section, I the intensity, and τ_h the recombination time for holes. With typical values for these parameters [17], f_h remains smaller than 5% for $I < 20$ MW/cm^2 and saturation stays negligible. This might be a consequence of the low typical value of S_h at 1.06 μm, as suggested by the low absorption at this wavelength.

The energies required to write a grating in the counterpropagating geometry are nearly equal in the cw and nanosecond regimes (100 J/cm^2). This value agrees with the conclusions of [24]. In Y32-B (45° cut), the maximum two-wave mixing gain has been measured in the copropagating geometry for a grating spacing of 4 μm for extraordinary polarized beams. The measured gain, corrected from the erasing effect of the pump beam reflected by the rear face of the sample, is 16 cm^{-1}. As predicted in [26], this is smaller than the value obtained with cw illumination (23 cm^{-1}).

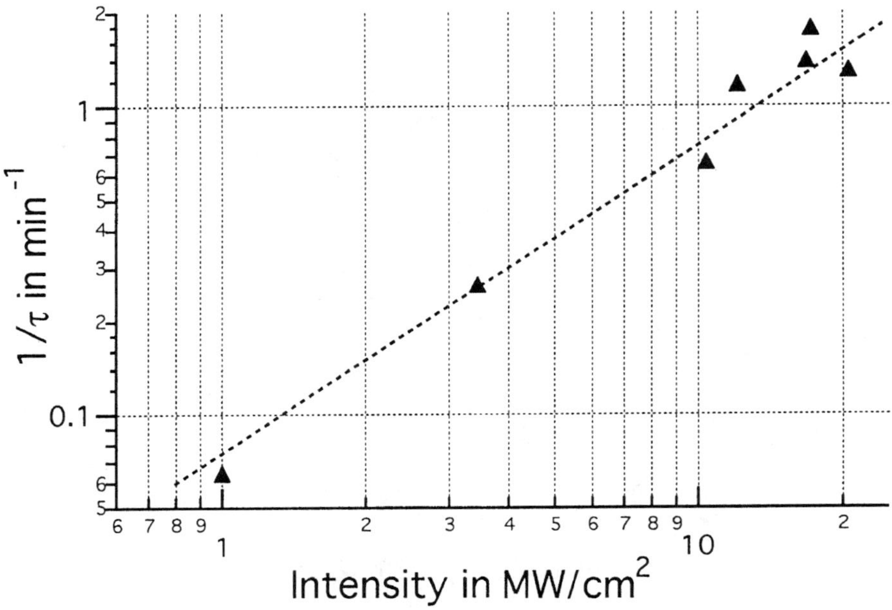

Figure 14.7: Inverse of the measured time constant (proportional to the density of free holes) versus pulsed intensity for sample FEE-Y32-B (repetition rate of 10 Hz). The dotted line is a guide for the eye.

To conclude, in the nanosecond illumination regime for pulsed average intensities up to 20 MW/cm^2 at 1.06 μm, the gain is intensity-independent, while the intensity × rise time product keeps constant. The maximum gain is marginally smaller than the one measured in the cw illumination regime. So, contrary to the results obtained in the visible range, nanosecond and cw illumination regimes are nearly equivalent for BaTiO$_3$:Rh at 1.06 μm for intensities up to 20 MW/cm^2, which is favorable for applications.

BaTiO$_3$:Rh is a photorefractive material with high gain and low absorption at 1.06 μm. It is accurately described by a three-charge state model, which agrees with spectroscopic and photorefractive studies. Oxidation of these crystals reduces significantly the rise time of the photorefractive effect, but, as it also increases the permanent absorption, it is not suited for dynamic wavefront correction of powerful lasers. Nevertheless, oxidation might be very useful for low power applications at laser diode wavelengths. For as-grown $BaTiO_3$:Rh at 1.06 μm, nanosecond and cw illuminations are nearly equivalent in terms of photorefractive gain and time constant for pulsed intensities up to 20 MW/cm^2. This latter result, which differs from these obtained in the visible range, is a key point for applications to Q-switched Nd:YAG lasers. Such a high photorefractive gain and low absorption material is attractive for high-reflectivity phase conjugation, as will be discussed in Section 14.2.

14.2 Self-Pumped Phase Conjugation

Phase conjugate mirrors fall into two main categories: phase conjugation by degenerate four-wave mixing and self-pumped phase conjugation. Degenerate four-wave mixing needs one signal beam and two externally provided pump beams. One major advantage of this configuration is that the phase conjugate reflectivity may be larger than unity if the signal intensity is small enough. This principle can be used in self-starting oscillators [30, 31]. The use of a phase conjugate mirror by degenerate four-wave mixing as the end mirror of a laser cavity has been studied theoretically [32] and experimentally [33], but this architecture needs an additional laser source to provide two pump beams, which is not favorable to applications. In self-pumped phase conjugation, this drawback is bypassed at the expense of a reflectivity limited to unity. Self-pumped phase conjugation has been widely studied since the early 1980s. In 1982, a self-pumped phase conjugator using $BaTiO_3$ was proposed [1]. In this self-starting geometry demonstrated at 514 nm with cw illumination, the two-pump beams were derived from the incident beam itself using beam-fanning and total internal reflection on the crystal faces. In the following discussion, this geometry is called "internal loop geometry." In 1984, various self-pumped phase conjugate mirrors were proposed, modeled, and experimentally tested [2]. Among them, the "linear phase conjugate mirror" requires two additional mirrors which form a linear resonator providing the two counterpropagating pump beams [2]. This phase conjugate mirror, along with its semilinear version (only one additional mirror), has been used in laser cavities as end mirrors [6]. One major drawback of this geometry is that coherence length is required to provide efficient phase conjugation. In the case of a Q-switched Nd:YAG laser, it is about 1 cm. This is why linear phase conjugate mirrors are not used for the application discussed in this chapter. The "ring phase conjugate mirror"

[2] uses two additional mirrors, which form a loop and feed the transmitted beam back to the photorefractive crystal. These two geometries (internal loop and ring) are often used for the improvement of laser sources such as laser diode injection and spectral narrowing [8] or self-corrected oscillators [34]. These two kinds of self-pumped phase conjugate mirrors with $BaTiO_3$:Rh are also currently being developed for the application to dynamic wavefront correction of nanosecond Nd:YAG lasers at 1.06 μm [35, 36]. They are discussed in the following sections.

14.2.1 Internal loop self-pumped phase conjugate mirror

As stated in the previous discussion, the internal loop phase conjugate mirror is widely used because the only required device is a photorefractive crystal. A theoretical description of such a phase conjugate mirror is difficult because the dimensions of the material and the incidence angle as well as its divergence affect the beam path in the photorefractive crystal and as a result, the numbers and positions of the interaction regions may vary from one case to another [37]. Moreover, the nature of the gratings (transmission or reflection gratings) involved in each interaction region is not easily determined. Experiments like low-coherence reflectometry are necessary [38]. Nevertheless, in all cases, this geometry is self-starting and the four-wave mixing develops from the beam-fanning. In a model considering plane waves interacting in two regions, the threshold in terms of gain times interaction length ($\Gamma\ell$) product is shown to be $\Gamma\ell = 8$, which is high compared to other geometries mentioned above [2].

At 1.06 μm, this internal loop geometry is used both with cw and nanosecond illumination to perform phase conjugation in $BaTiO_3$:Rh. It is depicted in Fig. 14.8 [35]. As mentioned in the literature for other wavelengths [7, 37], it is observed that the reflectivity is strongly dependent on the incidence angle on the photorefractive crystal (Fig. 14.9). When the adequate angle is chosen, the maximum measured reflectivity is 32% with nanosecond illumination at 10 Hz (the crystal was not antireflection coated). As discussed in Section 14.1, the sensitivity of $BaTiO_3$:Rh at 1.06 μm is low compared to the usual values in the visible range. So, the rise time of the phase conjugate reflectivity at 1.06 μm has to be optimized to be competitive. Indeed, the phase conjugate mirror is unable to compensate for aberrations which vary on a time scale that is shorter than the reflectivity rise time. With the internal loop geometry, the re-

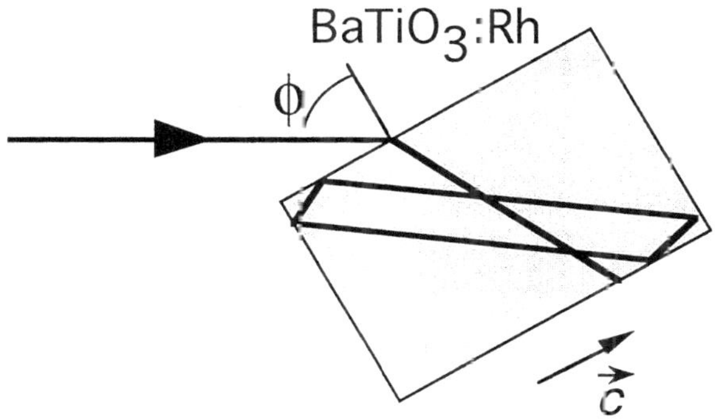

Figure 14.8: Arrangement of the internal loop self-pumped phase conjugate mirror.

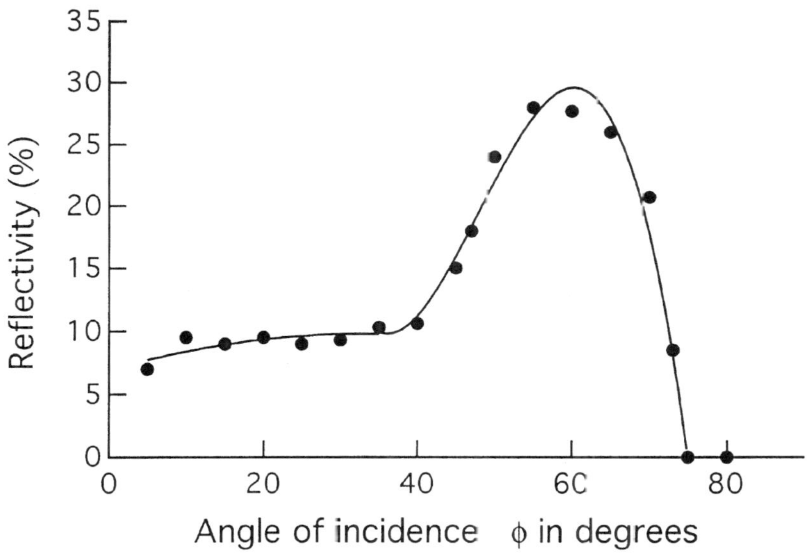

Figure 14.9: Steady-state reflectivity of the internal loop phase conjugate mirror versus incidence angle ϕ. The solid curve is a guide for the eye.

ported rise time from zero to 90% of the maximum reflectivity is 20 min for an incident fluence of 230 mJ/cm^2 at a repetition rate of 10 Hz. This leads to a required fluence of 2700 J/cm^2. At a repetition rate of 30 Hz, the required fluence goes down to 2000 J/cm^2 (5 min for an input fluence of 230 mJ/cm^2) [35].

Such a phase conjugate mirror is able to compensate for highly aberrated beams [34]. The fidelity of phase conjugation is 90%, even when a phase plate aberrator is introduced in the beam path. It was measured by the power-in-the-bucket technique: the fidelity is defined as the ratio of the phase-conjugate and incident pulsed energies that are transmitted through a pinhole in the focal plane of a lens. This technique, easy to set up, is widely used [35, 39, 40]. Other techniques, like shearing interferometry, enable to observe phase-difference gradients between the reference beam and the phase conjugate beam [41]. When analyzed by a Fourier transform method [42], phase difference can be evaluated in the transverse plane. But, as phase conjugation is a nonlinear process, the fidelity may strongly depend on the nature of the incident beam (Gaussian, speckled, etc.). Therefore, an extremely detailed analysis of the phase conjugate fidelity, only valid for a single kind of incident beam, seems useless. For the application to wavefront correction of laser sources, the focusing ratio is a relevant characteristic. This is why the power-in-the-bucket technique is employed.

14.2.2 Ring self-pumped phase conjugation

The ring self-pumped phase conjugate mirror consists of one photorefractive crystal and two additional mirrors which form a loop and feed the transmitted beams back to the crystal (Fig. 14.10). It is a self-starting phase conjugate mirror and the four-wave mixing is initiated by beam-fanning. It has been studied theoretically in a plane-wave model at steady state with pump depletion [2]. The threshold in terms of $\Gamma\ell$ product is found to be $\Gamma\ell = 2$, which is four times lower than for the internal loop geometry. Moreover, the reflectivity increases sharply with $\Gamma\ell$ and high reflectivities are easily obtained. A $\Gamma\ell$ of 3 is enough to generate a reflectivity of 90% for a lossless loop [2]. As will be shown in this section, in this geometry optical systems may be inserted in the loop to improve the fidelity of phase conjugation.

In the experiment shown in Fig. 14.10, a roof-cut photorefractive crystal referenced to as FEE-X16-45 is used. Indeed, under cw illumination, the total internal reflection on the crystal faces allows the oscillation of spurious beams and prevents beam-fanning from rising. Thus, no phase conjugate beam can be observed. Thanks to this roof cut, antireflection coatings, and a 45° orientation of the c-axis with respect to the input face, this oscillation is avoided as well as that of the beam-fanning between

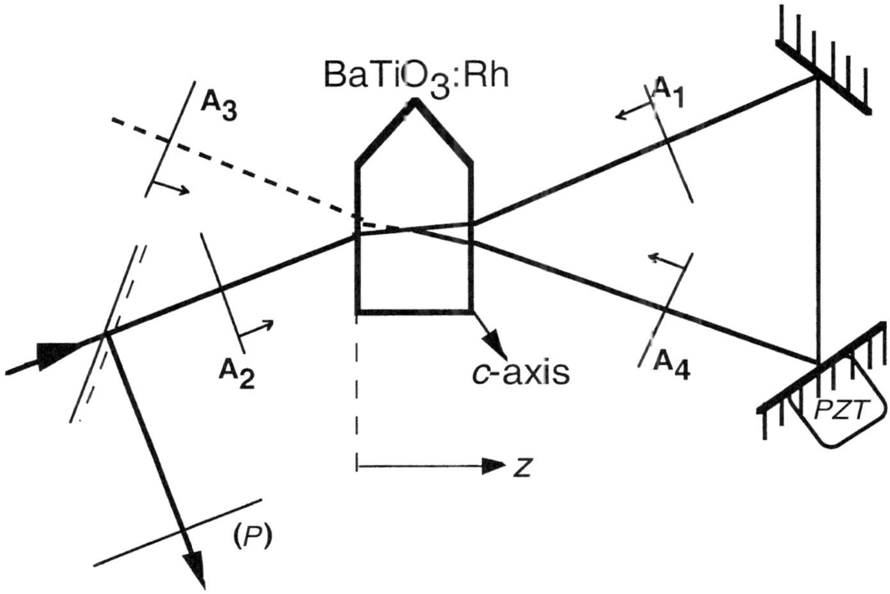

Figure 14.10: Schematic representation of the ring self-pumped phase conjugate mirror. Phase conjugate beams are observed in plane (P). (A_1, A_2, A_3 and A_4 are the complex amplitudes of interacting beams in a plane wave model. PZT: piezo mirror used to wash out the reflection gratings.)

input and output faces. The coherence length of the usual cw sources allows both the desired transmission grating (between A_2 and A_3, or A_1 and A_4) and unwanted reflection gratings (between A_1 and A_2, or A_1 and A_3, or A_4 and A_2, or A_4 and A_3) to be recorded. This is why a vibrating mirror is used under cw illumination [43]: the response time of the photorefractive material is longer than the time necessary to shift the reflection grating, which therefore cannot be recorded in the crystal. As a result, only the transmission grating remains. With nanosecond illumination, the coherence length of the source is too small to write reflection gratings and no vibrating mirror is needed.

As pointed out in [2], the reflectivity depends on the value of $\Gamma\ell$. As soon as $\Gamma\ell$ is high enough, the reflectivity saturates to its maximum value. The measured reflectivity with cw illumination is 70% for a transmission of the loop of 71%. With nanosecond illumination and other optical elements in the loop, a reflectivity of 79% is measured for a 81% loop transmission. Experimentally, the reflectivity proves to be limited to the transmission of the loop as soon as it reaches 5% (Fig. 14.11) [36]. A fit of experimental data by a plane-wave model at steady state including pump depletion leads to $\Gamma\ell = 8.2$ (Fig. 14.11). This is consistent with $\Gamma = 23\ \text{cm}^{-1}$ measured in [20] and a crystal thickness of 3.5 mm, as is the case in FEE-X16-45.

The second important feature in photorefractive phase conjugation is the reflectivity rise time. A typical time evolution of the reflectivity is given in Fig. 14.12. It shows a sharp increase characterized by the time needed for the reflectivity to grow from 10 to 90% of its maximum value $\tau_{90\%}\text{-}\tau_{10\%} \equiv \tau_{pc}$. To optimize this feature, a plane-wave model is used. Absorption may be neglected in BaTiO$_3$:Rh at 1.06 μm ($\alpha = 0.1\ \text{cm}^{-1}$). This model cannot take into account the spatial profiles of the interacting beams, like in three-dimensional numerical models [44], or the influence of beam-fanning [45]. But for $t > \tau_{10\%}$, one may reasonably consider that

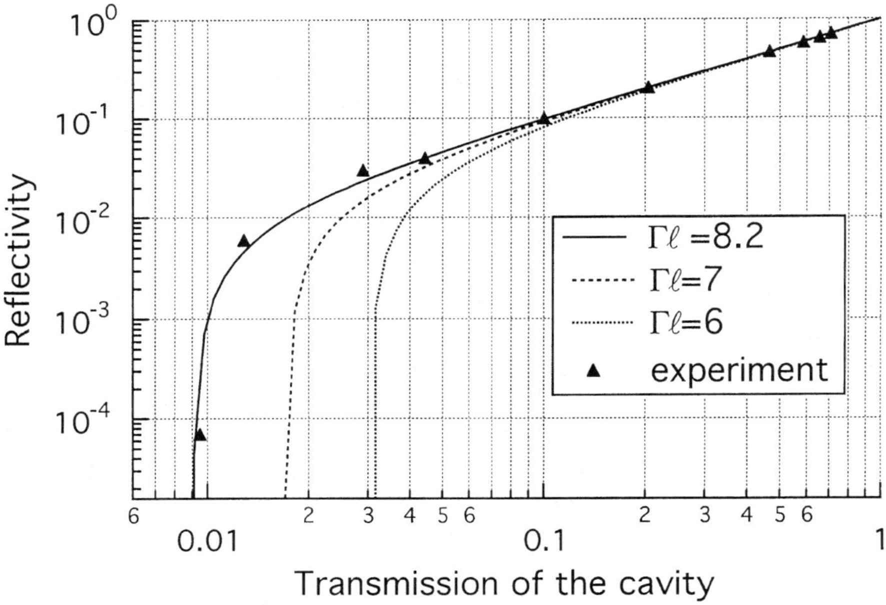

Figure 14.11: Reflectivity of the ring-phase conjugate mirror versus transmission of the loop. Experimental data are fitted by theoretical curves obtained in a plane-wave steady-state model with pump depletion for different values of the $\Gamma\ell$ product.

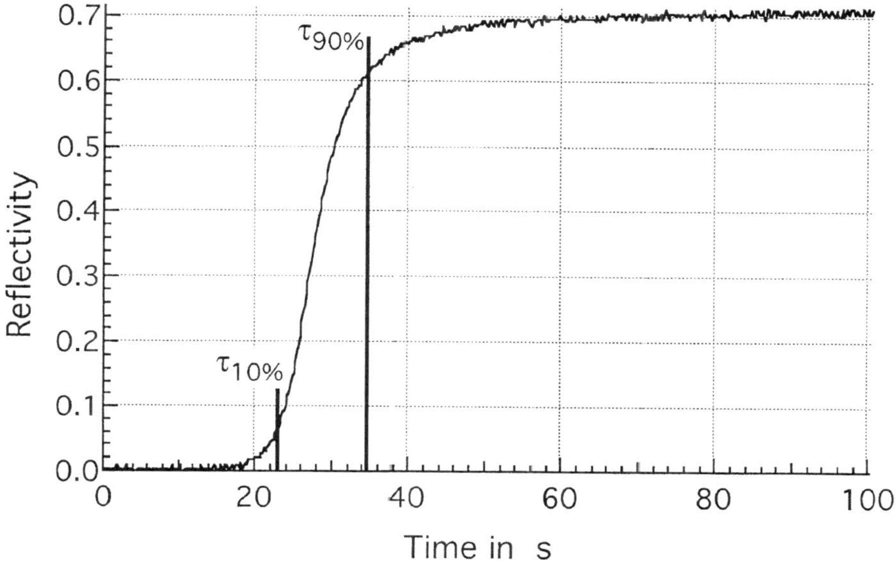

Figure 14.12: Experimental time evolution of the reflectivity of the ring self-pumped phase conjugate mirror for an incident cw intensity of 5 W/cm^2 at 1.06 μm.

the desired transmission grating is the only one remaining in the material. Consequently, this simplified model is sufficient to optimize the reflectivity rise time, using coupled-wave equations for a transmission grating only (Eq. 14.15), along with the time evolution of the space-charge field E_1 (Eq. 14.16).

$$\left.\begin{aligned}
\frac{\partial A_1(z,t)}{\partial z} &= i\frac{\Gamma}{4E_{sc}}E_1(z,t)A_4(z,t) \\
\frac{\partial A_2^*(z,t)}{\partial z} &= i\frac{\Gamma}{4E_{sc}}E_1(z,t)A_3^*(z,t) \\
\frac{\partial A_3(z,t)}{\partial z} &= -i\frac{\Gamma}{4E_{sc}}E_1(z,t)A_2(z,t) \\
\frac{\partial A_4^*(z,t)}{\partial z} &= -i\frac{\Gamma}{4E_{sc}}E_1(z,t)A_1^*(z,t)
\end{aligned}\right\} \tag{14.15}$$

where A_1, A_2, A_3, and A_4 are the complex amplitudes of the plane waves and E_{sc} is the steady-state space-charge field.

As detailed in Section 14.1, the photorefractive effect in BaTiO$_3$:Rh at 1.06 μm finds its origin in the three charge states of rhodium Rh^{3+},Rh^{4+}, and Rh^{5+} and a three-charge state band transport model is suitable to account for experimental photorefractive characterizations of this crystal. However, to describe the time evolution of the space-charge field, the simpler single-carrier, single-site model without applied electric field is used here. The time evolution of the space-charge field is then governed by:

$$\frac{\partial E_1(z, t)}{\partial t} = -\frac{1}{\tau_{ph}}[E_1(z, t) - im(z, t)E_{sc}] \tag{14.16}$$

where τ_{ph} is the photorefractive time constant for a given grating spacing and for a given $I_0 = \sum_i |A_i|^2$. $m(z, t)$ is the modulation of the interference pattern defined by

$$m(z, t) = 2(A_1 A_4^* + A_2^* A_3)/I_0 \tag{14.17}$$

We assume that the initial seeding amplitude A_3 results from a scattering at the input surface: A_3 $(z, t = 0) = \varepsilon$. This initial condition has no consequence on τ_{pc} as soon as ε is small enough.

For a given I_0, the important features for the kinetics of the phase conjugation are τ_{ph}, $\Gamma\ell$, and the transmission T of the loop, as in photorefractive oscillators [46]. For a given incidence angle θ on the crystal (i.e., a given grating spacing Λ), τ_{pc} is deduced by calculating the corresponding values of τ_{ph} and $\Gamma\ell$ and by inserting these values in Eqs. (14.15) and (14.16)and solving numerically. A theoretical plot of τ_{ph}, $\Gamma\ell$, and τ_{pc} is given in Fig. 14.13 for an incident intensity on the crystal of 5 W/cm^2, an effective density of traps of 5 10^{16} cm^{-3}, extraordinary polarized beams, an antireflection-coated 45°-cut crystal whose thickness is 3.5 mm, and a lossless loop. The rise time τ_{pc} appears to be minimized for incidence angles between 20° and 50°, corresponding to 0.7 μm $\leq \Lambda \leq$ 1.6 μm. This result is checked experimentally (Fig. 14.14). However, as the transmission T of the loop varies with the incidence angle θ, the rise time τ_{pc} for a given θ has to be compared with τ_{pc} at a different $\theta_{reference}$ for the same T. Here the value of $\theta_{reference}$ is chosen to be 20°. Experimental results are in agreement with the model: an incidence angle of $\theta = 32°$ leads to a value of τ_{pc} slightly smaller (20 s) than the one obtained at $\theta = 20°$ for the same T. At $\theta = 9°$, $\tau_{10\text{–}90\%}$ is much larger than the one obtained at $\theta = 20°$ for the same T. For reasons of comfort of work, an incidence angle

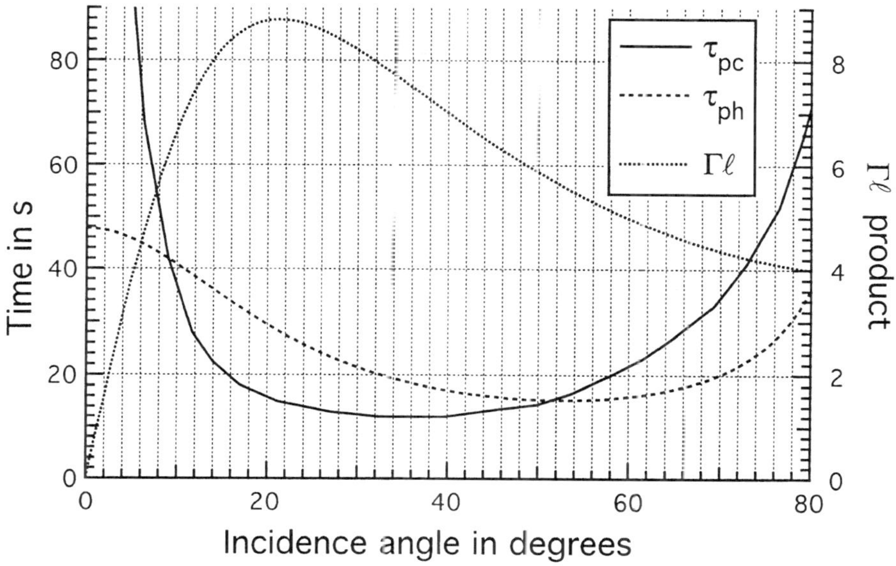

Figure 14.13: Theoretical variations of the $\Gamma\ell$ product, the photorefractive rise time τ_{ph} and the reflectivity rise time τ_{pc} as a function of the incidence angle θ, for an incident intensity of 5 W/cm^2, extraordinary polarized beams, a 45°-cut antireflection-coated crystal whose thickness is 3.5 mm, an effective density of traps of 5 10^{16} cm^{-3} and a lossless loop.

of 20° is chosen for further experiments. In this case, for a transmission of the loop of 71%, the measured rise time for an incident cw intensity of 5 W cm^2 is 12 s (Fig. 14.12), corresponding to an energy of $E_{10\text{–}90\%} = 60$ J/cm^2. With nanosecond illumination, at $\theta = 20°$, the required energy is $E_{10\text{–}90\%} = 90$ J/cm^2. The energy necessary to increase the reflectivity up to 90% of its maximum value is 300 J/cm^2. As the product intensity × rise time has been found to be constant, this can be extrapolated to give a response time of the phase conjugate mirror of 30 s for pulses of 10 mJ/cm^2 at a repetition rate of 1 kHz.

A further key point in phase conjugation is the fidelity of phase conjugation. This problem has been addressed theoretically and experimentally [44, 47]. Indeed, when the loop consists in two plane mirrors, the phase conjugate fidelity is poor. Two reasons may explain this.

First, the phase-matching condition:

$$\vec{k}_2 - \vec{k}_3 = \vec{k}_4 - \vec{k}_1 = \vec{k}_g \tag{14.18}$$

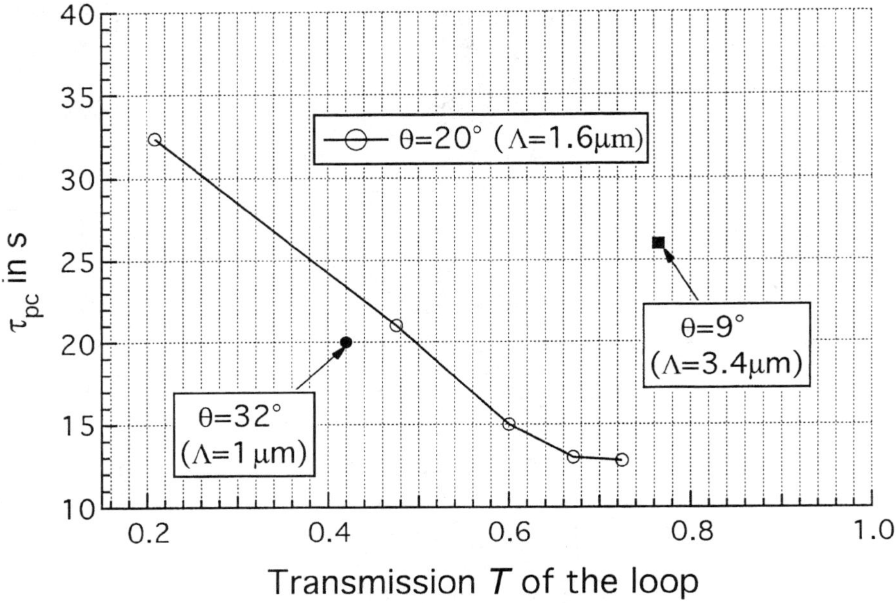

Figure 14.14: Experimental values of τ_{pc} for various incidence angles θ. As the transmission T of the loop varies with θ, the values of τ_{pc} for a given θ are to be compared with the value obtained at $\theta_{reference} = 20°$ for the same T.

where $\vec{k}_1$ are the wavevectors of beams A_i and where $\vec{k}_g$ is the grating wave vector, may be fulfilled on a circle called the degeneracy circle whose diameter is k_g (Fig. 14.15). This circle is clearly visible in Fig. 14.16 obtained at 1.06 μm when a screen is inserted in the loop before the four-wave mixing process has reached its steady-state.

Second, the quasi-Bragg matching condition:

$$|\vec{k}_2 - \vec{k}_4 - \vec{k}_3 + \vec{k}_1) \cdot \Omega\vec{\Omega}'| = |\delta\vec{k} \cdot \Omega\vec{\Omega}'| < \pi \qquad (14.19)$$

with Ω and Ω′ being two points in the interaction volume, allows wavevectors which point close to the degeneracy circle to exist (Fig. 14.15). This results in a roughly elliptical phase conjugate beam with a long axis along the y-axis [44, 47]. The insertion of optical elements in the loop improves the phase conjugate fidelity. With a three-prism system able to rotate the beam cross section by 90°, the phase conjugate beam is efficiently selected among the backscattered light. This device (Fig. 14.17), previously pro-

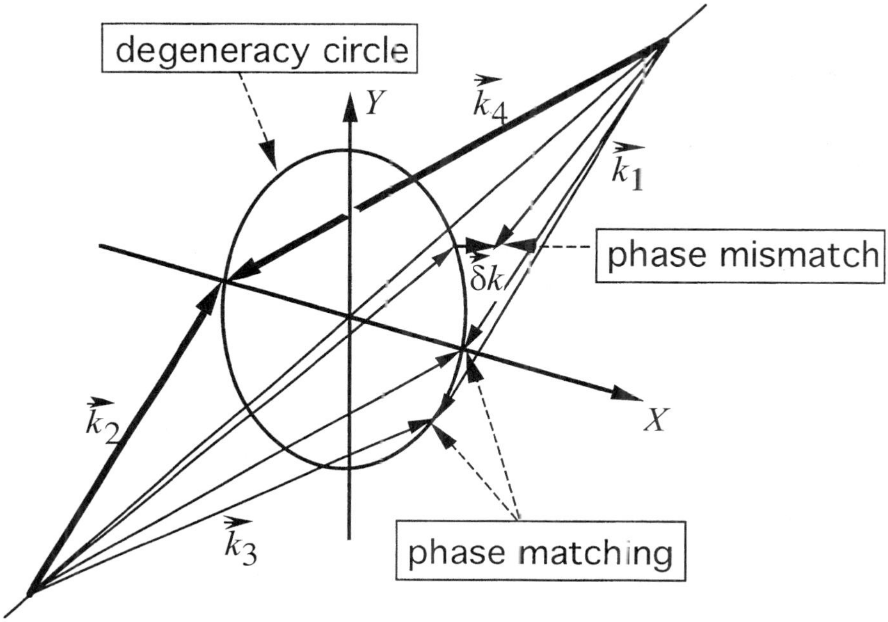

Figure 14.15: Representation of the wave vectors in the ring phase conjugate mirror. (k_1 = phase conjugate beam and k_3 = beam fanning) $\vec{k}_2$ (incident beam) and $\vec{k}_4$ *image of the incident beam through the loop) are fixed.*

posed [44] and tested in the visible range [47], is also used at 1.06 μm with nanosecond illumination and proves to be efficient [36]. Phase conjugation fidelities of 80% are measured by the power-in-the-bucket technique when lenses of focal length longer than 300 mm are inserted in the beam path. Because of size problems, the optical length of the loop cannot be made smaller than 12 cm. This prevents us from correcting highly aberrated beams. Indeed, after one lap in the loop, the propagation of the diverging aberrated beam affects the overlap of the beams inside the $BaTiO_3$:Rh crystal.

An effort has been done in compacting the ring phase conjugate mirror. A ring self-pumped phase conjugate mirror using total internal reflection on the faces of a crystal cut in a triangle shape was already proposed [48] but offered no control of the phase conjugate beam and of the gratings involved. Another solution is to use one spherical mirror in the loop (Fig.

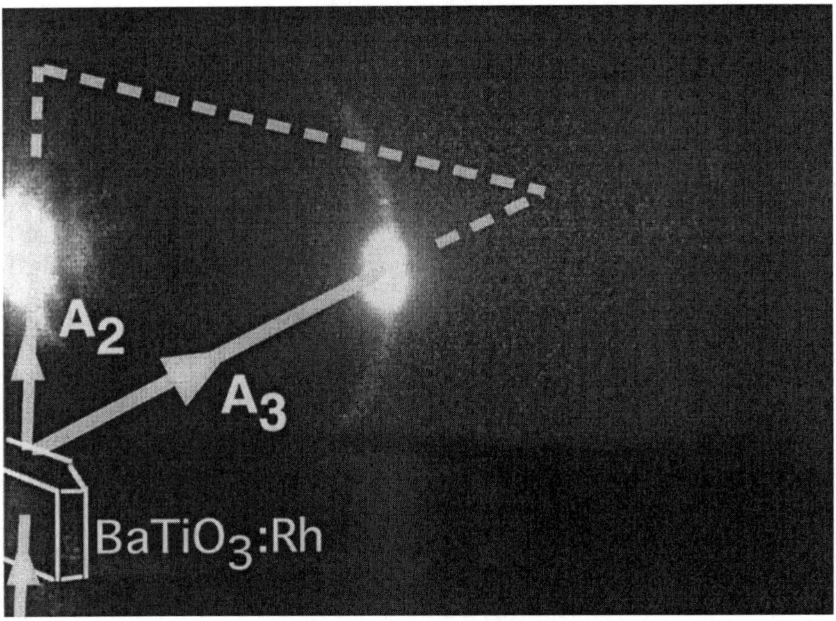

Figure 14.16: Observation of the degeneracy circle on beam A_3 at 1.06 μm on a screen inserted in the loop of the ring phase conjugate mirror, before the four-wave mixing has reached its steady-state. The position of the photorefractive crystal is represented with white lines. The beam path in the loop is also drawn in solid and dotted thick lines.

14.18). This configuration has been investigated theoretically. For doing this, an incident plane wave focusing near the crystal after one lap in the loop is considered (Fig. 14.18). It is also assumed that all waves are locally plane waves. The aberration considered is a tilt in the focusing beam ($\delta\alpha$ in the XZ-plane, $\delta\phi$ in the YZ'' plane). In these conditions, δk expressed in (X'', Y, Z'') is given by:

$$\delta\vec{k} = \vec{k}_2 - \vec{k}_4 - \vec{k}_3 + \vec{k}_1 = k \left| \begin{array}{c} \left[-\cos(2\theta - \alpha)\dfrac{f' - z}{f'}\dfrac{1}{\cos^2\alpha} \right]\delta\alpha \\ -\delta\alpha \sin(2\theta - \alpha)\dfrac{f' - z}{f'}\dfrac{1}{\cos^2\alpha} \\ \delta\phi\left[\dfrac{f' - z}{f'} - 1\right] \end{array} \right. \tag{14.20}$$

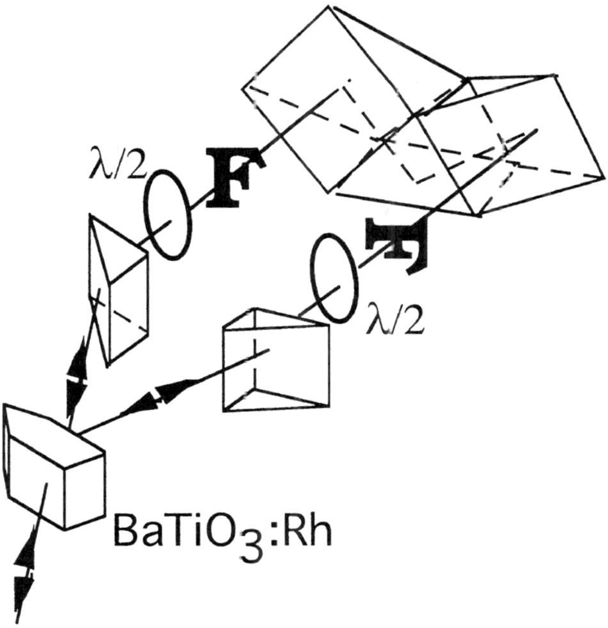

Figure 14.17: Ring self-pumped phase conjugate mirror with a three-prism system in the loop which performs a 90°-rotation of the beam cross section. Two half-wave plates maintain the extraordinary polarization of interacting beams in the crystal, which is necessary to provide high photorefractive gains and high phase conjugate reflectivities. The loop length is 12 cm.

where $2f'$ is the curvature of the spherical mirror; z, α and θ are defined in Fig. 14.18; and $k = 2\pi/\lambda$.

Introducing Eq. (14.20) into Eq. (14.19) leads to:

$\delta\phi = 0$

$$\delta\alpha < \frac{\pi}{\ell k\left[1 + \frac{f' - z}{f'}\frac{1}{\cos^2\alpha}(\cos(2\theta - \alpha) + \sin(2\theta - \alpha))\right]} \quad \text{if } \delta\phi = 0$$

$$\delta\phi < \frac{\pi}{\ell k\left[\frac{f' - z}{f'} - 1\right]} \quad \text{if } \delta\alpha = 0 \tag{14.21}$$

where, ℓ is the typical size of the interaction region. It is chosen here to be 2 mm.

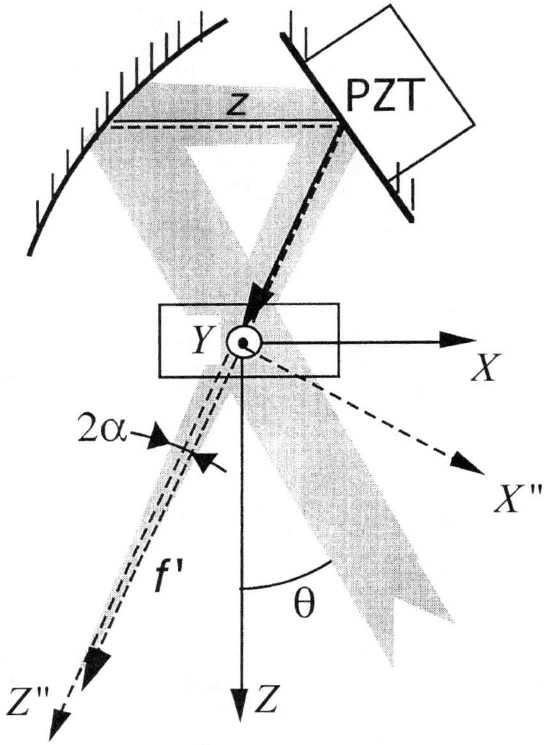

Figure 14.18: Compact ring-phase conjugate mirror including one curved mirror in the loop.

Equation (14.21) shows that the phase conjugate fidelity in the plane of incidence ($\delta\phi = 0$) is weakly dependent on the curvature $2f'$ of the spherical mirror. But in the *Y*-direction ($\delta\alpha = 0$), the fidelity strongly depends on the ratio $(f' - z)/f'$. For a Gaussian incident beam, the "phase conjugate" beam is elliptic with a long axis along the *Y*-direction if $(f' - z)/f' > 0.7$, whereas it remains Gaussian for $(f' - z)/f' < 0.7$ [49]. The ratio $(f' - z)/f'$ is also an expression for the change in the beam's diameter after one lap in the loop. The influence of this parameter on the phase conjugate fidelity is also pointed out in numerical simulations of [44]. The influence of $(f' - z)/f'$ has been tested experimentally by measuring the phase conjugate fidelity and monitoring the far-field profile of the phase conjugate beam. Even for small loop lengths close to the coherence length of the laser source, the vibrating plane mirror avoids reflection gratings in the crystal. The results reported in Fig. 14.19 are in agreement with the predictions. Using this simple configuration with a loop length of 6 cm, lenses of focal length from infinity down to 25 cm could be compensated for. Phase conjugation fidelities of 90% are obtained. Such a ring mirror may be adapted to any range of focal lengths of the aberrator, for instance from 80 down to 15 cm by increasing slightly the ratio $(f' - z)/f'$. The alignment of this device is not critical and the spherical mirror may even be replaced by a cylindrical mirror, as the fidelity in the incidence plane is weakly affected by the ratio $(f' - z)/f'$. With such a ring-phase conjugate mirror, the ability to correct highly aberrated beams has also been demonstrated (Fig. 14.20) along with the ability to restore images, even when lenses or strong aberrations are inserted in the beam path.

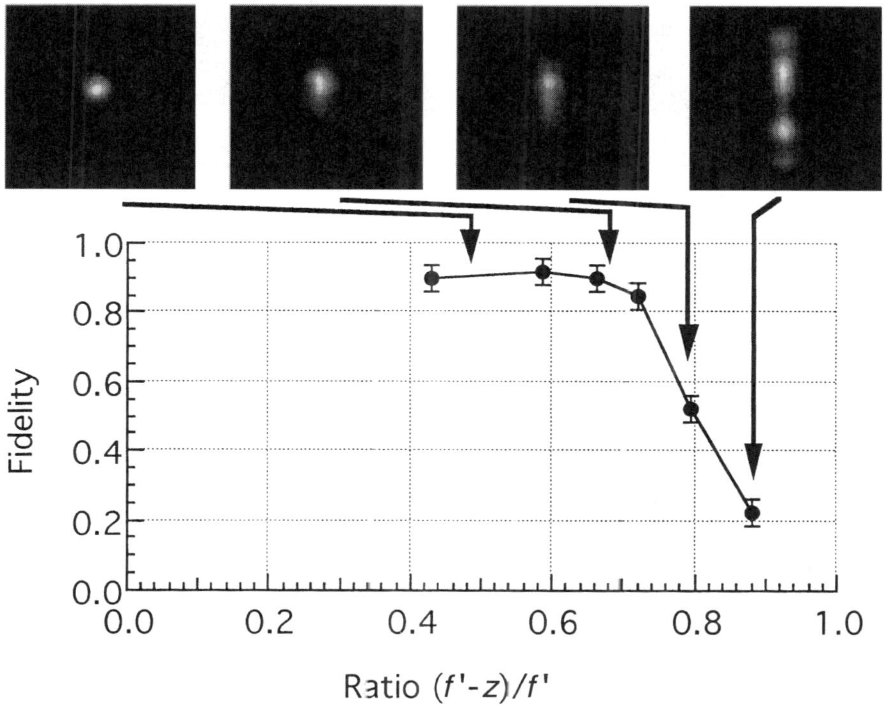

Figure 14.19: Phase conjugation fidelity of the compact ring mirror as a function of the ratio $(f' - z)/f'$ with nanosecond illumination at 1.06 μm. The corresponding far-field beam profiles are also given.

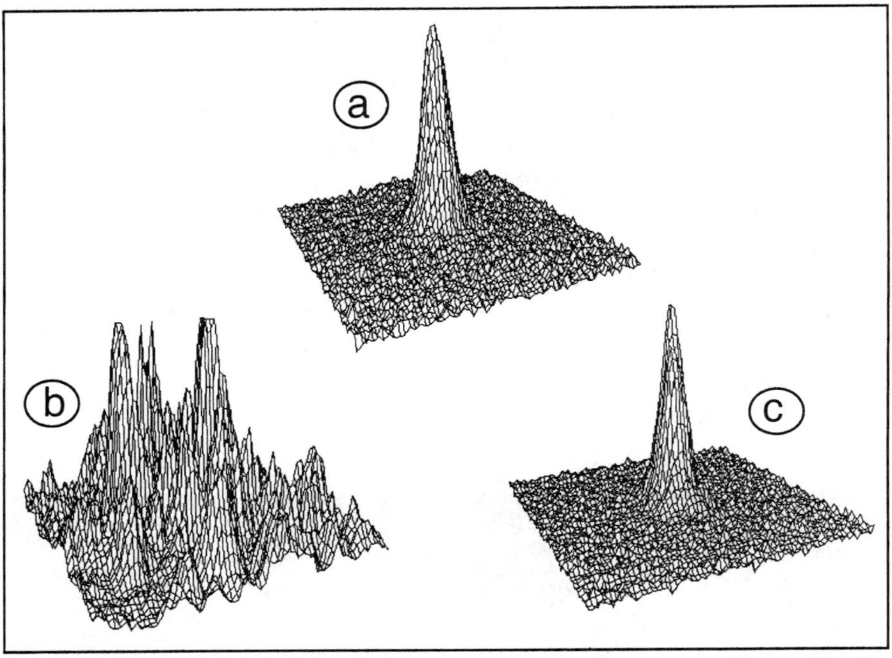

Figure 14.20: Correction of random aberrations at 1.06 μm with nanosecond illumination: (a) incident beam. Far-field profile after a double pass through a random aberrator and (b) reflection on a dielectric mirror, (c) reflection on the compact ring phase conjugate mirror.

Such a compact ring-phase conjugate mirror seems to put together advantages of both internal loop and ring geometries presented before. It is able to correct highly aberrated beams and the positioning is not critical. It thus appears to be a good candidate for all solid-state dynamic wavefront correction of MOPA laser sources which will be presented in the next section.

14.3 Dynamic wavefront correction of MOPA laser sources

BaTiO$_3$:Rh allows the phase conjugation of nanosecond pulses at 1.06 μm with high reflectivity and fidelity along with a reflectivity rise time which is compatible with laser manufacturer requirements. Therefore, the application of this technique to dynamic wavefront correction of Nd:YAG lasers is realistic. We will now describe the origin of aberrations present in Nd:YAG amplifier rods and present MOPA architectures including a photorefractive self-pumped phase conjugate mirror. Its experimental performances will be given. The results obtained with a photorefractive phase conjugator will be compared to the performances of other existing techniques.

14.3.1 Origin of aberrations in Nd:YAG amplifier rods

As soon as the repetition rate of the amplifier rod gets higher than the inverse of the thermal relaxation time (a few Hz), thermal effects occur in Nd:YAG rods [50]. These effects are caused by the pumping by flash lamps or laser diodes, because a significant amount of their energy is converted into heat. The rod is cooled only at its periphery, which leads to a significant radial gradient of temperature inside the material. This results in a nonuniform index variation. Indeed, three optical effects occur. First, the distribution of refractive index acts as a converging lens. Second, the nonuniform thermal expansion of the rod induces stresses in the material which, through the elastooptic effect, causes double focusing (bifocusing) according to the local polarization of the propagating wave. Third, thermal stresses locally modify the index ellipsoid and induce birefringence with a local fast axis in the radial direction and a local slow axis along the tangential direction, resulting in a depolarization of the propagating wave across the beam [51].

Birefringence may be partially compensated by using two amplifier rods separated by a 90° polarization rotator [52], or by adding optical imaging systems along with the 90° polarization rotator between the two rods [53]. These devices, easy to set up, may only provide a partial correction of the depolarization if the rods are slightly different. Other tricks have been proposed to avoid the use of two identical amplifier rods. A single amplifier followed by a Faraday rotator and a phase conjugate mirror proved to be efficient but bifocusing cannot be compensated [54]. This problem may be bypassed by splitting the depolarized beam into two orthogonal polarizations and performing a separate phase conjugation for each of these two polarizations and then passing through the amplifier again [55].

Thermal lensing in Nd:YAG rods has been widely studied. The thermal focal lens depends on the operating conditions of the amplifier, and especially on the absorbed average pump power [51, 56]. For a pulsed pumping, the higher the repetition rate, the shorter the thermal focal lens [56]. With typical values, the focal lens is about 50 cm for a flash lamp–pumped Nd:YAG amplifier at a repetition rate of 100 Hz. The thermal lens is hardly corrected by a compensative diverging lens having the same optical power as the laser rod, because transient effects occur in the rod, for instance during the warm-up of the laser. Moreover, for a strong pumping, the thermal lens is not spherical but rather highly aberrated [57]. Consequently, adaptative devices are required. Optical phase conjugation is one of them. Other techniques emerge and are discussed in Section 14.3.3. Optical phase conjugation at 1.06 μm can be achieved by various physical effects like stimulated Brillouin scattering (SBS), gain gratings or the photorefractive effect, and different materials are available for each effect at 1.06 μm. Section 14.3.2 is devoted to the use of photorefractive $BaTiO_3$:Rh crystals for this application. Section 14.3.3 details other techniques to perform dynamic wavefront correction. Photorefractive beam cleanup, gain gratings, adaptative optics, and SBS in various materials are presented and compared to the performances of $BaTiO_3$:Rh self-pumped phase conjugation.

14.3.2 MOPA laser sources including a photorefractive self-pumped phase conjugate mirror

The first compensation of a passive aberration by double pass after reflection on a phase conjugate mirror (using SBS) was performed in 1972 [58]. Less than 1 year later, a master oscillator-power amplifier (MOPA) structure using a Ruby amplifier was demonstrated [59]. Many laser geometries have been tested with various nonlinear media in which phase conjugation was performed. The MOPA structures demonstrated at 1.06 μm with either internal loop or ring $BaTiO_3$:Rh phase conjugate mirrors have been evaluated in a collaborative work between the Laboratoire Central de Recherches of Thomson-CSF (France) and the Laboratoire Charles Fabry de l'Institut d'Optique (France) [60]. They are depicted in Fig. 14.21. The laser beam experiences two (Fig. 14.21a) or four passes (Fig. 14.21b) in a Nd:YAG amplifier pumped by flash lamps at a repetition rate of 10 Hz. No care was taken to compensate for depolarization which is negligible at a repetition rate of 10 Hz. The oscillator is an intracavity-

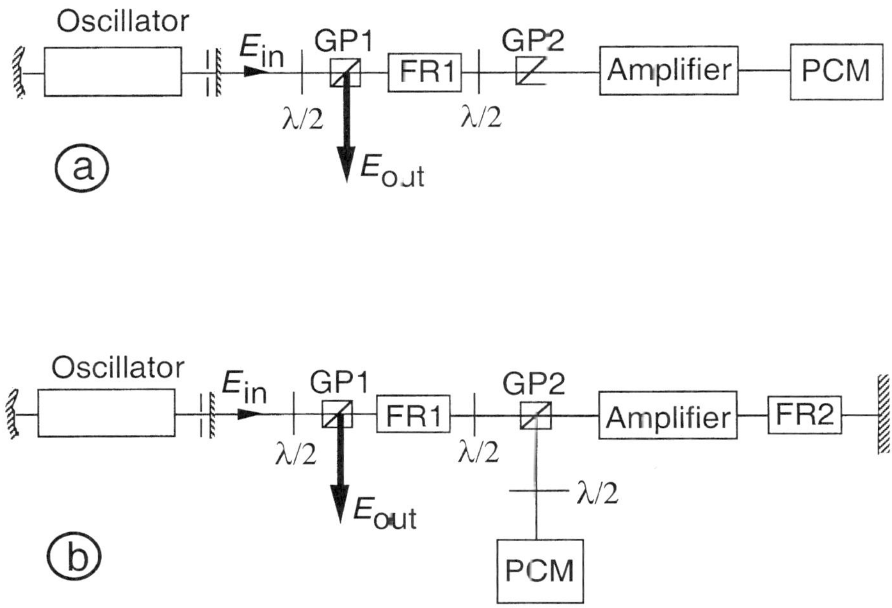

Figure 14.21: Nd:YAG master oscillator power amplifier architectures using $BaTiO_3$:Rh self-pumped phase conjugate mirrors. (a) Double-pass MOPA. (b) Four-pass MOPA. See text for a description of the beam path. (FR = Faraday rotator, GP = Glan polarizer, PCM = phase conjugate mirror.)

filtered Q-switched Nd:YAG laser delivering 20-ns pulses at a repetition rate of 10 Hz. The beam diameter at $1/e^2$ at the output of the oscillator is 2.5 mm. A first half-wave plate/Glan polarizer GP1 device is used to vary the incident energy in the amplifier rod. The device formed by GP1, the Faraday rotator FR1, the half-wave plate, and the Glan polarizer GP2, suppresses beams propagating back to the oscillator and performs the extraction of the output beam after a double pass in the amplifier rod (Fig. 14.21a). In the four-pass configuration, a Faraday rotator FR2 is inserted after the amplifier rod, so that the beam is directed to the phase conjugate mirror only after a double pass in the amplifier. A half-wave plate before the phase conjugate mirror changes the polarization from vertical to horizontal (extraordinary for the photorefractive crystal), which is necessary to provide high photorefractive gains in $BaTiO_3$:Rh. After a reflection on the phase conjugate mirror, the beam experiences a new double pass in the amplifier and its polarization is changed from vertical to horizontal by a double pass in FR2. The beam travels through polarizer GP2 and is extracted by polarizer GP1 after a rotation of its polarization induced by the half-wave plate and FR1 (Fig. 14.21b).

The energy per pulse at the output of the MOPAs is plotted in Fig. 14.22 versus energies per pulse at the output of the oscillator (after GP1) [60]. In the double-pass MOPA, the same output energy is obtained for an incident energy six times lower with the ring phase conjugate mirror than with the internal loop-phase conjugate mirror. In the four-pass MOPA, the difference in the incident energy for the same output energy reaches a factor of 40 in favor of the ring-phase conjugate mirror. This is a consequence of the larger reflectivity available with the ring-phase conjugate mirror, as pointed out in Section 14.2.

The two-pass MOPA rise time, defined as the time needed to increase the output energy from 0 to 90% of its maximum value with a ring-phase conjugate mirror, is given in Fig. 14.23 as a function of the incident energy

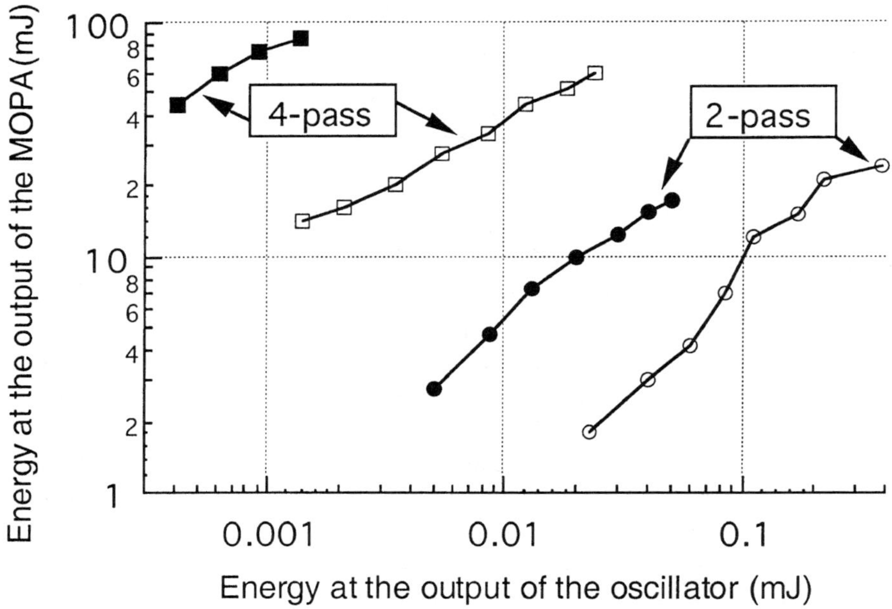

Figure 14.22: Output energy per pulse versus energy per pulse at the output of the oscillator for two- and four-pass MOPAs. Black markers: ring-phase conjugate mirror. White markers: internal loop-phase conjugate mirror.

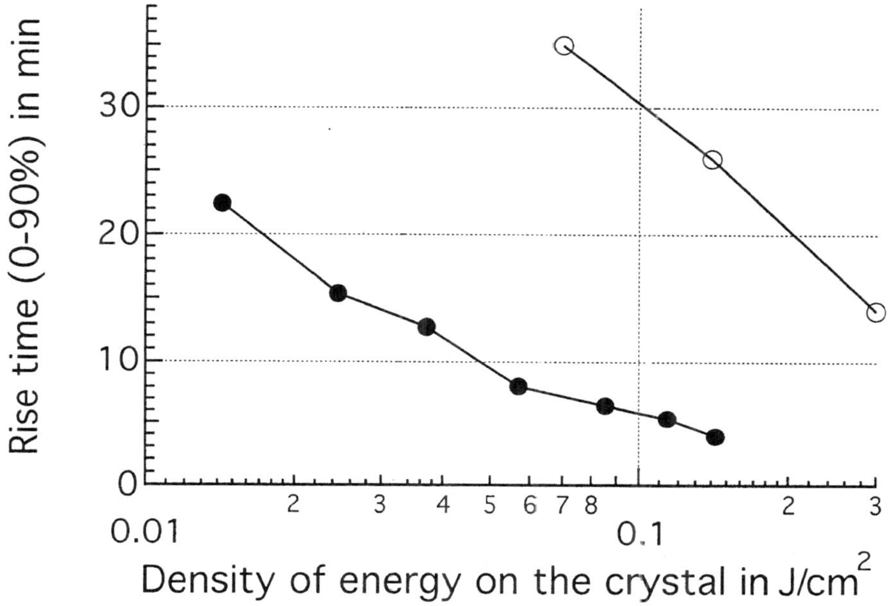

Figure 14.23: Rise time from 0 to 90% of the maximum output energy of the double-pass MOPA versus incident density of energy on the photorefractive crystal (repetition rate of 10 Hz). Black markers: ring-phase conjugate mirror. White markers: internal loop-phase conjugate mirror.

per square centimeter on the photorefractive crystal. It is compared to the two-pass-MOPA rise time with an internal loop phase conjugate mirror. We observe that the internal loop phase conjugate MOPA is five times slower than the ring phase conjugate MOPA. This is consistent with the experiments performed on both phase conjugate mirrors alone (see Section 14.2.3).

For both phase conjugate mirrors, the output beam is a quasi-diffraction-limited Gaussian beam. Anyway, it should be kept in mind that the focal length of the thermal lens at a repetition rate of 10 Hz is larger than 2 m. Such aberrations are easily compensated by both phase conjugate mirrors. To demonstrate the ability of correction of strong aberrations, an additional random-phase plate is inserted after the amplifier rod in the two-pass geometry, the output beam keeps TEM_{00} at steady state as shown in Fig. 14.24.

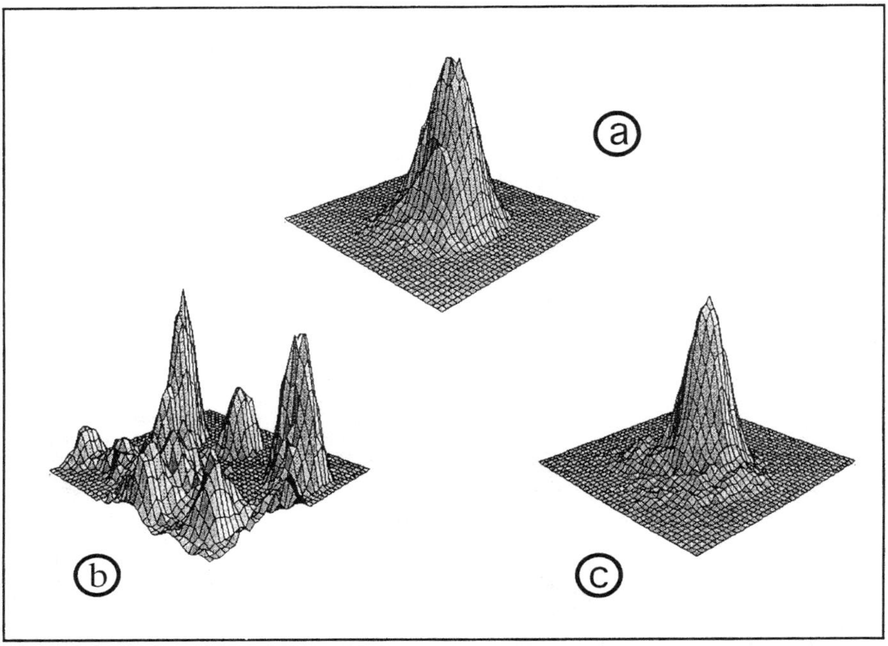

Figure 14.24: Compensation of strong aberration in a MOPA configuration. (a) Beam profile delivered by the oscillator. (b) Far-field output beam profile when a random-phase plate is inserted after the amplifier and when the phase conjugate mirror is replaced by a dielectric mirror. (c) Far-field output beam profile with the random phase plate after the amplifier and with the phase conjugate mirror instead of the dielectric mirror.

The feasibility of nanosecond Nd:YAG MOPA laser sources with photorefractive self-pumped phase conjugate mirrors using BaTiO$_3$:Rh has been demonstrated with success at a repetition rate of 10 Hz. Satisfying results have also been obtained at a repetition rate of 30 Hz [60]. Such all-solid-state phase conjugate mirrors are well suited for high repetition rates wavefront correction of diode-pumped MOPA, which may exhibit a high average power but a low energy per pulse. Recently, a high-repetition-rate all-solid-state diode-pumped MOPA was demonstrated using an internal loop BaTiO$_3$:Rh phase conjugate mirror. The repetition rate is 200 Hz. The output energy per pulse is 20 mJ in a nearly diffraction limited beam, and the rise time is 1.1 min [61]. The results obtained up to now suggest that the compact ring phase conjugate mirror would lead to an output energy of 30 mJ and a rise time of about 10 s under the same conditions.

14.3.3 Comparison of photorefractive self-pumped phase conjugation to other existing techniques

In the previous sections, we described wavefront correction by photorefractive self-pumped phase conjugation. Dynamic wavefront correction can also be achieved by two-wave mixing. This so-called beam cleanup technique (Fig. 14.25) consists in splitting the aberrated intense laser beam into two beams. One, the signal beam of extremely low power, is spatially filtered to provide a TEM_{00} beam. The other is a powerful and aberrated pump beam. Both interfere in the photorefractive crystal. The $\pi/2$ phase shift between the interference pattern and the index grating in the material allows energy transfer from the pump beam to the signal beam, without phase transfer. As a result, the signal beam is amplified while keeping its TEM_{00} structure. Beam cleanup has been studied with nanosecond illumination at 532 nm in $BaTiO_3$:Co where 41% of the pump beam is transferred to the signal beam, which corresponds to 90% of the maximum energy transfer achievable, taking absorption into account [27]. At 1.06 μm in $BaTiO_3$:Rh, the overall efficiency, measured as the ratio of the output amplified signal power to the power of the beam delivered by the laser, is 28% [62]. This is a relatively poor efficiency compared to phase conjugation reflectivities at the same wavelength [36]. Indeed, it was shown that high-phase conjugate reflectivities were achievable with relatively small gain time interaction length products [2]. High transfer efficiencies by two-wave mixing need higher $\Gamma\ell$ values. Moreover, beam-fanning is damageable in beam cleanup: in high-gain crystals, part of the pump energy is transferred to beam-fanning and not to the signal beam. To get rid of this beam-fanning, the available photorefractive gain has to be reduced by incoherent illumination or by crystal tilting [27]. Moreover, to perform beam cleanup with sources of short coherence length, the path lengths of the two arms have to be adjusted carefully, so that the beams may interfere on the whole thickness of the crystal. In self-pumped phase conjugation, this is done automatically. These results show that beam cleanup seems less adapted to applications than phase conjugation.

Photorefractive wave mixing is not the only way to perform wavefront correction. Indeed, the amplifier itself can be used as the nonlinear medium in which gain gratings by way of gain saturation can be photoin-

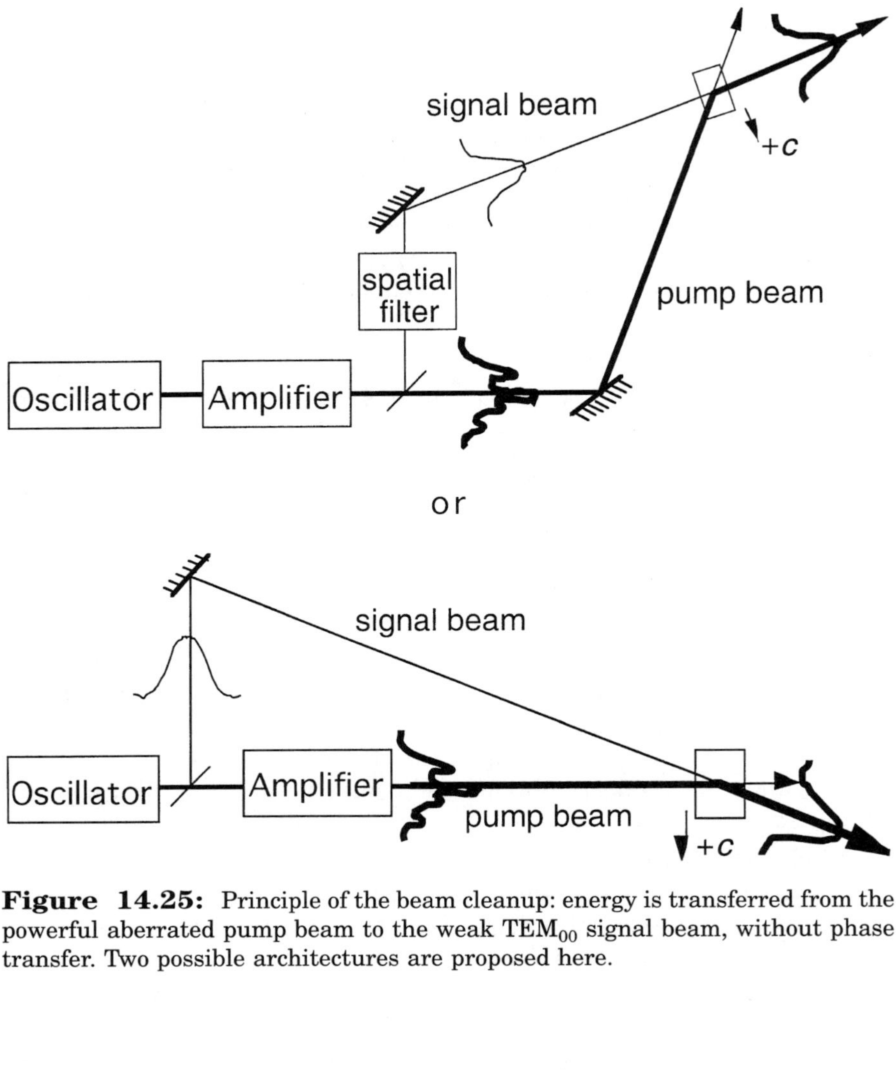

Figure 14.25: Principle of the beam cleanup: energy is transferred from the powerful aberrated pump beam to the weak TEM_{00} signal beam, without phase transfer. Two possible architectures are proposed here.

duced. This idea, initially proposed in 1979 [63], leads to the development of loop resonators and self-starting loops in which pulse-to-pulse phase conjugation is performed inside the Nd:YAG amplifier rod itself (Fig. 14.26) [64, 65]. Moreover, this geometry selects a single longitudinal mode and is self Q-switching. Nevertheless, it needs high-laser-gain amplifiers, which is not the case in high-repetition-rate operation regimes. This geometry is also greedy in energy as at least two amplifiers are often needed to

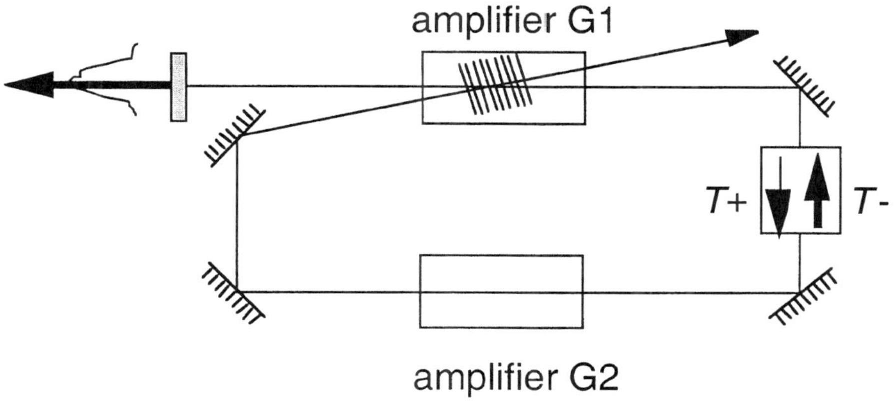

Figure 14.26: Scheme of the self-starting holographic laser oscillator. $T-$ and $T+$ are the transmissions of a nonreciprocal element, which ensures unidirectional lasing in the direction of the output coupler. Phase conjugation occurs in amplifier G1 by gain saturation.

implement a TEM_{00} self-starting source, one used as the active nonlinear medium for phase conjugation and the other as the amplifier medium.

Other active techniques use adaptative optics. The active mirror can be a deformable mirror or an optically addressed light valve (OALV). The OALV is based on a spatial light modulator using liquid crystal technology. An incoherent intensity modulation modifies the orientation of the birefringent liquid crystal molecules, which provides a coherent phase modulation. Such a system has been used to correct phase aberrations of pulsed nanosecond and femtosecond laser beams. Optically addressing is achieved by a mask imaged on the OALV and made by electrically addressing a liquid crystal matrix (Fig. 14.27) [66]. A deformable mirror placed in the laser cavity may also control the divergence of the output beam and increase its brightness. With such a technique, an increase of the beam radiant intensity of 10–15 is reported without any drop in the output power in a 25-W Nd:$YAlO_3$ laser [67]. Various technologies and devices using adaptative mirrors may be found in [68].

One of the most studied and well-developed techniques to perform phase conjugation is stimulated Brillouin scattering (SBS), by means of which a laser beam is focused in a liquid or gas cell. The stimulated scattering then results from the coupling of two processes: (1) the input and scattered waves interfere and produce an acoustic wave by electro-

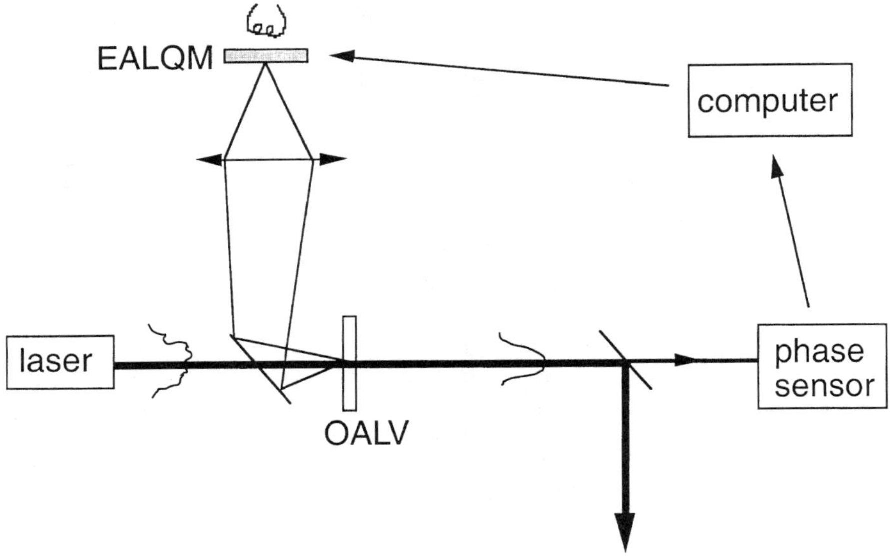

Figure 14.27: Schematic representation of active dynamic correction of laser beams with an optically addressed light valve (OALV). The electrically addressed liquid crystal matrix (EALQM) is imaged on the OALV which changes an incoherent illumination modulation into a phase modulation.

striction, and (2) the input wave is Bragg-diffracted from the acoustic wave and produces the scattered wave. Above a given threshold in intensity, a large amount of the input wave is converted into the scattered wave. SBS cells are widely used and prove to be efficient phase conjugate mirrors. Nanosecond MOPA Nd:YAG laser sources using a SBS phase conjugate mirror are used to obtain high nonlinear conversion processes because of the high brightness of the source. Indeed, with such a laser source, a 80% efficiency second harmonic generation and a 49% third-harmonic generation with BBO (beta-barium borate) are reported [69]. High-average-power nanosecond MOPA laser sources using SBS phase conjugation are also demonstrated: 200 W average power at a repetition rate of 100 Hz with a burst of 20 Q-switched pulses during each flash-lamp pulse is available with a TEM_{00} profile and depolarization compensation [70]. However, SBS phase conjugation in liquid or gas cells presents a few inconveniences. First, to exceed the intensity threshold, one must operate with high-energy pulses and focus the beam into the cell. So, high repetition rate amplifiers having a poor small-signal gain factor can hardly be employed. Second, in focused geometries with high input energies, instabilities occur in the SBS phase conjugation process. Third, liquid or gas SBS cells are often dangerous to manipulate due to their toxicity or high pressure. Quartz fibers are alternative nonlinear media that can produce SBS phase conjugation. The long interaction length and small cross section of the fiber contribute to a decrease of the threshold in intensity. Moreover, such fibers are easy to manipulate and the low Brillouin gain in quartz fibers is compensated by a large interaction length. At 1.06 μm with 30-ns pulses, SBS phase conjugation is reported with a reflectivity of 51% and a fidelity of 93% measured by the power-in-the-bucket technique [71]. This technique is efficient in a broad spectral range. At 532 nm with 30-ns pulses, a reflectivity of 51% and a fidelity higher than 70% are obtained. At 355 nm with 30 ns pulses, a reflectivity of 62% and a fidelity higher than 70% are measured in quartz fibers [72]. New materials presenting SBS phase conjugation effects are currently under investigation. For instance, organic crystals like *l*-arginine phosphate monohydrate (LAP) seem to be efficient nonlinear media in which SBS phase conjugation can be performed. With LAP, which presents a high Brillouin gain coefficient, reflectivities of 65% and fidelities of 90% are reported at 1.06 μm [73]. One of the main advantages of SBS compared to photorefractive self-pumped phase conjugation is that it performs a pulse-to-pulse phase conjugation. In other words, it is suitable for rapidly time-varying aberrations. But, unlike photorefractive materials, it is hardly used with millisecond illumination or high repetition rate (> 1 kHz) nanosecond pulses, because the threshold in intensity is difficult to exceed.

14.4 Conclusion

In this chapter, we have discussed $BaTiO_3$:Rh, a new infrared-sensitive photorefractive material. Its photorefractive properties at 1.06 μm are described under both cw and nanosecond illuminations. The main characteristics of this reproducible material are a high photorefractive gain (23 cm^{-1}) and a weak absorption (0.1 cm^{-1}). A three-charge state band transport model based on spectroscopic studies describes it accurately. This model also predicts a fourfold improvement of the time constant by an oxidation of the sample, along with an increase of absorption and of the dark conductivity. Contrary to the results obtained in $BaTiO_3$ with visible light, nanosecond and cw illuminations are equivalent for $BaTiO_3$:Rh at 1.06 μm in terms of gain and time constant. We have also described the performances of two self-pumped phase conjugate mirrors using $BaTiO_3$:Rh. Compact ring phase conjugate mirrors present a higher reflectivity and a shorter rise time than internal loop phase conjugate mirrors, and may compensate as well for highly aberrated beams. Moreover, the control of the involved gratings enables an accurate prediction of the performances of the mirror. Both types of phase conjugate mirrors were introduced in MOPA architectures at a repetition rate of 10 Hz. The output beam is TEM_{00} in both cases. As predicted, the MOPA using a ring mirror is more energy efficient and has a shorter rise time than MOPA using an internal loop mirror. The compact ring phase conjugate mirrors seem well suited for application to all-solid-state dynamic wavefront correction of high repetition rate (>1 kHz) diode-pumped nanosecond Nd:YAG MOPAs, which deliver a high average power with a relatively low energy per pulse. Such diode-pumped MOPAs have already been demonstrated at a repetition rate of 200 Hz with an internal loop phase conjugate mirror. High repetition rate operation may hardly be accessible to SBS phase conjugation because the energy per pulse may be too low to exceed the threshold. Nevertheless, other promising techniques are also emerging, like SBS in quartz fibers or active wavefront restoration with optically addressed light valves.

We should also mention that other photorefractive materials are now infrared-sensitive and could be an alternative to $BaTiO_3$:Rh. For instance, photorefractive strontium and barium niobate (SBN) doped with cerium and calcium is sensitive up to 850 nm [74], but the low value of the gain at this wavelength cannot lead to high reflectivity phase conjugation. Nevertheless, by a preillumination with green light, SBN:Ce can be activated at 1.06 μm [75]. This material may be interesting for wavefront correction.

Tin hypothiodiphosphate ($Sn_2P_2S_6$) has been studied for a long time for its ferroelectric properties. It also exhibits interesting photorefractive properties at 1.06 μm. This material appears as an intermediate between the high-gain but poor response time $BaTiO_3$:Rh and the short response time but poor gain semi-insulating materials [76]. At 10 W/cm^2, a two-wave mixing gain of 7 cm^{-1} and a response time of 10 ms has been reported [77]. Several problems still arise in this crystal, e.g., $Sn_2P_2S_6$ needs to be preilluminated with white light to be efficient at 1.06 μm.

Moreover, parasitic thermal excitation of charges progressively decreases the gain. To partially eliminate this problem and obtain permanent high gains, low-temperature operations and moving grating techniques are used. This contributes to improve the performances of this promising infrared-sensitive material [78].

Efforts have been made to increase the sensitivity of potassium niobate ($KNbO_3$) in the near-infrared spectrum. Several dopants have been tested (Ce, Cu, Co, Fe, Rh, Mn, and Ni) and a significant response at 1.06 μm has been observed in Rh-, Fe-, Mn-, and Mn-Rh-doped $KNbO_3$ crystals [79]. The sensitivity of $KNbO_3$:Rh in the near-infrared spectrum can be increased by a reduction of the crystal at high temperature, leading to a sensitivity of the same order as $BaTiO_3$:Rh [80]. The problem of the reproducibility of the performances of reduced samples remains, but such a crystal could be an alternative to $BaTiO_3$:Rh for various applications.

References

1. J. Feinberg, "Self-pumped, continuous-wave phase conjugator using internal reflection," *Opt. Lett.,* **7**, 486–488, 1982.
2. M. Cronin-Golomb, B. Fischer, J. O. White and A. Yariv, "Theory and applications of four-wave mixing in photorefractive media," *IEEE J. Quant. Electron.,* **QE–20**, 12–29, 1984.
3. B. Fischer and S. Sternklar, "New optical gyroscope based on the ring passive phase conjugator," *Appl. Phys. Lett.,* **47**, 1–3, 1985.
4. S. Weiss, M. Segev and B. Fischer, "Line narrowing and self-frequency scanning of laser diode arrays coupled to a photorefractive oscillator," *IEEE J. Quant. Electron.,* **24**, 706–708, 1988.
5. R. A. McFarlane and D. G. Steel, "Laser oscillator using resonator with self-pumped phase conjugate mirror," *Opt. Lett.,* **8**, 208–210, 1983.
6. M. Cronin-Golomb, B. Fischer, J. Nilsen, J. O. White and A. Yariv, "Laser with dynamic intracavity distortion correction capability," *Appl. Phys. Lett.,* **41**, 219–220, 1982.
7. B. A. Wechsler, M. B. Klein, C. C. Nelson and R. N. Schwartz, "Spectroscopic and photorefractive properties of infrared-sensitive rhodium-doped barium titanate," *Opt. Lett.,* **19**, 536–538, 1994.
8. M. Lobel, P. M. Petersen and P. M. Johansen, "Suppressing self-induced frequency scanning of a phase conjugate diode laser array with using counterbalance dispersion," *Appl. Phys. Lett.,* **72**, 1263–1265, 1998.

9. M. Kaczmarek, P. Hribek and R. W. Eason, "Near infrared incoherent coupling and photorefractive response time of 'blue' Rh:$BaTiO_3$," *Opt. Comm.,* **136**, 277–282, 1997.

10. S. MacCormack, G. D. Bacher, J. Feinberg, S. O'Brien, R. J. lang, M. B. Klein and B. A. Wechsler, "Powerful, diffraction-limited semiconductor laser using photorefractive beam coupling," *Opt. Lett.,* **22**, 227–229, 1997.

11. M. Kaczmarek and R. W. Eason, "Very high gain single pass two-beam coupling in blue $BaTiO_3$:Rh," *Opt. Lett.,* **20**, 1850–1852, 1995.

12. G. W. Ross, P. Hribek, R. W. Eason, M. H. Garrett and D. Rytz, "Impurity enhanced self-pumped phase conjugation in the near infrared in 'blue' $BaTiO_3$," *Opt. Comm.,* **101**, 60–64 1993.

13. N. V. Kukhtarev, V. B. Markov, S. G. Odoulov, M. S. Soskin and V. L. Vinetskii, "Holographic storage in electrooptic crystals," *Ferroelectrics,* **22**, 949–960, 1979.

14. P. Tayebati and P. Magherefteh, "Theory of the photorefractive effect for $Bi_{12}SiO_{20}$ and $BaTiO_3$ with shallow traps," *J. Opt. Soc. Am.,* **B 8**, 1053–1064, 1991.

15. Ph. Delaye, L. A. de Montmorillon, I. Biaggio, J. C. Launay and G. Roosen, "Wavelength dependent effective trap density in CdTe: Evidence for the presence of two photorefractive species," *Opt. Comm.,* **134**, 580–590, 1997.

16. K. Buse and E. Krätzig, "Three-valence charge-transport model for explanation of the photorefractive effect," *Appl. Phys.,* **B61**, 27–32, 1995.

17. N. Huot, J. M. C. Jonathan and G. Roosen, "Validity of the three-charge-state model in photorefractive $BaTiO_3$:Rh at 1.06 μm in the cw regime," *Appl. Phys.,* **B65**, 489–493, 1997.

18. H. Kröse, R. Scharfschwerdt, O.F. Schirmer and H. Hesse, "Light-induced charge transport in $BaTiO_3$ via three charge states of rhodium," *Appl. Phys.,* **B61**, 1–7, 1995.

19. M. Kaczmarek, G. W. Ross, R. W. Eason, M. J. Damzen, R. Ramos-Garcia and M. H. Garrett, "Intensity-dependent absorption and its modelling in infrared sensitive rhodium-doped $BaTiO_3$," *Opt. Comm.,* **126**, 175–184, 1996.

20. N. Huot, J. M. C. Jonathan, G. Pauliat, D. Rytz and G. Roosen, "Characterization of a photorefractive rhodium doped barium titanate at 1.06 μm," *Opt. Comm.,* **135**, 133–137, 1997.

21. A. Brignon, D. Geffroy, J. P. Huignard, I. Mnushkina and M. H. Garrett, "Experimental investigations of the photorefractive properties of rhodium-doped $BaTiO_3$ at 1.06 μm," *Opt. Comm.,* **137**, 311–316, 1997.

22. Ph. Delaye, J. M. C. Jonathan, G. Pauliat and G. Roosen, "Photorefractive materials: Specifications relevant to applications," *P. Appl. Opt.*, **5**, 541–559, 1996.

23. N. Huot, G. Pauliat, J. M. C. Jonathan, G. Roosen, R. Scharfschwerdt, O. F. Schirmer and D. Rytz, "Four fold improvement of the photorefractive time constant of $BaTiO_3$:Rh by oxidation," *Technical Digest CLEO Europe'98*, Paper CFE2, Glasgow, September 14–18, 1998.

24. G. C. Valley, "Short pulse grating formation in photorefractive materials," *IEEE J. Quant. Electron.*, **QE–19**, 1637–1645, 1983.

25. G. C. Valley, "Simultaneous electron/hole transport in photorefractive materials." *J. Appl. Phys.*, **59**, 3363–3366, 1986.

26. M. J. Damzen and N. Barry, "Intensity dependent hole/electron competition and photocarrier saturation in $BaTiO_3$ when using intense laser pulses," *J. Opt. Soc. Am.*, **B10**, 600–606, 1993.

27. L. Mager, G. Pauliat, D. Rytz, M. H. Garrett and G. Roosen, "Wavefront correction of nanosecond pulses at 532 nm by photorefractive two-wave mixing," *Nonlinear Opt.*, **11**, 135–144, 1995.

28. N. Barry, L. Duffault, R. Troth, R. Ramos-Garcia and M. J. Damzen, "Comparison between continuous-wave and pulsed photorefraction in barium titanate" *J. Opt. Soc. Am.*, **B11**, 1758–1763, 1994.

29. N. Huot, J. M. C. Jonathan, G. Roosen and D. Rytz, "Two-wave mixing in photorefractive $BaTiO_3$:Rh at 1.06 μm in the nanosecond regime," *Opt. Lett.*, **22**, 976–978, 1997.

30. J. Feinberg and R. Hellwarth, "Phase-conjugating mirror with continuous-wave gain," *Opt. Lett.*, **5**, 519–521, 1980.

31. A. Desfarges, V. Kermène, B. Colombeau, M. Vampouille and C. Froehly, "Wave-front reconstruction with a Fourier hologram in a phase-conjugating mirror oscillator," *Opt. Lett.*, **20**, 1940–1942, 1996.

32. P. A. Bélanger, A. Hardy and A. E. Siegman, "Resonant modes of optical cavities with phase-conjugate mirrors," *Appl. Opt.*, **19**, 602–609, 1980.

33. R. C. Lind and D. G. Steel, "Demonstration of the longitudinal modes and aberration-correction properties of a continuous-wave dye laser with phase-conjugate mirror," *Opt. Lett.*, **6**, 554–556, 1981.

34. A. Litvinenko and S. G. Odoulov, "Copper-vapor laser with self-starting $LiNbO_3$ nonlinear mirror," *Opt. Lett.*, **9**, 68–70, 1984.

35. A. Brignon, J. P. Huignard, M. H. Garrett and I. Mnushkina, "Self-pumped phase conjugation in rhodium-doped $BaTiO_3$ with 1.06-μm nanosecond pulses," *Opt. Lett.*, **22**, 215–517, 1997.

36. N. Huot, J. M. C. Jonathan, G. Roosen and D. Rytz, "Characterization and optimization of a ring self-pumped phase-conjugate mirror at 1.06 μm with $BaTiO_3$:Rh," *J. Opt. Soc. Am.*, **B 15**, 1992–1999, 1998.

37. A. V. Nowak, T. R. Moore and R. A. Fisher, "Observations of internal beam production in barium titanate phase conjugators," *J. Opt. Soc. Am.*, **B5**, 1864–1877, 1988.

38. P. Lambelet, R. P. Salathé, M. H. Garrett and D. Rytz, "Characterization of a photorefractive phase conjugator by low-coherence reflectometry," *Appl. Phys. Lett.*, **64**, 1079–1081, 1994.

39. W. T. Whitney, M. T. Duignan and B. J. Feldman, "Stimulated Brillouin scattering and phase conjugation of multiline hydrogen fluoride laser radiation," *J. Opt. Soc. Am.*, **B7**, 2160–2168, 1990.

40. B. W. Liby, J. K. McIver and D. Statman, "Beam quality measurements of a unidirectional self-pumped phase conjugate mirror," *Opt. Comm.*, **101**, 79–84, 1993.

41. B. Fleck, A. Kiessling, G. Notni and L. Wenke, "A simple interferometric method for measuring the fidelity of phase conjugated beams," *J. Mod. Opt.*, **38**, 495–502, 1991.

42. M. Takeda, H. Ina and S. Kobayashi, "Fourier-transform method of fringe-pattern analysis for computer-based topography and interferometry," *J. Opt. Soc. Am.*, **B72**, 156–160, 1982.

43. M. Cronin-Golomb, J. Paslaski and A. Yariv, "Vibration resistance, short coherence length operation, and mode-locked pumping in passive phase conjugators," *Appl. Phys. Lett.*, **47**, 1131–1333, 1985.

44. V. T. Tikhonchuk and A. A. Zozulya, "Structure of light beams in self-pumped four-wave mixing geometries for phase conjugation and mutual conjugation," *Prog. Quant. Electron.*, **15**, 231–293, 1991.

45. N. V. Bogodaev, L. I. Ivleva, A. S. Korshunov, A. V. Mamaev, N. N. Poloskov and A. A. Zozulya, "Geometry of a self-pumped passive ring mirror in crystals with strong fanning," *J. Opt. Soc. Am.*, **B10**, 1054–1059, 1993.

46. G. Pauliat, M. Ingold and P. Günter, "Analysis of the build-up of oscillations in self-induced photorefractive light-resonators," *IEEE J. Quant. Electron.*, **25**, 201–207, 1989.

47. L. Mager, G. Pauliat, D. Rytz and G. Roosen, "Two- and four-wave mixing in photorefractive materials for dynamic correction of pulsed laser beams," in I. A. Khoo, F. Simoni and C. Umeton eds., *Novel Optical Material & Applications*, (Wiley, New York, 1996, pp. 149–174. See also L. Mager, C. Lacquarnoy, G. Pauliat, M. H. Garrett, D. Rytz, G. Roosen, "High quality self-pumped phase

conjugation of nanosecond pulses at 532 nm using photorefractive $BaTiO_3$," *Opt. Lett.,* **19**, 1508–1510, 1994.

48. M. Cronin-Golomb and C. D. Brandle, "Ring self-pumped phase conjugator using total internal reflection in photorefractive strontium barium niobate," *Opt. Lett.,* **14**, 462–464, 1989.
49. N. Huot, J. M. C. Jonathan, G. Pauliat, D. Rytz and G. Roosen, "Self-pumped phase conjugate $BaTiO_3$:Rh ring mirror at 1.06 μm: optimization of reflectivity, rise time and fidelity," *SPIE,* **3470**, 8–15, 1998.
50. W. Koechner, *Solid-state Laser Engineering*, Springer Verlag, 1988.
51. J. D. Foster and L. M. Osterink, "Thermal effects in Nd:YAG laser," *J. Appl. Phys.,* **41**, 3656–3663, 1970.
52. W. C. Scott and M. de Witt, "Birefringence compensation and TEM_{00} mode enhancement in a Nd:YAG laser," *Appl. Phys. Lett.,* **18**, 3–4, 1971.
53. Q. Lü, N. Kügler, H. Weber, S. Dong, N. Müller and U. Wittrock, "A novel approach for compensation of birefringence in cylindrical Nd:YAG rods," *Opt. Quant. Electron.,* **28**, 57–69, 1996.
54. I. D. Carr and D. C. Hanna, "Performance of a Nd:YAG oscillator / amplifier with phase-conjugation via stimulated Brillouin scattering," *Appl. Phys.,* **B36**, 83–92, 1985.
55. N. G. Basov, V. F. Efimkov, I. G. Zubarev, A. V. Kotov, S. I. Mikhailov and M. G. Smirnov, "Inversion of wavefront in SMBS of a depolarized pump," *JETP Lett.,* **28**, 197–201, 1978.
56. H. J. Eichler, A. Haase, R. Menzel and A. Siemoneit, "Thermal lensing and depolarization in a highly pumped Nd:YAG laser amplifier," *J. Phys.,* **D26**, 1884–1891, 1993.
57. D. A. Rockwell, "A review of phase-conjugate solid state lasers," *IEEE. J. Quant. Electron.,* **24**, 1124–1140, 1988.
58. B. Y. Zel'dovich, V. I. Popovichev, V. V. Ragul'skii and F. SD. Faizullov, "Connection between the wave fronts of the reflected and exciting light in stimulated Mandel'shtam-Brillouin scattering," *JETP Lett.,* **15**, 109–113, 1972.
59. O. Y. Nosach, V. I. Popovichev, V. V. Ragul'skii and F.S D. Faizullov, "Cancellation of phase distortions in an amplifying medium with a 'Brillouin mirror' ," *JETP Lett.,* **16**, 435–438, 1972.
60. N. Huot, J. M. C. Jonathan, G. Pauliat , G. Roosen, A. Brignon and J. P. Huignard, "Nd:YAG oscillator power amplifier using $BaTiO_3$:Rh internal loop and ring self-pumped phase conjugate mirrors," *Technical Digest CLEO Europe'98,* Paper CWO2, Glasgow, September 14–18, 1998.
61. A. Brignon, S. Senac, J. L. Ayral and J. P. Huignard, "Rhodium-doped barium titanate phase-conjugate mirror for an all-solid-state, high-repetition-rate,

diode-pumped Nd:YAG master-oscillator power amplifier laser," *Appl. Opt.*, **37**, 3990–3995, 1998.

62. A. Brignon, J. P. Huignard, I. Mnushkina and M. H. Garrett, "Spatial beam clean-up of a Nd:YAG laser operating at 1.06 μm with two-wave mixing in Rh:BaTiO$_3$," *Appl. Opt.*, **36**, 7788–7793, 1997.
63. A. Tomita, "Phase conjugation using gain saturation of a Nd:YAG laser," *Appl. Phys. Lett.*, **34**, 463–464, 1979.
64. P. Sillard, A. Brignon and J. P. Huignard, "Loop resonators with self-pumped phase conjugate mirrors in solid-state saturable amplifiers," *J. Opt. Soc. Am.*, **B14**, 2049–2058, 1997.
65. K. S. Syed, G. J. Crots and M. J. Damzen, "Transient modelling of a self-starting holographic laser oscillator," *Opt. Comm.*, **146**, 181–185, 1998.
66. J. C. Chanteloup, B. Loiseaux, J. P. Huignard and H. Baldis, "Detection and correction of the spatial phase of ultrashort laser pulses using an optically addressed light valve," *Technical Digest, CLEO'97,* Paper CThD8, Baltimore, MD, 1997.
67. S. A. Chetkin and G. V. Vdovin, "Deformable mirror correction of a thermal lens induced in the active rod of a solid state laser," *Opt. Comm.*, **100**, 159–165, 1993.
68. Special Issue *Opt. Eng.*, **36**, 1997. See also M. A. Vorontsov, G. W. Carhart, D. V. Pruidze, J. C. Ricklin and D. G. Voelz, "Adaptative imaging system for phase-distorted extended source and mutliple-distance objects," *Appl. Opt.*, **36**, 3319–3328, 1997.
69. J. K. Timinski, C. D. Nabors, G. Frangineas and D. K. Negus, "High efficiency harmonic conversion of a Nd:YAG phase-conjugated master oscillator / power amplifier laser system," *Advanced Solid-State Lasers,* Paper MD2, Memphis, TN, 1995.
70. S. Seidel and N. Kugler, "Nd:YAG 200-W average power oscillator-amplifier system with stimulated-Brillouin-scattering phase conjugation and depolarization compensation," *J. Opt. Soc. Am.*, **B14**, 1885–1888, 1997.
71. H. J. Eichler, J. Kunde and B. Liu, "Quartz fibre phase conjugators with high fidelity and reflectivity," *Opt. Comm.*, **139**, 327–334, 1997.
72. H. J. Eichler, J. Kunde and B. Liu, "Fiber phase conjugators at 1064-nm, 532-nm and 355-nm wavelengths," *Opt. Lett.*, **22**, 495–497, 1997.
73. H. Yoshida, M. Nakatsuka, H. Fulita, T. Sasaki and K. Yoshida, "High-energy operation of a stimulated Brillouin scattering mirror in L-arginine phosphate monhydrate crystal," *Appl. Opt.*, **36**, 7783–7787, 1997.

74. R. A. Rakuljic, K. Sayano, A. Agranat, A. Yariv and R. R. Neurgaonkar, "Photorefractive properties of Ce- and Ca-doped SBN," *Appl. Phys. Lett.*, **43**, 1465–1467, 1988.

75. A. Gerwens, M. Simon, K. Buse and E. Krätzig, "Activation of cerium-doped strontium-barium niobate for infrared holographic recording," *Opt. Comm.*, **135**, 347–351, 1997.

76. S. G. Odoulov, A. N. Shumelyuk, U. Hellwig, R. A. Rupp and A. A. Grabar, "Photorefractive beam coupling in tin hypothiodiphosphate in the near infrared," *Opt. Lett.*, **21**, 752–754, 1996.

77. S. G. Odoulov, A. N. Shumelyuk, U. Hellwig, R. A. Rupp, A. A. Grabar and I. M. Stoyka, "Photorefraction in tin hypothiodophosphate in the near infrared," *J. Opt. Soc. Am.*, **B13**, 2352–2360, 1996.

78. S. G. Odoulov, A. N. Shumelyuk, G. A. Brost and K. M. Magde, "Enhancement of beam coupling in the near infrared for tin hypothiodophosphate," *Appl. Phys. Lett.*, **69**, 3665–3667, 1996.

79. C. Medrano, M. Zgonik, I. Liakatas and P. Günter, "Infrared photorefractive effect in $KNbO_3$ crystals," *J. Opt. Soc. Am.*, **B13**, 2657–2661, 1996.

80. M. Ewart, R. Ryf, C. Medrano, H. Wüest, M. Zgonik and P. Günter, "High photorefractive sensitivity at 830 nm in reduced rhodium-doped $KNbO_3$," *Opt. Lett.*, **22**, 781–783, 1997.

Chapter 15

Space-Time Processing with Photorefractive Volume Holography Using Femtosecond Laser Pulses

Yeshaiahu Fainman, Pang-chen Sun, and Yuri T. Mazurenko

Department of Electrical and Computer Engineering, University of California, San Diego, La Jolla, California

15.1 Introduction

The growing field of ultrashort-pulse laser technology has elevated conventional holography to new dimensions. These new holographic methods can process time-domain in addition to spatial-domain information, which includes temporal evolution of optical signals on nanosecond to femtosecond scale. Photorefractive volume holographic materials provide the medium necessary for recording and reconstruction of such space-time optical signals. New holographic techniques are currently being developed for mutual processing of optical signals representing different combinations

PHOTOREFRACTIVE OPTICS
Materials, Properties and Applications

ISBN 0-12-774810-5/$30.00

of space and time domains, e.g., space-space, time-time, space-time, and time-space. The photorefractive volume holography with ultrashort optical pulse signals is useful for a variety of novel applications such as optical space-time information processing, three-dimensional (3-D) optical storage, and optical interconnections. This chapter will review the basic principles of such holographic techniques, describing experimental techniques and present experimental demonstration results on storage and optical interconnection application examples, as well as critically evaluate the future research directions. The spatial and the temporal domain holography are discussed in Sections 15.2 and 15.3, respectively. Space-time photorefractive volume holography is introduced in Section 15.4 with a special emphasis on its application for free-space-to-guided-wave optical interconnections. Summary and future research directions are presented in Section 15.5.

15.2 Spatial-domain holography

Spatial-domain holography was first proposed by Gabor [1, 2] in 1948. Holography, originally called wavefront reconstruction, consists of recording and reconstruction of both amplitude and phase information of an optical field. A spatial hologram is obtained by recording an interference pattern between an object and a reference wavefront in a high-resolution holographic recording medium. Depending on the object field and the reference field distributions in the hologram plane, it is typically useful to distinguish among image plane, Fresnel transform, Frounhofer transform, and Fourier transform holograms. Each of these configurations has advantages and disadvantages depending on a specific application. For example, an image plane hologram has the least chromatic dispersion and therefore can produce a high-quality color image display when a broadband light source is used for reconstruction. In contrast, a Fourier transform hologram is usually recorded and reconstructed using monochromatic light, allowing implementation of numerous optical signal processing operations of two-dimensional (2-D) spatial image information. To achieve higher efficiencies in reconstruction, spatial domain holography uses recording in a volume of photorefractive materials. Spatial domain holography has been used for various optical spatial signal processing applications including spatial filtering [3, 4], signal representation [5], optical storage [4, 6–9], and optical interconnects [10–12]. Further details on recording and reconstruction of spatial holograms using photorefractive volume holographic materials as well as their applications for spatial information processing can be found in [13] and [14].

15.3 Temporal holography

The invention of ultrashort pulse lasers enabled development of techniques to determine instantaneous motion of light wave packets. In 1970, Duguay and Mattick [15] of Bell Laboratories created a new ultrahigh-speed photography technique that used an ultrafast shutter built of an optical Kerr cell. They used a 10-ps light pulse to trigger the optical Kerr cell and successfully recorded an image of a short section of a light beam traveling inside a bottle of milky water.

Simultaneously, Denisyuk and Staselko *et al.* [16, 17] discovered that the brightness distribution of an image reconstructed from a hologram is affected by the degree of temporal coherence of the light source, changing from point to point in the hologram plane depending on the pathlength difference between the object beam and the reference beam. They claimed that this method "allows recording of ultrashort light pulses of duration from 3×10^{-9} up to 3×10^{-13} s or less, and performs precise measurements of small temporal shifts of these pulses." In 1978, Abramson [18] named the method light-in-flight to indicate its unique ability "to produce a frameless motion picture of ultrafast phenomena."

Temporal-domain holography, analogously to its spatial-domain counterpart, can process temporally distributed optical signals in the spectral domain (or temporal Fourier transform domain) as well as directly in the time domain. Mazurenko [19, 20] introduced spectral holography that can reconstruct temporal wave packets by recording the temporal Fourier transform, i.e., temporal spectrum of these wave packets. This technique also allows one to perform temporal filtering of optical signals varying in time, similar to the spatial filtering of spatial optical signals.

15.3.1 Time-domain holography

Time-domain holography [16–18, 21] is based on direct recording and reconstruction of time- or space-time-dependent optical signals. Consider an object illuminated by an ultrashort laser pulse as shown in Fig. 15.1. The reflected signal wave has a complex space-time wave packet structure

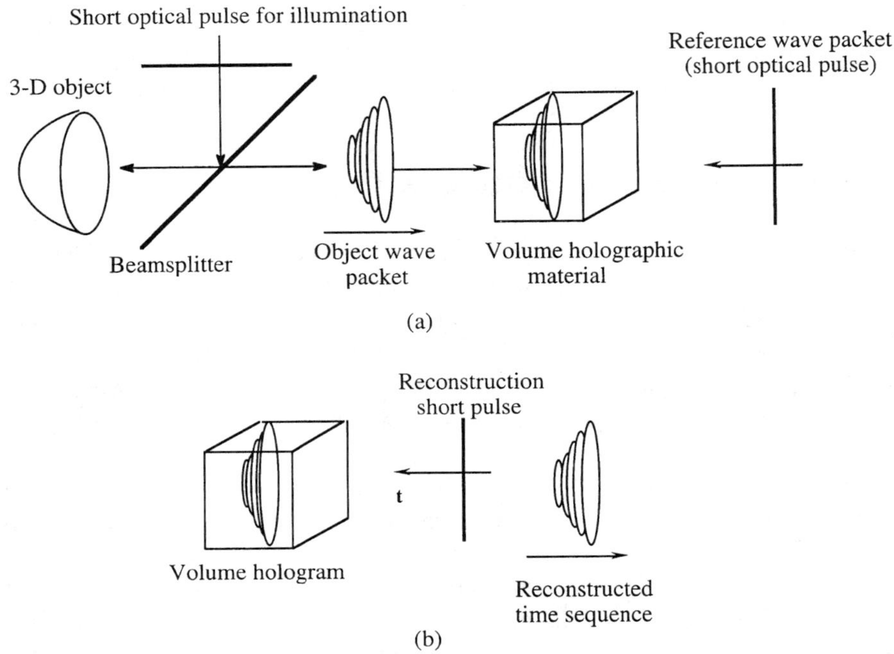

Figure 15.1: Schematic diagram of (a) recording and (b) reconstruction with time-domain holography.

determined by the shape of the object. When this optical signal wavefront interferes with a counterpropagating reference plane wave packet in a 3-D volume of a photorefractive material, a volume hologram is formed. We assume that the photorefractive material provides uniform spectral response for the spectral content of the short-pulse radiation. Since the photorefractive recording process is sensitive to the modulation of the interference pattern, the induced refractive index modulation will be distributed in the photorefractive volume only at spatial coordinates where both the signal and the reference waves coexist. Such temporal hologram recording process can be seen as conversion of spatiotemporal signal into a spatial 3-D volume. During the reconstruction process, a short-pulse reference wave packet is introduced into the volume of the photorefractive material. The resultant reconstruction will reproduce the recorded complex space-time signal. Note that this is valid only for image plane holographic recording since we have neglected the diffraction effects. In the following we will introduce several important applications of direct time-domain holography.

15.3.1.1 Light-in-flight holography

The principle of light-in-flight holography is based on recording the consecutive images of a traveling light wave packet in motion using an extremely fast shutter driven by ultrashort light pulses. This principle is similar to the high-speed photography technique, but at much higher rates. For example, high-speed photography can freeze the motion of a fired bullet, whereas the light-in-flight techniques can freeze the propagation of an ultrashort light pulse.

This concept has been demonstrated employing different techniques such as optical Kerr shutter [15], streak camera [22], and holography [16–18]. The exclusive requirement of coherence gives holography a critical advantage over other methods, because for holographic technique the timing of the "ultrafast shutter" is determined by the temporal coherence function between the object field and the reference field. The maximum of this coherence function occurs at spatial coordinates with zero pathlength difference. Therefore, a history of temporal evolution of the optical short-pulse propagation in time can be continuously monitored from a light-in-flight hologram. This process is enabled by localization of the temporal coherence function that encodes different time frames at different spatial coordinated in the hologram. Note that the width of the temporal coherence function which determines the "shutter time" (i.e., the time resolution) depends only on the spectral content of the light source and is independent of the time domain intensity distribution of the optical radiation. Therefore, for light-in-flight holography, a short coherence continuous wave (CW) laser can perform the same task as a short-pulse laser, provided that the spectral bandwidth of two sources is comparable. This advantage considerably extends the flexibility of the holographic method.

One of the most important applications of light-in-flight holography is imaging through highly scattering medium [23–27], such as, for example, imaging through biological tissue. The principle behind this imaging technique is based on utilization of the first arriving light. When a short pulse of light travels through a highly scattering medium in which some absorbing structures are embedded (Fig. 15.2), light waves will exhibit

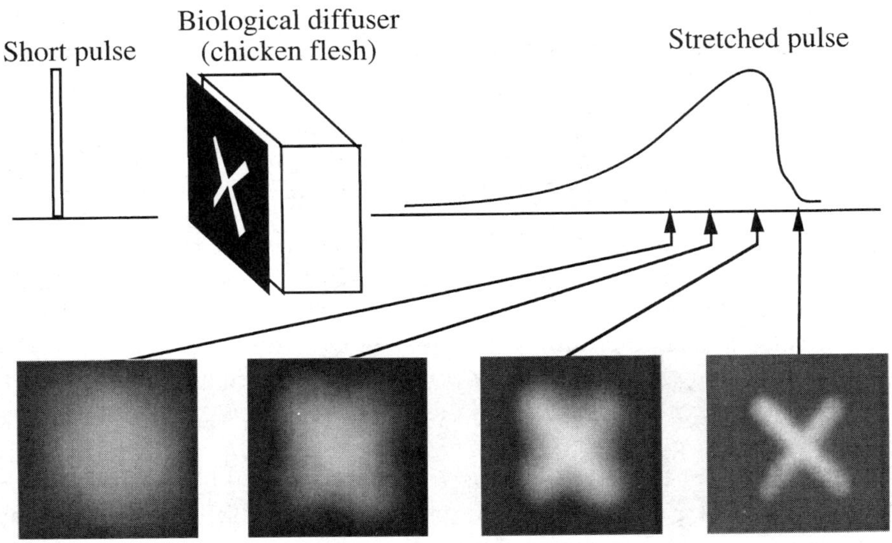

Figure 15.2: The experimental result demonstrating imaging through biological sample with light-in-flight holography. The best image of the target is obtained from the first-arriving light.

multiple scattering effects and diffuse in all directions. Thus, the light waves emerging from the opposite side of the scattering medium will have the shape of an elongated pulse for which the pulse-width will depend on the strength of the scattering and absorption, as well as the thickness of the medium. Light waves that emerge first experience the least scattering effects, and are capable of forming the best-quality image of the absorbers embedded in the material. Therefore, to obtain the best image, these first emerging lightwaves or first arriving lightwaves must be separated from lightwaves that emerge later and thus contain the scattered signals. As described in the preceding discussion, light-in-flight holography can divide the temporal information distribution of a wave packet into separate time zones, and then record them in successively distributed coherence cells on the hologram, through which we can see a continuous progression of information emerging from the time-elongated wave packet. The first-arriving light and the later-arriving light will be coded into different temporal time zones. Thus, at the position on the hologram where the first-arriving light is represented, a sharp image of the absorber can be seen (Fig. 15.2). This image will gradually degrade as the observer's eye moves toward the portion of the hologram representing the later-arriving lightwaves.

15.3.1.2 Three-dimensional storage

An additional important application of the time-domain holography is 3-D volume holographic storage. In existing volume holographic 3-D memories the pages of spatial information are multiplexed in the volume using phase- [6] and/or wavelength-multiplexed encoding [28, 29]. During readout, an appropriate code is set onto the reference beam, which reconstructs the corresponding page of information at the output of the memory system. In contrast to the existing photorefractive 3-D memories, a new approach uses true 3-D storage, where the pages of information are stored in separate subvolumes. Such partitioning of the volume can be obtained using a technique based on collision of ultrashort pulses (Fig. 15.3) [30, 31]. An ultrashort laser pulse is split into two, where one is modulated by the 2-D $N \times N$ spatial image information that we want to store, while the other is used as a reference wave to record a reflection-type hologram (Fig. 15.3a). The information-carrying pulse, imaged into the volume of the media, will collide with the counterpropagating uniform amplitude reference pulse in the volume of the holographic material centered at a single longitudinal coordinate z_i determined by the relative time delay between the two beams. The intensity distribution of the interference between the collided pulses will give rise to recording a reflection-type volume grating, confined in the longitudinal direction. The dimension of this confinement is determined by collision length, i.e., by the spatial pulse width $l_p = c\tau_p/n$, where c is the speed of light in a vacuum, τ_p is the temporal pulse width, and n is the refractive index of the volume holographic material. Alternatively, in a transmission hologram geometry recording with ultrashort laser pulses (Fig. 15.3b), the collision concept provides transverse hologram confinement of effective thickness $l_p = c\tau_p/n \sin\theta$, where θ is the angle between the two beams in air. We call such holograms quasi-infinite as will be further discussed in the following section.

For retrieval of the recorded information (memory readout), the reference pulse of pulse width τ_p is sent into the volume hologram which diffracts on the recorded subvolumes generating a sequence of time-

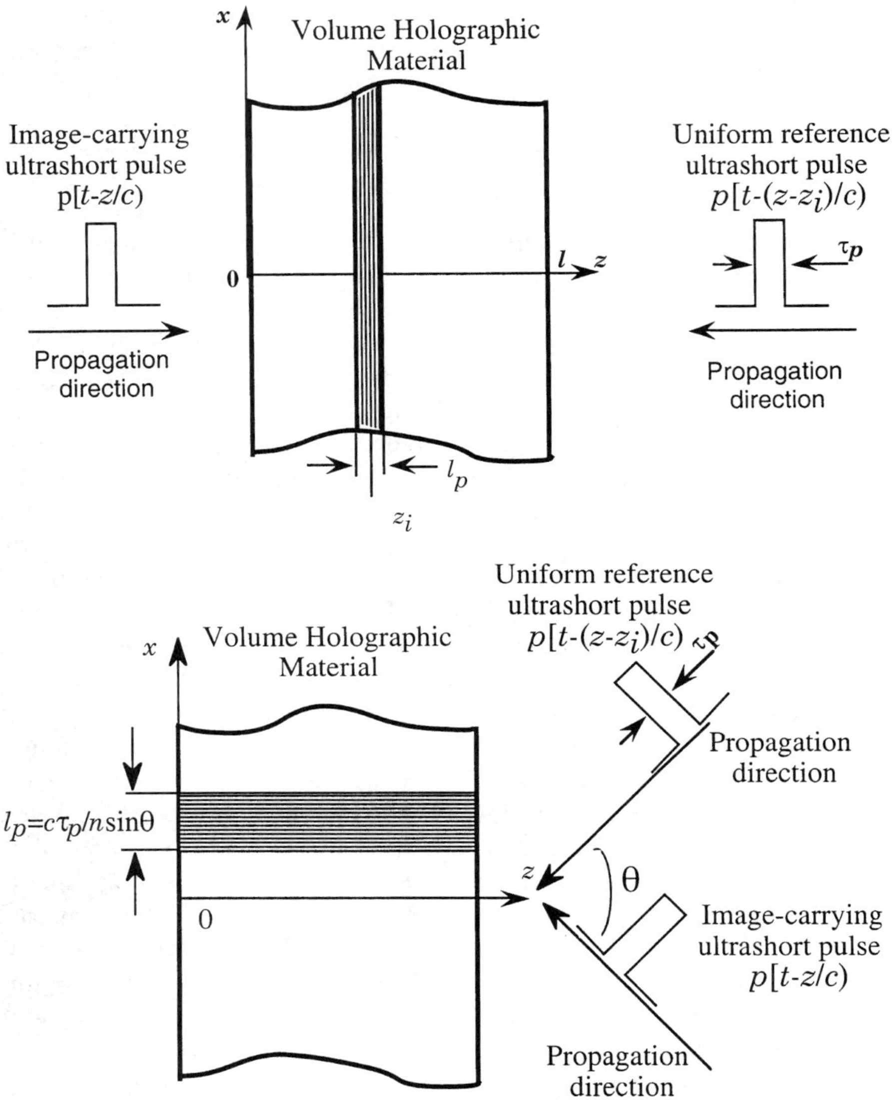

Figure 15.3: (a) Schematic diagram for true volume partitioning using ultrashort pulse collision recording in counterpropagation (i.e., recording reflection-type hologram) geometry. (b) Schematic diagram for true volume partitioning using ultrashort pulse collision recording in transmission geometry.

delayed two-dimensional (2-D) pages of the recorded information images. To access a single page from the reconstructed sequence one must use a decoding technique, which for the case of time-space coding will require to employ one of the existing time-gating techniques [25, 32]. With such a decoding method it will be possible to retrieve 2-D pages of stored information similar to those of the methods based on phase and/or wavelength orthogonal coding. The random access to the pages in the sequence can be accomplished using either acoustooptical or electrooptical devices integrated with discrete temporal delay lines. For example, with an electrooptical or acoustooptical deflector of $\tau_a = 1$ μs and pages of size $N^2 = 10^6$ we can obtain data transfer rates of 1 tbit/s. In the following discussion we investigate wavelength multiplexing of such quasi-infinite holograms.

15.3.1.3 Quasi-infinite wavelength-multiplexed volume holograms

We further analyze the recording process with ultrashort laser pulses for wavelength-multiplexed storage (Fig. 15.4). The spatial interference pattern is weighted by the mutual coherence function of these partially coherent fields. Let $g(t)\exp(j\omega_c t)$ be the mutual coherence function of the signal and the reference waves, where $g(t)$ is the envelope and ω_c is the center frequency. Volume holographic materials are sensitive to the spatial intensity modulation, and thus the mutual coherence function is recorded as a volume grating with amplitude varying along the longitudinal coordinate $z = vt/2$, where v is the light velocity in the medium and a factor of two encounters for the counterpropagation recording. This grating is described by $h(z) = g(2z/v)\exp(j\omega_c 2z/v)$. When interfering waves possess time limited mutual coherence function, the recorded grating will be of limited extend along the z-axis, and therefore can be completely recorded in the material of finite longitudinal dimension.

For the hologram reconstruction process, such a quasi-infinite hologram can be seen as a superposition of "monochromatic gratings" of infinite spatial extent in z, with grating amplitude weighted by the Fourier transform of $h(z)$. With such description, when a quasi-infinite hologram is reconstructed by a monochromatic plane-wave of frequency ω, only the Bragg-matched, infinite gratings will determine the amplitude diffraction efficiency. Consequently, the amplitude's diffraction efficiency dependence on frequency ω can be expressed by the Fourier transform of the mutual

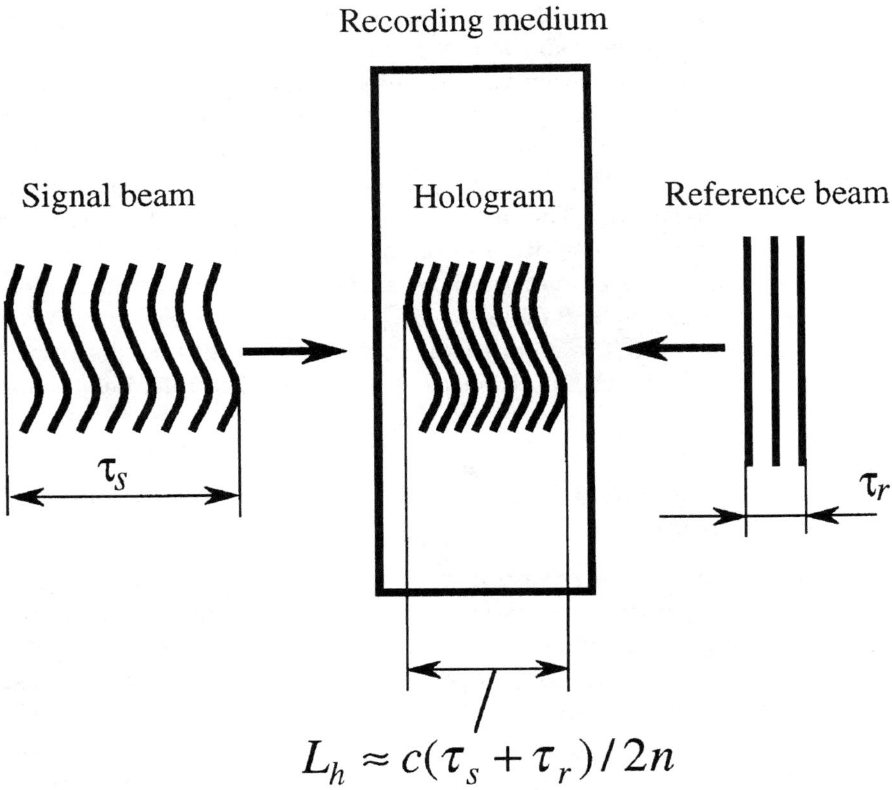

Figure 15.4: Quasi-infinite volume hologram of a wave packet.

coherence function, i.e., the normalized diffraction efficiency $G(\zeta)$ will be proportional to the mutual spectral density [33],

$$G(\zeta) = \int_{-\infty}^{+\infty} g(t)\exp[-j\zeta t]dt\,, \tag{15.1}$$

where $\zeta = \omega - \omega_c$. Furthermore, notice that even for recording wide angular bandwidth signals, the quasi-infinite holograms will not be truncated by the material boundaries, and Eq. (15.1) will be valid for all the components of the angular bandwidth. In contrast, recording such holograms with monochromatic light will result in gratings truncated by the material boundaries, introducing windows varying with angular components, which in turn lead to the variation of diffraction efficiency spectrum within the angular bandwidth.

We next examine wavelength-multiplexing of such quasi-infinite holograms. Consider wavelength multiplexing by using pulses with center frequencies $\omega_{cj} = \omega_0 + j\Delta$, where ω_0 is a center of the multiplexing frequency band, $j = 0, \pm 1, \ldots \pm J/2$ is the integer corresponding to a given information page, J is an even integer such that the total number of multiplexed holograms is $(J + 1)$, and Δ is the frequency increment. The jth information page is reconstructed with a monochromatic wave of corresponding frequency $\omega_{cj} = \omega_0 + j\Delta$. The multiplexing is performed by means of recording many quasi-infinite holograms in the same volume, causing cross talk noise during reconstruction. This cross talk can be diminished by a proper choice of mutual coherence function. For example, the mutual coherence function, $g_q(t)$, can be generated from rectangular functions $p(t) = (1/2T)\mathrm{rect}(t/2T)$ with unit area and parameter $T = \pi/\Delta$,

$$g_q(t) = p(t) * \cdots * p(t), \tag{15.2}$$

where * denotes convolution operation applied $q - 1$ times, and the function $g_q(t)$ has the boundaries $-qT$ and qT. The length of the interference pattern, $L_c = qT/v$, should be less than the material length L_h. The amplitude diffraction efficiency for the mutual coherence function of Eq. (15.2) follows from Eq. (15.1):

$$G_q(\zeta) = (\sin \zeta T / \zeta T)^q, \tag{15.3}$$

where $\zeta = \omega - \omega_{cj} = \omega - (\omega_0 + j\Delta)$. This function has its maximum at $\zeta = 0$ and zeros at $\zeta = n\Delta$ for any integer $n \neq 0$. For recall of the jth information page, the reconstruction wave is tuned to the jth central frequency $\omega_{cj} = \omega_0 + j\Delta$ and only the jth hologram is reconstructed without any cross talk.

In practice, since the frequency bandwidth of optical radiation used for hologram recording is always limited, the mutual-coherence function spreads to infinity, thus violating the condition of obtaining a quasi-infinite hologram and leading to a certain level of cross talk. To counter such physical limitations we choose to truncate the mutual spectral density of the reference and signal radiation $G_q(\zeta)$ (which is proportional to the diffraction efficiency) using a square window $\mathrm{rect}(\zeta/2\Omega)$. The introduced bandwidth 2Ω limits the number of multiplexed holograms to the value $J \approx 2\,\Omega T/\pi$. We also notice that since during the changing of the multiplexing frequency ω_{cj} the mutual spectral density of jth hologram is shifted by the value $j\Delta$, the total employed spectral bandwidth is 4Ω.

A new mutual spectral density $G_q(\zeta)\text{rect}(\zeta/2\Omega)$ can be generated in practice, and since it possesses the same zero-crossings as $G_q(\zeta)$, cross talk noise will not be generated. However, its corresponding mutual coherence function

$$g_q(t) = \frac{1}{2\pi}\int_{-\Omega}^{+\Omega} G_q(\zeta)\exp[j\zeta t]d\zeta = g_q(t) * \sin(\Omega t)/\pi t \qquad (15.4)$$

will need to be recorded in a material of infinite longitudinal extent (for $q = 1$ see Fig. 15.5). The boundaries of the ideal mutual coherence function, $-qT$ and qT have been extended to infinity due to the convolution with the $\sin(\Omega t)/(\pi t)$ (Eq. 15.2). The material boundaries introduce truncation of the mutual coherence function of Eq. (15.3) by a rectangular window $\text{rect}(t/2T_h)$, where $T_h = L_h/v$. Finally, the diffraction efficiency of such a truncated hologram is

$$G_q(\zeta) = \int_{-Th}^{+Th} g_q(t)\exp[-j\zeta t]dt = G_q(\zeta)\text{rect}(\zeta/2\Omega) * 2\sin(T_h\zeta)/\zeta\,. \qquad (15.5)$$

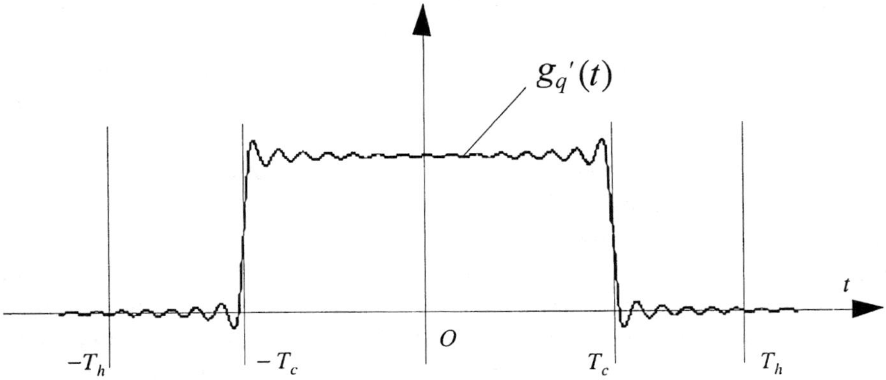

Figure 15.5: Recording of the nonideal coherence function in a medium with a finite longitudinal dimension.

Owing to the convolution operation, we can observe from Eq. (15.5) that the spectrum $G'_q(\zeta)$ is different from the spectrum $G_q(\zeta)\text{rect}(\zeta/2\Omega)$, causing cross talk.

We define a cross talk factor $Q(\zeta)$ which is the deviation of the practically achievable diffraction efficiency $G'_q(\zeta)$ from that which does not introduce the cross talk, $G_q(\zeta)\text{rect}(\zeta/2\Omega)$:

$$Q_q(\zeta) = G'_q(\zeta) - G_q(\zeta)\text{rect}(\zeta/2\Omega) = -\left(\int_{-\infty}^{-Th} + \int_{Th}^{+\infty}\right) g'_q(t)\exp[-j\zeta t]dt\,. \tag{15.6}$$

For estimating the cross talk noise calculation we estimate the last equation [31], yielding the value

$$|Q_q(\zeta)|^2 \approx \frac{1}{\pi^2} G_q^2(\Omega)\{si^2[(\Omega + \zeta)T_h] + si^2[(\Omega - \zeta)T_h]\}\,. \tag{15.7}$$

For characterization of the cross talk noise we next consider reconstruction of the hologram corresponding to $j = 0$, i.e., reconstruction with center frequency ω_0 The normalized diffraction efficiency for this hologram is $G'_q(0)$, while simultaneously the cross talk diffraction efficiencies $G'_q(-j\Delta)$ are also generated. The amplitude of the jth cross talk noise term is $N_j = S_j Q_q(-j\Delta)$, where S_j is the amplitude of the jth signal with $j \neq 0$. The total amplitude noise N due to cross talk is a coherent sum,

$$N = \left(\sum_{j=-J/2}^{-1} + \sum_{j=1}^{J/2}\right) S_j Q_q(j\Delta)\,, \tag{15.8}$$

where $J + 1 \approx J$ is the total number of multiplexed holograms. The power of the cross talk noise is

$$\langle|N|^2\rangle = \left(\sum_{j=-J/2}^{-1} + \sum_{j=1}^{J/2}\right) |Q_q(j\Delta)|^2\,, \tag{15.9}$$

where we assumed that S_j are statistically independent complex random quantities with average power $\langle|S_j|^2\rangle = 1$ and we also used the fact that the number of holograms is large such that the sum of the cross-products of the type $S_k Q(k\Delta) S_l^* Q^*(l\Delta)$ is close to zero.

We will estimate value $\langle|N|^2\rangle = 1$ using Eq. (15.7). The spectral boundaries of our recorded mutual spectral densities satisfy $J\Delta \leq 2\Omega \leq (J + 1)\Delta$.

The spectral holography approach [19, 20, 35–40] uses the temporal spectrum decomposition of the signal wave packets instead of direct time-domain recording method. The spectrum decomposition process can be accomplished by using an optical spectral device built of a grating or a prism in combination with a lens. The spectral decomposition waves (SDW) of a short wave packet (Fig. 15.6) from such a spectral device is linearly dispersed along the x-direction, where each subdivision of the wave in the x-direction contains a longer wave packet of narrower spectral bandwidth compared with the temporal extend of the original wave packet. If two SDWs, one from the signal wave packet and the other from the reference wave packet, are brought together in a volume of the photorefractive recording medium, a spectral hologram is formed. The setup for recording and reconstruction of spectral holograms is shown in Fig. 15.7. During the reconstruction process a spectral device, similar to that of Fig. 15.6, but in reverse, is used to perform the inverse transformation, i.e., from the temporal spectral domain back to the time domain. Due to narrower spectral bandwidth of the SDWs dispersed in space, the fringe pattern and the recorded spectral hologram will have much higher contrast in comparison to that obtained with direct time-domain holography. Improvement in the fringe contrast and more efficient photorefractive recording are achieved by generating and recording SDWs of the ultrashort pulses. However, the SDWs are generated at the expense of trading-in one of the two dimensions in space domain, thereby leaving only one-

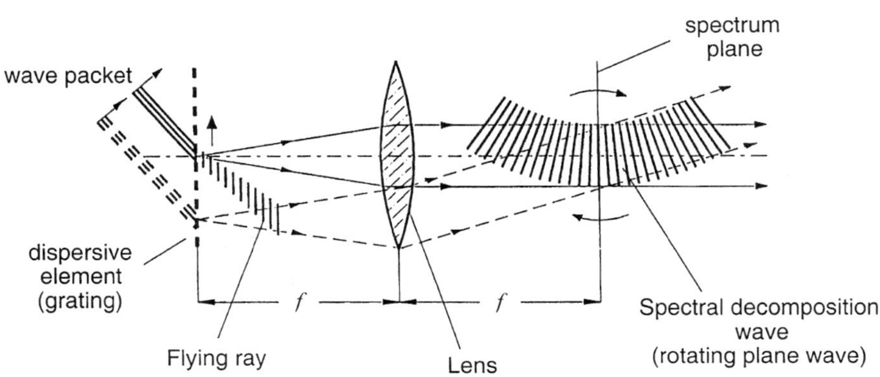

Figure 15.6: Schematic diagram of wave packet spectral decomposition by a spectral device.

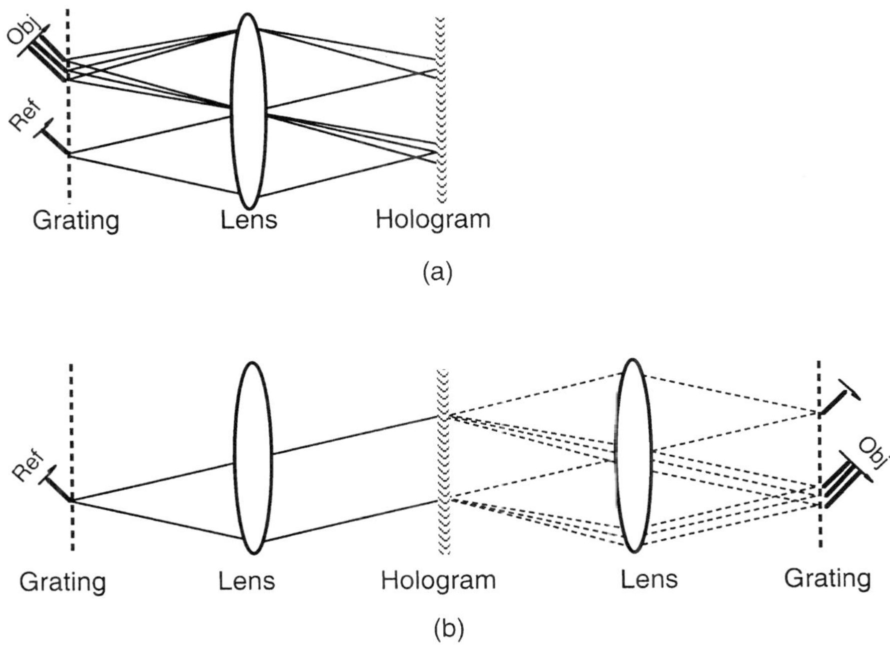

Figure 15.7: Optical system for (a) recording and (b) reconstruction of a spectral hologram.

dimensional (1-D) spatial information channel for optical information processing. The recording time of a spectral hologram depends on the spectral resolution of the optical spectral device. If a longer recording time is desired, then it is necessary to choose a higher resolution spectral device such as a Fabry-Perot etalon. An experimental result [39] (Fig. 15.8) shows that wave packet of nanosecond duration can be recorded and reproduced by spectral hologram with Fabry-Perot etalons as spectral decomposition devices. The wave packet of nanosecond scale time duration occupies a few tenths of centimeters in space, which makes them impractical for direct time-domain volume holographic recording.

15.3.2.1 Nonvolatile photorefractive spectral holography

Volatility is an important issue associated with using photorefractive materials for volume holographic storage applications. In other words, if

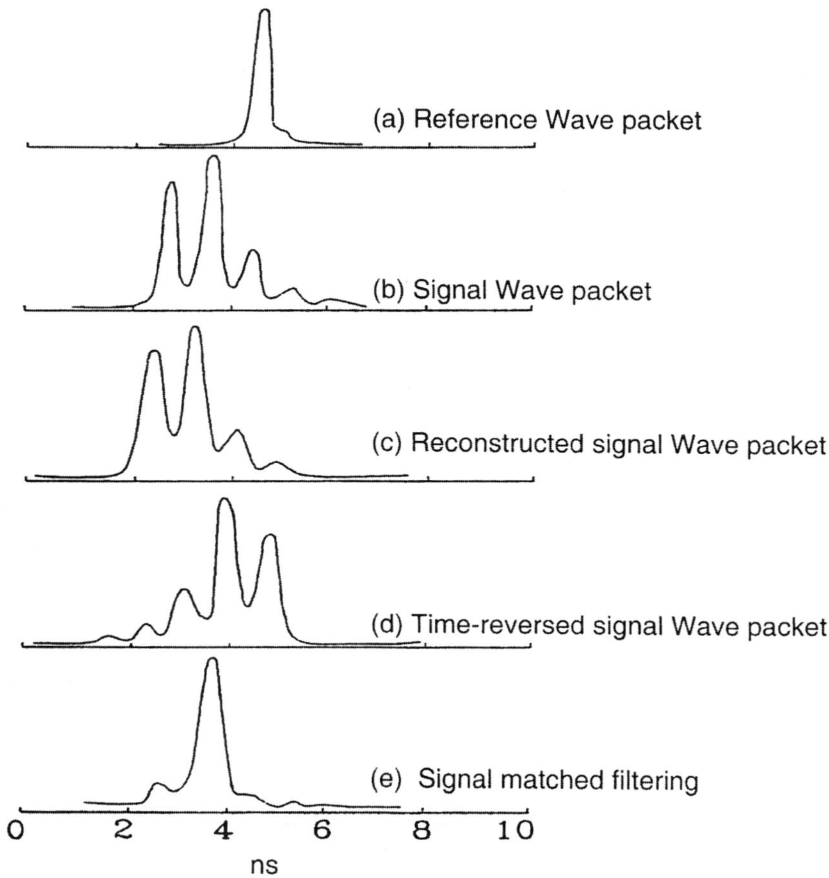

Figure 15.8: Experimental results from spectral holography with nanosecond range wave packages: (a) reference wave packet, (b) original signal wave packet, (c) reconstructed signal wave packet, (d) reconstructed time-reversed signal wave packet, and (e) cross-correlation function obtained from temporal matched filtering.

the same wavelength is used for recording as well as for reconstruction of photorefractive hologram, then the reconstruction process will necessarily cause erasure of the recorded information hologram. This issue has been addressed by using different wavelengths of radiation during the recording and the reconstruction stages of the information holograms. The recording of the hologram is performed at a wavelength of radiation for which the photorefractive material has high photorefractive sensitivity [13, 14], whereas the recorded hologram reconstruction is performed using radiation at a wavelength for which the photorefractive sensitivity is negligibly small, thereby allowing nonvolatile storage. However, to achieve efficient hologram reconstruction, the Bragg matching condition needs to be satisfied. Several elaborate solutions to the Bragg matching condition have been proposed for spatial domain page-oriented holographic memory systems where dual-wavelength recording and reconstruction have been employed [41]. To achieve nonvolatile storage, we use dual-wavelength recording and reconstruction of spectral holograms. With our approach the Bragg matching condition is easily satisfied over the entire spatial frequency range by adjusting the angle of incidence of the spectral decomposition wave derived from the readout wave. Such a simple adjustment is possible because holograms recorded with this method have a quasi-1D data format and we can choose the direction of the spatial frequency carrier of the interfering SDWs to be orthogonal to the direction of the wavelength decomposition.

The experimental setup [42] for spectral holographic storage is schematically shown in Fig. 15.9. We use a mode-locked Ti:sapphire laser producing 200-fs pulses at the center wavelength of 920 nm with a repetition rate of 77 MHz. The infrared radiation (920 nm) is divided into two beams, one of which is kept for the hologram reconstruction process while the other one is frequency-doubled using a second-harmonic generator producing pulses at a wavelength of 460 nm for hologram recording. The radiation at 460 nm is split into two beams, one used as a reference beam while the other one is transmitted through a pulse shaper to produce an object beam pulse sequence. The pulse sequence is tailored using spectral domain filtering of an incident transform-limited pulse introduced into a pulse shaper. The shape of the pulse sequence is proportional to the time-domain convolution between the transform-limited input pulse and the Fourier transform of the spectral filter in the pulse shaper.

The object beam from the pulse shaper and the reference pulse are introduced into the spectral holographic storage system of Fig. 15.9. The SDW of both the signal pulse and the reference pulse are identically spread in the horizontal direction while in the vertical direction they merge and overlap with each other. The spectral components from both beams coincide, producing an interference pattern that can be recorded as a spectral hologram. A 1-mm-thick $LiNbO_3$ crystal is placed in the

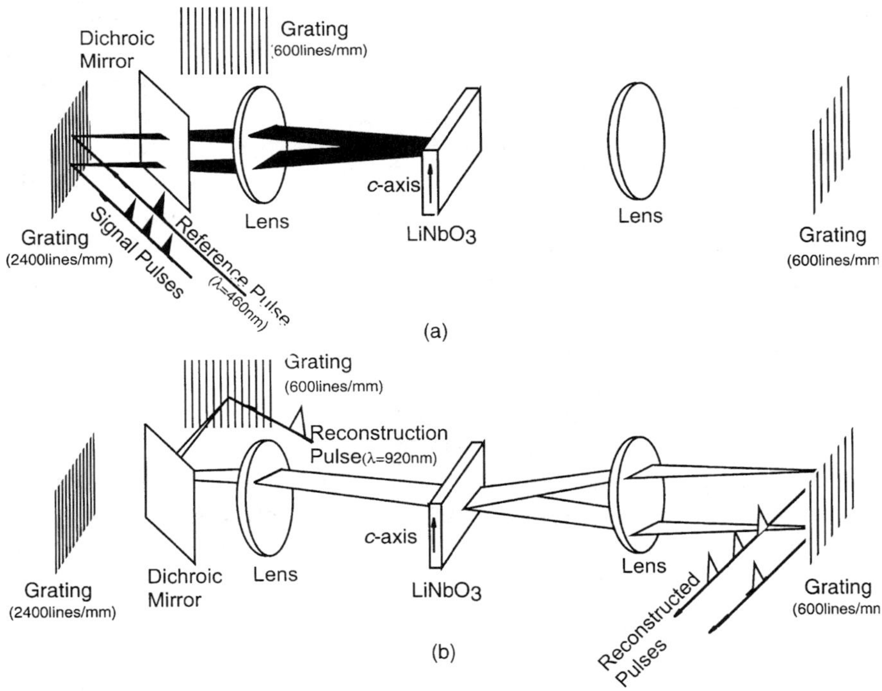

Figure 15.9: Nonvolatile spectral holographic setup from optical pulse storage: (a) recording and (b) reconstruction setup

Fourier transform plane for recording of this spectral hologram. Typical recording times vary in an interval of 60–90 s. As the recording process takes a relatively long time, the stability of the system is crucial.

During the reconstruction process, a readout pulse beam of 920 nm wavelength is introduced into the recording system via a dichroic beam-splitter, which is reflecting radiation at 920 nm while transmitting radiation at 460 nm. The readout beam, after diffracting from a reflection grating, is then Fourier transformed by a lens. To satisfy the Bragg matching condition, the readout beam is shifted vertically to achieve the best diffraction efficiency from the spectral hologram. In our experiment, we set the scaling factor to 1 by choosing gratings with appropriate spatial frequencies. The reconstructed SDW from the hologram is transmitted through the second Fourier transform lens and combined by the second reflection grating producing the reconstructed signal in the time domain. The reconstructed pulses are introduced into an autocorrelator for detection and analysis.

Figure 15.10 shows the experimental results of the reconstructed pulse sequence that was stored in the spectral hologram. The reconstruction result shows that each of the reconstructed pulses has a pulse width of about 200 fs and that the pulse separation is 1.97 ps, consistent with the calculated pulse separation of 2.1 ps with the scaling factor of about 1. The diffraction efficiency is measured to be 7%, and did not show any degradation for over 24 hours of continuous reconstruction. The second

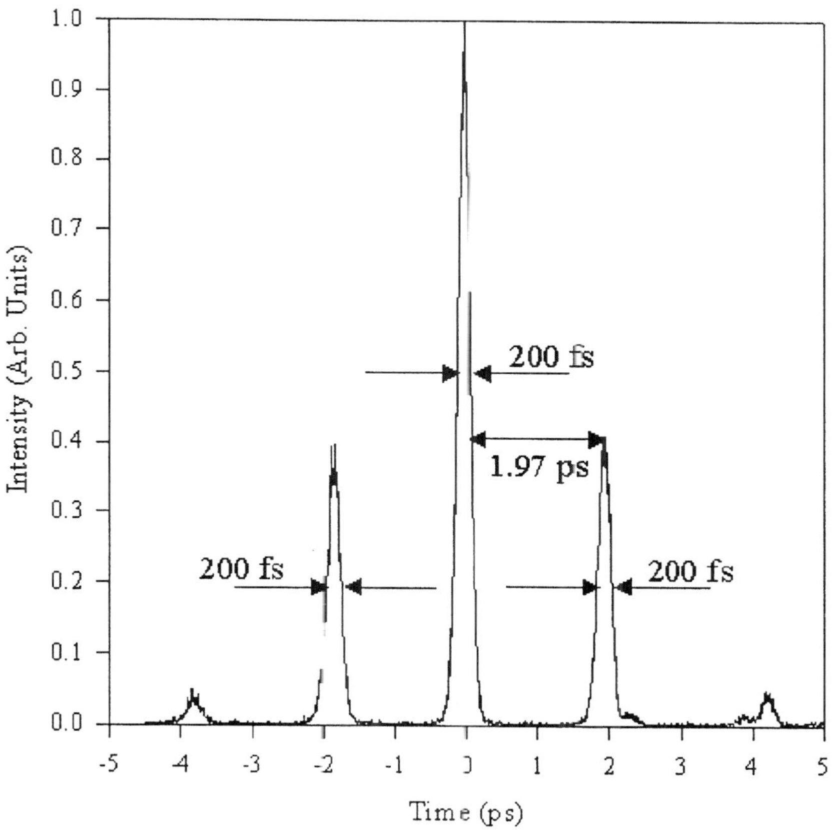

Figure 15.10: Experimental result of short-pulse signal reconstructed from nonvolatile spectral holographic recording.

order peaks appeared due to imperfections in the Ronchi grating used in our experiment.

For spectral holography with dual-wavelength writing and reconstructing the hologram, the time separation of the readout pulse train can be larger or smaller than that of the input recorded pulse train. This is similar to the effect of magnification in conventional spatial holography when different wavelengths are used during the writing and reading of the hologram. The reconstructed image can be magnified or demagnified depending on the ratio between the writing and reading wavelengths. In contrast to space domain holography, the magnification in spectral holography is also affected by the spectral resolving power of the spectral decomposition devices, e.g., the gratings used during the recording and reconstruction processes. For example, it is possible to record a spectral hologram of nanosecond pulse signals and reconstruct them with subpicosecond scale pulses, obtaining nearly 1000 times magnification. For this application, it will be necessary to use a spectral decomposition device with high spectral resolution, such as, for example, a Fabry-Perot interferometer .

15.3.2.2 Recall, time reversal, and temporal filtering of optical wave packets

Similar to spatial holography that allows reconstruction of the real and the virtual images of the recorded optical signal, the reconstruction of a spectral hologram can also reproduce, in general, two signals: the original recorded signal wave packet and the time-inverted signal wave packets (Figs. 15.8c and d). These two output signals represent the real and the phase-conjugate signals reconstructed from the hologram.

In addition to the ability to record and reconstruct optical wave packets, the spectral holography is also capable of performing time domain filtering of optical wave packets [38–40]. The concept is similar to that of Vander Lugt filtering for spatial domain information processing. Consider a set of target spatio-temporal signal wave packets that are prerecorded in a form of spectral holograms multiplexed in the volume of a photorefractive material for later use as filters for time-domain signals. Let a test spatio-temporal signal wave packet be introduced into a temporal Fourier transform processing system containing the hologram with the multiplexed temporal filters. Depending on the relative orientation of the filters and the incident signal, the system can produce at the output either the convolution or the cross-correlation (Fig. 15.8e) between the incident signal wave packet and one of the prestored target wave packet signals. Cross-correlation output is used for implementing the temporal matched filtering operation [39, 40], pulse compression [39, 40], and dispersion compensation [40]. The combination of the convolution and cross-correlation has been used for temporal signal encoding and decoding [40]. Finally, volume spectral holograms have been used to control the shape of ultrashort laser pulses for ultrafast spectroscopy applications [40].

15.4 Space-time holographic processing

The analogy between the spatial domain Vander Lugt filter and temporal domain spectral filtering with the spectral photorefractive volume holograms suggests the possibility of converting optical information between the spatial and temporal domains [43, 44]. In the following discussion, we introduce a holographic method that allows parallel-to-serial (i.e., space-to-time) optical signal conversion by encoding spatial frequency spectrum of the parallel optical signals in space domain onto the temporal frequency spectrum of ultrashort optical pulses. Moreover, by combining this technique with serial-to-parallel conversion methods [45, 46] we can demonstrate the possibility of transmitting parallel optical signals over a long distance optical fiber network. Such space-time and time-space domain interchange is critically important for ultrahigh-speed fiber-optic communication applications.

The all-optical parallel-to-serial conversion processor is shown schematically in Fig. 15.11a. The processor consists of two independent optical channels for carrying temporal signals and spatial signals. The temporal information carrying channel consists of a pair of gratings and a 4-F lens arrangement. The incident pulses are transformed by the input reflecting grating and the first lens into temporal frequency spectrum distributed in space of the focal plane of the first lens, while the second lens and the output reflecting grating are performing the inverse transformation of the temporal spectrum distribution back to the time domain signal. The spatial information carrying channel utilized with a CW laser source is a simple optical spatial Fourier transform arrangement consisting of the input image and the reference waves introduced via a beamsplitter to share the second lens of the temporal channel in the backward direction. To achieve interaction between the temporal and spatial frequency infor-

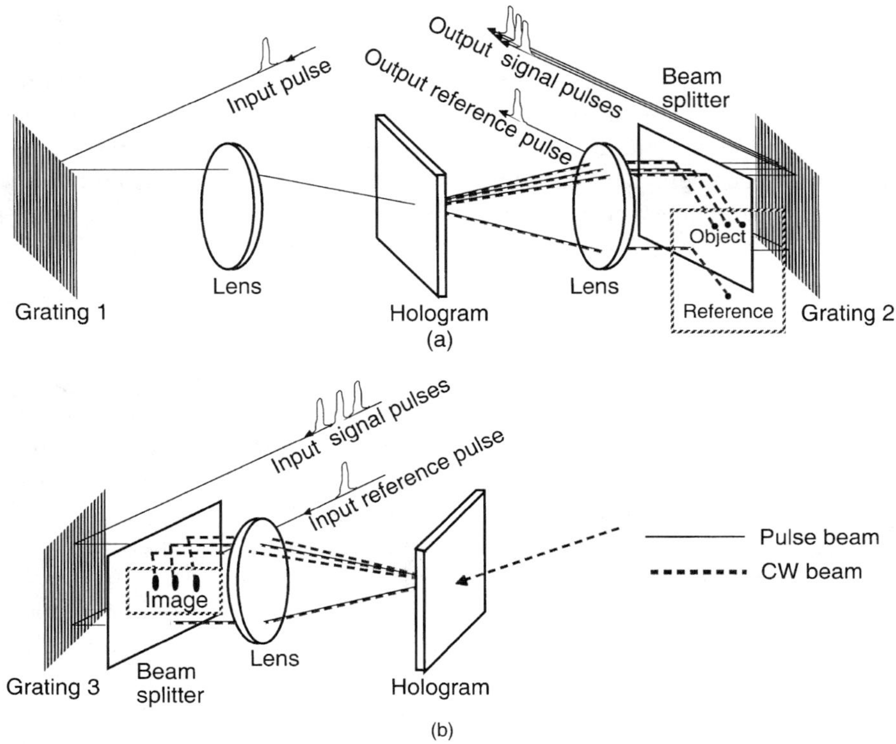

Figure 15.11: Schematic diagram of optical processors for (a) parallel-to-serial and (b) serial-to-parallel conversions.

mation we use a real-time holographic material in a four-wave mixing arrangement. In the following discussion we provide theoretical and experimental results that demonstrate all-optical parallel-to-serial conversion of one-dimensional spatial signals.

Let a spatially collimated and temporally transform-limited optical pulse propagating in z'-direction (Fig. 15.12) be described by

$$s(t) = p(t - t_0)\exp(j\omega_c t), \tag{15.15}$$

where $p(t)$ is the temporal envelop function of the pulse, t_0 is the initial time, and ω_c is the carrier frequency. In the frequency domain, each frequency component of the pulse is described by

$$S(\omega) = P(\omega - \omega_c)\exp[-j(\omega - \omega_c)t_0], \tag{15.16}$$

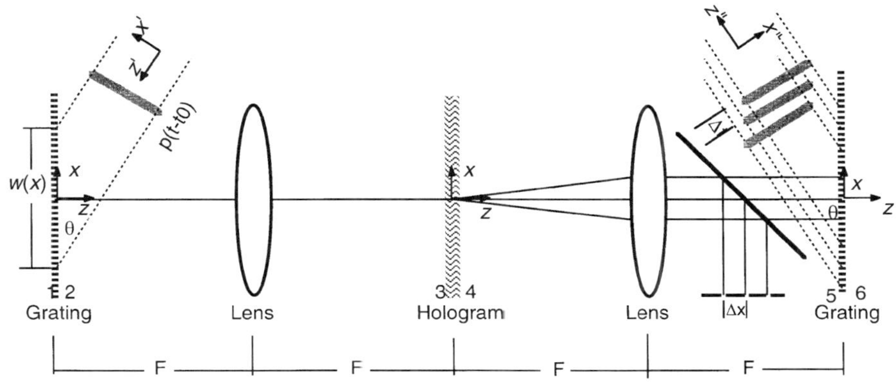

Figure 15.12: Schematic diagram of parallel-to-serial conversion using four-wave mixing in a real-time photorefractive recording material.

where $P(\omega)$ is the temporal Fourier transform of the function $p(t)$. Assuming that the system of Fig. 15.12 is linear and time invariant, the output of the system is determined by the convolution of the input pulse with the system impulse response (or equivalently, by the inverse Fourier transformation of the product between the spectrum of the input pulse with the system temporal transfer function). In the following discussion we derive the temporal transfer function of the system using conventional Fourier optics analysis.

Consider a monochromatic plane wave of frequency ω incident on the first reflecting grating at an inclination angle θ. The grating is arranged to diffract the carrier frequency component ω_c into the direction of the optical axis of the system. Thus, the diffracted optical field of frequency ω in plane 1 is given by

$$s_1(x;\omega) = \exp\left[-j\frac{(\omega - \omega_c)}{c}\alpha x\right]w(x), \tag{15.17}$$

where $\alpha = \sin\theta$ and $w(x)$ is the pupil function of the reflecting grating. The field in plane 2 is determined by the spatial Fourier transform of the field in plane 1, yielding

$$s_2(f_x;\omega) = W\left[f_x + \frac{(\omega - \omega_c)}{2\pi c}\alpha\right], \tag{15.18}$$

where $W(f_x)$ is the spatial Fourier transform of $w(x)$, $f_x = \omega x/2\pi cF$, and F is the lens focal length. Equation (15.18) shows that if an input optical pulse is introduced into the system it will be spatially dispersed in the Fourier transform plane where each spectral component occupies a width determined by the function $W(f_x)$. Let a spatial Fourier transform hologram in Fig. 15.12 be recorded using a spatial Fourier transform setup shown in Fig. 15.11a. The hologram contains information of the spatial Fourier transform of a sequence of equally spaced coherent point sources where each corresponds to a single bit of data from a spatially distributed data array. The hologram serves as a temporal frequency filter with transmittance

$$t(f_x) = \sum_n A_n \exp\left[j2\pi n\Delta\left(\frac{\omega_w}{\omega} f_x\right)\right], \tag{15.19}$$

where A_n is the amplitude of the nth bit in the spatial data array, Δ is the spatial separation between adjacent elements in the data array, ω_w is the optical frequency of the writing field used for recording the hologram, and the spatial carrier frequency term has been neglected. The ratio ω_w / ω accounts for the difference in spatial frequencies of the spatial and temporal optical channels. Thus, the field behind the hologram in plane 3 is $s_3(f_x;\omega) = s_2(f_x;\omega)t(f_x)$ with $s_2(f_x;\omega)$ and $t(f_x)$ from Eqs. (15.18) and (15.19), respectively. The second spatial Fourier transform yields the optical field on plane 4,

$$s_4(x;\omega) = \sum_n A_n w\left[-x + n\left(\frac{\omega_w}{\omega}\Delta\right)\right] \exp\left[j\frac{(\omega - \omega_c)}{c}\alpha\left[x - n\left(\frac{\omega_w}{\omega}\Delta\right)\right]\right], \tag{15.20}$$

where the minus sign in $w(-x)$ indicates that the image is inverted. The field in plane 4 is diffracted by the second reflecting grating yielding the output field propagating in the z''-direction,

$$s_5(x'';\omega) = \sum_n A_n w''\left[-x'' + n\left(\frac{\omega_w}{\omega}\Delta\right)\alpha\right] \exp\left[-j\frac{n(\omega - \omega_c)\omega_w\Delta}{c\omega}\alpha\right], \tag{15.21}$$

where a coordinate rotation from (x, z) to (x'', z'') is performed. Equation (15.21) represents the temporal transfer function of the system.

Finally, when a short optical pulse with a temporal spectrum described by Eq. (15.16) is introduced to the input, the system output is determined by the inverse Fourier transform of the product of Eqs. (15.16) and (15.21),

$$s_0(x'';t) = F_{\omega}^{-1}\{P(\omega - \omega_c)\exp[-j(\omega - \omega_c)t_0]s_5(x'';\omega)\}. \quad (15.22)$$

The function $P(\omega - \omega_c)$ is band-limited with a bandwidth of $\Delta\omega$. If this bandwidth is much smaller than the central frequency ω_c (for our case, the pulse of ~150 fs with central wavelength at 480 nm has $\Delta\omega/\omega_c$ of about 1%), then the values ω_w/ω in Eq. (15.21) can be approximated by ω_w/ω_c and Eq. (15.22) can be solved as

$$s_0(x'';t) = \left\{\sum_n A_n w''[-x'' + n\Delta'']p(t - t_0 - n\delta t)\right\}\exp(+j\omega_c t), \quad (15.23)$$

where $\Delta'' = \alpha\frac{\omega_w\Delta}{\omega_c}$, $\delta t = \alpha\frac{\omega_w\Delta}{\omega_c c}$, and t_0' is introduced to account for the time delay in the system. Equation (15.23) shows a temporal sequence constructed of original incident pulses. The pulses are separated from each other by the same distance δt and are modulated in amplitude with one-to-one correspondence to the data array from the spatial channel. Therefore, information is converted from space domain (parallel) to time domain (sequential) via the interaction of the corresponding frequency components. Note that the output pulses obtained with this technique propagate in the same direction, but with slightly different transverse extents due to the difference in projections of the entrance pupil $w(x)$ on the output grating. This may cause a slight variation in coupling efficiency when these pulses are coupled into an optical fiber.

The parallel-to-serial conversion experiments were implemented using a processor shown in Fig. 15.11a (transmitter node) with a 1-D binary data array of Fig. 15.13a. The output pulses (Fig. 15.13b) were transmitted to the input of the processor shown in Fig. 15.11b (receiver node) that implements serial-to-parallel conversion. A single reference pulse produced by the nondiffracted portion of the readout beam in Fig. 15.11a was also transmitted to the receiver for decoding. The serial-to-parallel conversion is based on recording a spectral hologram between the input signal pulses and the reference pulse. The CW monochromatic readout wave is used to read out the recorded hologram. It is diffracted and modulated by the spatial frequencies of the spectral hologram, and

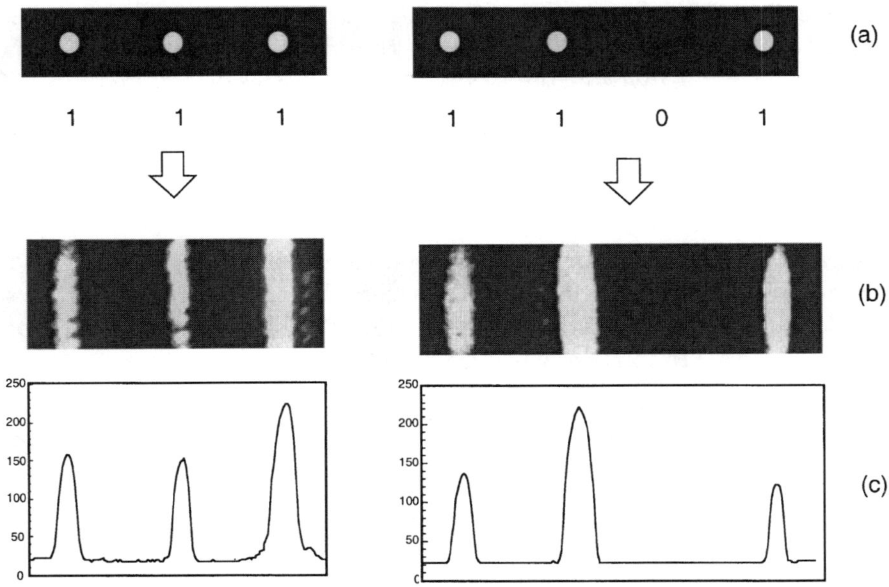

Figure 15.13: Experimental results of image transmission using parallel-to-serial and serial-to-parallel conversion: (a) photograph of the 1-D input data arrays, (b) photograph of the 1-D binary data array images reconstructed at the receiver, (c) plots of the intensity profile in (b).

Fourier transformed by a lens. The reconstructed 1-D spatial image shown in Fig. 15.13b exhibits one-to-one correspondence with the transmitted image of Fig. 15.13a.

The experiments were performed using 150-fs optical pulses of about 1 kW peak power at a wavelength of 480 nm generated from a frequency doubled mode-locked Ti:sapphire laser. The spatial Fourier transform hologram (Fig. 15.11a) of the 1-D binary data array shown in Fig. 15.13a was recorded in a 1-mm-thick photorefractive crystal of lithium niobate using the 488 nm line of an Ar^+ laser. The small angle between the reference and the object beams was used to ease the Bragg matching requirement for the reconstruction with optical pulse at a center wavelength of 480 nm. The resultant output pulses and the reference pulse were transmitted to the receiver via free space propagation. In practice, the signal and the reference pulses can be transmitted through two identical fibers or through a single fiber using polarization multiplexing. In our experiments the output pulses and the reference pulse were transmitted through a low-pass spatial filter to emulate the transmission through a fiber, i.e., to ensure that no spatial information is transmitted to the receiver. At the receiver (Fig. 15.11b) we recorded a spectral hologram between the input signal pulses and the reference pulse using another photorefractive lithium niobate crystal. The spectral hologram was read out using the 488 nm line of an Ar^+ laser, and then Fourier transformed by a lens to yield the serial-to-parallel-converted 1-D spatial output signal shown in Fig. 15.13b. The resultant images are stretched in the vertical direction due to the nature of the spectral holograms that are inherently 1-D. Note that the speed of transformations in our experiments was limited by the time response of photorefractive lithium niobate crystal (~5 s). Recent progress in multiple-quantum-well photorefractive material shows potential for high-speed data conversion [47, 48]. In the future we are planning to use fast nonlinear optical materials such as photorefractive semiconductor crystal [46, 48], composite nonlinear materials [49] and electronic nonlinear optical materials to provide high-speed and real-time operation.

If the input pulse contains wide bandwidth, the approximation for Eq. (15.23) becomes invalid because higher-order terms from the Taylor expansion of $\omega - \omega_c/\omega$ need to be taken into consideration, which will cause dispersed signal pulses. We have studied the influence of such effect on the information transformation of the optical signals in the spatial and the temporal domain [44, 50]. We have found that the pulses generated from the parallel to serial conversion system will experience dispersion due to the difference in optical frequencies used for recording and reconstruction of the hologram in the parallel to serial converter. The spatial Fourier transform hologram of the spatial channels information is recorded using quasi-monochromatic waves, whereas the reconstruction is performed using a short optical pulse, i.e., by a broadband wave, resulting in a temporal chirp of the readout pulse sequence. However, the dispersed information due to this frequency mismatch will be carried to the receiver node and used for recording the spectral hologram. This spectral hologram at the receiver will resemble the original spatial Fourier transform hologram at the transmitter. This phenomenon in spectral holography can be seen as temporal phase-conjugation. The demonstrated parallel-to-serial and serial-to-parallel conversions via holographic processors possess additional advantages: (1) they do not require transform-limited input pulses, since the system is self-referenced; (2) optical dispersion induced by the holographic materials, communication channel, as well as all optical components is self-compensated, because the reference beam propagates through the same material as the signal beam.

15.5 Summary and future directions

We have introduced and discussed spatial and temporal holography in photorefractive materials. These new holographic methods can process four-dimensional (4-D) information, which in addition to the three spatial coordinates also includes the temporal evolution of optical signals. Photorefractive volume holographic materials provide the medium necessary for recording and reconstruction of such 4-D optical information carrying signals in real time. We described the two disadvantages of direct time-domain holography, the low fringe contrast and the limited recording time. These disadvantages are shown to be overcome by using the method of spectral domain holography. Applications of direct time-domain and spectral-domain holography for image processing, temporal matched filtering, pulse shaping, 3-D optical storage, and optical interconnects have been discussed.

In addition, we have introduced space-time holographic processing that allows the conversion between the spatial and the temporal optical information carrying channels. This method was used to demonstrate experimentally parallel-to-serial and serial-to-parallel data conversion for 1-D images and image-format data transmission. The holographic processors demonstrated in this chapter provide the following advantages: (1) transform-limited input pulses are not required, since the system is self-referenced; and (2) optical dispersion induced by the holographic materials, communication channel, as well as all optical components are self-compensated, because the reference beam propagates through the same material as the signal beam.

Progress in the area of space-time processing with photorefractive holography will depend on the future developments in the supporting technologies such as compact and cost-effective laser sources, fast electrooptical modulators and photodetector arrays, passive and programmable diffractive optical elements, etc. Of particular importance is the development of fast photorefractive volume holographic materials such as bulk semiconductor materials and semiconductor microstructures [13, 47].

Acknowledgments

This work was supported in part by the National Science Foundation (NSF), Ballistic Missile Defense Organization (BMDO), Air Force Office of Scientific Research (AFOSR), and North American Treaty Organization (NATO). Y. Mazurenko would like to acknowledge the Russian Foundation for Fundamental Research.

References

1. D. Gabor, "A new microscopic principle," *Nature,* **161**, 777–779, 1948.
2. D. Gabor, "Microscopy by reconstructed wavefronts," *Proc. Roy. Soc. A,* **197**, 454–462, 1949.
3. J. E. Ford, Y. Fainman and S. H. Lee, "Time integrating interferometry using photorefractive fanout," *Opt. Lett.,* **13**, 856–858, 1988.
4. Y. Fainman, "Application of photorefractive devices for optical computing," *Proc. SPIE,* **1150**, 120–141, 1989.
5. P. C. Sun and Y. Fainman, "Real-time generation of the Wigner distribution of complex functions using phase conjugation in photorefractive materials," *Opt. Lett.,* **15**, 941–943, 1990.
6. Y. Taketomi, J. Ford, H. Sasaki, J. Ma, Y. Fainman, S. H. Lee, "Incremental recording for photorefractive hologram multiplexing," *Opt. Lett.,* **16**, 1774–1776, 1991.
7. H. Sasaki, Y. Fainman, J. Ford, Y. Taketomi, S. H. Lee, "Dynamic photorefractive memories," *Opt. Lett.,* **16**, 1874–1876, 1991.
8. J. E. Ford, J. Ma, Y. Fainman, S. H. Lee, Y. Taketomi, D. Bize and R. R. Neurgaonkar, "Multiplex holography in strontium barium niobate with applied field," *J. Opt. Soc. Am. A,* **9**, 1183–1185, 1992.
9. H. Sasaki, J. Ma, Y. Fainman, S. H. Lee and Y. Taketomi, "Fast update of dynamic photorefractive optical memory," *Opt. Lett.,* **17**, 1468–1470, 1992.
10. J. Ma, B. Catanzaro, J. E. Ford, Y. Fainman, and S. H. Lee, "Photorefractive holographic lenses and applications for dynamic focusing and dynamic image shifting," *J. Opt. Soc. Am.,* **11**, 2471–2480, 1994.
11. J. Ford, Y. Fainman, and S. H. Lee, "Reconfigurable array interconnection by photorefractive correlation," *Appl. Opt.,* **33**, 5363–5377, 1994.

12. B. Catanzaro, J. Ma, Y. Fainman, and S. H. Lee, "Wavelength multiplexed reconfigurable interconnect using a volume holographic lens array," *Opt. Lett.*, **19**, 1358–1360, 1994.

13. Y. Fainman, J. Ma and S. H. Lee, "Nonlinear optical materials and applications," *Material Science Reports,* **9**(2, 3), 53–139, 1993.

14. P. Gunter and J. P. Huignard (eds.), *Photorefractive Material and Their Applications II, Survey of Applications,* Springer-Verlag, New York, 1989.

15. M. A. Duguay and A. T. Mattick, "Ultrahigh speed photography of picosecond light pulses and echoes," *Appl. Opt.,* **10**, 2162–2170, 1971.

16. Y. N. Denisyuk, D. I. Staselko, and R. R. Herke, "On the effect of the time and spatial coherence of radiation source on the image produced by a hologram," in *Proceedings, Applications of Holography,* Besancon, France, July, 1970.

17. D. I. Staselko, Y. N. Denisyuk, and A. G. Smirnow, "Holographic recording of the time-coherence pattern of a wave train from a pulsed laser source," *Opt. Spectrosc.,* **26**, 413–420, 1969.

18. Nils Abramson, "Light-in-flight recording by holography," *Opt. Lett.,* **3**, 121–123, 1978.

19. Yu. T. Mazurenko, "Interference of spectrally dispersed light," *Opt. Spectrosc.,* **56**, 357, 1984.

20. Yu. T. Mazurenko, "Detection and reconstruction of ultrashort pulses by interference of spectrally dispersed light," *Sov. J. Quan. Elec.,* **15**, 815–818, 1985.

21. Yu. T. Mazurenko, "Reconstruction of a nonstationary wave field by holography in a 3-D medium," *Opt. Spectrosc.,* **57**, 343, 1985.

22. J. C. Hebden and R. A. Kruger, "Transillumination imaging performance: A time-of-flight imaging system," *Med. Phys.,* **17**, 351–356, 1990.

23. N. H. Abramson and K. G. Spears, " Single pulse light-in-flight recording by holography," *Appl. Opt.,* **28**, 1834–1841, 1989.

24. K. G. Spears, J. Serafin, N. Abramson, X. Zhu, and H. Bjelkhagen, "Chronocoherent imaging for medicine," *IEEE Trans. Biomed. Eng.,* **36**, 1210–1221, 1989.

25. E. Leith, H. Chen, Y. Chen, D. Dilworth, J. Lopez, R. Masri, J. Rudd, and J. Valdmanis, " Electronic holography and speckle methods for imaging through tissue using femtosecond gated pulses," *Appl. Opt.,* **30**, 4204–4210, 1991,

26. E. N. Leith, C. Chen, H. Chen, Y. Chen, D. Dilworth, J. Lopez, J. Rudd, P. C. Sun, J. Valdmanis, and G. Vossler, " Imaging through scattering media with holography," *J. Opt. Soc. Am. A,* **9**, 1148–1153, 1992

27. D. Huang, E. A. Swanson, C. P. Lin, J. S. Schuman, W. G. Stinson, W. Chang, M. R. Hee, T. Flotte, K. Gregory, C. A. Puliafito, and J. G. Fujimoto, "Optical coherence tomography," *Science,* **254**, 1178–1181, 1991.

28. A. Yariv, "Interpage and interpixel cross talk in orthogonal (wavelength-multiplexed) holograms," *Opt. Lett.,* **18**, 652–654, 1993.

29. K. Curtis, C. Gu, D. Psaltis, "Cross talk in wavelength multiplexed holographic memories," *Opt. Lett.,* **18**, 1001–1003, 1993.

30. K. B. Hill, K. G. Purchase, and D. J. Brady, " Pulsed-image generation and detection," *Opt. Lett.,* **20**, 1201–1203, 1995.

31. Y. Mazurenko and Y. Fainman, "Cross-talk of wavelength-multiplexed quasi-infinite holograms," *Opt. Lett.,* **23**, 963–965, 1998.

32. K. Yu, Q. Xing, R. Alfano, "Imaging objects hidden in highly scattering media using femtosecond using second harmonic generation cross-correlation time gating," *Opt. Lett.,* **16**, 1019–1021, 1991.

33. M. Born and E. Wolf, *Principles of Optics,* 10, Pergamon, New York, 1980.

34. C. E. Shannon, *Proc. IRE,* **37**, 10, 1949.

35. Y. T. Mazurenko, "Holography of wave packets," *Appl. Phys. B,* **50**, 101–114, 1990.

36. K. Ema and F. Shimizu, "Optical pulse shaping using a Fourier transform hologram," *Japan J. Appl. Phys.,* **29**, 631–633, 1990.

37. Yu. T. Mazurenko, S. E. Putilin, V. V. Kuznetsov, and L. M. Lavrenov, "Recording, reconstruction and time reversal of light pulses," *Opt. Spectrosc.,* **71**, 223–224, 1991.

38. Yu. T. Mazurenko, "Time-domain Fourier transform holography and possible applications in signal processing," *Opt. Eng.,* **31**, 739–749, 1992.

39. Yu. T. Mazurenko, S. E. Putilin, V. L. Bogdanov, G.V. Lukomskij, M. V. Smirnov and Yu. A. Cherkasov, "Spectral holography of pico- and nanosecond laser pulses," *SPIE,* **2097**, *Laser Application,* 380, 1993.

40. A. M. Weiner, D. E. Leaird, D. H. Reitze, E. G. Paek, "Femtosecond spectral holography," *IEEE J. Quantum Electron.,* **28**, 2251–2261, 1992.

41. D. Psaltis, F. Mok, H. Li, "Nonvolatile storage in photorefractive crystals," *Optics Letters,* **19** (3), 210–212, 1994.

42. K. Oba, P. C. Sun, and Y. Fainman, "Nonvolatile photorefractive spectral holography," *Optics Letters,* **23**, 915–917, 1998.

43. M. C. Nuss and R. L. Morrison, " Time-domain images," *Opt. Lett.,* **20**, 740–742, 1995.

44. P. C. Sun, Yu. T. Mazurenko, and Y. Fainman, "All-optical parallel-to-serial conversion by holographic spatial-to-temporal frequency encoding," OSA topi-

cal meeting on optical computing at Salt Lake City, *Tech. Digest,* **10**, 236–238, 1995. Also see P. C. Sun, Y. Mazurenko, W. S. C. Chang, P. K. L. Yu, and Y. Fainman, "All Optical Parallel-to-serial Conversion by Holographic Spatial-to-Temporal Frequency Encoding," *Optics Letters* **20**, 1728–1730 1995.

45. K. Ema, M. Kuwata-Gonokami, and F. Shimizu, "All-optical sub-Tbits/s serial-to-parallel conversion using excitonic giant nonlinearity," *Appl. Phys. Lett.,* **59**, 2799–2801, 1990.
46. M. C. Nuss, M. Li, T. H. Chiu, A. M. Weiner, and A. Patrovi, "Time-to-space mapping of femtosecond pulses," *Optics Letters,* **19**, 664–666, 1994.
47. Y. Ding, R. M. Brubaker, and D. D. Nolte, M. R. Melloch, and A. M. Weiner, "Femtosecond pulse shaping by dynamic holograms in photorefractive multiple quantum wells," *Optics Letters,* **22**, 718–720, 1997.
48. A. Partovi, A. M. Glass, D. H. Olson, G. J. Zydzik, H. M. O'Bryan, T. H. Chiu, and W. H. Knox, "Cr-doped GaSa/AlGaAs Semi-insulating multiple quantum well photorefractive devices," *Appl. Phys. Lett.,* **62**, 464–466, 1993.
49. R. W. Boyd and J. E. Spie, "Nonlinear optical susceptibilities of layered composite material," *J. Opt. Soc. Am. B,* **11**, 297–303, 1994.
50. D. M. Marom, P. C. Sun, and Y. Fainman, "Analysis of spatial-temporal converters for all-optical communication links," *Applied Optics,* **37**, 2858–2868, 1998.

Chapter 16

Dynamics of Photorefractive Fibers

Francis T. S. Yu and Shizhuo Yin

Department of Electrical Engineering, Pennsylvania State University, University Park, Pennsylvania

16.1. Introduction

From the first discovery of the photorefractive (PR) effect in 1966 [1], numerous efforts have been made for the use of PR materials to real-time optical processing, memories, interconnects and computing [2–5]. Recent years, with the rapid advent of semiconductor technology, have brought significant achievements in PR applications to electronic devices. These are primarily due to the availability of the compact-size semiconductor and solid-state lasers, fast speed and high resolution charge-coupled device (CCD) cameras, and high contrast ratio and large spatial bandwidth product liquid crystal spatial light modulators (SLMs). These are the major impetuses that drive us to seek for suitable recording materials so that a practical commercial memory system can be developed. One of the critical tasks is to improve the currently available volume holographic recording technology by which photorefractive crystals, photopolymers [6–7], and other exotic materials can be exploited.

It has been reported that as many as 5000 images have been stored in a single PR crystal by using the angular-multiplexing technique [8],

PHOTOREFRACTIVE OPTICS
Materials, Properties and Applications

ISBN 0-12-774810-5/$30.00

which represents a significant progress in the holographic storage. However, the high cost and low diffraction efficiency of the bulk materials limit their practical applications. An alternative to bulk material, single-crystal PR fibers has been reported in recent years [9–12]; such alternative fibers are more cost-effective and have a unique geometric shape that may be essential for some specific applications, such as for applications in fiber-optic communications, fiber sensors, and other uses.

Although there have been several articles reported in PR fiber fabrication, due to page limitation, we shall, however, concentrate our effort mainly on the recent work done at the Pennsylvania State University. First, we briefly introduce a laser-heated pedestal growth (LHPG) technology for single-crystal fiber fabrication. Second, a phase conjugate fiber holographic construction architecture will be presented, which has the advantage of minimizing the intramodal noise excitation [13]. Angular and wavelength selectivities for PR fiber hologram storage using coupled wave theory will be analyzed [14]. The calculated results have been shown to be consistent with the experimental data as reported by Hesselink *et al.* [11]. Since the reflection-type wavelength-multiplexed holograms offer a higher and uniform selectivity over a wider range of the construction angles, we will concentrate our attention on this type of fiber holograms. Cross talk noise between the wavelength-multiplexed channel will be evaluated, and we will show the narrower the laser linewidth will lower the cross talk between the holographic channels. Because successive recordings would partially erase the previous recorded holographic gratings, the recording-erasure dynamics of the fiber hologram will be evaluated. The storage dynamics of the PR fiber hologram will be experimentally demonstrated. We will show that a 1-cm length of PR fiber can store as much as over a hundred holograms. Toward the end of this chapter, several potential applications of the PR fiber such as applied to hologram memories, turnable filters, and true-time delay lines will be addressed.

16.2 Fabrication of photorefractive fibers

In this section we will discuss the fabrication of single-crystal fibers using laser-heated pedestal growth (LHPG) technology [9,10]. The LHPG technique is a relatively new crystal fiber growth technology that can grow 50 to 1000-μm diameter fiber, up to 10 cm long. The major advantages of growing single-crystal fiber using the LHP are:

1. It does not require any holding device (e.g., a crucible) so that the contamination from the holder can be avoided.
2. Some growing limitations, such as the melting point, shape, etc. can be alleviated.
3. Since the LHPG technique used a CO_2 laser as the heating source, the growing process is much easier to control. For example, the laser beam has an annular taper form, so that the crystal fiber can be grown in the vertical direction.
4. The LHPG technique can also be used in a variety of applications, e.g., as applied to high-temperature superconductor wire-rod and other novel devices.

In addition, high temperatures (as high as 2400°C) with steep gradients, can be achieved using the LHPG technique, so that a variety of single-crystal fibers, such as $LiNbO_3$, $Sr_{1-x}Ba_xNb_2O_6$, $BaTiO_3$, b-BaB_2O_5, and $Bi_2CaCu_2O_8$, can be grown using this technology. The LHPG system at Penn State is depicted in Fig 16.1, and the schematic diagrams describ-

Figure 16.1: Laser-heated pedestal growth (LHPG) system.

ing the growth procedure are shown in Figs. 16.1–16.4. For example, preformed Ce:Fe-doped $LiNbO_3$ is placed in a holder that maintains at a precise vertical position. A CO_2 laser of about 50 W is brought to focus by first forming an annular beam with a reflaxicon. By steering the annular beam with a 45° flat mirror, a highly intense focusing beam can be formed at the preformed crystal, using a concave parabolic mirror, as shown in Fig. 16.3. When the seed rod is brought in contact with the molten tip of the preformed crystal, the fiber growth can be initiated by slowly pulling the seed upward with a computer control driver, as illustrated in Fig. 16.4. Thus a single crystal fiber can be grown with the above-described LHPG system. A typical sample of the Ce:Fe–doped $LiNbO_3$ PR fiber is shown in Fig. 16.5. The doping levels of Ce:Fe: are about 300 and 500 ppm, respectively. To ensure that the fabricated PR fiber is indeed a single-domain crystal, the grown fiber is subjected to an x-ray Laue diffraction for the c- and a-crystal axes, respectively, as shown in Fig. 16.6, where it is confirmed that the grown fiber is indeed a single-crystal fiber.

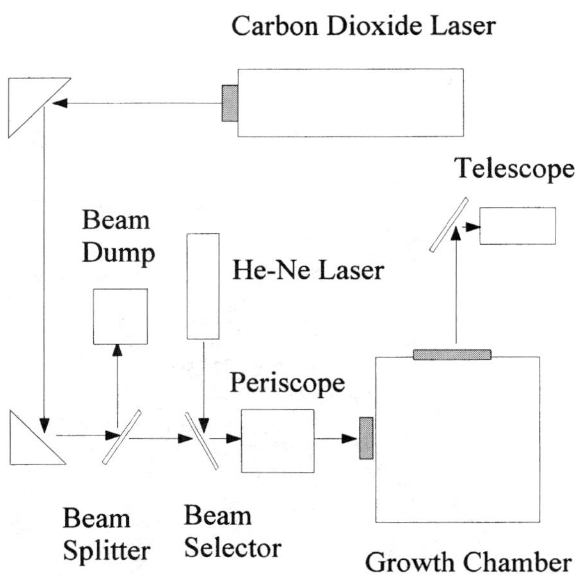

Figure 16.2: The LHPG system.

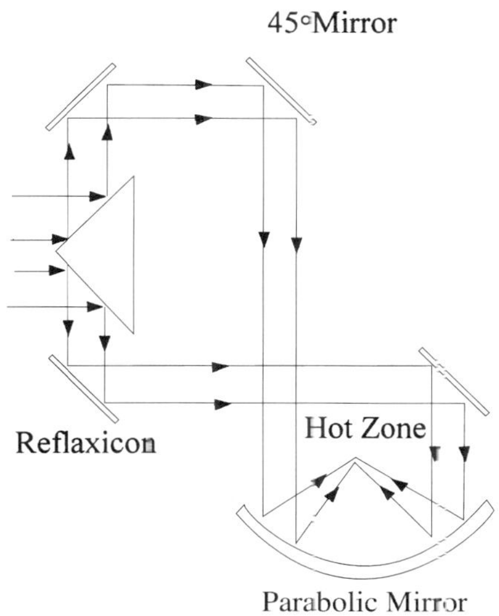

Figure 16.3: Growth chamber.

16.3 Constructing photorefractive fiber holograms

First we will describe a method for recording an axial fiber hologram in which the intermodal excitation can be compensated. Let a linearly polarized monochromatic plane wave be coupled into a PR fiber with a condensing lens. A cone of light beams having different polarizations, caused by the anisotropism and the Fresnel reflection within the fiber will eventually exit from the other end [15]. Since the light beam propagates along the fiber axis, the spatial information content of the incident light will be scrambled at the exit end. However, if the incident beam (called an object beam) interferes with a reference beam, which is assumed to be coupled into the fiber at the opposite end, a hologram (similar to that of a reflection-type hologram) can be formed. Nevertheless, if the length-to-diameter ratio of the PR fiber is large, say that in the order of hundreds (e.g., length/diameter > 100), the object beam cannot be faithfully reproduced merely by using the reference beam. This is primarily due to intermodal scrambling within the fiber. But if the fiber hologram

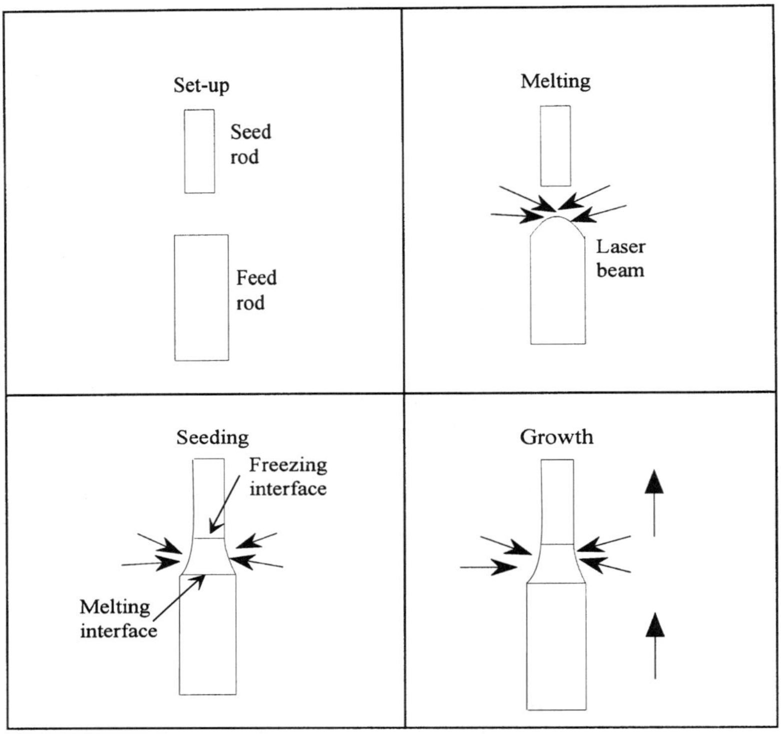

Figure 16.4: Fiber growth.

construction is using a conjugate reference beam instead, which can be derived from a polarization-preserving conjugating mirror shown in Fig. 16.7, a transmission-type (instead of reflection-type) fiber hologram can be made. Note that the writing reference beam R has been made to exceed the coherence length of the object beam O, so that the returned conjugate reference beam R^* will be coherent with the object beam O along the fiber. Thus we see that the recorded object beam can be actually read out by the reference beam R.

To illustrate the fiber hologram formation as described in the preceding discussion, an object transparency is launched into the PR fiber using an imaging lens. This forms an object beam O that propagates along the fiber. Because the reference beam R is launched from the other end of the fiber, the polarization preserving phase-conjugated mirror reflects a

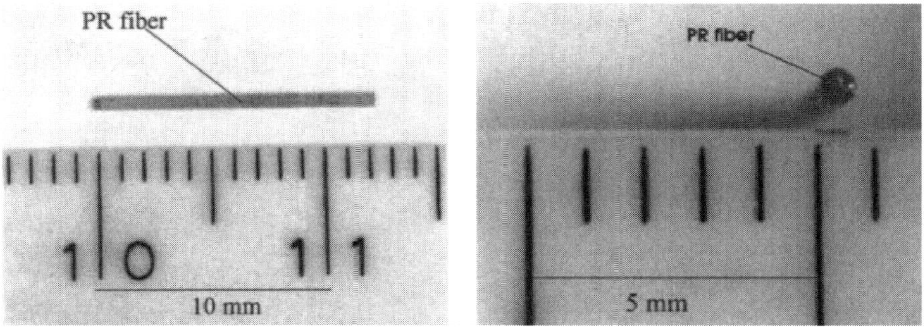

(a) (b)

Figure 16.5: Ce:Fe:$LiNbO_3$ PR crystal fiber. (a) Side-view, (b) Cross-sectional view.

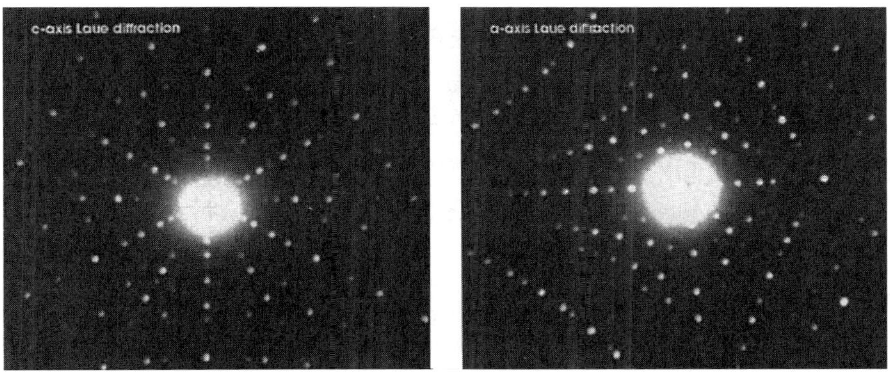

(a) (b)

Figure 16.6: (a) c-axis x-ray diffraction pattern. (b) a-axis x-ray diffraction pattern.

conjugate reference beam R^* that is within the coherence length of the object beam. Then a transmission-type fiber hologram can be constructed, as given by

$$H = |O|^2 + |R|^2 + O^*R^* + OR \tag{16.1}$$

It is trivial to see that if the recorded PR fiber hologram is read out by the reference beam R, a phase conjugated object beam O^* will be

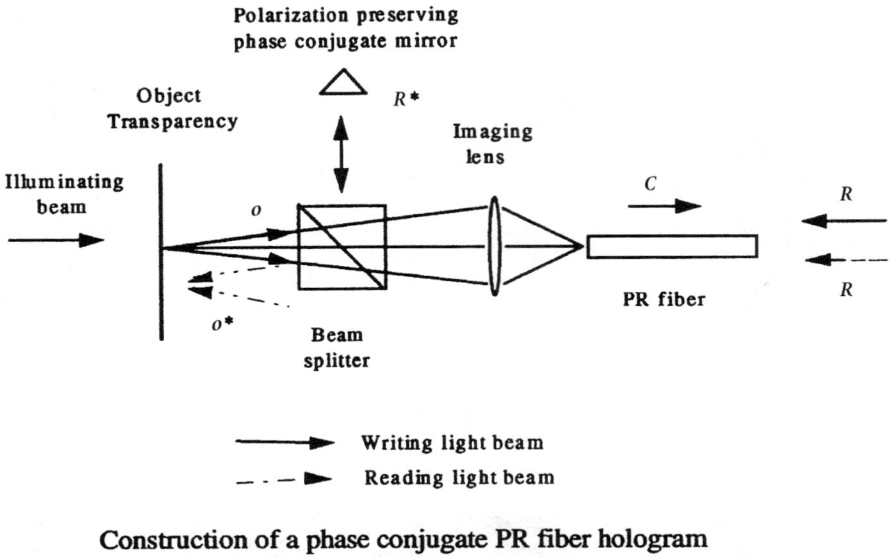

Figure 16.7: Construction of a phase-conjugate PR fiber hologram.

reconstructed as shown by the dashed lines in Fig. 16.7. Since the read-out conjugated object beam O^* retraces back along the fiber, the intermodal excitations within the fiber hologram can be compensated. Thus a faithful conjugated object beam O^* can be reconstructed.

On the other hand, if the length-to-diameter ratio is adequately small (e.g., $\ell/d \leq 10$), the PR fiber can be treated as a bulk crystal for which a reflection-type PR fiber hologram can be directly constructed. This is particularly true by using two directly counterpropagated beams for the fiber hologram construction, as will be seen in the subsequent sections.

16.4 Selectivities of fiber holograms

Since the PR fibers we are interested in are primarily for the application to photonic devices (such as optical memories, sensors, true-time delay lines, and others), the PR fibers that will be described from now on will be high-order mode fibers, and the length-to-diameter ratio is assumed sufficiently small (e.g., length/diameter < 10). Thus, the PR fibers we will be discussing can be treated as bulk materials instead of thin fibers. This assumption is particularly true for counterpropagated beams in reflection-type fiber holograms. By referring to the unslanted transmission and reflection-type holograms shown in Fig. 16.8, the normalized diffraction efficiencies can be shown as [16, 17],

$$\eta_t = \frac{\sin^2(\upsilon_t^2 + \xi_t^2)^{1/2}}{1 + \frac{\xi_t^2}{\upsilon_t^2}}, \tag{16.2}$$

$$\eta_r = \frac{1}{1 + \frac{1 - \xi_r^2/\upsilon_r^2}{[\sinh(\upsilon_r^2 - \xi_r^2)^{1/2}]^2}}, \tag{16.3}$$

where the subscripts t and r denote the transmission-type and the reflection-type holograms, respectively,

$$\upsilon_t = \frac{\pi \Delta n d}{\lambda \cos\theta}, \quad \xi_t = \frac{2\pi n d \sin\theta}{\lambda}\Delta\theta,$$

$$\upsilon_r = \frac{\pi \Delta n d}{\lambda \sin\theta}, \quad \xi_r = \frac{2\pi n d \cos\theta}{\lambda}\Delta\theta,$$

where λ is the recording wavelength, n and Δn are the refractive index and modulation index, d is the length of the fiber, and θ and $\Delta\theta$ are the internal construction angle and its variation. Notice that for weak coupling condition, i.e., $|\upsilon| << \xi|$, Eqs. (2) and (3) can be approximated by

$$\eta_t \approx \upsilon_t^2 \operatorname{sinc}^2 \xi_t, \tag{16.4}$$

$$\eta_r \approx \upsilon_r^2 \operatorname{sinc}^2 \xi_r. \tag{16.5}$$

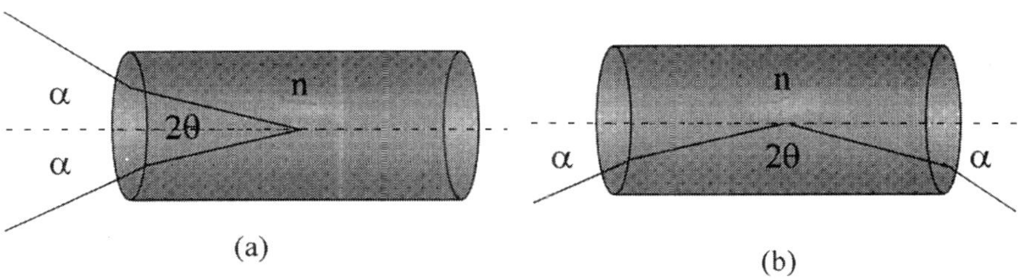

Figure 16.8: Fiber holographic constructions: (a) transmission fiber hologram, (b) reflection fiber hologram.

For example, typical values of n and Δn for a $LiNbO_3$ crystal would be $n = 2.28$ and $\Delta n_{max} = 10^{-3}$–10^{-5} [18]. Since Δn (for each multiplexed hologram) is sufficiently smaller than Δn_{max}, Eqs. (16.4) and (16.5) are considered to be as good estimations under the weak coupling assumption.

Since the angular selectivity is inversely related to the full width of the sinc factor, the internal (writing) angular sensitivity can be shown as

$$\Delta\theta_t = \frac{\lambda}{nd\sin\theta}, \ \Delta\theta_r = \frac{\lambda}{nd\cos\theta} \tag{16.6}$$

By referring to Snell's law of refraction (Fig. 16.8), we have

$$\sin\alpha = n\sin\theta, \tag{16.7}$$

$$\cos\alpha = n\cos\theta \tag{16.8}$$

where Eqs. (7) and (8) are for the transmission-type and the reflection-type fiber holograms, respectively. The angular selectivities (in radians) as expressed in terms of the external writing angle α can be written as [1, 2],

$$\left\{\frac{1}{\Delta\alpha}\right\}_t = \frac{\sin\alpha\cos\alpha}{\sqrt{n^2-\sin^2\alpha}}\frac{d}{\lambda}, \tag{16.9}$$

$$\left\{\frac{1}{\Delta\alpha}\right\}_r = \frac{\sin\alpha\cos\alpha}{\sqrt{n^2-\cos^2\alpha}}\frac{d}{\lambda}. \tag{16.10}$$

where the subscripts t and r denote the transmission and the reflection-type holograms, respectively.

To evaluate the wavelength selectivity for a fiber hologram, **k**-vector diagrams are used as shown in Fig. 16.9. By referring to the *law of momentum conservation*, a change in the diffraction beam angle by $\Delta\theta$ is needed. If the readout wavelength is deviated from the recording wavelength by $\Delta\lambda$, from Fig. 16.9a we see that

$$(|k + \Delta k|)\sin\theta + (|k + \Delta k|)\sin(\theta - \Delta\theta) = 2|k|\sin\theta, \tag{16.11}$$

where $|k| = 2\pi n/\lambda$. The wavelength selectivity for the transmission-type fiber hologram (in terms of internal writing angle θ) can be shown as

$$\left|\frac{\lambda}{\delta\lambda}\right|_t = 1 + \frac{2}{\dfrac{\sin\theta}{\sin(\theta - \Delta\theta)} - 1}. \tag{16.12}$$

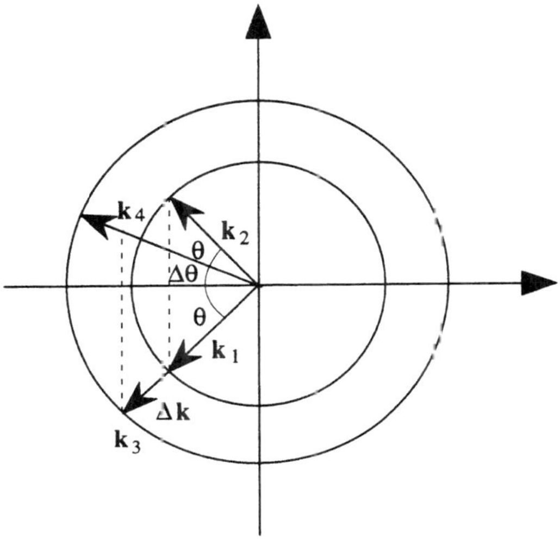

(a)

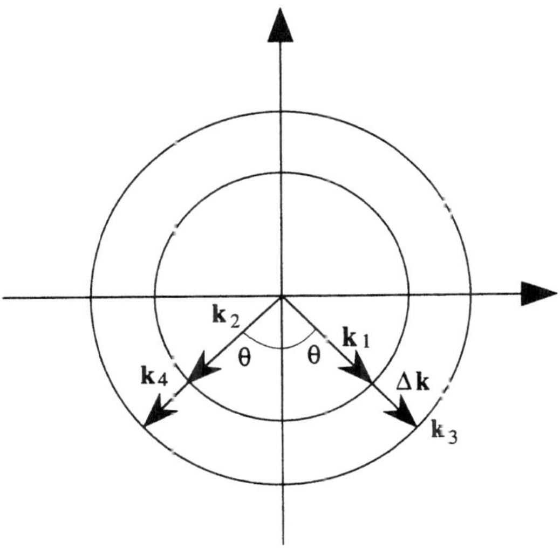

(b)

Figure 16.9: Momentum conservation principle. $\mathbf{k}_1$and $\mathbf{k}_2$, writing wave vectors: $\mathbf{k}_3$, readout wave vector: $\mathbf{k}_4$, diffracted wave vector: $\mathbf{\Delta k}$, reading wavelength deviation: (a) for transmission hologram, (b) for reflection hologram.

In view of Eqs. (4) through (8), the wavelength selectivity for the transmission-type fiber hologram can be written as

$$\left\{\left|\frac{\lambda}{\delta\lambda}\right|\right\}_t + 1 + \frac{2}{\dfrac{\sin\alpha/n}{\sin\left[\sin^{-1}\left(\dfrac{\sin\alpha}{n}\right) - \dfrac{\lambda}{d\sin\alpha}\right]} - 1}. \tag{16.13}$$

Similarly, by referring to the **k**-vector diagram of Fig. 14.9b, the wavelength selectivity for the reflection-type fiber hologram can be shown to be

$$\left\{\left|\frac{\lambda}{\Delta\lambda}\right|\right\}_r = \frac{\sqrt{n^2 - \cos^2\alpha}}{\lambda}d. \tag{16.14}$$

To test the validity of our derivations, a comparison with the experimental results obtained by Hesselink [11] are tabulated in Table 16.1, where we see that the rate of change is in the same order as the experimental data obtained. The consistency of our derivation can be interpreted in this way: as the light beams interacted along the fiber (due to internal reflections) they behaved in a similar manner as the light beams interacting within a thick crystal.

The angular and the wavelength selectivities for the transmission-type and the reflection-type PR fiber holograms, as a function of external writing angle, are plotted in Figs. 16.10 and 16.11, where we see that both selectivities increase as the fiber length increases. However, there is limited usable-range for angular selectivities (i.e., $15° \leq \alpha \leq 75°$ for

Recording angle (2 α)	**Angular selectivity calculate results (degrees)**	**Angular selectivity experimental data (degrees)**
0.7	2.41	1.43
3	0.64	0.86
5	0.39	0.57
10	0.19	0.44
20	0.098	0.26

Table 16.1: Calculated and Experimental Data for Angular Selectivity for the Fiber Hologram $\lambda = 515$ nm, $n = 2.28$, $d = 4$ mm.

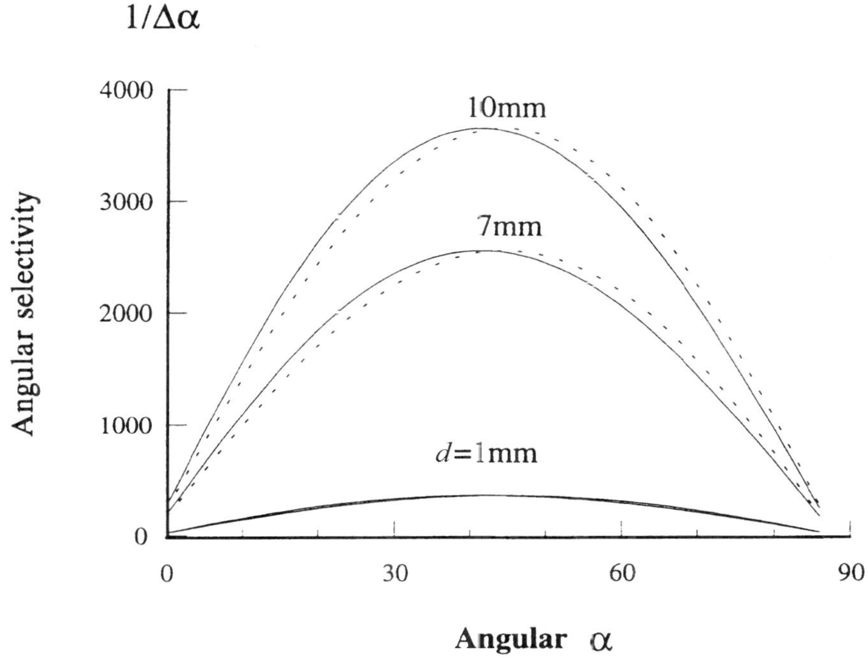

Figure 16.10: Angular selectivity (dashed line for transmission hologram and solid line for reflection hologram). Recording wavelength $\lambda = 0.7$ μm.

full-width half-maximum (FWHM)) for both the transmission and the reflection-type fiber holograms. On the other hand, the wavelength selectivity for the reflection-type fiber hologram has been shown severalfold higher than the transmission-type, and the selectivity is relatively uniformly distributed over the entire range of recording angles. Thus, the reflection-type wavelength-multiplexed PR fiber hologram would be a better choice for holographic data storage and other applications. Attention should be drawn to the fact that Rakuljic *et al.* [19] and Curtis *et al.* [20] have subsequently shown that the cross talk in a wavelength multiplexed volume hologram is less severe as compared with an angular multiplexed hologram. To have a feeling of magnitude, we assume the length of the PR fiber hologram is $d = 7$ mm, and the construction wavelength is $\lambda = 0.7$ μm. Then the wavelength selectivity for a reflection-type fiber hologram can be as high as the order of 10^4.

To conclude this section, we point out that the wavelength multiplexing technique [20, 21] is particularly essential for fiber hologram construction. To achieve high angular selectivity, one of the recording light beams (either object or reference beam) has to be away from the optical axis so that the external write-in angle can be arranged within 30°–150°, as shown in Fig. 16.12a. However, the off-axis beam will excite high-order modes, which will severely increase the speckle noise. As one can see, the higher-order mode excitation can be significantly reduced by using wavelength multiplexing technique as shown in Fig. 16.12b. In other words, by arranging a counterpropagating object and reference beam configuration for fiber hologram construction, the excited higher-order modes can be minimized. Needless to say, as can be seen in Fig. 16.11, the reflection-type holograms offer a much higher wavelength selectivity, which potentially can be applied to high-density optical fiber memories.

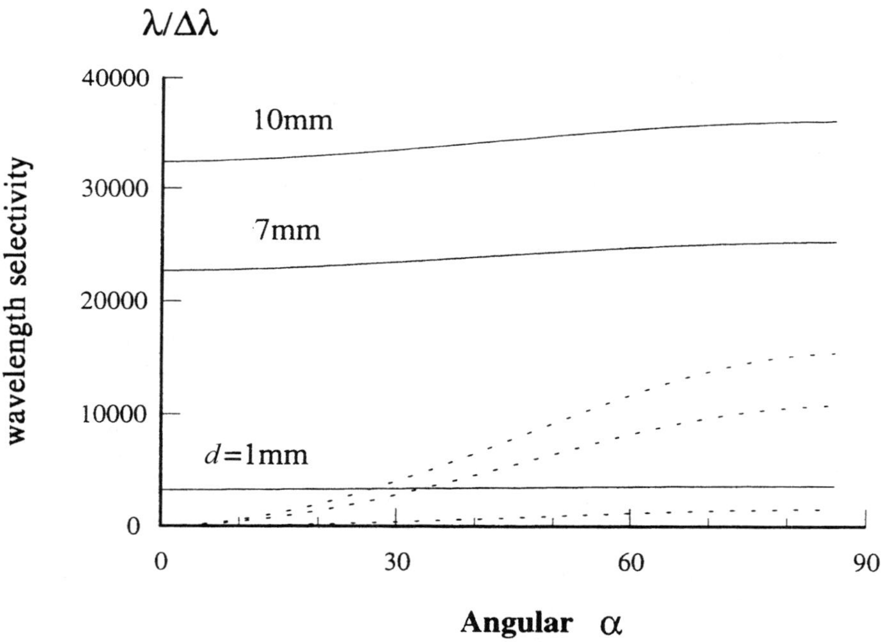

Figure 16.11: Wavelength selectivity (dashed line for transmission hologram and solid line for reflection hologram), recording wavelength $\lambda = 0.7\ \mu m$.

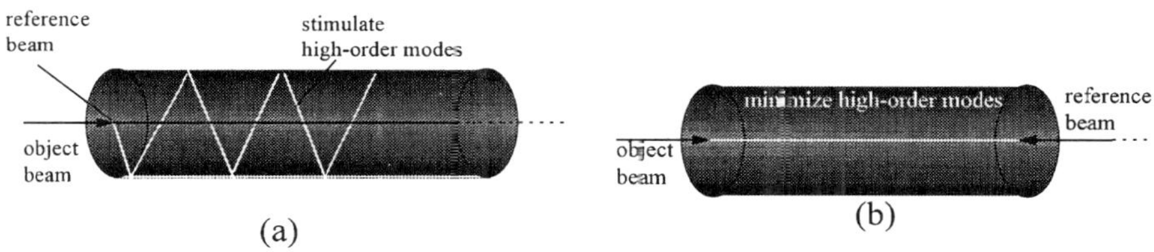

Figure 16.12: (a) Angular multiplexing fiber hologram construction. (b) Wavelength multiplexing fiber hologram construction.

16.5 Cross talk noise

Theoretically speaking, the upper-bound storage capacity for a volume hologram can be as high as V/λ^3 (where V is the volume of the recording material and λ is the wavelength) [22]. For example, the storage capacity for a 1 cm^3 crystal can in principle be as high as 10^{12} bits. However, there are practical limitations that prevent its reaching to this upper limit. Certainly, cross talk noise is one of the major limiting factors among them. Cross talk noise for angular multiplexed thick hologram has been discussed by Ramberg [23] and Gu *et al.* [24], who have shown that the cross talk noise is proportional to the number of stored holograms. Moreover, the effect of the wavelength spread has also been evaluated by Gu and Hong [25]; they have shown that the cross talk noise approaches a saturated level when the number of stored holograms is large. However, their study is based on the assumption of infinitesimal linewidth of the reading beam. In this section, we shall discuss the cross talk due to the wavelength spread over the multiplexed holograms. Because the reflection-type fiber hologram has been shown a better choice for the fiber holographic storage, we shall focus our attention primarily on this type of fiber hologram as follows.

We begin our discussion with the construction of a reflection-type fiber hologram as shown in Fig. 16.13, in which the object and the reference beams are assumed to be directly counterpropagated along the fiber axis during the holographic construction. Since the PR fiber is assumed to be sufficiently short (e.g., in the order of 0.5–1 mm in diameter and 3–20 mm in length), the length-to-diameter ratio is small (e.g., ~ 10). In other words, the writing beams can only reflect a few times from the boundary

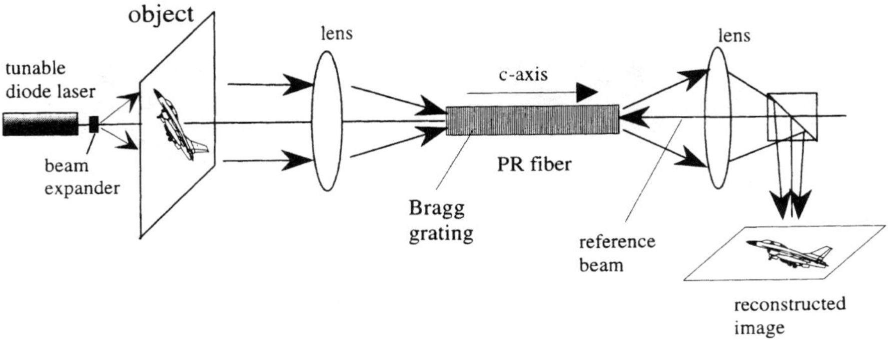

Figure 16.13: Wavelength-multiplexed reflection fiber holographic construction.

surface within the fiber. By using the bulky crystal approximation, the diffraction efficiency (assume no losses) can be shown as [26]

$$\eta = \left[1 + \frac{1 - \xi^2/v^2}{\sinh^2(v^2 - \xi^2)^{1/2}}\right]^{-1}, \tag{16.15}$$

in which we have used the direct counterpropagating beam approximation (i.e., $2\theta = 180°$), where

$$\xi = \frac{2\pi nd\,\Delta\lambda}{\lambda_0(\lambda_0 + 2\Delta\lambda)}, \quad v = \frac{\pi\Delta nd}{(\lambda_0 + \Delta\lambda)\sqrt{1 + 2\Delta\lambda/\lambda_0}}, \tag{16.16}$$

are for the unslanted case, d is the fiber length, n is the refractive index of the fiber crystal, Δn is the induced refractive index due to the writing beam intensity, and $\Delta\lambda$ is the mismatched wavelength of the reading beam with respect to the writing wavelength λ_0. The diffraction efficiencies as a function of mismatched wavelength $\Delta\lambda$ are plotted in Fig. 16.14, in which we see that the diffraction efficiency decreases rapidly as $\Delta\lambda$ increases. Nevertheless, as the fiber length increases, both the diffraction efficiency and spectral bandwidth improve. The peak efficiency can be shown as $\eta_{peak} = \eta\,(\Delta\lambda = 0) = \tanh^2(\pi\Delta nd/\lambda)$, which is a monotonically increasing function as the coupling strength $\pi n_1 d/\lambda_0$ increases. In other words, for a higher peak efficiency, it requires higher refractive index variation Δn, longer fiber length, and shorter writing wavelength λ_0. However, to have a larger dynamic storage capacity, as pointed out by Hong *et al.* [27], Δn should be maintained as small as possible. Thus by imposing the weak

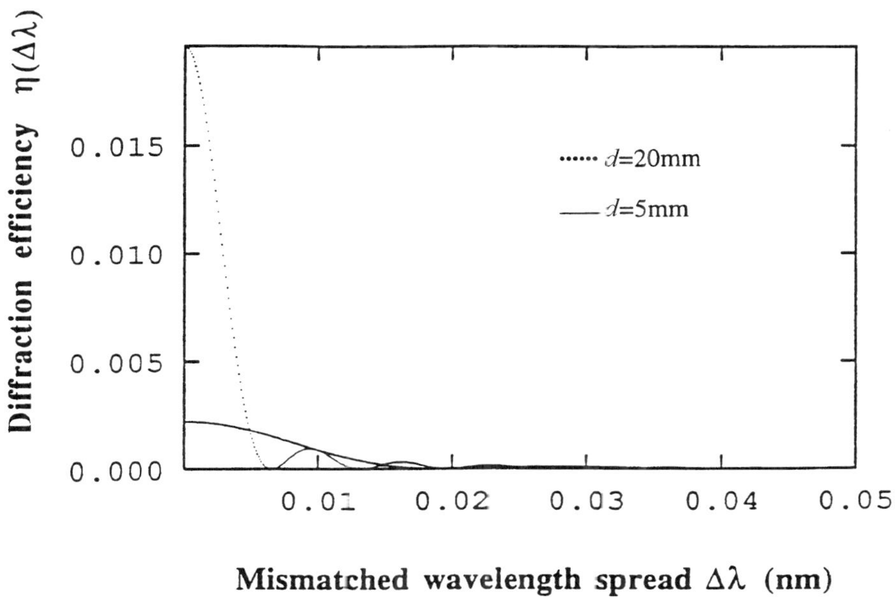

Figure 16.14: Diffraction efficiency η as a function of mismatched wavelength $\Delta\lambda$. Solid line: $\eta_{\text{peak}} = 0.2\%$ ($n_1 = 2 \times 10^{-6}$, $\lambda_0 = 670$ nm, $d = 5$ mm; dotted line: $\eta_{\text{peak}} = 0.2\%$ ($n_1 = 2 \times 10^{-6}$, $\lambda_0 = 670$ nm, $d = 5$mm).

coupling condition and allowing $\eta(\Delta\lambda) = 0$ the spectral bandwidth of the multiplexed channel can be shown to be

$$\Delta\lambda_{\min} \approx \frac{\lambda_0^2}{2nd}, \tag{16.17}$$

which is proportional to the square of the construction wavelength and inversely proportional to the refractive index and the fiber length.

To have a feeling of magnitude, we assume $\lambda_0 = 670$ nm, $n = 2.28$, and the fiber length $d = 5$ mm, the spectral bandwidth of the channel would be $\Delta\lambda_{\min} = 0.02$ nm. We note that Eq. 16.15 is derived under the assumption of zero linewidth for recording beams. However, if the recording beams have a finite spectral width $\Delta\lambda_1$, the profile of η would be broader, i.e., $\Delta\lambda_{\min} = \Delta\lambda_{\min} + \Delta\lambda_1$. Thus, to have a finer tuning channel, the narrower the spectral width of the light source needs to be.

If one regards the readout image from a given channel as the signal and the wavelength spread over into this channel as noise, then the signal-to-(cross talk)-noise ratio can be defined as

$$SNR_{\text{CT}}^{(j)} = \frac{\int \text{rect}\left[\frac{\Delta\lambda}{\Delta\lambda_1}\right] \eta d(\Delta\lambda)}{\sum\limits_{\substack{m=-(M-j) \\ m \neq j}}^{M+j} \int \text{rect}\left[\frac{\Delta\lambda - m\delta\lambda}{\Delta\lambda_1}\right] \eta d(\Delta\lambda)}, \tag{16.18}$$

where we assumed that the total number of holographic channels is $2M + 1$, $\delta\lambda$ is the separation between holographic channels, and $\Delta\lambda_1$ is the spectral width of the light source. For simplicity, we assume that the recording linewidth is negligibly small, so that the cross talk effect would be primarily due to the reading linewidth. Notice that similar results can also be obtained when recording linewidth is not assumed negligibly small. To minimize the cross talk noise between the holographic channels, we let the channel separation ($\delta\lambda$) equal the spectral bandwidth of the channels $\Delta\lambda_{\text{min}}$, i.e.,

$$\delta\lambda = \Delta\lambda_{\text{min}}. \tag{16.19}$$

Again, to have a sense of the magnitude involved, we assume the modulated refractive index is $\Delta n = 2 \times 10^{-5}$, the length of the fiber is $d = 5$ mm, and the central wavelength is $\lambda_0 = 670$ nm. Then the wavelength-multiplexed channel separation will be $\delta\lambda = \Delta\lambda_{\text{min}} = 0.02$ nm, and the total number of multiplexed holographic channels is $2M + 1 = 501$. By using the preceding data, the corresponding $\text{SNR}_{\text{CT}}^{(j)}$ are plotted in Fig. 16.15, for $\Delta\lambda_1 = 1.5 \times 10^{-4}$ nm ($\sim$ 100 MHz) and $\Delta\lambda_1 = 7.5 \times 10^{-3}$ nm ($\sim$ 5 GHz), respectively. In view of these plots, we note that the SNR_{CT} values are relatively uniformly distributed over the *j*th channel and one plot using a narrower spectral beam width has a higher signal-to-(cross-talk)-noise ratio.

The SNR_{CT} as a function of the number of multiplexed holograms is also plotted in Fig. 16.16, for $\Delta\lambda_\ell = 100$ MHz and 5 GHz, respectively. Again we see that by using narrower recording beam width it would have a higher signal-to-(cross talk)-noise ratio. As the number of stored holograms increases, the SNR_{CT} tends to be more independent as the stored holograms increase. The effects of SNR_{CT} caused by the recording bandwidth are also depicted in Fig. 16.17, where we see that as the

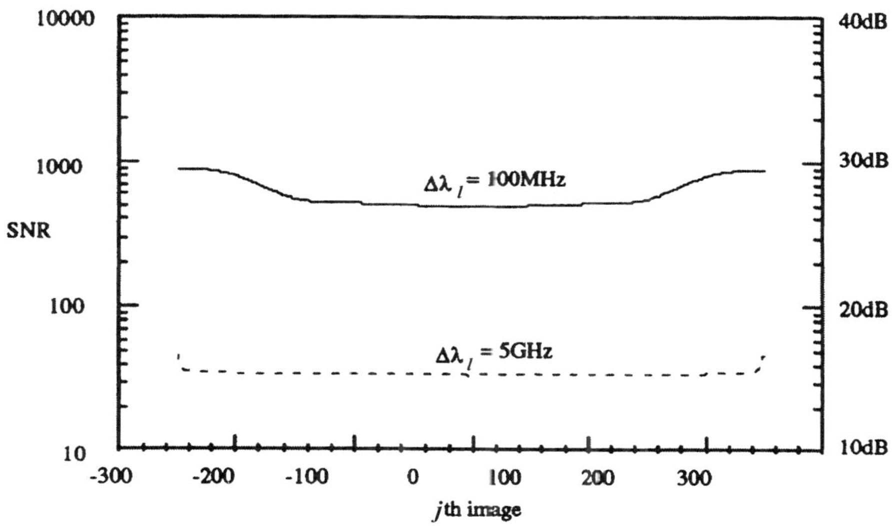

Figure 16.15: Distribution of signal-to-noise (cross talk) ratio $\mathrm{SNR}_{\mathrm{CT}}^{(j)}$ with respect to the jth multiplexed channel.

channels separate further, the noise performance tends to improve, and the $\mathrm{SNR}_{\mathrm{CT}}$ decreases rapidly as the bandwidth of the reading beam increases.

Since the cross talk noise in a fiber hologram is determined by the spectral bandwidth of the channel, it imposes the constraint of the storage capacity. For example, if the tuning range of the light source is on the order of 10–10^2 nm, the cross talk–limited storage capacity (C_{CT}) can be calculated as

$$C_{\mathrm{CT}} = \frac{\text{Tuning range}}{\text{Channel separation}}. \tag{16.20}$$

If we assume $\Delta\lambda_{\min}$ is on the order of 10^{-2}–10^{-3} nm, then C_{CT} would be on the order of 10^3–10^5 pages of holographic data with a readout $\mathrm{SNR}_{\mathrm{CT}}$ of 10–20 dB.

16.6 Recording erasure dynamics

Photorefractive materials offer promising applications to large capacity optical memories; however, in constructing the multiplexed holograms,

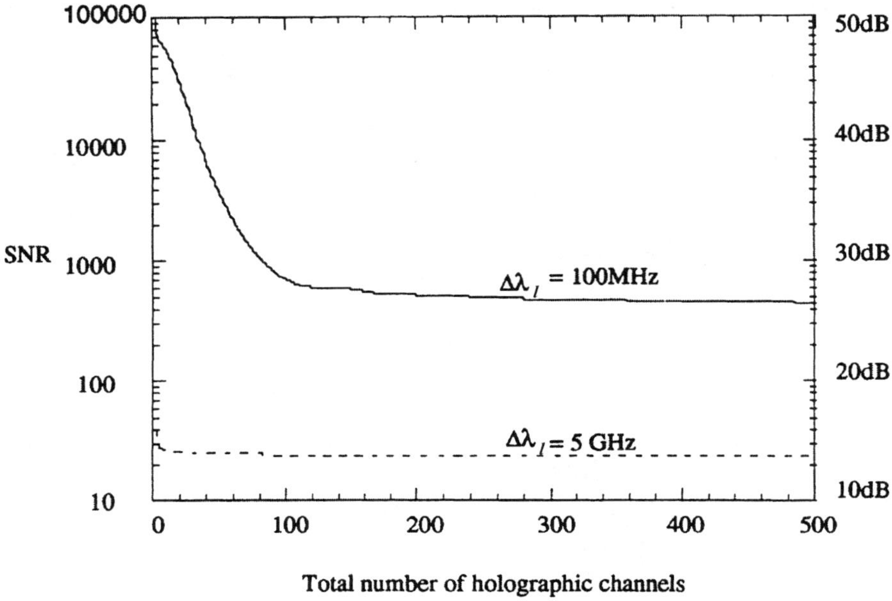

Figure 16.16: Signal-to-noise (cross talk) ratio $\mathrm{SNR}_{\mathrm{CT}}^{(0)}$ as a function of the number of multiplexed channels.

the successive recordings will partially erase the preceding recordings, which imposes stringent constraints on the storage capacity. We note that the recording-erasure dynamics, for angular-multiplexed transmission-type holograms, have been analyzed by Maniloff and Johnson [28]. However, the coupled-wave approach is valid only for uniform refractive index modulation and is not sufficient for the case of nonuniform index modulation. In this section, we will investigate the recording-erasure dynamics for a wavelength-multiplexed reflection-type PR fiber hologram. Since the transfer of energy depends on the relative sign of the writing wave vectors, the dynamic behavior of a reflection-type hologram would be different from a transmission-type hologram. In a PR fiber hologram recording, the counterpropagation of the object and the reference beams produces a nonuniform refractive index modulation along the fiber, by means of which successive recording causes an incoherent erasure of the preceding recorded holograms. In fact, the erasure does produce uniform decay in index-modulation depth, but has no effect on the recorded modulation

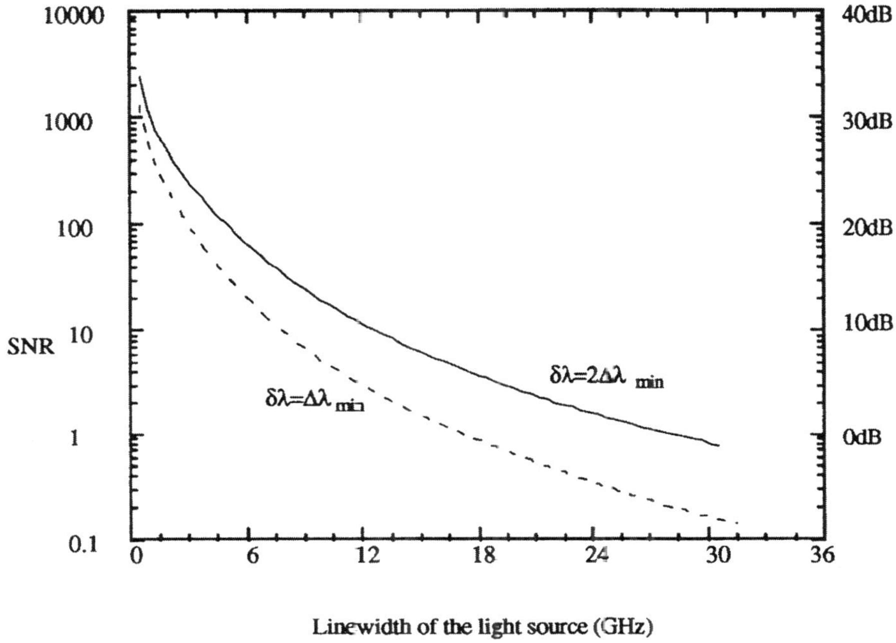

Figure 16.17: $\mathrm{SNR}_{\mathrm{CT}}^{(j)}$ as a function of source linewidth $\Delta\lambda$, for different channel separations.

ratio. Referring to the geometry of the fiber hologram construction shown in Fig. 16.18, the induced refractive index-modulation can be shown as

$$\Delta n = \Delta n_0(t_r) m(z) \cos[\varphi_R(z) - \varphi_s(z) + \varphi_g], \tag{16.21}$$

where $\Delta n_0(t_r)$ represents the refractive index perturbation, t_r is the recording time, ψ_g is the phase shift (assumed to be constant) relative to the interference fringes, $\psi_s(z)$ and $\psi_R(z)$ are the phase factors of the signal and the reference beams, respectively, and $m(t)$ is the modulation contrast ratio as given by

$$m(z) = \frac{2\sqrt{I_S(z) I_R(z)}}{I_S(z) - I_R(z)}. \tag{16.22}$$

Strictly speaking $\Delta n_0(t_r)$ is dependent upon a number of physical factors however, for simplicity we assume it is a monotonic function as given by

$$\Delta n_0(t_r) = \Delta n_s[1 - \exp(-t_r/\tau)], \tag{16.23}$$

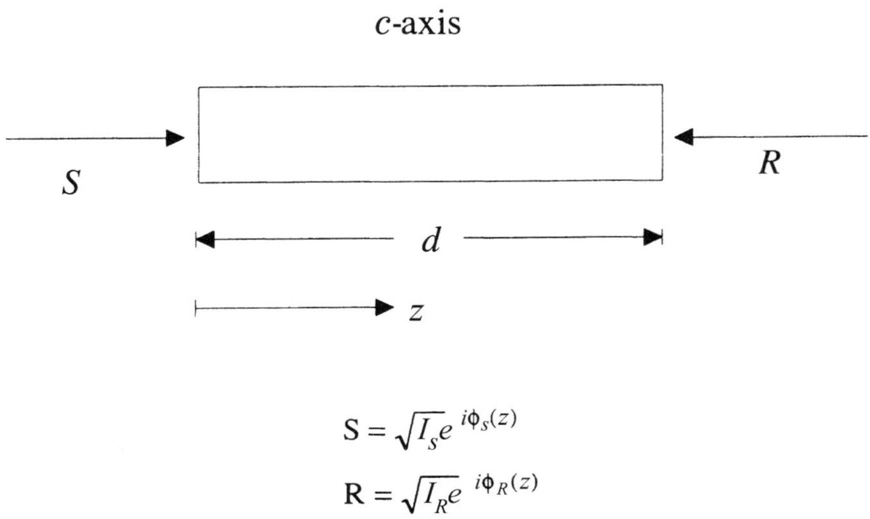

Figure 16.18: Geometry for direct counterpropagated beams fiber hologram construction.

where Δn_s represents the saturated value of the refractive-index change, and t_r and τ are the recording time and the recording time constant, respectively.

Although the photovoltaic effect would have influence on the two-wave mixing dynamics [29, 30] we assume this effect is negligibly small so it can be neglected. There is, however, the sign of the phase-shift produced by the coupling between the writing beams; it is dependent upon the direction of the c-axis of the crystal fiber relative to the wave propagation, which will cause a nonlinear refractive-index modulation ratio along the PR fiber.

By subjecting the scalar wave equation to the refractive index modulation described in Eq. (16.22) with the assumption of slowly varying field, a set of nonlinear coupling equations can be developed [31, 32], such as

$$\frac{\partial S}{\partial Z}(z,t_r) + i\Gamma_r(t_r)\frac{\sqrt{I_S(z)I_R(z)}}{I_S(z)+I_R(z)}\exp[-i(\varphi_R(z) - \varphi_S(z) + \varphi_g)R(z,t_r) = 0\,, \tag{16.24}$$

$$\frac{\partial R}{\partial Z}(z,t_r) - i\Gamma_r(t_r)\frac{\sqrt{I_S(z)I_R(z)}}{I_S(z)+I_R(z)}\exp[i(\varphi_R(z) - \varphi_S(z) + \psi_g)]S(z,t_r) = 0\,, \tag{16.25}$$

in which the modulation index is

$$m(z) = \frac{2\sqrt{I_S(z)I_R(z)}}{I_S(z) + I_R(z)} = \frac{1}{\sqrt{1 + m_1 \exp(2\Gamma_r z \sin\varphi_g)}}, \tag{16.26}$$

where

$$\varphi_R(z) - \varphi_S(z) = \Gamma_r z \cos\varphi_g ; \tag{16.27}$$

$$\Gamma_r(t_r) = \frac{\pi \Delta n_0(t_r)}{\lambda}, \tag{16.28}$$

$$m_1 = \frac{[m_0^2 - \exp(-2\Gamma_r d \sin\varphi_g)]^2}{4m_0(m_0 + 1)[m_0 + \exp(-2\Gamma_r d \sin\varphi_g)]}, \tag{16.29}$$

where λ is the recording wavelength, $m_0 = I_R(L)/I_s(0)$ is the reference-to-signal beam ratio, and d is the length of the fiber.

Notice that the PR fiber is assumed to be thick and to act like a bulk crystal. From Eq. (16.26), we see that the modulation index is highly nonuniform as a function of z, since Γ_r is a time-dependence factor. To evaluate the diffraction efficiency of the PR fiber hologram, a new set of coupling equations must be developed. By applying the weak reading beam assumption with the constraint of Bragg diffraction condition, we have

$$\frac{\partial s}{\partial z}(z) + f(z)r(z) = 0, \tag{16.30}$$

$$\frac{\partial r}{\partial z}(z) + f^*(z)s(z) = 0, \tag{16.31}$$

where the superscript asterisk represents the complex conjugate, $r(z)$ is the reading beam, and $s(z)$ is the readout beam;

$$f(z) = i\Gamma_e(t_e)m(z)\exp[i(\varphi_R(z) - \varphi_S(z) + \varphi_g)], \tag{16.32}$$

and

$$\Gamma_e(t_e) = \frac{\pi \Delta n_e(t_e)}{\lambda} \tag{16.33}$$

in which a new refractive index perturbation is defined,

$$\Delta n_e(t_e) = \Delta n_0(t_r)\exp(-t_e/\tau_e), \tag{16.34}$$

where t_e and τ_e are the erasure-time and the erasure time constant, respectively. We note that the losses, due to scattering, absorption, and reflection, have been disregarded. Strictly speaking, the recording and the erasure time constants are dependent upon the recording intensity and the spectral content of the light source. For simplicity, we shall treat them as independent time-constants.

It would be difficult to evaluate a general solution for the preceding coupled equations. However, the solution can be simplified if we assume $\varphi_g = \pi/2$. This corresponds to maximum beam coupling, and it is in fact the usual case for PR crystals such as $BaTiO_3$. Thus the coupling equations can be simplified as

$$s(z) = s_1 \exp[\beta_1(z)] + s_2 \exp[\beta_2(z)]\,, \tag{16.35}$$

$$r(z) = r_1 \exp[\beta_1(z)] + r_2 \exp[\beta_2(z)]\,, \tag{16.36}$$

in which $\beta_i(t)$ remains to be determined, where r_i and s_i are arbitrary constants depending upon the boundary conditions of the fiber. By letting the summed-coefficients that associate with $\exp[\beta_1(z)]$ and $\exp[\beta_2(z)]$ be equal to zero, we have

$$r_i \frac{\partial \beta_i(z)}{\partial t} = -f^*(z) s_i\,, \quad i = 1, 2\,, \tag{16.37}$$

$$s_i \frac{\partial \beta_i(z)}{\partial t} = -f(z) r_i\,, \quad i = 1, 2\,, \tag{16.38}$$

which can be shown as

$$\left[\frac{\partial \beta_i(z)}{\partial z}\right]^2 = ff^*\,, \quad i = 1, 2\,, \tag{16.39}$$

$$ff^* = \frac{\Gamma_e^2}{1 + m_1 \exp(2\Gamma_r z)}\,, \tag{16.40}$$

where the superscript asterisk represents the complex conjugate. Thus the solutions for $B_i(z)$ can be shown as [32]

$$\beta_{1,2}(z) = \beta(z) = \pm\left\{\Gamma_e z - \frac{\Gamma_e}{\gamma_r} \ln[\sqrt{1 + m_1 \exp(2\Gamma_r z)}]\right\}. \tag{16.41}$$

By further substituting into the coupling equations of Eqs. (16.35 and 16.36), we have

$$s(z) = c_1 \sinh(\beta(z) + c_2)\,, \tag{16.42}$$

where c_1 and c_2 are arbitrary constants. By further subjecting to the fiber boundary conditions, i.e., $s(0) = 0$ and $r(d) = 1$, the diffraction efficiency of the PR fiber hologram can be shown as

$$\eta = s(d)s^*(d) = \tanh^2\left\{\Gamma_e d - \frac{\Gamma_e}{\Gamma_r}\ln\left[\frac{1+\sqrt{1+m_1\exp(2\Gamma_r d)}}{1+\sqrt{1+m_1}}\right]\right\}. \quad (16.43)$$

In view of this result, two major conclusions can be drawn as given in the following:

First, the diffraction efficiency will never reach 100% even though the PR fiber is assumed lossless. As Γ_r approaches infinity, the achievable diffraction efficiency for a single exposure ($\Gamma_e = \Gamma_r$) fiber hologram approaches

$$\eta_\infty \approx \tanh^2\left(\ln\frac{\sqrt{m_{1,\infty}}}{1+\sqrt{1+m_{1,\infty}}}\right), \quad (16.44)$$

where

$$m_{1,\infty} = \frac{m_0^2}{4(m_0+1)}. \quad (16.45)$$

Thus, the maximum achievable diffraction efficiency, i.e., $\eta_\infty \approx 88.83\%$, can only be obtained by unity construction beam ratio, i.e., $m_0 = 1$. As can be seen, the existence of the grating phase shift would limit the diffraction efficiency, which sets the practical storage capacity of the fiber hologram.

Second, although high construction beam ratio ($m_0 \sim$ hundreds) is one of the prerequisites to ensure a higher signal-to-noise ratio and low hologram image distortion, the achievable diffraction efficiency decreases as the recording beam ratio increases.

The achievable diffraction efficiency of a PR fiber hologram for $m_0 = 10$, and $m_0 = 100$, as a function of recording time is plotted in Fig. 16.19. For comparison, the diffraction efficiencies obtained using Kogelnik's formula without taking the PR effect (i.e., $\varphi_g = 0$) are also plotted, in which we see that the achievable diffraction efficiency will ultimately reach 100%, as the recording time increases for non-PR material. However for PR fiber hologram, there is a significant drop in diffraction efficiency, which was affected by the construction beam ratio. For example, the (achievable) diffraction efficiency levels off to about 30% for a beam ratio of $m_0 = 10$, and reduces to 2% for $m_0 = 100$.

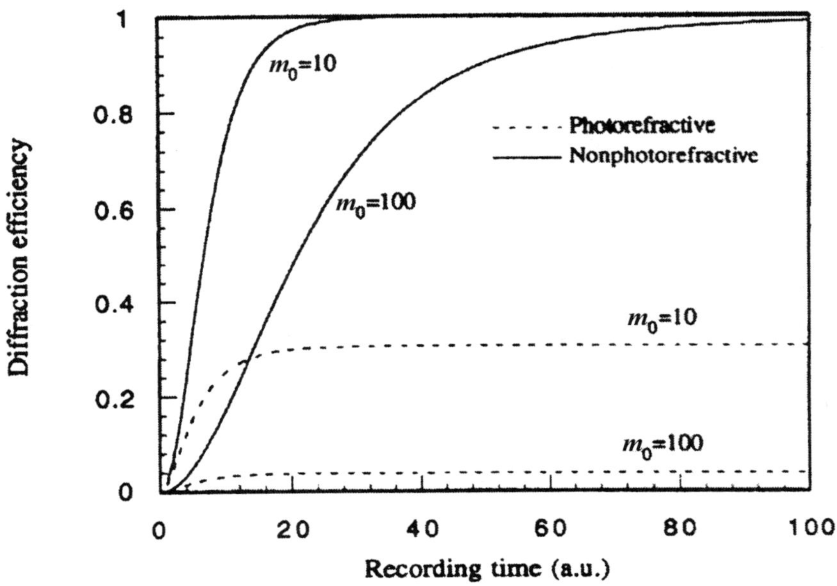

Figure 16.19: Achievable diffraction efficiency as a function of recording time. Dashed curves: based on Eq. (43). Solid curve: based on Kogelnik's formula.

16.7 Storage capacity

To evaluate the storage capacity of a PR fiber hologram, it is logical to equalize and to minimize the diffraction efficiencies for all the multiplexed holograms. For example, the required recording time of the nth hologram is determined by treating the overall recording time needed for the remaining successive recordings as the erasure time for the nth exposure—thus by equating the diffraction efficiencies for the nth and the $(n-1)$th holograms. As referred to in Eq. 16.43, we have

$$\left\{\frac{\pi\Delta n_0 t_{r,\mathrm{n}-1}}{\lambda}L - \ln\left[\frac{1+\sqrt{1+m_1\exp(2\Gamma_r t_{r,\mathrm{n}-1}L)}}{1+\sqrt{1+m_1}}\right]\right\}\exp\left(-\frac{t_{r,n}}{\tau_e}\right)$$

$$= \frac{\pi\Delta n_0 t_{r,n}}{\lambda}L - \ln\left[\frac{1+\sqrt{1+m_1\exp(2\Gamma_r t_{r,n}L)}}{1+\sqrt{1+m_1}}\right], \tag{16.46}$$

where $t_{r,n}$, $t_{r,n-1}$ are the recording times for the nth and $(n-1)$th holograms, respectively.

It is apparent that the recording time for each of the holograms can be determined from the total number of multiplexed holograms. Fig. 16.20 shows the variation of the required recording times as a function of a number of multiplexed holograms (for $m_0 = 1$, 10, and 100), in which is assumed that $\tau_r = 120$ s, $\tau_e = 210$ s, $\varphi_g = \pi/2$, $d = 2$ mm, $\lambda = 514.5$ nm, $\Delta n_s = 10^{-3}$, and $t_{r,l} = 20$ s. From these plots, we see that the required recording time decreases rapidly as the number of multiplexed holograms increases. Therefore, if one follows the prescribed recording recipes, a uniform diffraction efficiency may be obtained for all the recorded (multiplexed) holograms. From this figure, we also see that the required recording time does not appreciably alter if one uses different writing beam ratio (i.e., m_0). Nonetheless, diffraction efficiency is appreciably affected by the recording beam ratio, as shown in Fig. 16.21, where the diffraction efficiency decreases at the same rate as the number of storage holograms increases. There is, however, a 20 dB drop from $m_0 = 1$ to $m_0 = 100$, for which the storage capacity is anticipated to be lower for higher beam

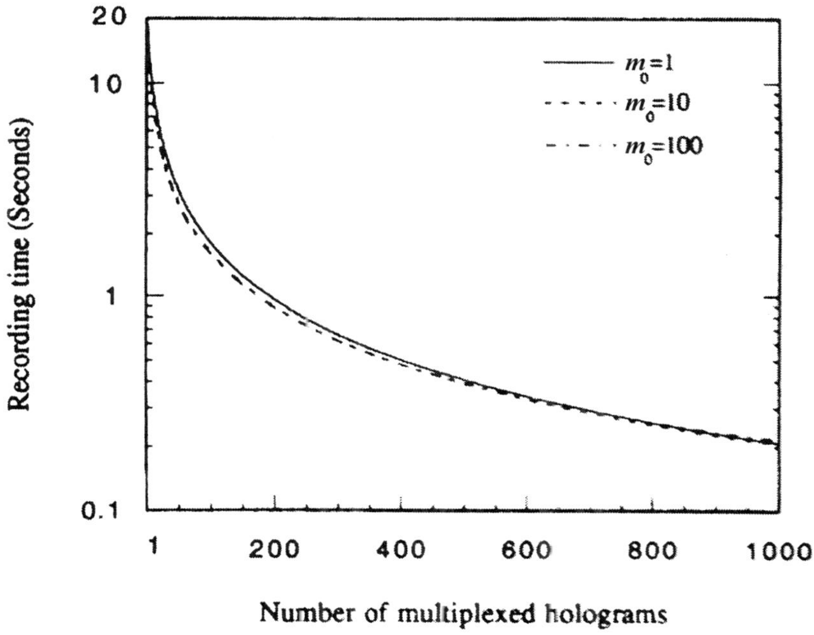

Figure 16.20: Recording times as a function of nth multiplexed hologram: $\tau_r = 120$ s. $\tau_e = 210$ s, $d = 2$ mm, $\lambda = 514.5$ nm, $\Delta n_\varepsilon = 10^{-3}$, $\tau_{r.1} = 20$ s.

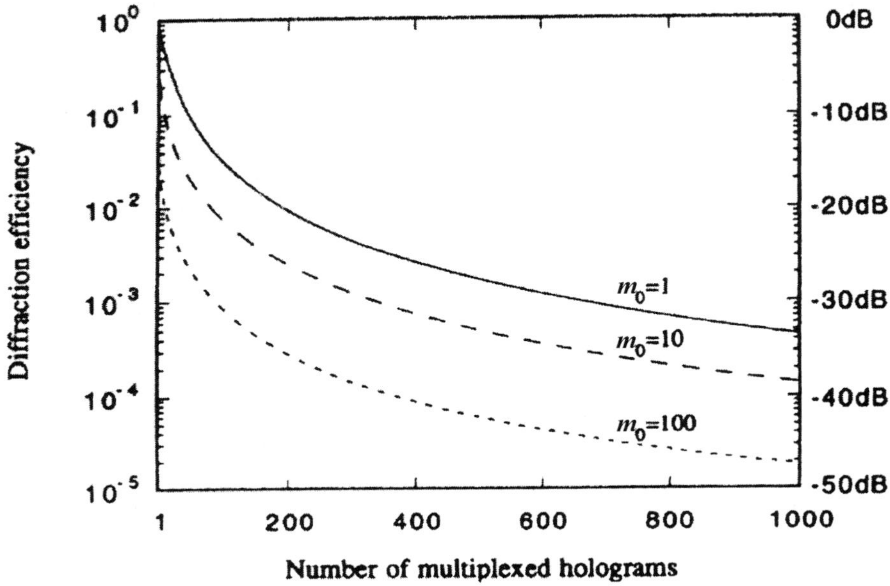

Figure 16.21: Diffraction efficiencies as a function of the number of multiplexed holograms (same parameters as for Fig. 16.20).

ratio. Nevertheless, in principle, a storage capacity of 2000 holograms is achievable in a 2-cm PR fiber using the unity writing beam ratio.

The evaluation of PR fiber storage capacity is based on the assumption that there is no cross talk noise among the multiplexed holograms, and no fanning noise (amplified scattering noise) in each subhologram. We stress that the cross talk noise is another major factor that limits the storage capacity, as has been shown in Section 16.5. The cross talk–limited storage capacity, for a wavelength-multiplexed reflection-type fiber hologram, has been shown as high as 10^3–10^5 pages of data. Since the fanning noise would affect the quality of the hologram image visibility [25], it would further limit the storage capacity. Another piece of circumstantial evidence is that the fanning noise could also cause the recording and the erasure time-constants to deviate from the true physical values, which would affect the actual exposure time schedule for obtaining the optimum storage capacity in a PR fiber hologram.

16.8 Application to photonic devices

Although there are numerous applications of PR fibers to photonic devices, due to page limitation, we focus here on a few recent findings from work done at Pennsylvania State University.

16.8.1 As applied to holographic memories

In this section we shall experimentally show that a high-density fiber hologram memory can be constructed. Because of long fiber axis structures, in the preceding sections we have shown that reflection-type holographic storage is feasible to realize in practice. As long as the length-to-diameter ratio of the PR fiber is adequately small, for example in the order of 10–30, the PR fiber hologram construction can be treated as a thick crystal. We will now utilize a specially doped PR fiber grown by the laser-heated pedestal growth system experimental demonstration, as depicted in Fig. 16.22.

To detail the specification of the experiment, the single-crystal fiber is grown from a preformed Ce:Fe-doped $LiNb0_3$ crystal, which contains 300 ppm Ce and 500 ppm Fe. The growth fiber is then subjected to a 700°C

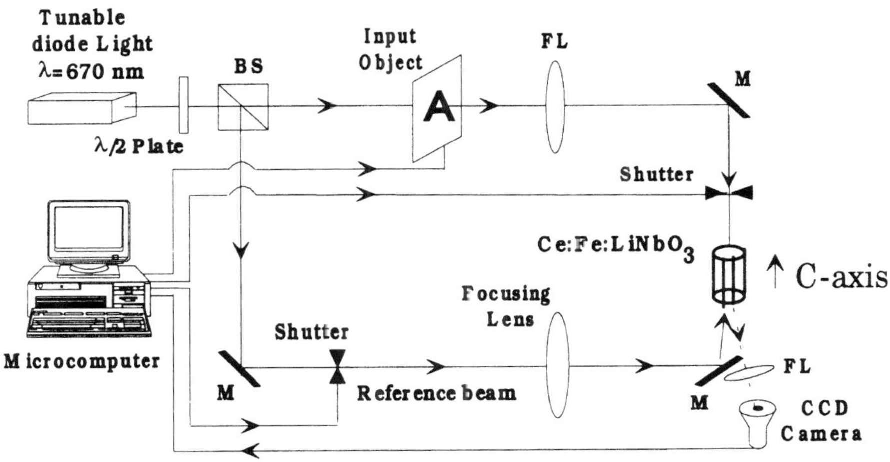

Figure 16.22: Setup for the construction of a wavelength-multiplexed reflection-type fiber hologram.

annealing process for about 5 h in an open-air environment. Although we have grown several fibers, the one we used for experimental demonstration is a 7-mm-long fiber with a diameter about 0.7 mm. This fiber has a length-to-diameter ratio of about 10. Since the Ce:Fe-doped $LiNb0_3$ fiber has a high photosensitivity in the red light region [21], a low-power, visible light tunable diode laser can be used for the hologram recordings. Moreover, the PR fiber we used has a high Curie point of about 1200°C, and it is robust against the ambient temperature variation.

In the construction of the fiber hologram, a 5-mW tunable laser diode (New Focus, Model 6102, $\lambda \simeq 670$ nm, with a tuning range of 12 nm) is used as the light source. The object beam is measured to about 2 mW/cm^2, and we have used counterpropagating beams for fiber hologram construction. By tuning the wavelength $\Delta\lambda = 0.1$ nm per multiplexed recording with a 5-s exposure time, a wavelength-multiplexed fiber hologram is constructed. If the fiber hologram is sequentially read by the tunable laser, a sequence of holographic images can be reconstructed, as shown in part in Fig. 16.23. In view of these reconstructed images, aside from the monitor scanning lines, the images show some distortions, which

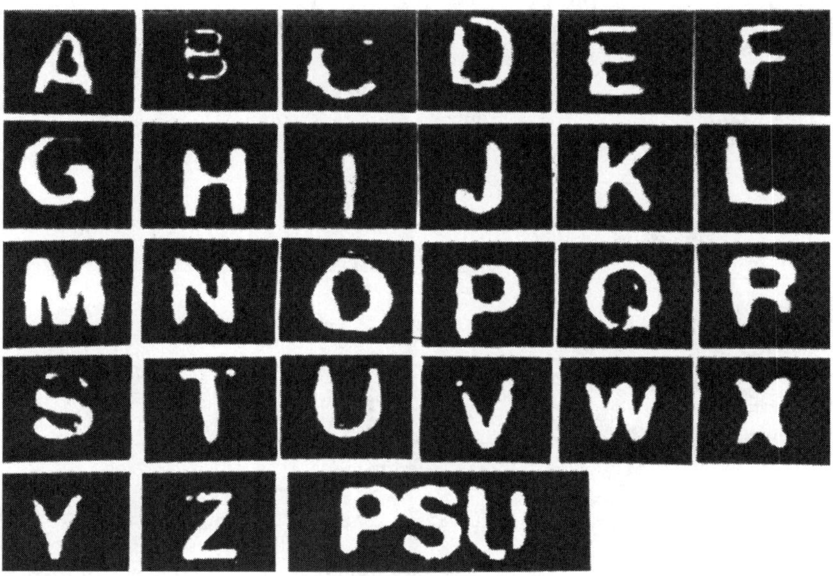

Figure 16.23: Holographic images obtained from a Ce-Fe-doped $LiNbO_3$ PR fiber hologram.

could result from the nonsmoothness of the fiber surface. Nevertheless, we have successfully recorded over 100 holograms by using this 7-mm-long fiber. This indicates that high-density fiber hologram storage can be realized in practice.

16.8.2 As applied to fiber sensors

Fiber sensors are among the most promising sensing devices [33]. However, most fiber sensors exploit the temporal content for sensing. At Pennsylvania State University, we have developed a new type of fiber sensor aspect called a fiber specklegram sensor (FSS), that exploits the spatial content for sensing [34, 35]. In other words, instead of using a single-mode fiber, we have used a multimode fiber so that the complex speckle field induced by intermodal phasings can be exploited for sensing.

As applied to the FSS, a small PR fiber can be directly coupled with a sensing fiber, as shown in Fig. 16.24. A step-index multimode silica fiber (about 50 μm core diameter) is used for the sensing fiber, and it is coupled with a specially doped Ce:Fe:$LiNbO_3$ PR fiber (about 7 mm in

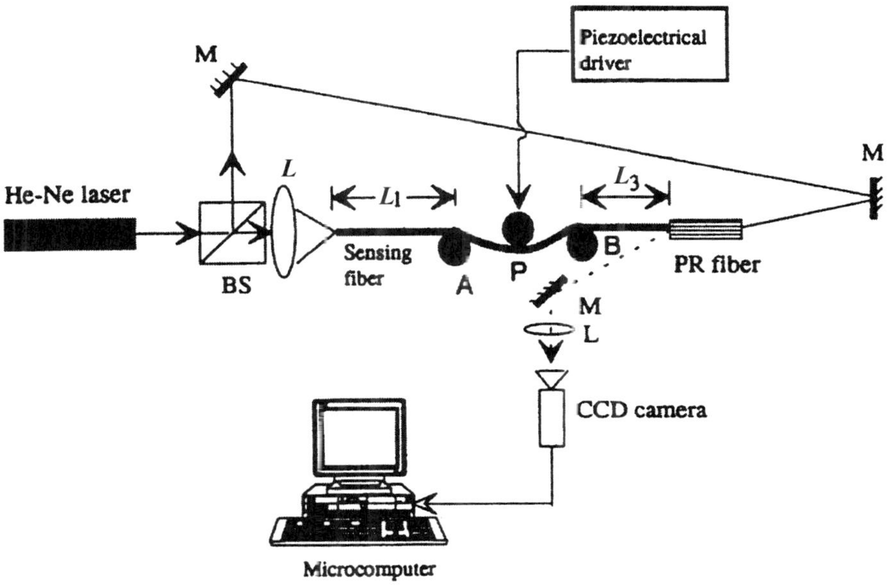

Figure 16.24: Experimental setup of the PR FSS system.

length and 0.7 mm in diameter). A reflection-type PR fiber specklegram can be made by using the counterpropagating beams. If the fiber specklegram is read by the same fiber speckle field, a reference field can be read out. To introduce the transversal displacement on the sensing fiber, a cylindrical rod affixed with a piezoelectric driver is used as the bending device [36], and the bending radius (i.e., the distance between the bending points A and B) is measured to about 2 mm. By plotting the normalized correlation peak intensity (NCPI) as a function of transversal displacement shown in Fig. 16.25, we see that a fairly linear (i.e., from 0 to 1.4 μm) transversal displacement curve is obtained. The dynamic range of this FSS setup is about 1.3 μm, and the corresponding sensitivity measurement is as high as 0.05 μm. It is needless to stress that the sensitivity can be further improved if a higher-mode sensing fiber is used, and the FSS system can be easily designed in compact form. Thus we see that the advantages of using the FSS for sensing must be the high sensitivity, compact size, and robustness to environmental perturbation. Because of simplicity and low-cost operation, we anticipate that the FSS will have

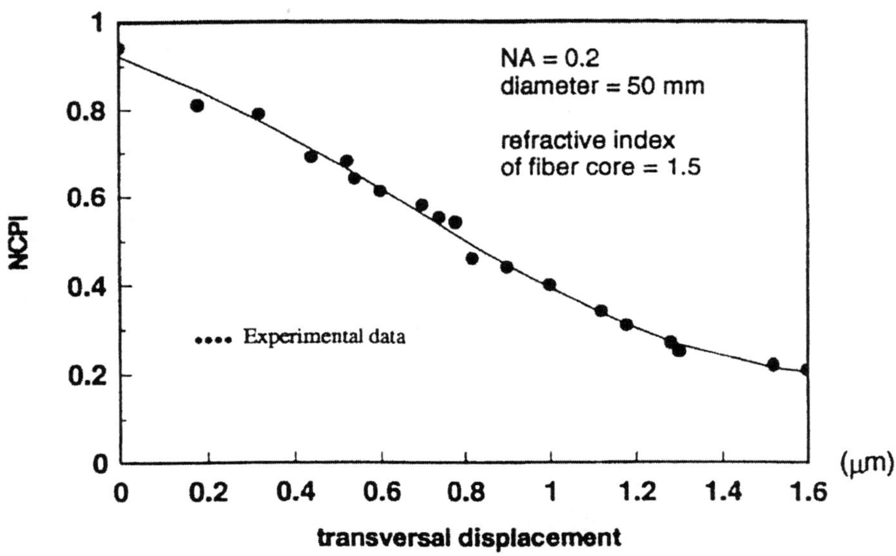

Figure 16.25: Normalized correlation peak as a function of transversal displacement.

significant practical applications, in applications such as acoustic wave detection, structural fatigue monitoring, and others.

16.8.3 As applied to tunable filters

A rapid and reliable communication system is one of the crucial information superhighway entities. Because of high-speed transmission and cost-effective operation, fiber optics and related technologies have been either employed or will be applied to the vast cutting-edge communication technologies field. In fact, fiber-optic communication is one of the 10 biggest inventions of the twentieth century. To exploit the extremely high optical carrier frequency (on the order of 10^{14} Hz) for communication, the wavelength division multiplexing (WDM) technique has been used [37]. Since multichannel signals (where each channel is encoded with different wavelength) can simultaneously propagate in a single optical fiber, transmission rates in the order of 10^{12} bit/s can potentially be realized without using expensive high-speed electronic circuitries. This has led to the design of large-scale information superhighway networks. One of the most critical components for the implementation of such a network must be the availability of high-speed narrow-band tunable filters. In other words, these filters are capable of carrying out functions of wavelength selection, switching, and routing. They can also serve as cross-interconnects for the reconfigurable networks.

Although acoustooptic (AO) cells have been used as the wavelength tunable filter [38–40], there are a number of limitations associated with using acoustooptical tunable filter (AOTF), such as low SNR and slow turning speed. For example, given that the sound speed in $LiNbO_3$ crystal in an AO cell is 106 nm/s. If we assume that the interaction length between the acoustic and light waves is about 10 nm, then the tuning speed of an AO cell would be in the order of a few milliseconds range. It is apparent that AOTF can not be applied for some high-speed communication networks, which require tuning speed in the order of a few nanoseconds or even higher.

Let us now describe the fabrication of a high-speed tunable filter. Since single-crystal PR fiber of 100 mm length is feasible to fabricate by the laser-heated pedestal growth technology, thus long grating structure can be recorded within the fiber. This corresponds to a wavelength selectivity as high as 10^5 order. For example, by using a recording wavelength of $\lambda = 1.5\ \mu$m, the 100-mm-long grating would have a bandwidth of about

0.02 nm. There are a few additional features of using the PR fiber for tunable filter as follows:

1. While the wavelength selectivity of a PR fiber filter is comparable to that of the Fabry-Perot etalon, one of the major advantages is that the *f*-number of the fiber filter can be designed to match with the optical system by simply controlling the curvature of the grating.
2. Physical structure of the filter makes it easier to couple with other fiber networks. This would not only reduce the size of the device but also increase the stability and robustness of the networking system.
3. Photorefractive materials have a strong electrooptic effect (EO) by which the refractive index can be tuned rapidly with an applied electric field. Since the EO moderator has been shown with a rise-time of about 1 ns, the tunable speed of a PR fiber filter within a few nanosecond range is feasible, in contrast with AO devices, which have a tuning speed of about ms range.

Mention should also be referenced that bulk material tunable filter has been developed by Rukuljic and Leyva [41]. However, the advantage of using tunable fiber filters is the long axial length, which corresponds to high wavelength selectivity and extremely narrow bandwidth, as can be seen in Figure 16.26. Since the counterpropagated beams are recorded within the PR fiber, a reflection-type grating can be made. We would note that the PR fiber we used for the tunable filter is a Ce:Fe-doped $LiNbO_3$ fiber. It is about 100 mm long with a diameter of about 1 mm. By using a 1-min recording time, about 20% diffraction efficiency from the fiber filter is obtained. By tuning the reading wavelength, the wavelength response of this fiber filter and spectral bandwidth are measured to about $\Delta\lambda_m = 0.022$ nm, for which we see that a narrow-bandwidth filter can indeed be developed. We stress that these data are about one order higher than the state-of-the-art AO tunable filter. To further demonstrate the reliability of this fiber filter, the fiber data (as given by $d = 100$ mm, $n = 2.3$, $\lambda = 670$ nm, and $\alpha = 180°$) were substituted in Eq. (14). The calculated bandwidth is $\Delta\lambda_r = 0.02$ nm, which is about 10% smaller than the experimental result.

We note that the crystal $LiNbO_3$ fiber is intentionally drawn along the *a*-axis, reserving the *c*-axis (which has a large EO coefficient) for the

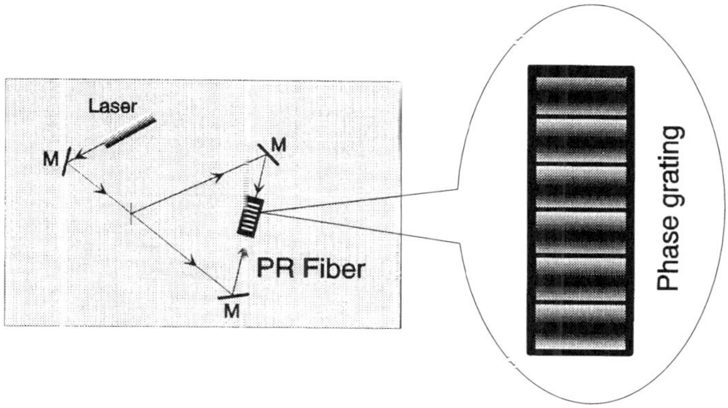

(a)

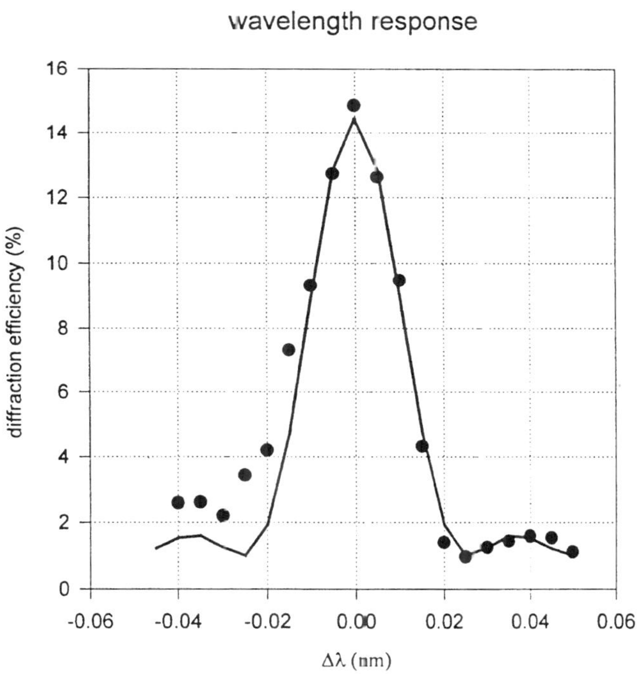

(b)

Figure 16.26: (a) Experimental setup. (b) Wavelength response.

transversal applied field. To ensure a uniform applied field across the fiber, the drawn fiber is first polished into rectangular form, as shown in Fig. 16.27. By coating the external electrodes on the upper and the lower surfaces of the rectangular fiber, a uniform electric field can be applied across the c-axis direction. Needless to say, the refractive index can then be tuned by the induced EO effect (i.e., Pockels effect). However, to improve the signal-to-(sidelobe) ratio, a Hamming-window Bragg-grating shown in Fig. 16.28 can be used as given by [42]

$$\Delta n(z) = \left[0.54 - 0.46\cos\left(\frac{2\pi z}{d}\right)\right]\cos\left(\frac{4\pi n}{\lambda}z\right), \tag{16.47}$$

which can be written as

$$\begin{aligned} n(z) &= n + n_1\left[0.54 - 0.46\cos\left(\frac{2\pi z}{d}\right)\right]\cos\left(\frac{4\pi n}{\lambda}z\right) \\ &= n + n_1 0.54\cos\left(\frac{4\pi n}{\lambda}z\right) - 0.23\cos\left[\left(\frac{4\pi n}{\lambda}\right)\right. \\ &\quad \left. - 0.23\cos\left[\left(\frac{4\pi n}{\lambda} - \frac{2\pi}{d}\right)z\right]\right. \end{aligned} \tag{16.48}$$

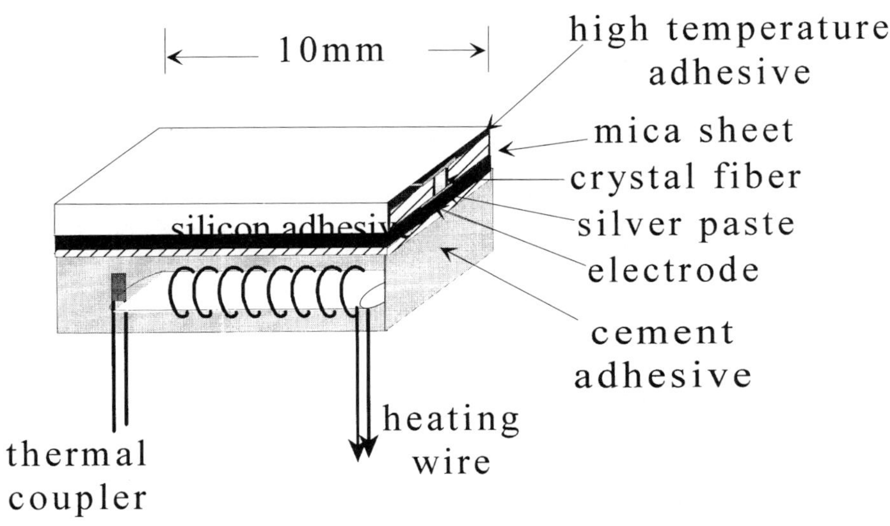

Figure 16.27: Structure of a PR fiber-based tunable filter.

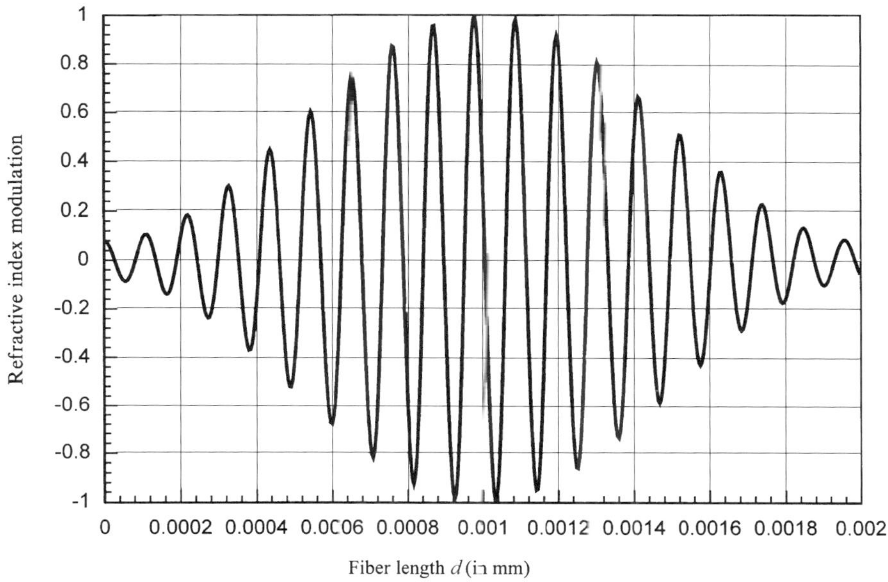

Figure 16.28: Hamming refractive index modulation profile.

where d and n are the length and the refractive index of the fiber, respectively, z is the axial coordinate, λ is the central wavelength, and n_1 is the modulation depth of the refractive index. We note that the (Hamming) refractive index distribution can be obtained by superimposing three sinusoidal gratings of the same spatial frequency but with different phase shifts. The wavelength responses for the rectangular and for the Hamming (Bragg) grating are plotted in Fig. 16.29, where we see that the signal-to-(sidelobe) ratio (SNR) for Hamming gratings performs better. It is about 30 dB higher as compared with the rectangular grating. There is, however, a small price to be paid, namely, by broadening the bandwidth of the filter. Nevertheless, the bandwidth can be made narrower by using a longer length of the PR fiber. We stress that higher SNR can be further extended by using other refractive index–modulated windows. For example, if one uses the Blackman-window which corresponds to a five-grating case [43], then the SNR can be made as high as 60 dB. However, the complexity of filter fabrication would be substantially increased.

To demonstrate the tunability, the preceding fiber filter is subjected to an applied electric field, and the corresponding wavelength response

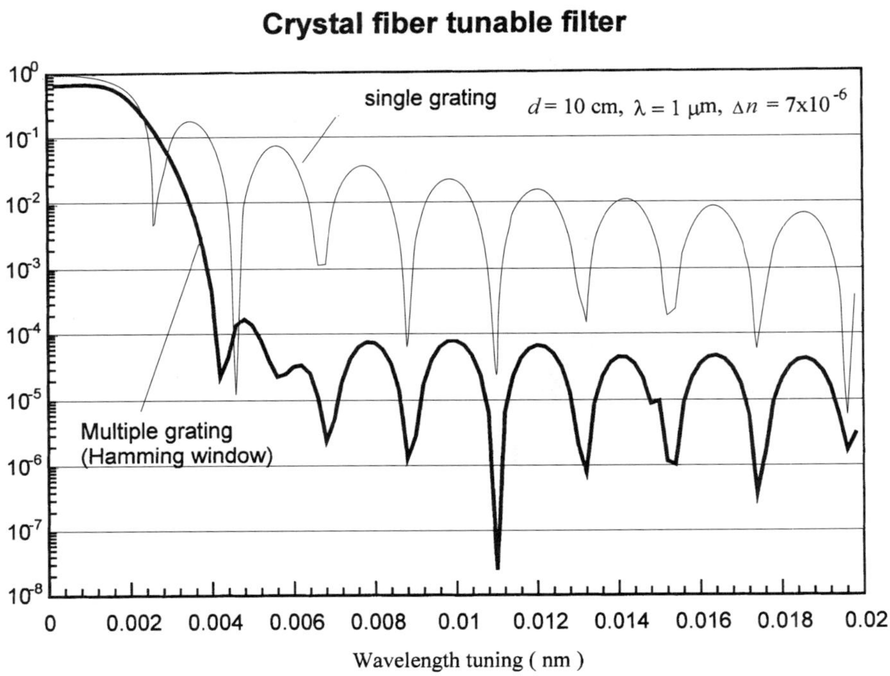

Figure 16.29: Wavelength response for crystal fiber tunable filter.

is plotted in Fig. 16.30. In light of this figure, we see that the wavelength response has been shifted toward the right side, although it suffers a minor diffraction efficiency loss. Nevertheless, we have demonstrated that high-speed high-SNR, narrow-band tunable filters can actually be developed in practice.

16.8.4 As applied to true-time delay lines

It is well known that phased array antennas (PAA) play an important role for the target detection [44]. To achieve good resolution and scanning range, typical parameters for required PAA are given as follows: (1) the RF frequency range is from 1 GHz to 25 GHz, (2) the azimuthal scanning range is between $-60°$ and $+60°$, (3) the altitude direction scanning range is from $-15°$ to $+75°$. Based on these data, the required radar beam angle is estimated to be as small as about 1°, and the time response is estimated

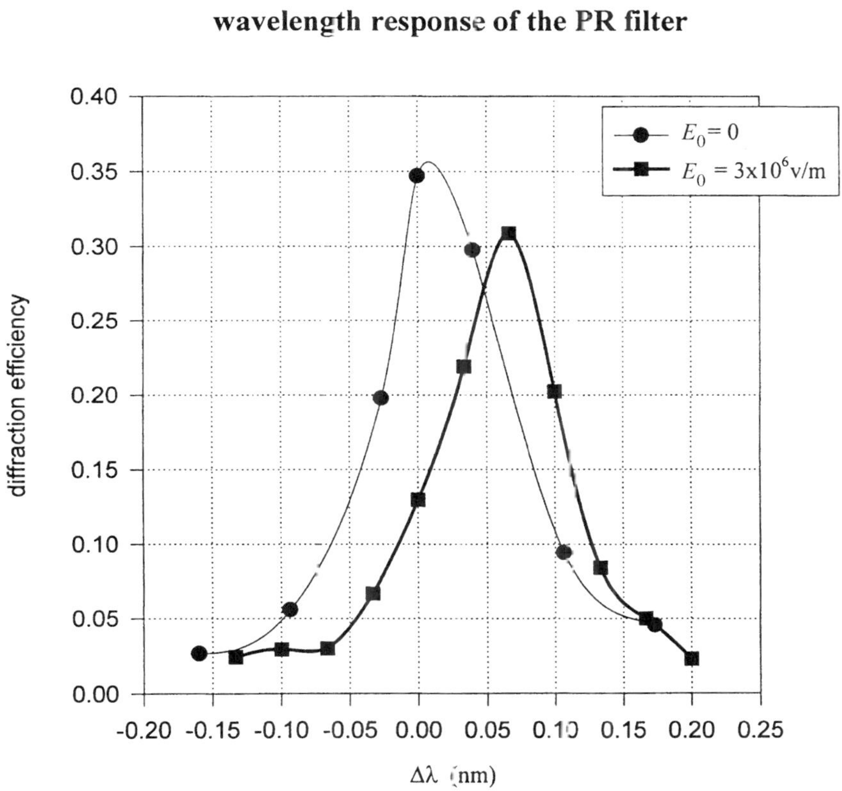

Figure 16.30: Fast tuning: shifted wavelength response by applied electric field.

between 40 and 1000 ps. Thus, for a hundred antenna elements the time resolution for a true time delay line would be about 0.4–10 ps. Obviously, it is a challenging task for the electronic technology to come up with these numbers. For instance, the speed of an analog to digital (A/D) converter has to be as fast as 0.4 ps (i.e., 25000 GHz). Unfortunately, such an A/D converter is not currently available, and it will not be available for some time. Even if we divide the antenna array into 12 × 12 subgroups, the required A/D conversion speed is within the range of 200 GHz, which is extremely difficult to develop with the current electronic technology.

On the other hand, fiber optic true-time delay (TTD) line is a potential application to phased array antennas [45]. For instance, a "squint" phenomenon of electronic phase shift systems, caused by the energy associated with the different frequencies by the array antennas, can be overcome by using TTD systems. In other words, TTD systems can radiate all frequency components in the same direction. One type of fiber optic TTD is based on fiber-optic Bragg gratings fabricated in silica glass fiber as illustrated in Fig. 16.31. A set of Bragg gratings of different central frequencies is implemented at different locations in the fiber. Since different wavelengths have different time delays from each grating, the time resolution of the TTD line can be estimated by $2nd/c$, where nd is the optical distance between adjacent gratings, and c is the speed of the light. However, using this type of delay line entails one major drawback, namely, tunability. In other words, the tuning speed should be in the order of 10 ms or even higher. It is, therefore, trivial to see that the proposed tunable PR fiber filters can be used for true-time delay lines as illustrated in Fig. 16.32a. Since large tuning range and high-speed tuning are not easy to come by from a single tunable diode laser, an array of diode lasers of different wavelengths can be used for this purpose as shown in the figure.

The system operation is described in Fig. 16.31b, in which the first row shows the wavelength distributions from the diode lasers, and the second row shows the responses from the Bragg gratings. Notice that the spectral distributions of the diode lasers are intentionally set apart from the respective grating responses. Thus by tuning a filter grating, a selected wavelength signal can be made reflected. For example, by tuning the central wavelength of the first grating from λ_{10} to λ_1, a signal of λ_1 from the first grating will be reflected. It is therefore trivial to see that by

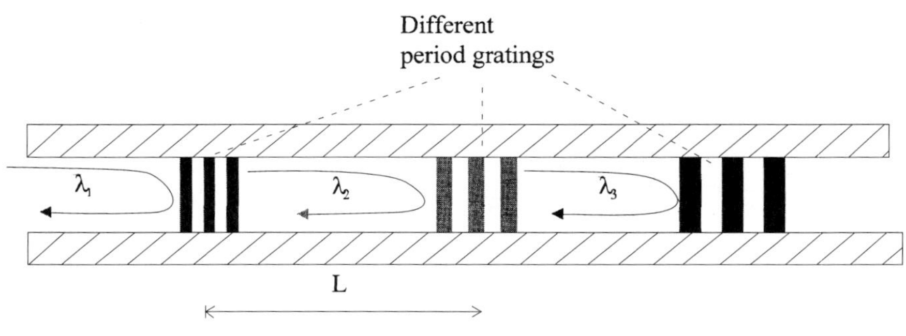

Figure 16.31: Fiber-optic TTD based on multiple Bragg gratings.

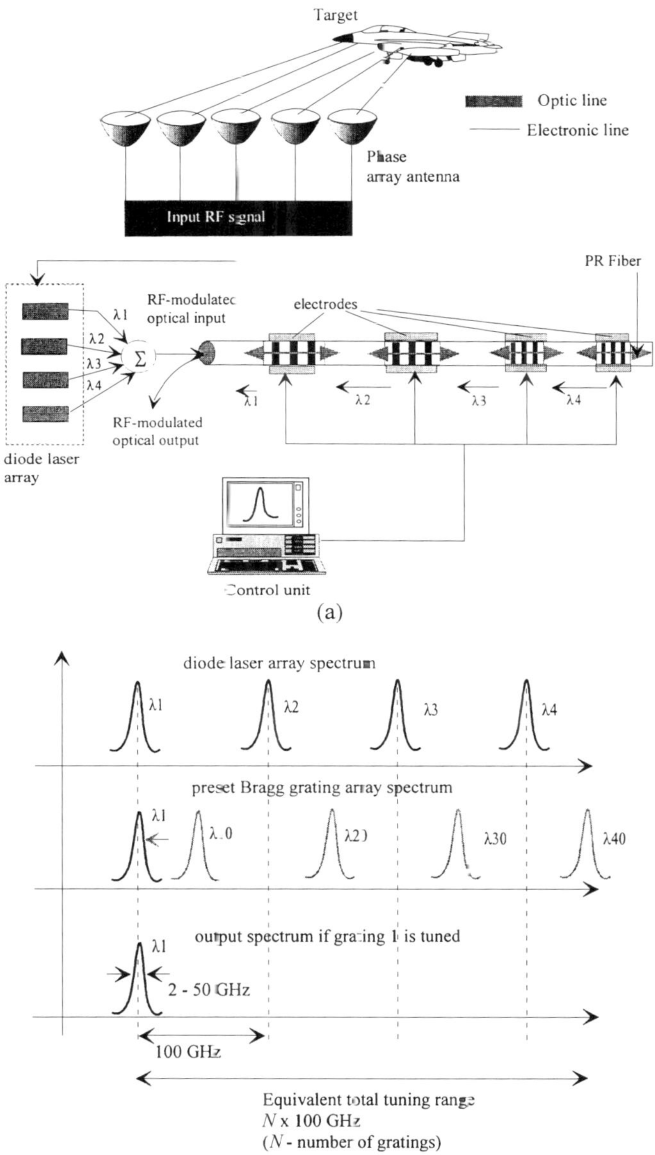

(b)

Figure 16.32: (a) Application of tunable PR fiber filter to PAA. (b) True-time delay signals in a PAA system.

simply tuning the fiber gratings it is possible to reflect different time-delay signals for different wavelengths.

To end up this section, we mention a few advantages of using the tunable fiber filter as applied to true-time delay lines:

1. It has a high tuning speed, since the grating filters can be parallel addressed.
2. It is flexible to operate, since each grating filter can be independently tuned.
3. It is robust to environment factors, since each grating filter can be readjusted to compensate the perturbations (e.g., due to temperature).
4. It has a large tuning range, since a large number of tunable Bragg gratings can be implemented.

16.9 Conclusion

In this chapter, recent work on the dynamics of photorefractive fibers has been discussed. The fabrication of single-crystal PR fiber using a laser-heated pedestal growth technique is illustrated. To compensate for the modal distortion caused by PR fiber, particularly for transmission-type fiber holograms, phase-conjugate recording techniques can be used. Analysis of angular and wavelength multiplexings in a PR fiber hologram is calculated, and we have found that reflection-type fiber holograms offer higher and more uniform wavelength selectivity. In view of the axial length of PR fibers, channel cross talk caused by wavelength spread has been evaluated; we have shown that high signal-to-(cross talk) noise can be achieved by using a narrower linewidth of the light source. Because successive recordings can partially erase the preceding recordings, recording-erasure dynamics have been investigated. The result shows that reflection-type fiber holograms would have higher storage capacity. To confirm our claim, we have experimentally demonstrated that a large-capacity-memory fiber hologram can be realized in practice. By further taking advantage of the fiber axial structure, we have shown that PR fibers are suitable for developing new photonic devices. To name a few such devices, we note parallel addressable fiber-array memories, sensitive fiber sensors, high-speed tunable filter, true-time delay lines, and others. We believe that the research and development of PR fibers is at the threshold of practical reality. However, several major tasks remain to be done before such applications become widespread.

References

1. A. Ashkin, G. D. Boyd, J. M. Dziedric, R. G. Smith, A. A. Ballman, J. J. Levinstein, and K. Nassau, "Optically induced refractive index inhomogeneities in $LiNbO_3$ and $LiTaO_3$," *Appl. Phys. Lett.*, **9**, 72–74, 1966.
2. P. Gunter and J. P. Huignard, *Photorefractive Materials and their Applications I & II*, Springer-Verlag, Berlin, 1988.
3. M. P. Petrov, S. I. Stepanov, A. V. Khomenko, *Photorefractive Crystals in Coherent Optical Systems*, Springer-Verlag, Berlin, 1991.
4. P. Yeh, *Introduction to Photorefractive Nonlinear Optics*, Wiley, New York, 1993.
5. D. D. Nolte (ed.), *Photorefractive Effects and Materials*, Boston, Kluwer Academic Publishers, Orlando FL, 1995.
6. V. V. Krongauz and A. D. Trifunac (eds.), *Processes in Photorefractive Polymers*, Chapman & Hall, New York, 1995.
7. R. A. Lessard and G. Manivannan, *Selected Papers on Photopolymers: Physics, Chemistry, and Applications*, Washington, DC, SPIE Optical Engineering Press, 1995.
8. F. H. Mok, "Angle-multiplexed storage of 5000 holograms in lithium niobate," *Opt. Lett.*, **18**, 915–917, 1993.
9. R. S. Feigelson, *Growth of Fiber Crystals, Crystal Growth of Electronic Materials*, North-Holland, Amsterdam, 1985, pp.127–144.
10. M. Saifi, B. Dubois, E. M. Vogel, and F. A. Thiel, "Growth of tetragonal $BaTiO_3$ crystal fibers," *J. Mater. Res.*, **1**, 452–456, 1986.
11. L. Hesselink and S. Redfield, "Photorefractive holographic recording in strontium barium niobate fibers," *Opt. Lett.*, **13**, 877–879, 1988.
12. J. K. Yamamoto and A. S. Bhalla,"Growth of $Sr_{1-x}Ba_xNb_2O_6$ single crystal fibers," *Mater. Res. Bull.*, **24**, 761–765, 1989.
13. S. Wu, A. Mayers, S. Rajan and F. T. S. Yu, "Use of photorefractive fibers to optical interconnections and switching," *Appl. Opt.*, **29**, 1059–1061, 1990.
14. H. Kogelnik, "Coupled wave theory for thick holograms," *Bell Sys. Tech. J.*, **48**, 2909–2947, 1969.
15. W. A. Gambling, D. N. Payne, and H. Matsumura, "Mode conversion coefficients in optical fibers," *Appl. Opt.*, **14**, 1538, 1975.
16. S. Wu, Q. Song, A. Mayers, and F. T. S. Yu, "Reconfigurable interconnections using nonlinear photorefractive crystals," *Appl. Opt.*, **29**, 1118–1125, 1990.
17. F. T. S. Yu and S. Jutamulia, *Optical Signal Processing, Computing and Neural Networks,* Wiley-Interscience, New York, 1992.

18. P. Gunter and J. P. Huignard, *Photorefractive Materials and Their Applications*, Vol. I, Springer-Verlag, Berlin, 1988.
19. G. A. Rakuljic, V. Leyva, and A. Yariv, "Optical data storage by using orthogonal wavelength-multiplexed volume holograms," *Opt. Lett.,* **17**, 1471–1473, 1992.
20. K. Curtis, C. Gu, and D. Psaltis, "Cross talk in wavelength-multiplexed holographic memories," *Opt. Lett.,* **18**, 1001–1003, 1993.
21. F. T. S. Yu, S. Yin, and A. S. Bhalla, "Wavelength-multiplexed holographic construction using a Ce:Fe: doped PR fiber with a tunable visible-light laser diode," *IEEE Phot. Tech. Lett.* **5**, 1230–1233, 1993.
22. P. J. Van Heerden, "Theory of optical information storage in solid," *Appl. Opt.,* **2**, 239–400, 1963.
23. E. G. Ramberg, "Holographic information storage," *RCA Rev.,* **33**, 5–53, 1972.
24. C. Gu, J. Hong, I. McMichael, R. Saxena, and F. Mok, "Cross talk limited storage capacity of volume holographic memory," *J. Opt. Soc. Am. A,* **9**, 1978–1983.
25. C. Gu and J. Hong, "Capacity of wavelength multiplexed holographic memory," in *OSA Annual Meeting Technique Digest Series,* **23**, paper WT2, OSA, Washington, DC, 1992.
26. F. T. S. Yu, F. Zhao, H. Zhou and S. Yin, "Cross-talk noise in a wavelength-multiplexed reflection-type photorefractive (PR) fiber hologram," *Opt. Lett.,* **18**, 1849–1851, 1993.
27. J. H. Hong, P. Yeh, D. Psaltis and D. Brady, "Diffraction efficiency of strong holograms," *Opt. Lett.,* **15**, 344–346, 1990.
28. E. S. Maniloff and K. M. Johnson, "Maximized photorefractive holographic storage," *J. Appl. Phys.,* **70**, 4702–4707, 1993.
29. A. M. Glass, D. Von Der Linde and T. Negran, "High-voltage bulk photovoltaic effect and the photorefractive process in $LiNbO_3$," *J. Appl. Phys. Lett.,* **25**, 233–235, 1974.
30. C. Gu, J. Hong., H.–Y. Li, D. Psaltis and P. Yeh, "Dynamics of granting formation in photovoltaic media," *J. Appl. Phys.,* **69**, 1167–1172, 1991.
31. P. Yeh, "Two-wave mixing in nonlinear media," *IEEE J. Quantum Electron.,* **QE23**, 484–519, 1989.
32. H. Zhou, F. Zhao and F. T. S. Yu, "Diffraction properties of a reflection-type photorefractive hologram," *Appl. Opt.,* **33**, 4345–4352, 1994.
33. E. Udd., *Fiber Optic Sensors*, Wiley, New York, 1991.
34. S. Wu, S. Yin and F. T. S. Yu, "Sensing with fiber speckle holograms," *Appl. Opt.,* **30**, 4468–4470, 1991.
35. S. Wu, S. Yin, S. Rajan and F. T. S. Yu, "Multiple channel sensing with fiber specklegram," *Appl. Opt.,* **31**, 5975–5983, 1992.
36. F. T. S. Yu, S. Yin, J. Zhang and R. Guo, "Application of fiber speckle hologram to fiber sensing," *Appl. Opt.,* **33**, 5201–5202, 1994.

37. C. Brackett *et al.,* "A scalable multiwavelength multihop optical network: A proposal for research on all optical networks," *IEEE J. Lightwave Tech.,* **LT-11**, 736–753, 1993.
38. M. Gottlieb, C. L. M. Ireland and J. M. Ley, *Electro-optics and Acoustooptic Scanning and Deflection,* Marcel Dekker, New York, 1983.
39. D. Smith *et al.,* "Integrated optics acoustically tunable filters for WDM networks," *IEEE J. Select. Areas Comm.*, **8**, 1151–1159, 1990.
40. W. Trunta *et al.,* "Anamalous sidelobes and birefringence apodization in acoustooptic tunable filters," *Opt. Lett.,* **18**, 28–30, 1993.
41. G. A. Rakulic and V. Leyva, "Volume holographic narrow-bandoptical filter," *Opt. Lett.,* **18**, 459–460, 1993.
42. S. Yin, B. D. Guenther and F. T. S. Yu, "Narrow-band filter using multiple photorefractive gratings," *SPIE,* **2529**, 196–203, 1995.
43. N. K. Bose, *Digital Filters* North-Holland, Amsterdam, 1985.
44. R. C. Hansen, *Phased Array Antennas*, Wiley-Interscience, New York, 1998.
45. S. Yin, "Fabrication of $LiNbO_3$ photorefractive single crystal fibers and their applications," *SPIE,* **2849**, 156–167, 1996.

Index